CAUSE OF DEA[TH]

M000233224

YOU BET YOUR LIFE

YOU BET YOUR LIFE

Your Guide to Deadly Risk

Sheila Buff & Joe Buff, MS, FSA

New York

CONTENTS

Causes of Death
NATURAL CAUSES

EVERYTHING ELSE

Accidental Harm and Injury

Judicial and State Actions

Intentional Harm and Injury

The Starting Point: Baseline Data on Life and Death

The information in this book is based on the most recent data available as of December 1, 2021, except for the COVID-19 section, which uses data available as of May 2, 2022. Because complex demographic data takes time to gather and analyze, most available US population statistics are from the period 2017 to 2020. The baseline data below is to give perspective on causes of death and death rates for the overall US population.

The Global Picture

Top 10 Worldwide Causes of Death, 2019

	WORLDWIDE	HIGH-INCOME COUNTRIES	LOW-INCOME COUNTRIES
1.	Ischemic heart disease	Ischemic heart disease	Neonatal conditions
2.	Stroke	Alzheimer disease and other dementias	Lower respiratory infections
3.	Chronic obstructive pulmonary disease (COPD	Stroke	Ischemic heart disease
4.	Lower respiratory infections	Trachea, bronchus, lung cancers	Stroke
5.	Neonatal conditions	Chronic obstructive pulmonary disease (COPD)	Diarrheal diseases
6.	Trachea, bronchus, lung cancers	Lower respiratory infections	Malaria
7.	Alzheimer disease and other dementias	Colon and rectum cancers	Road injury
8.	Diarrheal diseases	Kidney disease	Tuberculosis
9.	Diabetes	Hypertensive heart disease	HIV/AIDS
10.	Kidney disease	Diabetes	Liver cirrhosis

NOTE The World Bank defines high-income economies as those with a gross national income (GNI) per capita of $12,696 or more in 2020. Low-income economies are those with a gross national income (GNI) per capita of $1,045 or less in 2020.

SOURCE WHO, who.int/data/mortality

The United States

The US Population, 2019

According to the 2020 decennial census as reported by the US Census Bureau, the total US population as of April 1, 2020, was 331,449,281. Of these, 50.8 percent are female. The age distribution as of 2019 was:

AGE	NUMBER	PERCENT OF POPULATION
0–18	75,307,800	23.6
19–25	27,799,100	8.7
26–34	39,817,700	12.5
35–54	81,478,600	25.5
55–64	42,061,700	13.2
65+	52,784,400	16.5

The racial/ethnic breakdown for the US in 2019 by percent of population was:

RACE/ETHNICITY	% POPULATION
White	76.3%
Black	13.4%
Hispanic/Latino	18.5%
Asian/Pacific Islander	6.1%
Two or more races	2.8%
Native American	1.3%

SOURCE   US Census Bureau, census.gov/topics/population.html

Life Expectancy

Life expectancy is the average number of years of life a person who has attained a given age can expect to live. The CDC's National Center for Health Statistics is the main source for reliable life expectancy data.

In 2019, life expectancy at birth was 78.8 years for the total US population, an increase of 0.1 year from 78.7 years in 2018. For males, life expectancy changed 0.1 year from 76.2 in 2018 to 76.3 in 2019. For females, life expectancy increased 0.2 year from 81.2 years in 2018 to 81.4 in 2019. The difference in life expectancy between females and males was 5.1 years.

Life expectancy at age 65 was 19.6 years for the total population. For males, life expectancy was 18.2 years; for females, life expectancy was 20.8 years. The difference between females and males was 2.6 years.

The average age of death in 2019 was 73.8 years. Average age of death is calculated by taking a weighted average of the total number of people who died in each single age group annually.

The COVID-19 pandemic caused US life expectancy to drop. In 2020, life expectancy at birth was 77 years, declining by 1.8 years from 78.8 years in 2019. Life expectancy at birth for males was 74.2 years, a decline of 2.1 years. For females, life expectancy at birth was 79.9 years, a decline of 1.5 years.

In 2020, life expectancy at birth was at its lowest level since 2000 for both the total population and for males. It was at its lowest level for females since 2003.

SOURCE   CDC NCHS, cdc.gov/nchs/nvss

The life expectancy at birth by race/ethnicity in 2019 was:

RACE/ETHNICITY	LIFE EXPECTANCY AT BIRTH
All	78.8 years
White	78.8
Black	74.8
Hispanic/Latino	81.9
Asian American	85.6
Native American	71.8

SOURCE   CDC NCHS, cdc.gov/nchs/nvss

Top 10 Leading Causes of Death in the US, 2020

CAUSE OF DEATH	NUMBER OF DEATHS
1. Heart disease	690,882
2. Cancer	598,932
3. COVID-19	345,323
4. Unintentional injuries	192,176
5. Stroke	159,050
6. COPD	151,637
7. Alzheimer disease	133,382
8. Diabetes	101,106
9. Flu and pneumonia	53,495
10. Kidney disease	52,260
Total deaths	3,313,980

SOURCES   Ahmad FB, Anderson RN. The Leading Causes of Death in the US for 2020. *JAMA*. 2021;325(18):1829–1830. doi:10.1001/jama.2021.5469

COVID-19 IMPACT ON LIFE EXPECTANCY, 2019–2020

In 2020, COVID-19 was the 3rd leading cause of death in the US. The other causes remained in the same order. Of the approximately 3.36 million US deaths in 2020, COVID-19 was the underlying or a contributing cause for nearly 378,000, or about 11.3 percent of all deaths.

	LIFE EXPECTANCY, 2019	LIFE EXPECTANCY, 2020	DECLINE IN YEARS
All	78.8 years	77	1.8
Males	76.3	74.2	2.1
Females	81.4	79.9	1.5
White	78.8	77.6*	1.2
Black	74.8	71.8*	3
Hispanic/Latino	81.8	78.8*	3

* provisional

SOURCES Andrasfay T, Goldman N. Association of the COVID-19 Pandemic With Estimated Life Expectancy by Race/Ethnicity in the United States, 2020. JAMA Network Open, June 1, 2021; CDC NCHS, cdc.gov/nchs/nvss/covid-19.htm

Deaths by Age, 2019

<1	20,921
1–14	9,173
15–24	29,771
25–44	142,164
45–64	535,330
65–74	555,559
75–84	688,027
85+	873,746

SOURCE CDC NCHS, cdc.gov/nchs/nvss

Deaths by Race/Ethnicity, 2019

White	2,400,232
Black	354,349
Hispanic	212,397
Asian/Pacific Islander	79,139
Native American	21,118

SOURCE CDC NCHS, cdc.gov/nchs/nvss

Best and Worst States

The overall US mortality rate is 723.6 per 100,000 population.

As of 2017, the states with the highest age-adjusted death rates per 100,000 population are:

Alabama	918.1
Kentucky	920.0
Mississippi	934.8
Oklahoma	893.2
West Virginia	953.8

The states with the lowest age-adjusted death rates per 100,000 population were:

California	609.0
Connecticut	644.2
Hawaii	572.5
Minnesota	648.0
New York	626.7

The average age-adjusted death rate for the 5 states with the highest rates (926.8) was 49 percent higher than the rate for 5 states with the lowest rates (624.0). The age-adjusted death rates for chronic lower respiratory diseases and unintentional injuries for the states with the highest rates (62.0 and 65.5, respectively) were almost double those of the states with the lowest rates (31.0 and 35.8, respectively).

The average death rate for the states with highest rates was 27 percent higher than the national rate (731.9). For the states with the lowest rates, the average death rate was 15 percent lower than the national rate.

In 2017, the 5 leading causes of death were the same for both groups. However, the average age-adjusted death rates for all 5 top causes overall were 47 percent higher for the states with the highest death rates compared to the states with the lowest: 568.5 per 100,000 population compared with 387.3.

SOURCE   CDC NCHS, cdc.gov/nchs/nvss

Premature and Preventable Deaths

Premature death is a measure of years of potential life lost (YPLL) due to death occurring before the age of 75. Deaths at younger ages contribute more to the premature death rate than deaths occurring closer to age 75. For example, a person dying at age 70 loses 5 years of potential life; a child dying at age 5 loses 70 years of potential life.

Preventable death is death from a cause that could have been avoided through preventive interventions. For example, tobacco-related deaths are preventable by giving up or never smoking. Many car crash deaths are preventable by wearing seatbelts. Many deaths from colorectal cancer could have been avoided by colonoscopy to removed polyps before they became malignant.

Up to half of all US premature deaths are due to behavioral and other preventable factors, but precise statistics are hard to find. Cause of death data from states doesn't always identify underlying causes of death. Someone who dies of a heart attack might not be identified as a smoker, for example, and it is difficult to assess the relative risk of death associated with particular behaviors, such as a sedentary lifestyle. On the other hand, suicides and gunshot victims can all be said to have died prematurely from a preventable cause.

Overall, tobacco use, diet/activity patterns, alcohol, infectious disease, medical errors, toxic agents, firearms, sexual behavior, motor vehicle accidents, and drug overdoses are the leading preventable causes of premature death. For details on the lethal risks of these causes of death, see the rest of this book.

In 2014 (most recent available data), potentially preventable deaths for the top 5 leading causes of death were:

CAUSE OF DEATH	ACTUAL DEATHS	POTENTIALLY PREVENTABLE
Heart disease	614,348	87,950
Cancer	591,699	63,209
Chronic lower respiratory disease	147,101	29,232
Accidents	136,053	45,331
Stroke	133,103	15,175

SOURCE   CDC NCHS, cdc.gov/nchs/ndi

At-Risk Populations

Populations at higher risk for premature death include:

Native American populations. Compared to White Americans, Native Americans and Alaska Native populations have higher premature mortality rates. The infant mortality rate is almost 2 times higher. Risk factors for higher rate of death from accidents include higher rates of alcohol-related incidents, rural environments, and lack of traffic safety.

Black populations. The risk of premature mortality is 2 times as high for Black Americans as for Whites. The infant mortality rate is 3 times higher. Disparities in deaths from heart disease and homicides are also drivers of premature mortality.

Rural versus urban. White adults in rural areas have the highest mortality and premature mortality rates. The disparity is largely due to suicides and drug overdoses.

SOURCE   United Health Foundation, *Health Disparities Report, 2020*, americashealthrankings.org/learn/reports/

Day of Death

In 2017, an average of 7,708 deaths occurred each day. January, February, and December were the months with the highest average daily number of deaths (8,478, 8,351, and 8,344, respectively). June, July, and August were the months with the lowest average daily number of deaths (7,298, 7,157, and 7,158, respectively).

On any given day, the leading cause of death is a heart attack, but Mondays are the worst for this cause of death. Suicides also peak on Mondays. Sunday sees a peak in firearms deaths. Tuesday is the safest day for car crashes but sees the most deaths from flu and pneumonia. Wednesday is neutral—no cause of death is higher than on any other day. The fewest firearms death occur on Thursdays. Fridays are near the peak for car crashes and drug overdoses, but these causes of death reach a peak on Saturdays.

SOURCE   CDC NCHS

Historic Top 10 Causes of Death in the US

	2020	1950	1920	1900
1.	Heart disease	Heart disease	Influenza/ pneumonia	Influenza/ pneumonia
2.	Cancer	Cancer	Heart disease	Tuberculosis
3.	COVID-19	Stroke	Tuberculosis	Gastrointestinal infections
4.	Accidents	Accidents	Stroke	Heart disease
5.	Stroke	Neonatal conditions	Kidney disease	Stroke
6.	COPD	Influenza/ pneumonia	Cancer	Kidney disease
7.	Alzheimer disease	Tuberculosis	Accidents	Accidents
8.	Diabetes	Arteriosclerosis	Diarrheal disease	Cancer
9.	Influenza/ pneumonia	Kidney disease	Premature birth	Senility
10.	Kidney disease	Diabetes	Death in childbirth	Diphtheria

NOTES   2020 mortality data is statistically distorted due to the impact of the COVID-19 pandemic. Accurate data for cause of death before 1900 is not available. COPD is chronic obstructive pulmonary disease. Neonatal conditions include premature birth. Arteriosclerosis today is classified as hypertensive disease. Senility today is classified as Alzheimer disease and other dementias.

SOURCES   CDC NCHS, cdc.gov/nchs/ndi; WHO, who.int/data/ mortality

An Actuary's Introduction

Joe Buff, MS, FSA

Since the dawn of humanity, civilization has advanced because men and women have a natural appetite, or are forced by necessity, to take risks, sometimes very deadly risks. We confront this issue almost every moment of our lives, from birth to death and while awake or while sleeping. Our Stone Age ancestors needed to figure out *which* risks, big and small, were worth taking and which ones should be avoided with great care. These constant evaluations, trade-offs, and gambles challenged (and excited or horrified) every human adult or child. Seeing a neighbor killed while hunting a saber-tooth tiger (whether for fur, trophies, or food) taught people that large carnivores were extremely dangerous and that herbivores were a much safer bet—and their meat more delicious and tender, too. Seeing a loved one drown while swimming in a lake during a mastodon-steak cookout or at a clan gathering taught them that even family recreation and sports activities can be fatal. Watching a husband, wife, child, or parent slowly die of cancer or drop dead from a heart attack—even before these common causes of death were properly understood—showed that dying could sometimes occur for no obvious external reason.

Then there were the risks of being killed by other people, whether murder or in war.

In modern times, many activities have become a lot less hazardous, and medical science advances constantly, but the age-old causes of death remain stubbornly with us and, with them, the drive to assess risk.

In the 1600s, the famed French mathematician and philosopher Blaise Pascal played a big role in inventing modern probability theory. The story goes that he was asked by a rich aristocrat friend to solve a technical problem in which, quite literally, there was a whole bunch of money on the table. How can a group of high-stakes gamblers fairly

Blaise Pascal, the father of actuarial science

divvy up the gigantic pot, when their card game gets interrupted in the middle of a hand?

It was a very practical question but hardly a simple one to answer. With considerable effort and subtle reasoning, Pascal derived a mathematically rigorous solution. Centuries of further analysis then ensued by some of humanity's best minds, who carried out all sorts of theoretical probability research and practical statistics sampling. The goal was to quantify, clearly and objectively, the risks and rewards of all sorts of human endeavors.

The most daunting risk of all is the risk of dying before one's time. A companion risk, however, is that of living too long. Both these calamities can cause serious financial distress and standard-of-living discomfort for the individual and their dependents/survivors. Seeking ways for big groups to collectively take on and evenly spread out these stresses, thereby somewhat mitigating the suffering of those unfortunate group members randomly struck by overly short or excessively long lifespans, mathematically inclined businesspeople invented life insurance and pension annuities. Then, to find the ideal but tricky compromise between making participation affordable and keeping the program solvent, careful study of life and death data was called for along with disciplined mathematical modeling of cashflow ins and outs over long time frames. Actuarial science grew out of this need for sound management of corporate programs such as life insurance and private pension plans and government programs such as Social Security.

Actuaries use complex formulas to overcome 2 problems. First, mortality rates by individual age generally increase the older you get, but not many people want to buy an insurance policy (or fund a pension plan) with charges that also go up (and up, and up) every year. Second, although in the aggregate the proportion of people in a large group who will die (or not) in the next year is predictable, no one can tell which given individuals will be the unlucky ones. The first problem is solved by procedures to derive "level premium life insurance" costs (or pension plan "level normal costs"), supported by "reserve" monies insurers must hold back out of the cash received in early years to then release to help pay benefits in later years. The second is solved by relying on the Law of Large Numbers, which says that the more people who join a risk pool, the more the actual annual death toll experienced among them will closely mimic the expected statistical mortality rate. So, while nobody has a crystal ball for knowing when any one person will die, the math lets businesses offer—and society enjoy—protective coverages for wage earners and their families.

Still, nobody knows when and, especially, how any individual will die. It takes extra work to figure out, objectively, exactly how one given person ought to make all their different daily decisions and choices and most wisely conduct their various affairs and activities to mitigate the risk of death in the best-informed way. Individuals don't get direct benefit from the Law of Large Numbers, and the aggregate pooling of financial risk doesn't do you much good if you're dead.

Thinking explicitly about death isn't laugh-out-loud fun, but it undeniably holds a grim fascination. The entertainment industry offers us an endless flow of horror movies, homicide dramas, disaster documentaries, and more. Rambo movies are famous (and infamous) for their immense body counts. But, as actuaries will tell you, thinking about death can also have great *utility*. It can help us make decisions and choices about what we and our family and friends ought to do or ought *not* to do to succeed at our business enterprises or at our pastimes or in love, to escape unnecessary monetary loss, illness or injury, and to avoid untimely death. You don't have to be an actuary or even be any good at math to make such decisions and choices wisely.

I'm not saying that all you need is this book. But it is a start.

When charting a course through life, and while planning daily activities for yourself and your loved ones, it is wise to "Know before you go." If you're thinking of getting some fresh air and exercise by going skiing or swimming, what are the probability odds of getting killed in an avalanche or eaten by a shark? If you're planning to travel much near or far, which type of transportation is safest or most dangerous (while you also trade off speed, cost, and discomfort versus luxury): hiking or hitchhiking, biking or riding a horse, motorcycle, car, bus, truck, passenger train, freight train, or commercial or private airplane or helicopter? If you're planning to re-

Actuary at work

tire soon to a different part of the country, which regions are more or less safe from earthquakes, floods, hurricanes, tornadoes, mudslides, wildfires, or tsunamis? And what, really, is the chance in the US of getting killed by a volcano? (I'm looking at you, Mount St. Helens, May 18, 1980.)

Being able to compare the odds of different ways to die can help us to put off that inevitable but regrettable moment when, sooner or later, the life of every human being is ended—by something or other. In short, how can we each best preserve life and limb while at the same time each have what we consider a healthy, active, emotionally rewarding time on this Earth? How can we most wisely make sure to have a truly fine and fun time, full of positive social interactions, beneficial mental and physical engagement, enjoyable recreation and sports, and memorable travel experiences–all to our personal tastes? How can this game of life, this ultimate gamble against the Grim Reaper, be based on best-informed ways to pick and choose how we do spend however long each of us does have?

One practical problem with making the wisest choices is that, in modern society, accurate and objective data can be hard to come by. Information is power. But in today's Information Age, popular perceptions of relative dangers can easily get misshaped by mouthpieces manipulating the data: lobbyists for different products (tobacco, alcohol) and industries (Big Pharma, Big Auto), Hollywood storytellers trying to sell movie tickets, distorted reporting in the media trying to sell newspapers and get website clicks ("Dog Bites Man" isn't news, but "Man Bites Dog" makes the front page), and even by malicious "fake news" Internet trolls, both foreign and domestic. On a day-to-day basis we can also be misled by our own selective misperception, a cognitive process of not quite registering all the goings-on around us clearly and accurately. This is where actuaries and other statistical analysts can offer verifiable facts, to help eliminate guesswork and subjectivity, and to empower you to work around these various biases.

There's an old saying, shared by cops and other first responders, physicians and morticians, county coroners, public health and municipal safety practitioners, and, yes, actuaries: "The bodies in the morgue don't lie." Many different government agencies, nonprofit organizations, trade and industry associations, and private companies make it their business to tally up the body count accurately, along with their reported causes of death. This vast storehouse of Big Data is the basis of the carefully researched entries in this book. So, other than simply for the excellent, rewarding purposes of entertainment and general edification, how can readers best harness the 1,000+ entries here to lead longer, safer, healthier lives?

Probabilities and Death Statistics: A Pocket Primer

You don't need to understand the intricate processes involved in determining death statistics to be able to make practical use of them. In short, the probability (odds, chances, likelihood, rate, risk) of any event occurring is usually expressed in 1 of 3 ways:

- *Decimal form*, i.e., a number between 0 and 1
- *Percentage form*, i.e., between 0 percent and 100 percent
- *Gambling odds form*, i.e., as a relative proportion of winning versus losing a bet, such as 2 to 1.

In the decimal form, 0 means something never happens, whereas 1 means it is certain always to happen. In the percentage form, 0 percent and 100 percent have the same meanings as 0 and 1. Something that sometimes occurs and sometimes doesn't will have a probability somewhere between 0 and 1, between 0 percent and 100 percent.

Converting from one form to another is pretty simple. Gambling odds such as "W to L" can be converted into decimal probability using the formula $W/(W + L)$. For instance, if the odds to win in a horse race are given as 3 to 7, then the probability of winning is $3/(3 + 7)$ which gives 0.3. Any decimal value between 0 and 1 can be converted into a percentage between 0 percent and 100 percent just by multiplying it by 100. So, in this horse race example, the winning probability of 0.3 can be expressed in percent as $0.3 \times 100 = 30$ percent.

But death rates, among various event probabilities, like death among all calamities, are in a class by themselves, with their

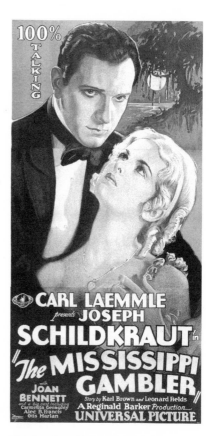

Figuring the odds is at the heart of intelligent risk taking.

own documentary conventions. For the rarest causes of death, the total number killed in a year is informative: the death *count*. But this isn't good for comparing the relative risk of the same cause of death in different places with different total populations. Death *rates* are needed for this. Death rates are sometimes expressed by saying how many people die in a period of 1 year out of 100,000 people alive at the start of the year. They can also be given as a percent or a decimal.

But not every *specialized* death statistic would mean much if the total deaths from that cause of death (COD) were simply compared to the total US population. This is true, for instance, for data comparing the degree of deadly danger involved in a given industry or profession. These numbers could be of keen interest to those engaged in that line of work; to job seekers contemplating that line of work; and to business owners, investors, executives, foremen, shop stewards, union reps, supervisors, risk managers and insurance consultants, and occupational safety regulators. For instance, the probability of being killed in a coal mine, to be most informative, should compare the number of such deaths in a year to the total number of people who work in coal mines during that year. The odds of being killed doing farm work should compare the number killed to the number of people who work on farms. Ditto for loggers, say, and for first responders.

Such statistics gain the broadest meaning when they are provided along with recent census counts of how many people *actually do work* in that industry (in coal mines, on farms, at logging, or as first responders) out of the total national population. For perspective, the total number of people living in the United States at year-end 2020 was 331 million.

It gets a bit more complicated when we talk about death involving transportation. For instance, death rates in automobile accidents can be calculated *per mile that cars travel* (vehicle miles), *per mile that people drive or ride* (passenger miles), or *per trip that people (or vehicles) take*. To measure the life-prolonging benefit *of wearing a seat belt*, we can compare the death rates among all vehicle mishaps divided into 2 subgroups: those in which seat belts were or weren't worn.

For death by disease, a very important statistic is each disease's *case fatality rate*. This is the percent of *all people who get the disease* who, sooner or later, *die specifically from that disease*. As we've seen with COVID-19, such stats (for any disease) can be especially hard to know accurately, for a pair of reasons: Not everyone who has a mild case was ever tested and diagnosed as positive, and not all those who die from it had that cause identified correctly on their medical records. Public health authorities have good ways to adjust for the biases due to such underreporting of cases and/or of deaths; recorded deaths divided by *diagnosed* cases will give a higher death rate than recorded deaths divided by an estimate of *total* (including undiagnosed) cases.

These different sorts of more specialized death statistics all provide interesting and informative data. The essential things are, first, to be clear which basis you're looking at, and, especially, to make sure to use the right basis for your given purpose.

The most important thing of all to know is that 2 *different* causes of death (CODs) can be compared and ranked as to their relative likelihoods (their relative *dangers*), by comparing their 2 death rates or probabilities, according to whichever particular documentary convention (decimal, percent, per 100,000) was used to tally the data. To be a meaningful comparison, though, it is *essential* that the 2 numbers be based on the *same* form of expression, such as "probability .03 compared to .01" or "probability 3 percent compared to probability 1 percent" or "3,000 deaths per 100,000 population compared to 1,000 deaths per 100,000." Whichever way they're expressed, what such a proper apples-to-apples comparison tells you is that the COD with the *higher* number is *more likely to occur* than the COD with the *lower* number. *How many times more likely* can be figured out simply by dividing the higher number by the lower number. In the example here, the COD with 3 percent is 3 times as likely as the COD with 1 percent. Such reliable statistical data lets you make wiser choices and take wiser precautions, when considering different work or leisure activities, health habits, or lifestyle choices. Good data comparisons give helpful guidance about deaths that, whatever the cause (accident, disease, act of nature, homicide), might be preventable or avoidable or inevitable.

Let's look now at a couple of examples taken from this book, to show how, by using some real-world specifics, you can choose between different ways to bet your (or someone else's) life. How can you choose between job offers in 2 different states where you might or might not want to live, say, Alabama and Alaska? If you look at the entry for "Population Mortality: Overall US, and Worst and Best States," you can see that the recent death rate for Alabama was 11.1 per 1,000 people, whereas for Alaska it was 6.0 per 1,000 people. Since Alaska's 6.0 is just over half of Alabama's 11.1, with both numbers on a comparable annual-deaths-per-thousand-total-population basis, you can see that Alaska would be the wiser choice from a life expectancy perspective.

Let's say you're on a public health task force, and you want to allocate the available anti-substance-abuse education budget between programs that separately target binge drinking and illicit drugs. Your task force decides to divvy up the spending in proportion to the national death rates from alcohol abuse versus drug abuse. You see in the

"Alcohol Abuse: Overview" entry that about 95,000 people die in the US every year as a result of abusing alcohol. Then you see in the "Drug Overdose: Overview" entry that drug abuse killed 70,630 people in the US in 2019. This says that you should allocate 95,000 divided by 95,000 + 70,630 to alcohol education, or about 58 percent, and the rest (42 percent) to drug education.

A Note on Data Quality

Actuaries know that any very large database can have gaps in its completeness, accuracy, and timeliness. Some of the statistics can be wrong, or missing altogether, or out of date. Some can even be intentionally misstated, due to political agendas or cultural biases. You'd think that stats about dead people would be straightforward.

Well, yes and no.

Challenges begin at the moment of death. Why someone died is sometimes very obvious: They were shot in the head, or run over by a garbage truck, or jumped off a bridge, or died of cancer. But sometimes it's not. If somebody is found dead, with no evident wounds or external causes, it might require an autopsy to figure out what killed them. Autopsies are expensive and take time; coroners make mistakes. Sometimes the decedent's family objects to an autopsy, perhaps for religious reasons. Sometimes the family brings pressure to conceal a suicide. Sometimes a helpful witness is actually a dissembling murderer. And sometimes a death crosscuts different causes or categories: A fatal car accident, say, might or might not result from a road rage incident, or vehicular suicide, or mechanical failure, or bad road conditions, or DUI or speeding or other driver error, or the death of the driver from

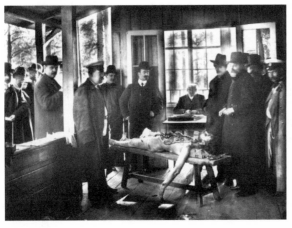

Forensic autopsies seek to determine the *cause of death* (specific injury or disease that caused the death) and *manner of death* (how the injury or disease led to death). Russian authorities examine the body of priest turned informant Grigory Gapon, 1906. Cause of death: asphyxiation by hanging. Manner of death: homicide.

a sudden stroke. Official accident investigators don't always have the time or funding resources, and all the technical forensic evidence, to sort out what really went on. They write down their best guess—which, though informed, is still only a guess.

US-wide death statistics begin with data from the over 3,000 separate counties in the 50 states, plus DC and other territories and possessions. The sources are likely to use different procedures, definitions, and standards, which can lead to some noise in the raw numbers, which then all have to get tallied up by some central office (think the FBI or the CDC). This analysis takes serious time, so a complete picture of US-wide death rates from different specific causes is often a few calendar years out of date by the time it gets revealed to the public.

In the real world, as opposed to on television shows such as *CSI* or *Law & Order*, homicides and suicides are especially challenging to classify and explain completely and correctly. Murders/suicides, suicides by cop, active shooter incidents (mass killings), not to mention gang-related murders and road rage killings, can easily be mischaracterized, misunderstood, even missed altogether—thus going undercounted or double-counted. The perpetrator is usually dishonest/uncooperative and often ends up dead, mute, and intestate. Sometimes a homicide goes unsolved (a cold case), or the suicide is mistaken as natural causes or murder. Any one homicide could be ambiguous or a hybrid, defying clear-cut attribution of motive and/or circumstances. Examples include missing persons (disappearances), officer-involved deaths (justified or excessive force?), drive-by shootings, deaths from random or stray gunfire, killings due to insanity/psychosis, hunting "accidents," and homicides in which both killer and victim might or might not be under the influence of drugs and/or alcohol. Such data can only be classified as "Unknown" or "Other."

Then there is intentional misreporting, usually underreporting, due to political pressures, personal prejudices, or even "fake news" disinformation. For example, because high crime rates are bad for business, tourism, and real estate values, murders, hate crimes, and road rage killings might get understated. Sometimes out-of-court plea deals lessen the severity of a reported crime where someone died. High suicide rates don't look good either. Such comparisons between different countries are particularly suspect. Would you, for instance, trust the repressive government in China to report accurately on citizens killing themselves?

Professional statisticians are well aware of all these potential distorting factors. They have sophisticated methods to try to account for such data errors and make corrections. What's important is that the figures that do get finally published serve as useful guides about how to live a safer and better life, and better protect our loved ones and friends from tragic calamity.

Interpreting the Data in This Book

For some entries, a recent and very representative annual average death toll can be calculated by averaging over just a few recent years of good data, such as 2014 through 2019. Sometimes those annual (year by year) death counts don't even vary a lot from their overall average. For other entries, such as the entry on meteorite strikes, the probability of dying is based on a 3,000-year-long average of 0 deaths in almost every historical year and 2 or 3 sudden disasters that each killed in the tens of thousands.

Whenever possible, this book uses recent (within the past 5 years whenever possible) national annual data from primary sources such as government agencies, trade and industry associations, nonprofit organizations, and medical studies. Secondary sources such as academic studies, insurance industry consumer information, and even studies compiled by personal injury law firms are used to supplement or update primary source data or when primary sources aren't available or are out-of-date. News media reports and history books are used for some discussions of historic and current events.

Past data cannot accurately predict future experience. Nor can overall national population death odds, given as fractions of a percent or as deaths per 100,000 people exposed to a risk, ever be counted on to happen in one given person's life. You, as one individual data point, are at any moment either all dead or all not dead. Even fatal phenomena that have recently had steady, stable annual US death tolls could in the near future vary tremendously and have either much higher mortality rates or much lower. This book is no crystal ball. Another deadly pandemic, or in contrast, a cure for cancer, could have very significant impact. Human existence is a great game. We each proceed at our own risk!

Glossary of Probability and Death Rate Terms

A good way to explain more of the working concepts underpinning probability theory and practice is via the following glossary of foundational terms. This specialized lingo shows you what *basic ideas* are useful to know about. Reading through the definitions will help you gain the needed knowledge.

Accidental death. An accidental death results from trauma caused in an accidental event, such as a car crash, an oil rig explosion, carbon monoxide poisoning, or a fall off a cliff (as opposed to one caused by disease/illness, homicide, or suicide). For life insurance purposes, a person killed in a homicide—depending on policy terms, the precise circumstances of the killing, and criminal law of the jurisdiction—might be covered as a type of accidental death.

Age-adjusted death rate. Death rate statistically modified to remove the effect of different age makeups among different populations, to allow more meaningful comparisons.

Age-specific death rate. The death rate applying only to people whose ages fall into a specified age range.

Brain dead. The 3 essential findings in brain death are coma, absence of brainstem reflexes, and cessation of breathing. Someone determined to be brain dead is legally and clinically dead even if their heart is still beating.

Case fatality rate. The percentage, among all people who are diagnosed with a disease, who actually die of it.

Cause of death. The *underlying* cause of death listed on a death certificate is the condition or injury that sets in motion the chain of events leading to death. This chain of events may include one or more *intermediate* causes, and ends with the *immediate* cause. For example, an HIV infection (underlying) may develop into AIDS (intermediate), leading to a fatal pneumonia (immediate). There may also be multiple *contributing* causes of death in addition to this main chain of events; for example, an AIDS patient whose immediate cause of death is pneumonia may also have been weakened by other opportunistic infections. Except as noted, the death rates cited in this book are based on underlying cause of death.

Compound probability. The probability that 2 or more events will *all* occur. When they are independent, this compound probability equals the multiplicative product of each event's own separate probability. For instance, if Event A has a probability of 0.3 and Event B has a probability of 0.4, then the probability that both A and B occur is $0.3 \times 0.4 = 0.12$.

Conditional probability. The probability that 1 event will occur given that another event has already occurred.

Death. The permanent cessation of all vital functions of the body, including heartbeat, respiration, and brain activity.

Death rate. The fraction of a population who die in a specific period of time, such as 1 year.

Expectation: The average amount or value of the loss or damage occurring from an event.

Full-time equivalent. Death statistics for a specific occupation are most meaningful when the number killed on the job in 1 year, in that line of work, is correlated with (normalized to) the total number of people who had that job during that year. This is done using the annual US deaths per 100,000 Full-Time Equivalent (FTE) employees. FTE takes proper account that exposure to an occupation's risk of death only happens while each individual worker is "on the clock." That summing together of all the various individual exposures is derived by adding together (a) the number of actual full-time employees, each counted as "1" person, (b) people who worked all year but only part-time, counting for each the appropriate fraction of the time they did work, and (c) people who worked full-time but only seasonally, also using the appropriate fraction of 1 person. For instance, if agricultural production employed 1,000,000 people full-time year round, and another 1,000,000 seasonal workers full-time but only for the 3 months of summer, plus another 1,000,000 school-age youngsters only half-time all year, then the FTE for agricultural workers would be $1,000,000 + 1,000,000 \times \frac{3}{12} + 1,000,000 \times \frac{1}{2} = 1,750,000$.

Incidence. The rate of occurrence of *new* cases of a specific disease, in a specific population during a specific period.

Independent events. Two or more events that do not share any cause-and-effect connections.

Infant mortality. Death occurring between live birth and attaining age 1.

Legally dead. A jurisdiction's court system has ruled favorably on a motion, usually filed by next of kin, that someone who has been missing with no signs of life for a mandatory years-long waiting period, is to be treated for all legal purposes as deceased.

Life expectancy. The average number of years that are lived by all individuals in a specified population, often measured only between attaining age 1 and later date of death, so as to leave out the effects of infant mortality.

Life span. How many years an individual actually lives or did live.

Morbidity. A synonym for "disease."

Mortality. A synonym for "death."

Normalization. The general process of deriving and adjusting mortality statistics so that they are properly comparable between different populations or subpopulations. For example, age-adjusted death rates, or number of deaths per 100,000 of population or subpopulation, use normalization.

Population mortality. The death rate actually experienced by the overall population of a specified place or region, including all causes of death and without adjustment for age and sex makeup.

Population probability. The probability of an event as measured across the entire population of people to which it can happen, such as the whole population of a country, or the totality of people employed in a particular occupation or who participate in a particular activity.

Premature death. Colloquially, premature death is when someone dies "before their time." Actuarially, it means someone whose attained age, at time of death, is less than the average life expectancy applicable to members of their particular demographic cohort, meaning all Americans born in the same year and having the same race and the same sex. When a death is premature, the statistic known as Years of Life Lost (YLLs) can be calculated (see that Glossary entry).

Prevalence. The total number of cases of a specific disease in a specific population at a given moment.

Preventable death. A preventable death is one whose underlying primary cause was some risk factor that could have been avoided through healthier and safer lifestyle, habits, and behaviors. Examples of preventable deaths are those due to smoking, morbid obesity, or impaired driving. Population statistics on preventable deaths are usually approximate/estimated, because death certificates often omit such information in favor of an immediate cause of death, such as lung cancer, heart attack, or going through the windshield and hitting the pavement in a car crash.

Probability. The odds or chances that a specified random or unpredictable event (such as cancer deaths or car accidents) will actually occur. Probability is usually measured and reported for a specified time period, such as 1 calendar year, or 1 year of a person's life starting when they first reach a specified age (age-specific).

Risk. The possibility or danger of something bad happening.

Risk management. Steps taken to reduce the probability of something bad happening.

Risk mitigation. Steps taken to reduce the damages caused if something bad does happen.

Sample probability. An estimate of population probability, obtained by polling or sampling a randomly chosen (unbiased) subset out of the total population. Usually done to get some approximate data while saving time, money, and work.

Statistical mean. The average result for an outcome, event, or phenomenon.

Subpopulation. Part of an overall population, in which all members of the subpopulation share a particular characteristic of interest, which they do not share with any of the rest of that overall population. The characteristic might be all American residents working in a particular occupation (such as firefighters or coal miners) or living in a particular smaller region (a city like Las Vegas or a state like Montana), or being diagnosed with a particular disease (such as lung cancer or COVID-19).

Years of life lost. For an individual, or for a larger group within an overall population, Years of Life Lost (YLLs) is calculated as the summed total of each affected individual's actuarial Life Expectancy (see that Glossary entry), minus their actual age at their premature death (see Glossary entry). Years of Life Lost is of interest because it gives an estimate of how many more years the dead people in the group under discussion could have collectively lived, if their deaths had not been premature.

Lies, Damned Lies, and Statistics: A Note on Sources

In compiling the statistics in this book, we used only sources that are widely accepted among actuaries as reliable, objective, accurate, current, and publicly available. Almost all the data comes from 3 types of primary sources: government agencies, industry associations, and health advocacy associations. The information from these sources is constantly being updated and changed, and much of it is in the form of statistical databases.

Secondary sources such as academic studies, peer-reviewed journal articles, insurance industry consumer information, and even studies compiled by personal injury law firms are used to supplement or update primary source data or when primary sources aren't available. News media reports and history books are used for discussions of historic events.

For each entry in this book, we provide the data source(s), usually with a URL. Interested readers can learn more and see the latest data by visiting the individual websites and exploring the resources they provide. The websites for these sources are constantly being updated, sometimes daily; much of the statistical information is in the form of databases or reports. We give the URLs for only the home pages and major subsections to avoid the problem of very lengthy, outdated, or broken links.

Two government agencies are the best starting points for population and health statistics: the Centers for Disease Control (CDC) and the National Institutes of Health (NIH).

The CDC (cdc.gov) is the national public health agency of the United States. It is one of the major operating components of the Department of Health and Human Services. The agency's mission is the protection of public health and safety through the control and prevention of disease, injury, and disability in the US and around the world. The National Center for Health Statistics (NCHS) division is the primary source for vital records such as the National Death Index. The CDC also publishes the *Mortality and Morbidity Weekly Report* (MMWR), a fascinating source of updated statistics. The National Institute for Occupational Safety and Health (NIOSH) is a valuable source for workplace death and morbidity information.

Data from the CDC is most easily accessed through CDC WONDER (Wide-ranging ONline Data for Epidemiologic Research) at wonder.cdc. gov. This valuable resource manages nearly 20 collections of public use data for US births, deaths, cancer diagnoses, tuberculosis cases, vaccinations, environmental exposures, and population estimates, among many other topics. The online databases are updated annually; some collections are updated monthly or weekly. The databases we drew on the most for this book are:

> CDC NCHS: National Center for Health Statistics, cdc.gov/nchs
> CDC NIOSH: National Institute for Occupational Safety and
> Health, cdc.gov/niosh
> CDC NVSS: National Vital Statistics System, cdc.gov/nchs/nvss
> CDC USCS: US Cancer Statistics, cdc.gov/cancer/uscs

The NIH (nih.gov), part of the Department of Health and Human Services, is the nation's medical research agency. The NIH has 27 separate institutes and centers that conduct research that leads to better health. Each individual institute and center provides massive amounts of information, including detailed statistics on disease incidence and prevalence within their specialties. This book draws on data from the following NIH institutes:

NIH: National Institutes of Health
> GARD: Genetic and Rare Diseases Information Center,
> rarediseases.info.nih.gov
> NCATS: National Center for Advancing Translational Sciences,
> ncats.nih.gov
> NCI: National Cancer Institute, cancer.gov
> NHLBI: National Heart, Lung, and Blood Institute, nhlbi.nih.gov
> NIA: National Institute of Aging, nia.nih.gov
> NIAAA: National Institute on Alcohol Abuse and Alcoholism,
> niaaa.nih.gov
> NIAID: National Institute of Allergy and Infectious Diseases,
> niaid.nih.gov
> NIAMSD: National Institute of Arthritis and Musculoskeletal and
> Skin Diseases, niams.nih.gov
> NICHD: *Eunice Kennedy Shriver* National Institute of Child
> Health and Human Development, nichd.nih.gov
> NIDA: National Institute on Drug Abuse, drugabuse.gov
> NIDDK: National Institute of Diabetes and Digestive and Kidney
> Diseases, niddk.nih.gov
> NIMH: National Institute of Mental Health, nimh.nih.gov
> NINDS: National Institute of Neurological Disorders and Stroke,
> ninds.nih.gov

The best starting points for statistics about most other causes of death are the numerous other federal agencies that collect this information as part of their larger overall responsibilities. The US Coast Guard (USCG) collects information about deaths at sea, for example, while the Federal Aviation Administration (FAA) collects data on aviation deaths. The Consumer Product Safety Commission (CPSC) collects data on fatalities from the many, many products that pose a fire, electrical, chemical, or mechanical hazard.

The data from one agency division often overlaps with another or with a different agency. The National Oceanic and Atmospheric Administration (NOAA) and the National Weather Service (NWS), both divisions of the Department of Commerce, both collect data on weather-related fatalities. So does the Federal Emergency Management Agency (FEMA), a division of the Department of Homeland Security. The Bureau of Labor Statistics (BLS) and the Occupational Safety and Health Administration (OSHA) are both divisions of the Department of Labor; both gather information on work-related mortality and morbidity. The information is complemented by data from the CDC's NIOSH division.

Industry associations are another valuable source of death data drawn from a range of state and federal sources—an approach that often provides further insight. Deaths from railroad accidents, for example, are reported by the Federal Railroad Administration, part of the Department of Transportation. This information is augmented by data from the Association of American Railroads and Operation Lifesaver, Inc. Similarly, data on recreational boating deaths from the USCG is augmented by the National Association of Boating Law Administrators; the FAA collects data on aviation deaths, as does the Experimental Aircraft Association (EAA). The National Highway Traffic Safety Administration (NHTSA), part of the Department of Transportation, collects data on traffic fatalities. So do 2 important insurance industry associations: the Insurance Institute for Highway Safety-Highway Loss Data Institute (IIHS-HLDI) and the Insurance Information Institute (III).

Health advocacy associations such as the American Heart Association (AHA), the American Society for Clinical Oncology (ASCO), and the National Organization for Rare Disorders (NORD) collect death data on a vast range of diseases and conditions, often in greater and more nuanced depth than the NIH or CDC.

Major sources we used are:

AACR: American Association for Cancer Research, aacr.org

AAP: American Academy of Pediatrics, aap.org

AAPCC: American Association of Poison Control Centers,
aapcc.org

AAR: Association of American Railroads, aar.org

ACS: American Cancer Society, cancer.org

AHA: American Heart Association, heart.org

AHQR: Agency for Healthcare Quality and Research, ahrq.gov

ALA: American Lung Association, lung.org

AMA: American Medical Association, ama-assn.org

ASCO: American Society of Clinical Oncology, asco.org

BATFE: Bureau of Alcohol, Tobacco, Firearms, and Explosives,
atf.gov

BLS: Bureau of Labor Statistics, bls.gov

CMS: Centers for Medicare and Medicaid Services, cms.gov

CPSC: Consumer Product Safety Commission, cpsc.gov

DOJ: US Department of Justice, justice.gov

EAA: Experimental Aircraft Association, eaa.org

FAA: Federal Aviation Administration, faa.gov

FBI: Federal Bureau of Investigation, fbi.gov

FMCSA: Federal Motor Carrier Safety Administration,
fmcsa.dot.gov

FRA: Federal Railroad Administration, railroads.dot.gov

FTA: Federal Transit Administration, transit.dot.gov

GVA: Gun Violence Archive, gunviolencearchive.org

IARC: International Agency for Research on Cancer, iarc.who.int

IIHS-HLDI: Insurance Institute for Highway Safety/Highway
Loss Data Institute, iihs.org

III: Insurance Information Institute, iii.org

NASA: National Aeronautics and Space Administration, nasa.gov

NASBLA: National Association of State Boating Law Administrators , nasbla.org

NATA: National Athletic Trainers Association, nata.org

NCCSIR: University of North Carolina National Center for
Catastrophic Sports Injury Research, nccsir.unc.edu

NCI SEER: National Cancer Institute Surveillance, Epidemiology,
and End Results, seer.cancer.gov

NFPA: National Fire Protection Association, nfpa.org

NHTSA: National Highway Traffic Safety Administration,
nhtsa.gov

NKF: National Kidney Foundation, kidney.org

NLSC: National Lightning Safety Council,
lightningsafetycouncil.org

NOAA: National Oceanic and Atmospheric Administration, noaa.gov

NORD: National Organization for Rare Disorders, rarediseases.org

NPS: National Park Service, nps.gov

NSAA: National Ski Area Association, nsaa.org

NSC: National Safety Council, nsc.org

NTSB: National Transportation Safety Board, ntsb.gov

NWS: National Weather Service, weather.gov

OLI: Operation Lifesaver, Inc., oli.org

OPTN: Organ Procurement and Transplantation Network, optn. transplant.hrsa.gov

UNOS: United Network for Organ Sharing, unos.org

USCG: United States Coast Guard, uscg.mil

USGS: United States Geological Survey, usgs.gov

USPA: United States Parachute Association, uspa.org

USPSTF: US Preventive Services Task Force, uspreventiveservicestaskforce.org

WHO: World Health Organization, who.int

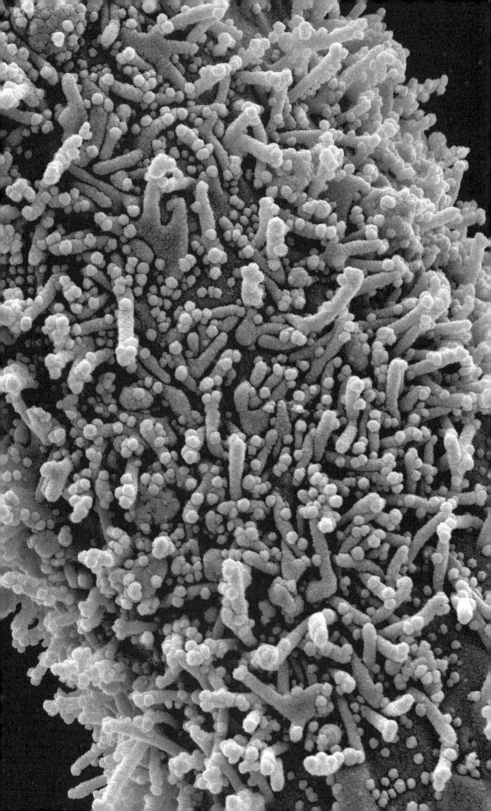

NATURAL CAUSES

Acute Conditions and Diseases

An acute condition generally develops suddenly and lasts a short time, often only a few days or weeks. Acute conditions are often caused by a viral or bacterial infection but can also be caused by an injury (a fall or car accident, for example) or by drug side effects and interactions. Acute conditions can be medical emergencies and some can be fatal, especially in the very young and the elderly as well as those with some preexisting medical conditions, whether related or unrelated to the acute issue.

The acute conditions discussed in this section are primarily caused by infections and adverse drug reactions/interactions. Deaths from acute injuries from falls, car crashes, and other sources are discussed throughout this book in the relevant sections (e.g., Motor Vehicle Crashes, Boating, and so on). The lethal risk is presented as the risk for people suffering from that particular cause, and in some cases, also as the risk for the overall population.

Acetaminophen (Tylenol) Overdose

ANNUAL US DEATHS   Acetaminophen overdose is the leading cause for calls to Poison Control Centers (more than 100,000 per year) and accounts for more than 56,000 emergency room visits, 2,600 hospitalizations, and an estimated 458 deaths due to acute liver failure each year.

Acetaminophen is an over-the-counter drug helpful for treating pain, inflammation, and fever. It is safe if taken as directed, but if too much is ingested in too short a period, it can cause severe liver damage and even death from liver failure. Acetaminophen poisoning leading to acute liver failure is estimated to kill 458 people in the US every year.

STAY SAFE   Use all over-the-counter medications only as directed on the label. Do not combine acetaminophen with alcohol.

SOURCE   AAPCC, aapcc.org/national-poison-data-system

Acute Cholecystitis (Gallstones)

ANNUAL US DEATHS   Gallstone disease causes approximately 100,000 deaths, some 7,000 of these due to acute complications of gallstones, such as perforation of the gallbladder.

Cholecystitis is inflammation of the gallbladder, usually because a gallstone is blocking the duct that carries bile from the gallbladder to the rest of the bile duct system. Acute cholecystitis is sudden inflammation characterized by severe pain, jaundice, and fever. It is a medical emergency usually treated by removing the gallbladder surgically.

Each year, in the United States, approximately 500,000 people develop symptoms or complications of gallstones and need to have the organ removed (cholecystectomy). The mortality rate for emergency cholecystectomy for acute cholecystitis is in the range of 3 to 5 percent. Complications such as gallbladder perforation occur in up to 10 percent of patients with acute cholecystitis. Gallbladder perforation is a severe complication with a mortality rate of up to 70 percent. In all, about 7,000 people a year die from complications of acute cholecystitis.

STAY SAFE   If you have severe abdominal pain, seek emergency medical treatment.

SOURCE   L.R. Soto et al. Fatal abdominal hemorrhage associated with gallbladder perforation due to large gallstones. *Proceedings (Baylor University Medical Center)*, April 2014.

Adverse Drug Events

ANNUAL US DEATHS   About 7,000 people die each year from ADE, most of them over age 65.

An adverse drug event (ADE) is an injury or dangerous effect caused by taking 1 or more medications. ADEs can be adverse drug reactions (ADRs), severe drug side effects, or allergic reactions. ADEs are often preventable because they mostly involve just 4 types of drugs: anticoagulants (e.g., warfarin), insulin, oral diabetes agents (e.g., sulfonylurea), and opioid painkillers. About two-thirds of hospital admissions for ADEs among older adults involved these drugs.

In the hospital setting, a very ill patient may have an ADE but die from something else or the ADE may not be recognized as the cause of death. The situation is similar in nursing homes, where patients are usually elderly and take numerous prescription and nonprescription medications. Many ADEs in all settings are simply not reported.

Each year ADEs comprise an estimated one-third of hospital adverse events, cause approximately 280,000 hospital admissions, affect approximately 2 million hospital stays, and prolong hospital length of stay by approximately 1.7 to 4.6 days. They account for two-thirds of postdischarge complications. In outpatient settings, ADEs account for more than 3.5 million physician office visits, an estimated 1 million emergency department (ED) visits, and approximately 125,000 hospital admissions each year. The FDA estimates that about 7,000 people die each year from adverse drug events, most of them over age 65.

STAY SAFE   Discuss all medications carefully with your doctor and take them as prescribed. Wear a medical alert bracelet if you have drug allergies.

SOURCE   US Department of Health and Human Services, Office of Disease Prevention and Health Promotion, National Action Plan for Adverse Drug Event Prevention, 2014.

Anaphylaxis

ANNUAL US DEATHS Estimated at less than 1 per 1,000,000 population

Anaphylaxis, also called anaphylactic shock, is a serious allergic reaction that is rapid in onset and may cause death. The most common causes are reactions to foods (especially peanuts), medications, stinging insects, and latex. Symptoms include hives, sneezing, swelling of the lips or tongue, swelling of the throat and trouble swallowing, shortness of breath, weak pulse, vomiting, and dizziness. Anaphylaxis is a serious medical emergency.

Between 1.6 and 5.1 percent of people in the US are estimated to have experienced anaphylaxis. It is rarely fatal; only an estimated 0.1 percent of emergency room visits end in death. Overall, anaphylaxis is estimated to be fatal in 0.7 to 2 percent of cases, with an annual US deaths that is estimated at less than 1 per 1,000,000 population.

STAY SAFE Administer an auto-injector (EpiPen) immediately if available. Call 911 immediately. Transport the victim immediately to an emergency room if 911 help will be delayed or is unavailable. If you are at risk of anaphylaxis from known allergies, carry an auto-injector, a wallet card, and/or a medical alert bracelet.

SOURCE P.J. Turner et al. Fatal Anaphylaxis: Mortality Rate and Risk Factors. *Journal of Allergy and Clinical Immunology: In Practice*, September 2017.

Burns

ANNUAL US DEATHS About 3,400

Burn injuries are thermal burns from contact with flames and other sources of high heat such as hot liquids (scalds), chemical burns, and electric burns. This section discusses the risk of death for people who survive the burn event itself and get medical treatment. A major burn is one that involves at least 25 percent of the body surface.

Every year in the US, about 1.1 million burn injuries require medical attention. In 2016, some 486,000 people with burn injuries received medical treatment, most in a hospital emergency room. About 40,000 were admitted to a hospital, including some 30,000 who were admitted to a hospital burn center. There are about 128 of these specialized units in the US. The survival rate for patients in burn centers is almost 97 percent overall.

Every day, over 300 children under age 19 are treated in emergency rooms for burn-related injuries. Two children a day die from burn injuries. Most (65 percent) of the children hospitalized for burns have scalds; about 20 percent have contact burns.

Approximately 3,400 people die of burns every year, including about 500 children under the age of 14. Most die from sepsis. The risk of death increases with the extent of the burn injury. Overall burn mortality ranges from 3 to 55 percent, depending on factors including age, extent of the burn injury, presence of inhalation injury, presence of other injuries such as blunt trauma (from a car crash, for example), and presence of serious comorbidities such as heart disease. When the burn area is less than 10 percent of total body surface area, the mortality rate is less than 1 percent. When the burn area is 90 percent or more of total body surface area, or 60 percent or more in elderly people, the mortality rate is near 100 percent. Worldwide, an estimated 180,000 deaths a year are caused by burns.

STAY SAFE   The majority (73.2 percent) of burn injuries happen in the home. Install smoke detectors and make sure they work. Install antiscald devices on tub faucets and showerheads. Water at the faucet should be less than 100 degrees F. Use safe cooking practices, such as never leaving food unattended on the stove and child-proofing burner controls. Supervise or restrict children's use of stoves, ovens, and especially microwaves.

SOURCES   American Burn Association National Burn Repository, ameriburn.org/research/burn-dataset; CDC NVSS, cdc.gov/nchs/nvss

Constipation

ANNUAL US DEATHS   About 900 deaths from diseases associated with or related to constipation. In 2019, 202 people died from fecal impaction, a severe complication of constipation.

Constipation is a condition causing fewer than 3 bowel movements in a week, stools that are hard, dry, or lumpy, or stools that are difficult or painful to pass. Constipation isn't a disease, but frequent or prolonged constipation may be symptomatic of an underlying medical problem. Prolonged constipation can cause fecal impaction, where the stool is so hard it cannot pass with a normal bowel movement. In rare cases, fecal impaction causes bowel perforation, gastrointestinal infection, peritonitis, and other serious complications that can result in death. In 2019, 202 people died of fecal impaction, for a death rate of 0.1 per 100,000 population.

Frequent constipation is a risk factor for death from stroke and cardiovascular disease. People with frequent constipation may have a 12 percent higher risk of all-cause mortality, an 11 percent greater risk of developing heart disease, and a 19 percent greater risk of ischemic stroke. Death caused by "straining at stool" may result from use of the Valsalva maneuver, voluntary contraction of chest muscles on a closed glottis while simultaneously contracting the abdominal muscles, causing a lethal drop in blood pressure.

STAY SAFE Prevent constipation by eating a high-fiber diet, drinking enough fluids, and getting regular exercise. Avoid straining on the toilet.

SOURCES CDC NCHS, cdc.gov/nchs/nvss/deaths.htm; NIH NIDDK, niddk.nih.gov/health-information/digestive-diseases/constipation; K. Sumida et al. Constipation and risk of death and cardiovascular events. *Atherosclerosis,* February 2019.

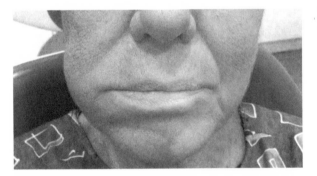

The face of dental abscess

Dental Disease

ANNUAL US DEATHS Estimated 8.25

Statistics related to death from an abscessed tooth are scarce. A 2013 study in the *Journal of Endodontics* found that between 2000 and 2008, a total of 61,439 hospitalizations for periapical abscess (pus at the tooth root) were reported; 66 patients died in the hospital.

STAY SAFE Brush your teeth and get regular dental checkups and cleanings.

SOURCE A.C. Shah et al. Outcomes of hospitalizations attributed to periapical abscess from 2000 to 2008. *Journal of Endodontics*, September 2013.

Diverticulitis

ANNUAL US DEATHS 3,400

Diverticulosis is a condition that occurs when small pouches, or sacs, form and push outward through weak spots in the wall of the sigmoid colon (terminal portion of the large intestine before reaching the rectum). Diverticulosis is very common. About 35 percent of US adults below age 50 have diverticulosis, while about 58 percent of adults over age 60 have it.

Most people with diverticulosis will never develop symptoms or problems. But approximately 5 percent of those with diverticulosis eventually develop diverticulitis, an inflammatory complication. In the US, about 200,000 are hospitalized for diverticulitis each year, 70,000 of whom are hospitalized for diverticular bleeding, which can be life-threatening. Other life-threatening complications include abscess, perforation, peritonitis, fistula, and intestinal obstruction. In all, about 3,400 people die each year from complications of diverticulitis.

STAY SAFE To prevent diverticulitis, eat a healthy diet with moderate amounts of fiber. Maintain a healthy weight, get regular exercise, and don't smoke.

SOURCE NIH NIDDK, niddk.nih.gov/health-information/digestive-diseases/diverticulosis-diverticulitis

Gangrene

ANNUAL US DEATHS Varies greatly with type and associated comorbidities

Gangrene is an overall term for dead or dying body tissues caused by an inadequate blood supply. Gangrene is one of the oldest recognized diseases—the ancient Greeks used the term *gangraina* to describe dead tissue. The most common body parts affected by gangrene are the fingers, toes, hands, arms, feet, and legs. Internal organs, such as the gallbladder, and muscles can also become gangrenous. Symptoms include discolored skin, severe pain followed by numbness, and foul discharge. Tissue damaged by gangrene is dead and cannot be saved.

Quick treatment to remove the dead tissue can keep the gangrene from spreading further.

Dry Gangrene

Dry gangrene is caused by conditions that reduce or block arterial blood flow to parts of the body. Trauma, frostbite, and injuries can cause dry gangrene. So can impaired circulation caused by diabetes, arteriosclerosis, and smoking. The skin over the affected area becomes hard and blackened. Treatment for dry gangrene is usually surgical removal of the dead tissue, often including amputation of the affected extremity. Because the tissue is dead but not infected, dry gangrene rarely causes death.

Wet Gangrene

Wet gangrene is dry gangrene that has become infected by *Clostridium* bacteria species. The stagnant blood from the restricted circulation provides an ideal environment for these bacteria to grow. The affected area is swollen and has oozing blisters; the area is usually warm and red and has a foul odor. Before antibiotics, wet gangrene affected as many as 5 percent of battlefield injuries. Symptoms of wet gangrene develop within 6 to 48 hours and progress rapidly. Toxins produced by the bacteria enter the body, causing sepsis and death unless treated quickly, usually with emergency amputation. About 1,000 cases of wet gangrene are reported annually in the US. The mortality rate of untreated wet gangrene is about 80 percent without treatment and about 20 percent with treatment.

Gas Gangrene

Gas gangrene is a more dangerous form of wet gangrene. It causes severe pain and fever. Gas produced by the bacteria make the skin above the affected area crackle when pressed. Approximately 1,000 cases of clostridial gas gangrene are reported annually in the US. The mortality rate for untreated gas gangrene is 100 percent. For treated gas gangrene, the mortality rate is 20 to 30 percent.

Diabetes and Gangrene

People with uncontrolled type 2 diabetes are at risk of foot ulcers that can become gangrenous. Every year in the US, about 73,000 lower limb amputations are performed on people with diabetes. About 50 percent of patients who have foot amputations for diabetic foot infections die within 5 years.

Fournier's Gangrene

Fournier's gangrene is gangrene with necrotizing fasciitis. It occurs mainly in male genitalia. Risk factors for Fournier's gangrene include diabetes, morbid obesity, alcohol abuse, and a compromised immune system. This form of gangrene is rare, with an incidence rate of 1.6 cases per 100,000 population. An estimated 20 to 40 percent of people with Fournier's gangrene die due to complications such as sepsis.

STAY SAFE Get prompt treatment for frostbite, injuries, and infections of the extremities. Careful blood glucose control can help prevent diabetic foot ulcers that can become gangrenous.

SOURCES Medicinenet.com/gangrene; Y.Y. Huang et al. Survival and associated risk factors in patients with diabetes and amputations caused by infectious foot gangrene. *Journal of Foot and Ankle Research*, January 2018.

Guillain-Barré Syndrome

ANNUAL US DEATHS 275 reported in 2019

Guillain-Barré syndrome (GBS) is a rare neurological disorder caused by an autoimmune attack on the body's peripheral nervous system (the nerves outside of the brain and spinal cord). Symptoms can range from very mild brief weakness to near total paralysis. The trigger for the mistaken attack is unknown, although most cases usually start within a few weeks of a viral or bacterial infection. About 70 percent of patients make a full recovery within a few months; almost all the rest recover but with some long-term muscle weakness from nerve damage. Each year in the US, an estimated 3,000 to 6,000 people develop GBS. The incidence may be higher because many cases are so mild they go unnoticed or unreported.

In 2019, 275 people died from GBS, for a mortality rate of 0.17 percent.

STAY SAFE If you notice muscle weakness within a few weeks of being sick, get medical attention.

SOURCES CDC NCHS, cdc.gov/nchs/nvss/deaths.htm; NIH NINDS, ninds.nih.gov/Disorders

Heat-Related Deaths

ANNUAL US DEATHS   505 reported in 2019

Extreme heat, defined as summertime temperatures that are much hotter and/or more humid than average, caused 505 deaths in 2019, for a mortality rate of 0.15 per 100,000 population. The highest yearly total of heat-related deaths was 1,050 in 1999; the lowest was 295 in 2004. The worst heat wave in US history occurred in the summer of 1980, when prolonged heat over the Midwest and Southern Plains caused approximately 1,700 deaths, many related to excess heat exposure. From 2014 to 2018, an average of 702 heat-related deaths (415 with heat as the underlying cause and 287 as a contributing cause) occurred in the US annually. Approximately 90 percent of heat-related deaths occur between May and September. Although Arizona, California, and Texas account for only approximately 23 percent of the US population, these 3 states accounted for approximately 37 percent of heat-related deaths.

Approximately 70 percent of heat-related deaths are among males. Babies, children under age 4, pregnant women, the elderly, and those with chronic illness are more vulnerable to heat-related illness and death.

STAY SAFE   To stay cool during a period of extreme heat, remain in air-conditioned buildings as much as possible. If you don't have air-conditioning, contact your local health authorities to find a nearby cooling center. Stay hydrated, wear loose clothing, take cool showers or baths to cool down, and don't engage in strenuous activities. Check in on vulnerable family, friends, and neighbors.

Heatstroke

Heatstroke, also called sunstroke, occurs when prolonged exposure to high temperatures causes the body temperature to rise to 104 degrees F (40 degrees C) or above. Heatstroke is a medical emergency that can kill directly by fatally damaging the brain, heart, or kidneys, but it is more likely to cause death by exacerbating underlying heart, lung, and kidney conditions. The total number of heatstroke deaths can't really be separated from the total number of heat-related deaths. However, the NSC estimates the lifetime odds of death from sunstroke at 1 in 8,248.

STAY SAFE   Call 911 immediately for symptoms of heatstroke or heat-related illness such as muscle cramps, headaches, nausea, vomiting, altered mental state, passing out. Move the person to shade if possible and immediately cool them by spraying with a garden hose, putting them into a cool tub or shower, sponging with cool water—anything that lowers the body temperature. Get them to drink lots of cool water.

SOURCES CDC NCHS, cdc.gov/nchs/nvss/deaths.htm; NWS, weather.gov/safety/heat; NSC nsc.org/road/safety-topics

Hypothermia

ANNUAL US DEATHS Average 1,301

When your body loses heat faster than you can produce it, hypothermia, or dangerously low body temperature (below 95 degrees F or 35 degrees C), can occur. The usual cause is prolonged outdoor exposure to cold weather, but hypothermia can occur at warmer temperatures in wet conditions.

In 2019, 1,115 people died of hypothermia in the US, for a mortality rate of 0.34 per 100,000 population. On average, 1,301 people die of hypothermia in the US each year. The highest yearly total was 1,536 in 2010; the lowest was 1,058 in 2006. Approximately 67 percent of deaths were among males. Hypothermia from exposure to excess natural cold kills about twice as many people each year as hyperthermia from exposure to excess natural heat.

Babies, small children, and the elderly are most at risk of hypothermia—about half of all deaths each year are among people aged 65 and up. People experiencing homelessness are also very vulnerable. Passing out in cold weather from drinking too much alcohol is also a risk factor. People in an outdoor setting, including hunters, skiers, climbers, boaters/rafters, and swimmers, can get caught by unexpected cold weather and die of hypothermia. At an ultramarathon race in China in May 2021, several runners were trapped by extreme weather that caused freezing rain and a sudden drop in air temperature. Of the 172 participants, 21 died of hypothermia.

Warning signs of hypothermia include shivering, exhaustion (or feeling very tired), confusion, fumbling hands, memory loss, slurred speech, and drowsiness. In babies, signs include cold, bright red skin and lethargy.

Hypothermia is a medical emergency. Call 911. Immediately move the person out of the cold or to a sheltered area, remove any wet clothing, and cover with dry blankets. If further warming is needed, do it gradually by applying warm compresses or an electric blanket to the center of the body. Offer warm drinks; do not give the person alcohol.

STAY SAFE Check the weather forecast and plan travel and outdoor activity appropriately. Dress in layers and wear a warm hat and appropriate footwear when going outside in the cold. Avoid getting wet. During severe cold weather, check in on vulnerable family, friends, and neighbors. If you are stranded in your car by snow or ice, stay with the vehicle. Cover yourself with any extra clothing or blankets. Run the motor and heater for only 10 minutes each hour, opening the window slightly to let in air. Make sure the tailpipe isn't blocked by snow.

SOURCES CDC NCHS, cdc.gov/nchs/nvss/deaths.htm; NWS, weather.gov/safety/cold

Hypovolemic Shock

ANNUAL US DEATHS 811 in 2019

Hypovolemic shock is caused by severe blood or other fluid loss that makes the heart unable to pump enough blood to the body. This type of shock can cause many organs to stop working. Symptoms include weak, rapid pulse; cold, clammy skin; faintness or dizziness; nausea. Hypovolemic shock is a very serious medical emergency.

An adult human has about 10 pints (1.2 to 1.5 gallons) of blood, or about 8 percent of body weight. Losing 20 percent or more of the blood volume causes hemorrhagic hypovolemic shock. Blood loss can be due to cuts, other external injuries, or internal bleeding. Loss of large amounts of body fluid can also cause hypovolemic shock, as from burns, severe diarrhea, and severe vomiting.

More than 3 million Americans experience shock every year, but most are saved by prompt emergency treatment using blood product transfusions or intravenous fluids. In 2019, 811 people died of hypovolemic shock, for a mortality rate of 0.01 per 100,000 population.

STAY SAFE Call 911 immediately if you suspect someone is in shock. If the bleeding is external, apply direct pressure to stop or slow it and hold in place until help arrives. If shock is from excess fluid loss, keep the patient warm and comfortable until help arrives. Begin CPR if necessary.

SOURCES J.W. Cannon, Hemorrhagic Shock. *New England Journal of Medicine*, 2018; CDC NCHS, cdc.gov/nchs/nvss/deaths.htm

Intestinal Obstruction

ANNUAL US DEATHS   6,943 in 2019

An intestinal obstruction is any blockage that keeps food and stool from moving through the intestinal tract. About 65 to 75 percent of cases are adhesions (scar tissue from previous surgery). Most other cases are caused by hernias and cancer. A full intestinal obstruction is a medical emergency. Surgery is almost always needed if the intestine is completely blocked or if the blood supply is cut off (ischemia).

If surgery is performed promptly, survival rates are good; about 5 percent of patients die. If untreated, small intestine obstructions cause death in 100 percent of patients; when surgery is performed within 36 hours, the mortality rate decreases to 8 percent. The mortality rate is 25 percent if the surgery is postponed beyond 36 hours. In 2019, 6,943 people died of an intestinal obstruction, for a death rate of 2.12 per 100,000 population.

STAY SAFE   If you have had abdominal or pelvic surgery, be aware of your increased risk of intestinal obstruction from adhesions.

SOURCES   F. Catena et al. Bowel obstruction: a narrative review for all physicians. *World Journal of Emergency Surgery*, 2019; CDC NCHS, cdc.gov/nchs/nvss/deaths.htm

Pancreatitis

ANNUAL US DEATHS   2,811 in 2019

Pancreatitis is inflammation of the pancreas, the gland that produces digestive enzymes and the hormone insulin. Backed-up digestive enzymes start digesting the pancreas itself—very painfully. Pancreatitis can be acute or chronic and is often related to gallstones, extremely high triglycerides, or heavy alcohol consumption. Either form is a serious condition that can lead to complications and death

The reported annual incidence of pancreatitis in the US is 40 to 50 cases per 100,000 population. The incidence of acute pancreatitis ranges from 4.9 to 35 per 100,000 population. Acute pancreatitis is the leading cause of gastrointestinal-related hospitalization in the US.

The overall mortality in patients with acute pancreatitis is 10 to 15 percent. In 2019, 2,811 people died of pancreatitis in the US, for a mortality rate of 0.86 per 100,000 population.

Racial Disparities

The annual incidence of acute pancreatitis in Native Americans is 4 per 100,000 population; in Whites, 5.7 per 100,000 population; and in Blacks, 20.7 per 100,000 population The hospitalization rates of patients with acute pancreatitis are 3 times higher for Black individuals than White individuals. The risk for Black Americans aged 35 to 64 years is 10 times higher than for any other group.

STAY SAFE　If you have gallstones, be aware of the increased risk of pancreatitis. Drink alcohol in moderation.

SOURCES　NIH NIDDK, niddk.nih.gov/health-information/ digestive-diseases/pancreatitis; American Pancreatic Association Pancreapedia, pancreapedia.org

Peptic Ulcer Disease

ANNUAL US DEATHS　1,137 (gastric) and 1,558 (duodenal) in 2019

Peptic ulcers are open sores in the lining of the stomach or small intestine. Inflammation from the bacterium *Helicobacter pylori* is the most common cause of peptic ulcers. Gastric ulcers develop inside the stomach; duodenal ulcers develop in the upper section of the small intestine.

Peptic ulcer disease (PUD) affects about 4.6 million people in the US each year. Most people respond well to acid blockers and treatment to eradicate the *H. pylori* bacterium. Untreated peptic ulcers can lead to perforation, where a hole in the lining of the stomach or small intestine leads to acute infection. The lifetime prevalence of perforation in patients with PUD is about 5 percent.

Peptic ulcer perforation is a medical emergency. Surgery has a mortality risk of 6 to 30 percent; the risk is higher for older people and

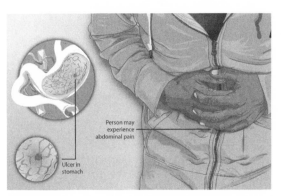

Untreated, a gastric ulcer can be lethal.

Person may experience abdominal pain

Ulcer in stomach

those with comorbidities. The mortality risk for perforated gastric ulcer is twice that of perforated duodenal ulcer. Postoperative long-term mortality for perforated peptic ulcers is high—15.2 percent of patients die within 30 days, 19.2 percent die at 90 days, 22.6 percent die at 1 year, and 24.8 percent die at 2 years.

In 2019, 1,137 Americans died from a gastric ulcer, for a mortality rate of 0.35 per 100,000 population. Another 1,558 Americans died from a duodenal ulcer, for a mortality rate of 0.47 per 100,000 population.

STAY SAFE Follow your doctor's advice for treating a peptic ulcer, including medication, dietary changes, and alcohol avoidance. Don't smoke. Get immediate medical attention if you have severe burning abdominal pain or see blood in your vomit or stool.

SOURCE K.T. Chung, V.G. Shelat. Perforated peptic ulcer: An update. *World Journal of Gastrointestinal Surgery*, January 2017.

Rhabdomyolysis

ANNUAL US DEATHS About 1,300

Rhabdomyolysis, often just called rhabdo, is the breakdown of muscle tissue that leads to the release of muscle fiber contents, including a protein called myoglobin and an enzyme called creatine kinase into the bloodstream. The breakdown products damage the kidneys as they attempt to filter them out. Causes can include overexertion during exercise and crush injuries. Severe cases can cause death from acute kidney injury. About 26,000 people a year in the US are diagnosed with rhabdomyolysis; about 5 percent, or approximately 1,300 people, die from it.

STAY SAFE Work up to high levels of exercise gradually. Go to the emergency room if you notice brown urine after a hard workout or other severe exertion.

SOURCE J.M. Sauret et al. Rhabdomyolysis. *American Family Physician*, March 2002.

Sepsis

ANNUAL US DEATHS 270,000

Sepsis occurs when an existing infection (for example, in the urinary tract or lungs) triggers an extreme response throughout the body.

Sepsis is usually caused by a bacterial infection, but it can also occur as a complication of viral infections, such as flu or COVID-19.

Sepsis is a medical emergency. Every year, approximately 1.7 million Americans develop sepsis, and nearly 270,000 die from it. One in 3 patients who die in a hospital succumb to sepsis. People with chronic health conditions such as diabetes, lung disease, kidney disease, and cancer are at higher risk of sepsis; so is anyone over age 65. In 2019, Maine had the lowest sepsis death rate in the country, with 2.7 deaths per 100,000 population. Louisiana had the worst rate, at 20.2 deaths per 100,000 population.

STAY SAFE   Wash your hands. Know the signs and symptoms of sepsis and act fast if you suspect you or a loved one has a bad infection that's not improving or is getting worse.

SOURCE   CDC NCHS, cdc.gov/nchs/nvss/deaths.htm and cdc.gov/nchs/pressroom/sosmap/septicemia_mortality

Spinal Cord Injuries

ANNUAL US DEATHS   About 1,250

A spinal cord injury (SCI) is severe damage caused by direct injury to the spinal cord itself or from damage to the vertebrae that surround the spinal cord. The damage interferes with transmission of nerve impulses and can result in temporary or permanent loss of sensation, movements, strength, and body functions below the site of the injury.

Statistics on death from SCIs are patchy and unreliable—often the cause of death on the scene of a car wreck or other event is not determined in detail. Every year, about 17,900 people in the US survive a traumatic spinal cord injury long enough to get to a hospital, or about 54 cases per 1,000,000 population. Of those, about 7 percent die within 24 hours.

Motor vehicle accidents are the leading cause of spinal cord injury, at about 38 percent, followed by falls, at about 32 percent. Acts of violence, primarily gunshot wounds, cause about 14 percent; sports and recreation cause about 8 percent. About 78 percent of new SCI cases are male.

Traumatic spinal cord injury leads to chronic impairment and disability and shortens life expectancy significantly. Mortality rates are significantly higher in the first year after the injury that in later years. Life expectancy for a 20-year-old without an SCI is estimated at 79 years, but for someone with an SCI affecting motor function at any level, life expectancy is 72 years. For a 20-year-old who is ventilator-dependent, life expectancy is only an additional 10 years. The most common causes of death are pneumonia and septicemia.

Actor Christopher Reeve of Superman fame suffered a severe spinal cord injury when he was thrown from a horse in 1995. He was paralyzed from the shoulders down and used a ventilator. He died from septicemia in 2004.

STAY SAFE   If you suspect a back or neck spinal injury, do not move the injured person—this can cause permanent paralysis and other serious complications. Call 911 immediately and take measures to keep the person still. Provide as much first aid as possible without moving the person's head or neck. Begin CPR if there is no pulse.

SOURCE   National Spinal Cord Injury Database, https://www.nscisc.uab.edu

Spontaneous Human Combustion

ANNUAL US DEATHS   0

Spontaneous human combustion (SHC) is the death by burning of a living (or recently deceased) human body without an apparent external source of ignition. The fire is thus believed to originate in the body of the victim, perhaps related to alcohol consumption. Although the concept is found in literature (most famously, Mr. Krook in *Bleak House*, by Charles Dickens) and popular culture (the on-stage cause of death for 2 drummers in the movie *This Is Spinal Tap*), no reports of real cases are medically or forensically documented. Supposed cases can all be traced to external ignition sources and usually involve a victim with impaired mobility.

STAY SAFE   Don't worry about this.

SOURCE   B. Radford. Spontaneous human combustion: facts & theories. *LiveScience.com*, December 18, 2013.

Traumatic Brain Injury

ANNUAL US DEATHS   61,000 in 2019

A traumatic brain injury (TBI) is an injury to the head that affects brain function. Causes of TBIs are bumps, blows, jolts, or a penetrating injury, such as a gunshot, to the head. Severe TBI is a major cause of death and disability in the US. In 2019, 61,000 people died from a TBI, about 166 TBI-related deaths every day, or a death rate of 17.3 per 100,000 population.

Suicide and accidental falls were the most common fatal injuries in 2017. Suicides accounted for 21,225 deaths, for 34.7 percent of all TBI-related deaths. Accidental falls resulted in 17,408 deaths, or 28.0 percent of all TBI-related deaths. The 3rd most common cause was motor vehicle crashes, which caused 11,098 of all TBI-related deaths.

By Age

Adults aged 75 and older have the highest rate of TBI-related deaths: 77.0 per 100,000 population in 2017, or 16,284 deaths, for 26.6 percent of all TBI deaths. Most—11,452—were from accidental falls. Those aged 65 to 74 years were 24.3 percent of all TBI-related deaths, and those aged 55 to 64 were 19.5 percent of all TBI-related deaths. Children aged 0 to 17 years accounted for 2,095 deaths, including 940 from car crashes and 665 from homicide.

By Sex

In 2017, males had significantly higher TBI death rates compared with females:

- Suicide: 18,436 male versus 2,789 female, or 13.0 versus 1.9 per 100,000 population (more than 6 times higher)
- Accidental falls: 10,180 male versus 7,228 female, or 6.1 versus 3.2 per 100,000 population
- Motor vehicle crashes: 8,036 male versus 3,062 female, or 4.9 versus 1.8 per 100,000 population (more than 2 times higher)
- Homicide: 4,316 male versus 1,665 female, or 2.7 versus 1.0 per 100,000 population (more than 2 times higher)

Rural versus Urban

People living in rural areas have a greater risk of dying from a TBI compared to people living in urban areas. They must travel further for emergency medical care, have less access to a Level I trauma center, and have less access to services such as specialized TBI care.

STAY SAFE Fall-proof your home and workspace. Wear your seat belt while driving. If you or someone you love is suicidal, call the National Suicide Prevention Lifeline at 1-800-273-8255 (TALK). Help is available 24/7/365.

Venous Thromboembolism (VTE)

ANNUAL US DEATHS 60,000 to 100,000

Venous thromboembolism (VTE) is a disorder that includes deep vein thrombosis (DVT) and pulmonary embolism. In VTE, a blood clot forms in a vein and blocks the circulation in the area. In DVT, the blood clot forms in a deep vein, usually in the lower leg, thigh, or pelvis, causing swelling, redness, and pain. If a portion of the DVT clot breaks off and travels through the bloodstream to the lungs, it causes a blockage called PE, causing pain and shortness of breath. A large PE stops blood from reaching the lungs and is quickly fatal.

As many as 900,000 people (1 to 2 per 1,000) each year in the US have a VTE. Exact death numbers are unknown, but anywhere from 60,000 to 100,000 people a year die of a pulmonary embolism that began as a DVT. Sudden death is the first symptom in about 25 percent of people who have a PE. An additional 10 to 30 percent die within 1 month of diagnosis. PE is often the immediate cause of death in people near the end of life from cancer, lung disease, and heart disease.

For reasons possibly related to increasing rates of obesity and sedentary lifestyle, the death rate from lung clots has been rising. Between 2008 and 2018, the death rate among people ages 25 to 64 increased 23 percent.

Risk factors for VTE include major surgery, confinement to bed, sitting for prolonged periods, especially with crossed legs, pregnancy, and cancer.

STAY SAFE Move around as much as possible when sitting for more than 4 hours, as when traveling. Do sitting leg exercises and wear loose clothing. If you are at risk for VTE, discuss the use of compression stockings and anticoagulant drugs with your doctor.

SOURCE CDC, cdc.gov/ncbddd/dvt/data.html

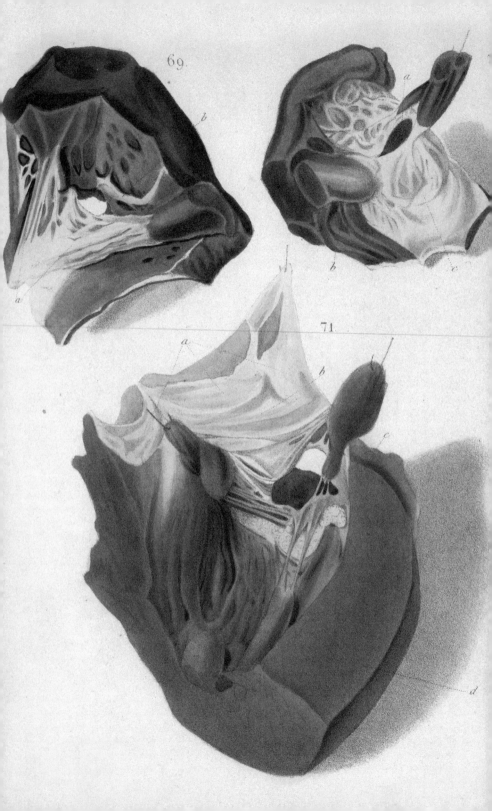

69.

71.

Chronic Diseases

C hronic diseases are defined broadly as conditions that last 1 year or more and require ongoing medical attention or limit activities of daily living or both. Six in 10 adults in the US have a chronic disease and 4 in 10 have 2 or more. Chronic diseases such as heart disease, cancer, lung disease, stroke, Alzheimer's disease, diabetes, and kidney disease are the leading causes of death and disability in the US.

The broad categories of chronic diseases in this section are listed in descending order as causes of death—Heart Disease, Cancer, Lung Disease, Stroke, Arterial Disease, Diabetes, and Kidney Disease. Within the major categories, the subcategories are listed alphabetically. In addition, the major category "Other Chronic Diseases" lists important chronic diseases that do not fall under the other major categories.

STAY SAFE Most chronic diseases can be avoided or delayed with lifestyle changes. Don't smoke, get 30 minutes of exercise most days, eat a healthy diet, maintain a healthy weight, limit alcohol, and get enough sleep. Get screened for chronic conditions as your doctor recommends and know your family history.

Heart Disease

> **ANNUAL US DEATHS** 659,041 in 2019

Heart disease is the leading cause of death in the United States. In 2019, 659,041 Americans died from heart disease, for a death rate of 200.8 per 100,000 population, or approximately 23.4 percent of all deaths, or almost 1 in every 4 deaths. Every 60 seconds, just over 1 person in the US dies from heart disease.

Heart disease is an overall term for several types of heart conditions. In the US, the most common type of heart disease is coronary heart disease (CHD), which affects the blood flow to the heart. Decreased blood flow due to a blocked coronary artery can cause a heart attack (myo-

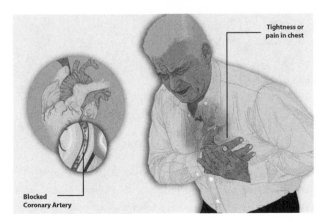

Tightness or pain in chest

Not every heart attack announces itself with chest pain, but many do.

Blocked Coronary Artery

cardial infarction, or MI). Other common types of heart disease include heart failure and heart valve disease. The age-adjusted prevalence of all types of heart disease is 11.2 percent.

Nearly half of all Americans have at least 1 of the top 3 risk factors for heart disease: high blood pressure, high cholesterol, and smoking. The annual cost of heart disease, including healthcare services, medicines, and lost productivity due to death, is about $219 billion a year.

By Age

Your risk of heart disease increases with age and jumps sharply after age 45. Heart disease, especially coronary artery disease, is on the rise in younger adults, however, due to the increase in obesity and type 2 diabetes among this age group.

For men, the age-related risk of death from all forms of heart disease is:

Age 25 to 34	10.2 deaths per 100,000 population
Age 35 to 44	35.5 deaths
Age 45 to 54	112.5 deaths
Age 55 to 64	271.3 deaths
Age 65 to 74	536.4 deaths
Age 75 to 84	1,312.0 deaths
Age 85 and over	4,403.5 deaths

For women, the age-related risk of death from all forms of heart disease is lower in general but starts to catch up during the menopause years as the protection from estrogen wanes. The numbers are:

Age 25 to 34	5.2 deaths per 100,000 population
Age 35 to 44	16.4 deaths
Age 45 to 54	47.4 deaths
Age 55 to 64	113.5 deaths
Age 65 to 74	266.0 deaths
Age 75 to 84	826.4 deaths
Age 85 and over	3,589.7 deaths

People with HIV are more likely to experience heart disease before age 60 than uninfected people. Their cumulative lifetime risk is 65 percent for men and 44 percent for women. This is similar to the risk for people with diabetes: 67 percent for men and 57 percent for women.

By Sex

The age-adjusted prevalence of heart disease is higher in males at 12.6 percent; in females it's 10.1 percent.

Among men, heart disease caused 24.4 percent of all deaths; among women, heart disease caused 22.3 percent of all deaths. Heart disease is the leading cause of death in men across almost all racial and ethnic groups: 357,761 men died from it in 2019, or about 1 in every 4 male deaths.

Heart disease isn't just a male problem. It is the leading cause of death for both White and Black women: 299,578 women died from it in 2017, or about 1 in every 5 female deaths. Structural gender bias in cardiology research and treatment means that women are often undertreated for heart disease. Even though 50 percent of heart disease occurs in women, treatment guidelines are based on studies that mostly looked at men.

By Race/Ethnicity

From 1999 through 2017, death rates for heart disease decreased for all racial and ethnic groups. In 1999, the death rate for Blacks was 337.4 per 100,000 population; in 2017, it was 208.0.

Among White people, the age-adjusted prevalence of heart disease is 11.5 percent. Among Black people, it's 10.0 percent, among Hispanic people it's 8.2 percent, among Asian people it's 7.7 percent, and among American Indian/Alaska Native people it's 14.6 percent.

About 1 in 13 (7.7 percent) White men and 1 in 14 (7.1 percent) Black men have coronary heart disease. About 1 in 17 (5.9 percent) Hispanic men have CHD. About 6.1 percent of White women, 6.5 percent of Black women, 6.0 percent of Hispanic women, and 3.2 percent of Asian women have CHD.

Geographic Trends

Death from heart disease varies considerably by state. The worst states for death from heart disease as of 2019 are:

1. Oklahoma, 228.5 deaths per 100,000 population, or 10,960 deaths
2. Alabama, 224.7 deaths per 100,000 population, or 13,448 deaths
3. Mississippi, 222.1 deaths per 100,000 population, or 7,997 deaths
4. Arkansas, 217.4 deaths per 100,000 population, or 8,669 deaths
5. Louisiana, 212.2 deaths per 100,000 population, or 11,302 deaths
6. Tennessee, 202.4 deaths per 100,000 population, or 16,814 deaths
7. Kentucky, 198.3 deaths per 100,000 population, or 10,742 deaths
8. West Virginia, 196.4 deaths per 100,000 population, or 5,087 deaths
9. Michigan, 195.0 deaths per 100,000 population, or 25,547 deaths
10. Ohio, 191.1 deaths per 100,000 population, or 29,160 deaths

The best states for death from heart disease as of 2019 are:

1. Minnesota, 116.7 deaths per 100,000 population, or 8,401 deaths
2. Hawaii, 120.3 deaths per 100,000 population, or 2,503 deaths
3. Massachusetts, 127.2 deaths per 100,000 population, or 11,761 deaths
4. Colorado, 127.7 deaths per 100,000 population, or 7,762 deaths
5. Alaska, 129.7 deaths per 100,000 population, or 843 deaths
6. Oregon, 131.0 deaths per 100,000 population, or 7,128 deaths
7. Arizona, 134.0 deaths per 100,000 population, or 12,587 deaths
8. Washington, 134.8 deaths per 100,000 population, or 11,862 deaths
9. California, 136.9 deaths per 100,000 population, or 62,394 deaths
10. Florida, 140.1 deaths per 100,000 population, or 47,144 deaths

Trends in Heart Disease Mortality

In 1900, heart disease was the 4th most common cause of death in the US. By 1930, as public health measures reduced deaths from infectious disease, heart disease moved up to be the commonest cause of death—and has remained there ever since. Heart disease deaths continued to increase until the mid-1960s, cresting at 466 deaths per age-adjusted 100,000 of population. By 1980, better treatment decreased the death rate to 345 deaths per 100,000, a 26 percent relative decrease. From 1980 to 2008, the prevalence decreased further, to 123 per 100,000, a 64 percent relative decrease. The downward trend is attributed to improvements in treatment and a continuing decrease in smoking. Death from heart disease reached a peak of 771,169 in 1985 but decreased to 596,577 in 2011. The downward trend has been reversing since then, however. The number of deaths from heart disease increased by 3.0 percent during the period 2011 to 2014, rising from 596,577 to 614,348. The reversal in the positive trend is probably due to the increasing prevalence of obesity and type 2 diabetes.

STAY SAFE   Don't smoke. If you smoke, quit. Free resources for quitting are available from BeTobaccoFree.gov, from the National Cancer Institute quitline at 1-877-44U-QUIT, and from the American Lung Association at lung.org/quit-smoking.

SOURCES   AHA, S.S. Virani et al. Heart Disease and Stroke Statistics-2021 Update: A Report From the American Heart Association. *Circulation*, February 2021; CDC NCHS, cdc.gov/nchs/ndi/index.htm; G.S. Tajeu et al. Black-White Differences in Cardiovascular Disease Mortality: A Prospective US Study, 2003–2017. *American Journal of Public Health*, May 2020.

Atrial Fibrillation

ANNUAL US DEATHS   In 2018, contributed to 175,326 and was the underlying cause of 25,845

Atrial fibrillation (AF or AFib) is the most common type of heart arrhythmia. During an episode of AF, the atria (the 2 upper chambers of the heart) beat irregularly. Blood doesn't flow as well as it should from the atria into the ventricles (the 2 lower chambers of the heart). The ventricles don't fill completely or pump enough blood to your lungs and body. AF is caused by a problem in the heart's electrical system. Atrial fibrillation can make blood pools in the heart, increasing the risk of clots that can cause a stroke or other complications.

In 2018, atrial fibrillation contributed to 175,326 deaths and was the underlying cause of death for 25,845 people. The age-adjusted mortality rate attributable to AF was 6.4 per 100,000 population in 2018. Although men and women have AF in approximately equal numbers, the adjusted risk of death is higher for women, with a relative risk 1.12, or 12 percent higher.

The death rate from AF as the primary or a contributing cause of death has been rising for more than 2 decades.

STAY SAFE Risk factors for atrial fibrillation are the same as for all other heart disease. Don't smoke. People who have had heart surgery are slightly more likely to develop AF later on.

SOURCES AHA, S.S. Virani et al. Heart Disease and Stroke Statistics-2021 Update: A Report From the American Heart Association. *Circulation*, February 2021; CDC, cdc.gov/heartdisease/atrial_fibrillation.htm

Cardiogenic Shock

ANNUAL US DEATHS 10,472 in 2019

In cardiogenic shock, the heart can no longer pump well enough to get enough blood and oxygen to the brain and other organs. This life-threatening condition is usually caused by a severe heart attack (myocardial infarction). Emergency treatment to restore circulation is needed.

Cardiogenic shock occurs in between 5 and 10 percent of all cases of acute myocardial infarction and is the leading cause of death after an MI, with an overall death rate of about 50 percent. The overall in-hospital mortality rate is about 39 percent. Roughly half of all people who survive cardiogenic shock after an MI will die in the next 6 to 12 months. In 2019, 10,472 people died of cardiogenic shock, for a mortality rate of 0.2 per 100,000 population.

BY RACE/ETHNICITY

Asians/Pacific Islanders have a higher incidence of cardiogenic shock (11.4 percent) than Hispanic (8.6 percent), White (8.0 percent), and Black (6.9 percent) patients. However, the death rate for cardiogenic shock among hospitalized patients is highest for Hispanic patients (40.6 percent), compared to Black (39.9 percent), White (38.9 percent), and Asian/Pacific Islander (37.6 percent) patients.

STAY SAFE Don't delay! Get immediate emergency medical help if you think you might be having a heart attack.

SOURCES AHA, S.S. Virani et al. Heart Disease and Stroke Statistics-2021 Update: A Report From the American Heart Association. *Circulation*, February 2021; NIH NHLBI, nhlbi.nih.gov/health-topics/cardiogenic-shock

Coronary Heart Disease and Heart Attack

ANNUAL US DEATHS 365,744 in 2018

Coronary heart disease (CHD), also known as coronary artery disease, is a narrowing of the small blood vessels that supply blood and oxygen to the heart. The primary cause is atherosclerosis (hardening of the arteries) caused by the formation of cholesterol plaques in the arteries. CHD is the leading cause of death in the United States for men and women.

Many people with CHD will go on to have a heart attack, also called a myocardial infarction (MI). An MI happens when blood flow to the heart through the coronary arteries suddenly becomes blocked, usually by a blood clot or ruptured plaque. Without blood flow to nourish it, the heart muscle begins to die. An American has a heart attack once every 40 seconds and 1 dies from a heart attack every minute.

Because CHD and MI are so closely linked, statistics for them are often combined. Most deaths from CHD are caused by an MI, but other forms of ischemic heart disease, where the heart muscle doesn't receive enough blood, are usually also included CHD statistics. As many as 50 percent of all cardiac deaths due to disease in the heart's vessels occur in individuals with no prior history or symptoms of heart disease.

In 2018, 20.1 million, or 7.2 percent, of adults over 20 had coronary heart disease; 365,744 people died from it. In 2006, CHD was the underlying cause of death for 425,425 people; the age-adjusted mortality rate was 135.0 deaths per 100,000 standard population.

In 2018, 8.8 million Americans, or 3.1 percent of the population, had a heart attack. Of those, 108,610, or approximately 14 percent, died.

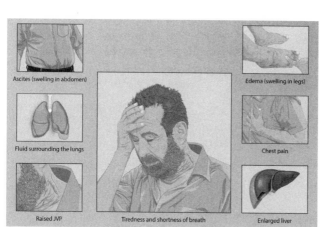

Ascites (swelling in abdomen)

Fluid surrounding the lungs

Raised JVP

Tiredness and shortness of breath

Edema (swelling in legs)

Chest pain

Enlarged liver

Heart failure is all too easy to neglect until it becomes disabling or deadly.

BY SEX

Men have more coronary heart disease than women. In 2018, the prevalence of CHD in men over 20 was 11.0 million, or 8.3 percent; in women it was 9.1 million, or 6.2 percent. Mortality from CHD for men was 215,032, or 58.8 percent; for women, it was 150,712, or 41.2 percent. In 2006, the death rate for men was 176.5 per 100,000 population; for women, it was 103.1. The death rate for men was 41.6 percent higher than for women.

Men have more heart attacks and die from them more often than women. However, women under 50 who have a heart attack are twice as likely to die from it than men the same age. In 2018, the prevalence of heart attack for men was approximately 5,800,000, or 4.3 percent, while the prevalence for women was approximately 3,000,000, or 2.1 percent. The mortality for men was 64,079, or 59.0 percent, while the mortality for women was 44,531, or 41.0 percent.

The median survival time after a first MI is 8.2 years for males and 5.5 years for females aged 45 or older. The risk of death after a heart attack continues for years after the event. Within 1 year after a first heart attack, 18 percent of males and 23 percent of females aged 45 or older will die. Within 5 years after a first heart attack, 36 percent of males and 47 percent of females aged 45 or older will die.

BY AGE AND RACE

Death from coronary heart disease is rare before age 45. The risk of death rises with age and differs between men and women. The average age of first myocardial infarction, for example, is 65.6 for men and 72.0 for women. Racial disparities in death rates are significant.

BY SEX: WOMEN

In 2006, 4,316 White women ages 45 to 54 died of CHD, for a rate of 24.1 per 100,000 population; in the same age group, 1,564 Black women died, for a rate of 56.0.

For White women ages 55 to 64, 10,137 died, for a rate of 73.8; in the same age group, 2,636 Black women died, for a rate of 147.8.

For White women ages 65 to 74, 19,287 died, for a rate of 221.0; in the same age group, 3,859 Black women died, for a rate of 367.2.

For White women ages 75 to 84, 50,538 died, for a rate of 740.4; in the same age group, 6,114 Black women died, for a rate of 940.8.

For White women age 85 and over, 89,442 died, for a rate of 2,761.6; in the same age group, 7,111 Black women died, for a rate of 2,599.5.

In 2006, 15,294 White men ages 45 to 54 died of CHD, for a rate of 86.2 per 100,000 population; in the same age group, 3,140 Black men died, for a rate of 13.9.

For White men ages 55 to 64, 27,772 died, for a rate of 212.7; in the same age group, 4,890 Black men died, for a rate of 340.1.

For White men ages 65 to 74, 36,434 died, for a rate of 483.8; in the same age group, 5,300 Black men died, for a rate of 704.9.

For White men ages 75 to 84, 60,452 died, for a rate of 1,275.5; in the same age group, 5,384 Black men died, for a rate of 1,456.9.

For White men ages 85 and over, 51,632 died, for a rate of 3,396.0; in the same age group, 2,973 Black men died, for a rate of 2,656.7

STAY SAFE Know the symptoms of a heart attack. Women often have less obvious symptoms and may not recognize they are having a heart attack. If you think you're having a heart attack, don't delay! Get emergency medical help immediately. Don't smoke. Free resources to help you quit smoking are available by calling 1-800-QUIT-NOW.

SOURCES AHA, S.S. Virani et al. Heart Disease and Stroke Statistics-2021 Update: A Report From the American Heart Association. *Circulation*, February 2021; CDC NCHS, cdc.gov/nchs/ndi/index.htm; NIH NHLBI, nhlbi.nih.gov/health-topics/heart-attack

Heart Failure

> **ANNUAL US DEATHS** 83,616 deaths in 2018

Heart failure, also known as congestive heart failure (CHF), means the heart muscle is weakened to the point where it cannot pump enough blood to meet the body's needs. One or both sides of the heart may be affected. Heart failure is usually a long-term consequence of previous medical conditions such as high blood pressure, coronary artery disease, heart attack, valvular heart disease, or arrhythmia. Heart failure rarely occurs without a preexisting condition, but sometimes it has no known cause (idiopathic heart failure).

In 2018, 83,616 people died of heart failure in the US. Of these, 38,487, or 46.0 percent, were male; 45,129, or 54.0 percent, were female. Heart failure was mentioned as an underlying cause of death on 379,800 death certificates, or 13.4 percent, in 2018. Heart failure is the primary cause of hospitalization in the elderly.

The age-adjusted rate for heart failure–related deaths decreased from 105.4 deaths per 100,000 population in 2000 to 81.4 in 2012. The rate then increased to 83.4 in 2013 and to 84.0 in 2014. The number of heart failure–related deaths declined from 2000 through 2009, but then increased steadily through 2014.

BY RACE/ETHNICITY

In 2014, the age-adjusted death rate for heart failure–related deaths for Black people was 91.5 deaths per 100,000 population; for White people, the death rate was 87.3; for Hispanic people the death rate was 53.3.

BY AGE

In 2014, the death rate was highest for adults aged 85 and over, with 2,842.9 deaths per 100,000 population for men and 2,333.5 deaths for women. That was followed by adults aged 75 to 84, for death rates of 720.0 deaths for men and 504.7 deaths for women. Among adults aged 65 to 74 the death rate was 201.8 for men and 124.6 for women. Among adults aged 45 to 64, the death rate was 41.3 for men and 24.0 for women.

The long-term outlook for heart failure patients is poor. More than half of those who develop CHF die within 5 years of diagnosis. Mortality following hospitalization for patients with heart failure is 10.4 percent at 30 days, 22 percent at 1 year, and 42.3 percent at 5 years.

STAY SAFE   Don't smoke. If you smoke, quit. Free resources to help you quit are available from BeTobaccoFree.gov, the National Cancer Institute quitline at 1-877-44U-QUIT, and the American Lung Association at lung.org/quit-smoking. Follow a heart healthy lifestyle, especially if you have already been diagnosed with heart disease.

SOURCES   AHA, S.S. Virani et al. Heart Disease and Stroke Statistics-2021 Update: A Report From the American Heart Association. *Circulation*, February 2021; CDC NCHS, cdc.gov/nchs/ndi/index.htm; NIH NHLBI, nhlbi.nih.gov/health-topics/heart-failure

Heart Inflammation: Endocarditis, Myocarditis, and Pericarditis

ANNUAL US DEATHS   Varies with type

Three types of heart inflammation cause almost all deaths from this condition: endocarditis, myocarditis, and pericarditis. Endocarditis is inflammation of the inner lining of the heart's chambers and valves. Myocarditis is inflammation of the heart muscle itself. Pericarditis is inflammation of the pericardium, the tissue that forms a sac around the heart. In all 3 conditions, a viral or bacterial infection can be a cause, but in many cases, the cause is unknown.

Almost all cases of fatal heart inflammation occur in people with preexisting heart problems, such as valvular heart disease, or in IV drug users or immunocompromised people. Heart inflammation is uncommon, but it has a high fatality rate and is a leading cause of cardiac death in young adults.

The incidence of infective *endocarditis* is low, with only 7.7 cases per 100,000 persons each year, or about 15,000 new cases each year. The mortality rate is high. Among hospitalized patients, mortality is 15 to 20 percent; the 1-year mortality rate is 40 percent.

Myocarditis has a mortality rate ranging from 25 to 45 percent, depending on the infectious agent. Myocarditis is most frequently diagnosed in younger adults between the ages of 20 and 40 years. It causes sudden cardiovascular death in between 4 and 14 percent of young adults. In 2019, acute myocarditis caused 5,278 deaths, for a mortality rate of 0.08 per 100,000.

Pericarditis is usually triggered by a viral infection. Approximately 1 per 1,000 hospital admissions are for acute pericarditis. Severe pericarditis has a high mortality rate of at least 12 percent and possibly as high as 40 percent of treated patients. In 2019, 1,235 people died from acute pericarditis, for a mortality rate of 0.02 per 100,000 population.

Statistics for death from heart inflammation are unreliable because many patients are in poor health and their death is attributed to an underlying cause, such as cancer or kidney disease.

STAY SAFE Alcohol abuse, intravenous drug abuse, and poor dental health all increase your risk of inflammatory heart disease by allowing germs to enter your bloodstream and travel to the heart. If you're at risk, brush and floss your teeth regularly, and have regular dental checkups.

SOURCES AHA, S.S. Virani et al. Heart Disease and Stroke Statistics-2021 Update: A Report From the American Heart Association. *Circulation*, February 2021; CDC NCHS, cdc.gov/nchs/ndi/index.htm; Myocarditis Foundation, myocarditisfoundation.org; NIH NHLBI, nhlbi.nih.gov/health-topics/heart-inflammation

Heart Transplant

ANNUAL US DEATHS About 284 in 2018

In 2019, 3,597 people had a heart transplant, surgery that removes a diseased heart and replaces it with a healthy heart from a deceased donor: 3,088 in adults and 509 in infants and children. The rate of heart transplants was 10.9 per million population.

Posttransplant mortality rates for adult recipients were 6.4 percent at 6 months and 7.9 percent at 1 year for transplants in 2018; 14.4 percent at 3 years for transplants in 2016; and 20.1 percent at 5 years for transplants in 2014. The number of survivors of heart transplants continues to increase. In 2019, 35,713 recipients of donated hearts were alive, versus 23,315 in 2008.

WAITING FOR TRANSPLANT

ANNUAL US DEATHS Rate of death while waiting for a heart transplant is calculated in deaths per 100 "wait-list years"; average pretransplant mortality is 8.3 deaths per 100 wait-list years.

In 2019, 7,562 people in the US were on a waiting list for a heart transplant. From 2008 to 2019, the number of new patients on the heart transplant wait-list increased by 42.5 percent, from 2,687 to 4,086. In that same time frame, the total number of patients waiting for a heart transplant increased by 42.6 percent, from 5,304 to 7,562. The most common diagnosis was cardiomyopathy, at 59.7 percent of candidates. The median wait-time for a transplant was 5.1 months. The average pretransplant mortality rate was 8.3 per 100 wait-list years. However, patients in the highest urgency category (status 1) had a pretransplant mortality rate of 113.5 deaths per 100 wait-list years.

STAY SAFE Consider postmortem organ donation to help others.

SOURCES NIH NHLBI, nhlbi.nih.gov/health-topics/heart-transplant; Organ Procurement and Transplantation Network (OPTN), optn.transplant.hrsa.gov/data

Sudden Cardiac Death

ANNUAL US DEATHS 94.8 per 100,000

Sudden cardiac arrest (SCD) happens when the heart suddenly and unexpectedly stops beating, usually due to a problem with heart rhythm called ventricular fibrillation. Blood stops flowing to the brain and other vital organs, causing death within minutes in most cases. Nine out of 10 victims die. Immediate treatment with hands-only CPR and defibrillation if available is the only treatment.

SCD is the 3rd leading cause of death in the US. Approximately 356,000 people of all ages experience EMS-assessed out-of-hospital nontraumatic SCD each year. In 2015, 347,322 adults died of SCD; the incidence of EMS-assessed SCD in people of any age was 110.8 individuals per 100,000 population. In 2018, SCD appeared among the multiple causes of death on 13.3 percent of death certificates, suggesting it killed 1 of every 7.5 people who died in the US. The any-mention age-adjusted annual rate of death from SCD in 2018 was 94.8 per 100,000 population.

Better treatment with drugs and implanted cardiac defibrillators brought the age-adjusted death rates for any mention of SCD from 137.7 per 100,000 person-years in 1999 to 94.8 per 100,000 person-years in 2018.

BY SEX

In 2018, SCD killed 10,273 men; it was an underlying cause in 195,723 deaths. SCD killed 8,716 women in 2018; it was an underlying cause in 182,040 cases.

BY RACE/ETHNICITY

In 2018, 7,679 White males died of primary SCD; 6,461 White females died. SCD killed 1,814 Black men and 1,653 Black females; 504 Hispanic males and 365 Hispanic females; and 194 Asian/Pacific Islander males and 170 Asian/Pacific Islander females.

CPR SURVIVAL

About 90 percent of people who experience an out-of-hospital cardiac arrest (OHCA) die, but immediate cardiopulmonary resuscitation can improve the chances of survival. Unlike the dramatic rescues seen on TV shows, however, the chances of survival from OHCA are still low. Data on bystander (not EMT) CPR is sketchy, but about 27 percent of the time, bystander CPR leads to return of a heartbeat. About 20 percent of those individuals survive long enough to be admitted to a hospital. About 8 percent of those who make it to the hospital survive long enough to be discharged, often with significant neurological damage. When CPR is administered by EMTs, a heartbeat is restored about 36 percent of the time. About 25 percent of those individuals survive long enough to be admitted to a hospital. About 11 percent of those who make it to the hospital survive long enough to be discharged, often with significant neurological damage.

Survival from an OHCA is sharply improved when bystanders apply an automated external defibrillator (AED) device. Cardiac arrest victims who quickly receive a shock from a publicly available AED administered by a bystander have 2.62 times higher odds of survival to hospital discharge and 2.73 times more favorable outcomes for functioning compared to victims who first received an AED shock after emergency responders arrive. The AHA estimates that about 1,700 lives are saved in the US per year by bystanders using an AED.

Survival from in-hospital cardiac arrest (code blue) is higher. In-hospital cardiac arrest occurs in over 290,000 adults each year in the US, for an incidence rate of 9 to 10 in-hospital cardiac arrests per 1,000 admissions. Among these patients, 25 percent survived to hospital discharge; 85 percent were discharged with a favorable neurological outcome.

STAY SAFE   Many individuals who die of SCD were unaware of having any risk factors for heart disease. Taking a statin drug does not prevent or reduce the risk of SCD.

SOURCES   AHA, S.S. Virani et al. Heart Disease and Stroke Statistics-2021 Update: A Report From the American Heart Association. *Circulation*, February 2021; CDC NCHS, cdc.gov/nchs/ ndi/index.htm; L.W. Andersen et al. In-Hospital Cardiac Arrest: A Review. *JAMA*, March 2019; R.A. Pollack et al. Impact of Bystander Automated External Defibrillator Use on Survival and Functional Outcomes in Shockable Observed Public Cardiac Arrests. *Circulation*, May 2018; S. Yan et al. The global survival rate among adult out-of-hospital cardiac arrest patients who received cardiopulmonary resuscitation: a systematic review and meta-analysis. *Critical Care*, February 2020.

SUDDEN ARRHYTHMIC DEATH SYNDROME

ANNUAL US DEATHS   Data is scarce; 1.2 SADS deaths per 100,000 person-years in 2011

Sudden arrhythmic death syndrome (SADS) is unexpected death from cardiac arrest, usually among apparently healthy adolescent or young men, and usually during sleep. This puzzling syndrome was first reported among South Asian refugees in the US in 1977. Accurate incidence data is scarce. In 1982, among this group the annual rate in the United States was 92 per 100,000 among Laotians-Hmong, 82 per 100,000 among other Laotian ethnic groups, and 59 per 100,000 among Cambodians. By 2011, for unknown reasons, the incidence of sudden unexplained death for all ethnicities had dropped to 1.2 per 100,000 person-years for persons under age 35.

STAY SAFE   Since the causes of this rare diagnosis are unknown, there are no specific recommendations for staying safe.

SOURCES   CDC NCHS, cdc.gov/nchs/ndi/index.htm; SADS Foundation, sads.org/What-is-SADS

Takotsubo Syndrome

ANNUAL US DEATHS   About 5.5

Takotsubo syndrome, also called broken heart syndrome or stress cardiomyopathy, occurs as the result of sudden, acute psychological or physical stress that rapidly weakens the heart muscle. The symptoms resemble a heart attack but are not caused by a blockage in a coronary artery. The symptoms are almost always temporary and reversible, with full recovery within days or weeks. Takotsubo syndrome is rare and death from it is very rare. In 2012, 5,480 patients were hospitalized with stress cardiomyopathy. The mean age was 67.2. The vast majority—91.6 percent—were White women. For patients with a primary diagnosis of takotsubo syndrome, in-hospital mortality was 1.1 percent. About 20 percent of patients will develop heart failure later on.

The syndrome was first identified in Japan in 1990. The name comes from the Japanese word for a type of octopus trap. The trap has a bulge at the tip that resembles the ballooning at the apex of the heart caused by stress cardiomyopathy.

STAY SAFE   If you think you're having a heart attack, don't delay! Get emergency medical help immediately.

SOURCE   R. Khera et al. Trends in hospitalization for takotsubo cardiomyopathy in the United States. *American Heart Journal*, 2016.

Valvular Disease

ANNUAL US DEATHS   3,046 (rheumatic valvular disease) and 3,046 (nonrheumatic valvular disease) in 2017

Valvular heart disease applies to a condition when any valve in the heart (mitral, tricuspid, aortic, pulmonary) is damaged or diseased.

Some heart valve disease is congenital, but, in most cases, it is the result of aging—the heart valves thicken and become stiff and calcified—or is a result of a previous heart problem such as a heart attack or heart failure. You are also at risk if you have risk factors for coronary heart disease.

Valvular heart disease affects about 2.5 percent of the US population. Rheumatic valvular disease, caused by infection with the same bacterium that causes strep throat, is rare, but about 13 percent of people born before 1943 and the advent of antibiotics have it. In 2017, rheumatic valvular heart disease caused 3,046 deaths; nonrheumatic valvular heart disease caused 24,811 deaths. Most people with valvular heart disease die from heart failure or atrial fibrillation, not from the valve disease itself.

In 2017, 15 percent of nonrheumatic valvular heart disease deaths were from mitral valve disease; 61 percent were from aortic valve disease; 24 percent were from disease in other valves.

In 2019, the prevalence of aortic valve disease was 116.34 per age-adjusted 100,000 population; the death rate was 1.76. For men, the prevalence was 242.69 and the death rate was 0.39. For women, the prevalence was 341.30 and the death rate was 0.49.

In 2019, the prevalence of mitral valve disease was 246.06 per age-adjusted 100,000 population; the death rate was 0.45. For men, the prevalence was 133.38 and the death rate was 1.85. For women, the prevalence was 341.30 and the death rate was 0.49.

STAY SAFE   Don't smoke. If you smoke, quit. Free resources to help you quit are available from the BeTobaccoFree.gov, the National Cancer Institute quitline at 1-877-44U-QUIT, and the American Lung Association at lung.org/quit-smoking. Follow a heart healthy lifestyle, especially if you have already been diagnosed with heart disease.

SOURCES   AHA, S.S. Virani et al. Heart Disease and Stroke Statistics-2021 Update: A Report From the American Heart Association. *Circulation*, February 2021; CDC NCHS, cdc.gov/nchs/ndi/index.htm

Racial and Ethnic Disparities in Heart Disease

> **ANNUAL US DEATHS** 22,699 Black men and 18,118 Black women died of coronary heart disease in 2018; heart attacks killed 6,650 Black men and 5,476 Black women.

The risk of death from heart disease differs by race and ethnicity. In 2017, 11.5 percent of White adults had heart disease; 9.5 percent of Black adults had heart disease; 7.4 percent of Hispanic adults had heart disease; and 6.0 percent of Asian adults had heart disease. The age-adjusted death rates for heart disease, however, show disparities by race and ethnicity. In 2017, the age-adjusted death rate per 100,000 persons for White adults was 168.9, but it was 208.0 for Black adults. For Hispanic adults it was 114.1; for Asian and Pacific Islander adults, it was 85.5. A Black person was more than twice as likely as an Asian or Pacific Islander to die of heart disease. The decreasing trend in heart disease deaths for Black adults since 2000 has reversed, with deaths increasing by 6.4 percent from 2011 to 2014.

In general, Black adults have more fatal heart disease than any other ethnic group. Based on 2015 to 2018 data, among Black adults over age 20, 60.1 percent of men and 58.8 percent of women had cardiovascular disease. In 2018, CVD killed 56,945 Black men and 53,641 Black women.

Coronary heart disease killed 22,699 Black men and 18,118 Black women in 2018. Heart attacks killed 6,650 Black men and 5,476 Black women. The age-adjusted death rate per 100,000 for CHD was 141.4 for Black men and 79.7 for Black women.

STAY SAFE Black and Hispanic adults are more likely than White adults to have hypertension, obesity, and diabetes, all risk factors for heart disease. Know your numbers and work to keep them in the healthy range. Don't smoke. If you smoke, quit. Free resources to help you quit are available from BeTobaccoFree.gov, the National Cancer Institute quitline at 1-877-44U-QUIT, and the American Lung Association at lung.org/quit-smoking.

SOURCES CDC NCHS, cdc.gov/nchs/ndi/index.htm; M. Van Dyke et al. Heart Disease Death Rates Among Blacks and Whites Aged ≥35 Years — United States, 1968–2015. *MMWR Surveillance Summary*, March 2018.

Brain and Cognitive Diseases

Stroke

> **ANNUAL US DEATHS** 150,005 in 2019

A stroke occurs when the blood flow to the brain is interrupted or reduced, keeping oxygen and nutrients from getting to brain cells. Without blood flow, brain cells start to die within a few minutes. This can cause lasting brain damage, long-term disability, or death.

Ischemic stroke is caused by a blood clot that blocks a blood vessel in the brain and keeps blood from flowing. About 80 percent of strokes are ischemic. Hemorrhagic stroke is caused by a blood vessel that breaks and bleeds into the brain.

Stroke is the 5th leading cause of death in the US. In 2019, 150,005 people died of a stroke, or 45.7 per 100,000 population, or approximately 1 of every 19 deaths. In 2018, on average someone died of a stroke every 3 minutes 33 seconds. The probability of death within 1 to 5 years after a stroke is highest in individuals 75 years of age or older.

Major risk factors for stroke include high blood pressure (the primary risk factor), diabetes, heart disease (especially atrial fibrillation), smoking, and age. Black Americans have a higher risk of stroke than any other ethnic group.

STAY SAFE Know the FAST signs of stroke: **F**ace drooping, **A**rm weakness, **S**peech slurred, **T**ime to call 911. Immediate treatment may save a life and increases the chances for successful rehabilitation and recovery.

SOURCES CDC NCHS, cdc. gov/nchs/ndi/index.htm; S.S. Virani et al. Heart Disease and Stroke Statistics-2021 Update: A Report From the American Heart Association. *Circulation*, February 2021; NIH NINDS, ninds. nih.gov/Disorders/All-Disorders/ Stroke-Information-Page

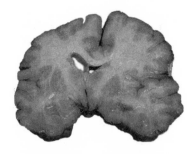

Section of an autopsied brain showing tissue damage caused by a fatal stroke

BY RACE/ETHNICITY

In 2018, Black males and females had higher age-adjusted death rates for stroke than any other ethnic group. Among all ages, stroke killed

8,851 Black males and 10,622 Black females. The age-adjusted rate for stroke was 58.2 per 100,000 for Black males and 47.5 for Black females.

In 2018, the age-adjusted death rate for stroke per 100,000 population for White males was 35.6; for White females it was 36.5. Black males and females had higher age-adjusted death rates for stroke than any other ethnic group. For Black males the rate was 58.3; for Black females the rate was 47.5. For Hispanic males the rate was 34.2; for Hispanic females it was 29.9. For Asian males, the rate was 31.7; for Asian females it was 27.8. For Native American males, the rate was 30.2; for Native American females, the rate was 30.2.

STAY SAFE Know the FAST signs of stroke: **F**ace drooping, **A**rm weakness, **S**peech slurred, **T**ime to call 911. Immediate treatment may save a life and increases the chances for successful rehabilitation and recovery.

BY SEX

Each year, approximately 55,000 more females than males have a stroke. The lifetime risk of stroke for women between the ages of 55 and 75 is 1 in 5. Stroke is the 3rd leading cause of death for women and kills twice as many women each year as breast cancer.

STROKE BELT

The Stroke Belt covers 11 states in the Southeast where deaths by stroke are 34 percent higher than they are for the rest of the country. The Stroke Belt states are Alabama, Arkansas, Georgia, Indiana, Kentucky, Louisiana, Mississippi, North Carolina, South Carolina, Tennessee, and Virginia.

The risk of death by stroke in these states may be higher because their residents have higher rates of obesity, cigarette smoking, and high blood pressure. In addition, the Stroke Belt region corresponds closely to higher incidence of type 2 diabetes, lung cancer, and cognitive decline.

Historically, the overall average stroke mortality has been approximately 30 percent higher in the Stroke Belt than in the rest of the nation and approximately 40 percent higher in the Stroke Belt "buckle" (North Carolina, South Carolina, and Georgia). The worst state of the Stroke Belt for death is Mississippi. In 2019 the state had an age-adjusted stroke death rate of 52.1 per 100,000 and 1,851 deaths out of a total population of 2.976 million. By comparison, the best state for stroke death in 2019 was New York, with an age-adjusted stroke death rate of 23.8 and 6,192 deaths out of a total population of 19.4 million.

SOURCE   Heart Attack and Stroke Prevention Center, thepreventioncenter.com/cardiovascular-disease/stroke-belt

Dementia

ANNUAL US DEATHS   271,872 in 2019

Dementia, or progressive decline in brain function, primarily affects older adults. The most common form of dementia is Alzheimer's disease, affecting about 80 percent of all people with dementia. In 2019, dementia was the 6th leading cause of death in the US. Among adults age 65 and up, Alzheimer's disease is the 5th leading cause of death. Worldwide, dementia is now the 7th leading cause of death. In 2019, some form of dementia was recorded as the underlying cause of death for 271,872 people in the US, including the 121,499 people who died of Alzheimer's disease.

Death from dementia is usually caused by something else, with dementia as the underlying cause. Aspiration pneumonia, bed sore complications, urinary tract infections, and dehydration are common causes of death. The percentage of deaths each year attributable to dementia is probably a serious undercount. Research suggests that the underlying cause of recorded on death certificates may underestimate the contribution of dementia by a factor of at least 2.7. When deaths attributable to cognitive impairment without dementia are added, the undercount gap is even greater, possibly as high as a factor of 4.8. Among Americans aged 70 and older, the research suggests about 14 percent of deaths are attributable to dementia.

In 2020, the COVID-19 pandemic caused an estimated 42,000 more deaths from Alzheimer's disease and other dementias compared to the 5-year average before 2020, or about 16 percent higher than expected.

STAY SAFE   Smoking and alcohol use contribute to dementia. Be alert for signs of dementia in older adults, such as personality changes, withdrawal, memory problems, sleep changes, confusion, and hallucinations. Early medical intervention can help manage the problem more effectively.

SOURCES   Alzheimer's Association 2021 Alzheimer's Disease Facts and Figures. *Alzheimer's and Dementia*, March 2021; CDC NCHS, cdc.gov/nchs/ndi/index.htm; NIH, alzheimers.gov; NIH, NIA, nih.nia. gov; A. Stokes et al. Estimates of the Association of Dementia with US Mortality Levels Using Linked Survey and Mortality Records. *JAMA Neurology*, August 2020.

ALZHEIMER'S DISEASE

> **ANNUAL US DEATHS** 121,499 deaths from Alzheimer's complications in 2019

In 2019, 121,499 Americans died of complications from Alzheimer's disease (AD). The most common complication was aspiration pneumonia, recorded in about half of all deaths. Because many people have undiagnosed or unrecorded Alzheimer's disease, the true death toll is almost certainly higher. Complicating the true death count is that Alzheimer's disease can only be verified by autopsy. Studies show that between 15 and 30 percent of those diagnosed with AD in life actually had some other form of dementia.

Most people with AD live an average of 4 to 8 years after diagnosis. About 66 percent of people with AD die in a nursing home, compared with about 20 percent of people with cancer and 28 percent of people dying from all other conditions.

In 1999, the death rate from Alzheimer's disease was 128.8 per 100,000 population. Reflecting an aging population, in 2019 the death rate was 233.8. Because women generally live longer than men, more women die of AD. The death rate in 2019 for women was 263.0; for men, it was 186.3. Mortality from AD increases sharply with advancing age. In 2019, the annual US deaths per 100,000 population for people aged 45 to 54 was 0.3; for people aged 55 to 64, it was 3.0; for people aged 65 to 74, it was 24.9; for people aged 75 to 84, it was 210.2; for people aged 85 and up, it was 1,191.3.

STAY SAFE Smoking and alcohol use contribute to dementia. Be alert for signs of dementia in older adults, such as personality changes, withdrawal, memory problems, sleep changes, confusion, and hallucinations. Early medical intervention can help manage the problem more effectively.

SOURCES Alzheimer's Association 2021 Alzheimer's Disease Facts and Figures. *Alzheimer's and Dementia*, March 2021; CDC NCHS, cdc.gov/nchs/ndi/index.htm; NIH, alzheimers.gov; A. Stokes et al. Estimates of the Association of Dementia with US Mortality Levels Using Linked Survey and Mortality Records. *JAMA Neurology*, August 2020.

CHRONIC TRAUMATIC ENCEPHALOPATHY

ANNUAL US DEATHS   No available data

Chronic traumatic encephalopathy (CTE) is a progressive degenerative disease of the brain found in people with a history of repetitive brain trauma. Athletes who experience both concussions that cause symptoms and subconcussive hits to the head that cause no symptoms are most affected. A 2017 study of brains obtained from NFL, college, and high school football players found CTE in 99 percent of the NFL players, 91 percent of the college players, and 21 percent of the high school players.

The repeated brain trauma triggers progressive degeneration of the brain tissue, including the buildup of an abnormal protein called tau. Symptoms of CTE usually appear gradually years later and include memory loss, confusion, impaired judgment, impulse control problems, aggression, depression, suicidality, parkinsonism, and progressive dementia.

Not everyone with a history of repeated brain trauma develops CTE; other factors, including genetics, may play a role. CTE can only be definitively diagnosed through postmortem analysis of the brain.

Junior Seau, a 12-time Pro Bowl selection, killed himself in 2012 at the age of 43, 3 years after his retirement from the NFL. His lifetime career as a linebacker with the Chargers, Dolphins, and Patriots included 268 games and nearly 1,900 tackles. Postmortem analysis confirmed severe CTE. In 2021, Phillip Adams, a retired defensive back who played for several NFL teams, shot and killed 6 people and then himself. His autopsy revealed stage 2 CTE, an abnormally severe diagnosis for someone in their 30s. Although numerous other former NFL players and other professional athletes have killed themselves because of CTE, no firm numbers for total deaths are available.

STAY SAFE   Wear appropriate head protective gear. Treat concussions promptly.

SOURCE   CTE Center at Boston University School of Medicine, Clinicopathological Evaluation of Chronic Traumatic Encephalopathy in Players of American Football. *JAMA*, 2017; 318(4):360–70.

VASCULAR DEMENTIA

> **ANNUAL US DEATHS**   No reliable accurate statistics; most patients die within 3 to 5 years of diagnosis.

Vascular dementia is the 2nd most common cause of dementia in the US. It's caused by the narrowing or blockage of arteries supplying blood to the brain and usually occurs after a stroke or a series of silent strokes. The symptoms include slowed thought, reduced concentration, and difficulty with executive function such as planning and making decisions.

About 20 percent of people who have a stroke develop vascular dementia within 6 months. Within 4 years of a stroke, the relative risk of vascular dementia is 5.5 times that of people the same age who have not had a stroke. Most people with vascular dementia die within 3 to 5 years of diagnosis. Most die from another stroke or a heart attack.

FRONTOTEMPORAL DEMENTIA

> **ANNUAL US DEATHS**   No reliable statistics; patients die within an average of 8 years of diagnosis.

Frontotemporal dementia (FTD) is an overall name for a group of disorders that cause neurodegenerative changes in the brain. The degeneration is caused by the buildup of misfolded proteins in the frontal and temporal lobes of the brain. Signs of FTD include changes in behavior and personality, progressive language decline, and progressive loss of motor function. Frontotemporal dementia usually begins in the late 50s. After Alzheimer's disease, it is the 2nd most common form of dementia in people under the age of 65. FTD probably affects about 50,000 to 60,000 people in the US, but the total prevalence is probably higher because many cases go undiagnosed or are misdiagnosed. People with FTD live on average for about 8 years after diagnosis.

DEMENTIA WITH LEWY BODIES

> **ANNUAL US DEATHS**   No available data

People with dementia with Lewy bodies (DLB) have symptoms similar to Parkinson's disease, along with cognitive impairment, sleep disorders, and hallucinations. Accurate data on incidence and prevalence are scarce, but DLB probably accounts for about 5 percent of all dementia cases in older populations. People with DLB usually live for 5.5 to 7.7 years after diagnosis.

Cancer

ANNUAL US DEATHS   606,520 in 2019

Cancer is the 2nd leading cause of death in the US, for a death rate of 158.3 per 100,000 men and women per year (based on 2013 to 2017 deaths). In 2019, 606,520 people died of cancer.

Worldwide, cancer is the 2nd leading cause of death. In 2018, 9.5 million people died from cancer-related causes, or approximately 16 percent of all deaths out of an estimated world population of approximately 7.592 billion. Worldwide, more than 1 in 6 deaths is due to cancer.

From 1991 to 2022, the overall cancer death rate in the US dropped 32 percent. That works out to approximately 3.5 million fewer cancer deaths than would have been expected if death rates had remained at their peak.

Cancer Terminology

Cancer is broadly defined as a disease caused by the uncontrolled growth of abnormal cells in a part of the body. Cancers are usually defined by their primary anatomical location, such as breast cancer or skin cancer. A cancer that has metastasized (spread) to a distant part of the body is still defined by the primary anatomical site.

Cancer has 2 main classifications: solid tumors (cancer in an organ such as the lung) or blood cancers (cancers that originate in the bone marrow, blood, or lymphatic system). The tumor can then be further defined by its tissue type, based on how the cancerous cells appear under a microscope. For example, lung cancer broadly refers to the primary site of the tumor, but it has 2 major tissue subtypes: non-small cell lung cancer and small cell lung cancer. Non-small cell lung cancer is further divided into tissue types named for the type of cells in which the cancer develops: squamous cell carcinoma, adenocarcinoma, and large cell carcinoma. In addition to common tissue types, some cancers can have numerous rare

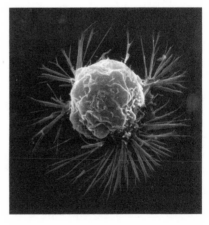

Electron photomicrograph of a single breast cancer cell

subtypes. Because statistics on cancer deaths don't always distinguish among tissue types and subtypes, for the purposes of this book, death statistics by cancer tissue type are included only where reliable information is available.

Types of Cancer

More than 100 different types of cancer are known, some very rare (see **Rare Cancers** below).

The top 10 leading causes of cancer death in 2020 were:

TYPE OF CANCER	ESTIMATED DEATHS, 2020	% OF ALL CANCER DEATHS
Lung cancer	135,720	22.4
Colon and rectum cancer	53,200	8.8
Breast cancer	42,170	7.0
Prostate cancer	33,330	5.5
Leukemia	23,100	3.8
Non-Hodgkin lymphoma	19,940	3.3
Bladder cancer	17,980	3.0
Kidney cancer	14,830	2.4
Uterine cancer	12,590	2.1
Melanoma of the skin	6,850	1.1

Approximately 40 percent of all men and women will be diagnosed with cancer at some point during their lifetimes, based on 2015 to 2017 data. Estimated projections of US cancer death to 2040 suggest that lung cancer will still be the leading cause of cancer death, but pancreatic cancer and liver cancer will become the 2nd and 3rd most common causes of cancer death. Colorectal cancer will be the 4th most common cause, and breast cancer will be the 5th.

The cancer mortality rate is higher among men than women (189.5 per 100,000 men and 135.7 per 100,000 women). Cancer mortality is highest in Black men (227.3 per 100,000) and lowest in Asian/Pacific Islander women (85.6 per 100,000).

Cancer and Age

Cancer is largely a disease of aging. In 2021, only 4.6 percent of all new cancer cases occurred among people ages 15 to 39. Deaths from cancer in this age group were 9,130, or 1.5 percent of all cancer deaths at any age.

In 2017, deaths from cancer by age group were:

AGE	RATE PER 100,000	DEATHS
<1	1.5	57
1–4	2.0	325
5–9	2.1	418
10–14	2.1	437
15–19	2.7	560
20–24	3.7	814
25–29	5.6	1,301
30–34	10.6	2,315
35–39	19.4	4,104
40–44	34.7	6,796
45–49	62.7	13,108
50–54	122.5	26,158
55–59	217.0	47,646
60–64	336.8	67,164
65–69	485.7	81,606
70–74	677.0	86,852
75–79	938.7	81,955
80–84	1240.9	73,962
85+	1599.9	103,521

SOURCES CDC USCS, cdc.gov/cancer/uscs/public-use/index.htm; NIH NCI, seer.cancer.gov

By Sex

For all anatomical sites combined, the cancer incidence rate is 20 percent higher in men than in women; the cancer death rate is 40 percent higher for men. Across different cancer types, the male-to-female mortality rate ratio (MRR) of primary cancer types in the United States was 1.17, meaning 1.17 men die of the cancer compared to just 1 woman. Some mortality rate ratios are much higher. The highest male-to-female MRR is for lip cancer at 5.51. For esophageal cancer, the MRR is 3.77; for bladder cancer, the MRR is 2.38.

Modifiable Risk Factors

Some cancer risks can't be changed: your age, your genetics, your exposure to environmental factors such as UV light, simple bad luck. However, nearly half of all cancer deaths are caused by factors that can be changed: smoking, obesity, sedentary lifestyle, poor diet, and other unhealthy behaviors.

- Smoking. Tobacco use is the leading preventable cause of death in the US. Cigarette smoking and secondhand smoke is estimated to cause more than 480,000 deaths annually. Nearly 9 out of 10 lung cancer deaths are caused by smoking cigarettes or secondhand smoke exposure.

- Obesity. Overweight and obesity are associated with increased risk of 13 types of cancer: Breast cancer (in women past menopause), colorectal cancer, endometrial cancer, esophageal cancer, gallbladder cancer, kidney cancer, liver cancer, ovarian cancer, pancreatic cancer, stomach cancer, thyroid cancer, multiple myeloma, and meningioma. Excess body weight is thought to be responsible for about 11 percent of cancers in women and about 5 percent of cancers in men, and for about 7 percent of all cancer deaths. Obesity is a better predictor of cancer risk than any other marker.

- Lack of exercise. Higher levels of physical activity are linked to lower risk of bladder cancer, breast cancer, colorectal cancer, endometrial cancer, esophageal cancer, kidney cancer, and stomach cancer. Mortality from cancer may be 82 percent higher among very sedentary people compared to very active people.

- Poor diet. In 2015, an estimated 80,110 new cancer cases among adults in the US were attributable to eating a poor diet—about 5.2 percent of all new cancer cases.

- Alcohol. An estimated 18,200 to 21,300 cancer deaths, or 3.2 to 3.7 percent of all US cancer deaths, are attributed to alcohol consumption. Most of the alcohol-attributable female cancer deaths were from breast cancer (56 to 66 percent). For men, upper airway and esophageal cancer deaths were more common (53 to 71 percent).

SOURCES   AACR, cancerprogressreport.aacr.org; ACS, cancer.org/healthy.html; ASCO, cancer.net/cancer-types; CDC USCS, cdc.gov/cancer/dcpc/data/index.htm; IARC, iarc.who.int/branches-csu; NIH NCI, seer.cancer.gov; USPSTF, uspreventiveservicestaskforce.org/uspstf/recommendation-topics

Anal Cancer

> **ANNUAL US DEATHS**   Estimated 1,350 in 2020

In 2020, an estimated 8,590 people were diagnosed with anal cancer, or 0.5 percent of all new cancer cases. There were an estimated 1,350 deaths, or 0.2 percent of all cancer deaths. The rate of new cases of anal cancer is 1.9 per 100,000 men and women per age-adjusted year.

Most anal cancers are related to human papillomavirus (HPV) infection. Smokers are about 8 times more likely to develop anal cancer than nonsmokers.

STAY SAFE   Don't smoke. Practice safe sex and get vaccinated for HPV.

SOURCES   ASCO, cancer.net/cancer-types; CDC USCS, cdc.gov/cancer/dcpc/data/index.htm; NIH NCI, seer.cancer.gov

Appendix Cancer

> **ANNUAL US DEATHS**   Reliable statistics unavailable

Appendix cancer, also called appendiceal cancer, is so rare that reliable statistics aren't available. Appendix cancer occurs at the rate 0.12 cases per 1,000,000 population per year and accounts for only about 0.5 percent of all tumors that start in the gastrointestinal tract, including the esophagus, pancreas, stomach, colon, rectum, anus, liver, biliary system, and small intestine. Actor Audrey Hepburn died of appendix cancer in 1993.

STAY SAFE   Getting older is the only known risk factor for appendix cancer.

SOURCES   ASCO, cancer.net/cancer-types; CDC USCS, cdc.gov/cancer/dcpc/data/index.htm; NIH NCI, seer.cancer.gov

Bile Duct Cancer (Cholangiocarcinoma)

> **ANNUAL US DEATHS**   8,000 cases, most resulting in death

Primary bile duct cancer is uncommon in the United States, with only an estimated 8,000 cases per year. This cancer is very lethal. The 5-year survival rate is 24 percent for cancer diagnosed at an early stage. If the cancer has spread to a distant part of the body, the 5-year survival rate is 2 percent.

STAY SAFE Avoid alcohol abuse, which can lead to liver cirrhosis and bile duct cancer.

SOURCES ASCO, cancer.net/cancer-types; CDC USCS, cdc.gov/cancer/dcpc/data/index.htm; NIH NCI, seer.cancer.gov

Bladder Cancer

ANNUAL US DEATHS Estimated 17,980 in 2020

In 2020, an estimated 81,400 new cases of bladder cancer were diagnosed, for 4.5 percent of all new cancer cases and a rate of 20.0 per 100,000 men and women per age-adjusted year. Bladder cancer is the 6th most common cancer in the US.

Bladder cancer is the 10th leading cause of cancer death in the US. An estimated 17,980 people died of bladder cancer in 2020, or 3.0 percent of all cancer deaths. The death rate was 4.3 per 100,000 men and women per age-adjusted year.

Bladder cancer affects and kills far more men than women. In 2018, the incidence rate for men was 34.9 per 100,000 persons, but only 8.6 for women. The death rate for men was 7.4 per 100,000 persons, but only 2.1 for women. For White men, the death rate was 7.8; for Black men, it was 5.3. For White women, the death rate was 2.2; for Black women, the death rate was 2.3.

Risk factors for bladder cancer include smoking and exposure to workplace chemicals such as paints, dyes, metals, and petroleum products.

STAY SAFE Don't smoke. Take appropriate precautions to avoid workplace exposure to toxins.

SOURCES ASCO, cancer.net/cancer-types; CDC USCS, cdc.gov/cancer/dcpc/data/index.htm; NIH NCI, seer.cancer.gov

Bone and Joint Cancer

ANNUAL US DEATHS Estimated 2,060 in 2021

Cancers affecting bones and joints are rare—as a group (osteosarcoma, Ewing sarcoma, malignant fibrous histiocytoma, and chondrosarcoma) these cancers are the 29th most common cause of cancer death. In 2021, only an estimated 3,610 new cases were diagnosed, or 0.2 percent of all new cancer cases. The rate of new cases of bone and joint cancer was 1.0 per 100,000 men and women per year.

Estimated deaths for 2021 were 2,060, or 0.3 percent of all cancer

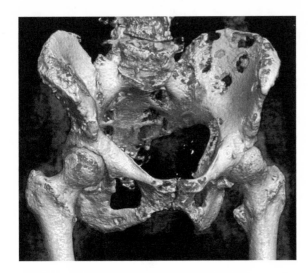

Metastasized bone cancer in pelvis and hips as rendered by a 3D CT scan

deaths. The death rate was 0.5 per 100,000 men and women per year. Based on 2018 data, approximately 0.1 percent of men and women will be diagnosed with bone and joint cancer at some point during their lifetime.

STAY SAFE Past treatment with chemotherapy or radiation therapy is a risk factor.

SOURCES ASCO, cancer.net/cancer-types; CDC USCS, cdc.gov/cancer/dcpc/data/index.htm; NIH NCI, seer.cancer.gov

Brain Cancer

ANNUAL US DEATHS 18,020 in 2020

Brain cancer is the 16th most common cancer in the US. In 2020, an estimated 23,890 adults (13,590 men and 10,300 women) were diagnosed with primary cancerous tumors of the brain and spinal cord, or 1.3 percent of all new cancer cases. Brain tumors account for 85 to 90 percent of all primary central nervous system tumors. The lifetime risk of developing this type of tumor is less than 1.0 percent.

About 3,540 children under the age of 15 were diagnosed with a brain or spinal cord tumor in 2020.

Brain and central nervous system cancer is the 9th leading cause of cancer death in the US. In 2020, 18,020 people died of brain cancer, or 3.0 percent of all cancer deaths. The death rate was 4.4 per 100,000 men and women per age-adjusted year; the rate was 5.4 per 100,000 for men and 3.2 per 100,000 for women. For White men, the death rate was 5.9 per 100,000; for Black men, the death rate was 3.2 per 100,000. For White women, the death rate was 3.9; for Black women it was 2.3.

The causes of brain and spinal cord cancer are largely unknown. No evidence shows that electromagnetic fields from power lines, cell phones, microwaves, and other sources are linked to an increased risk of brain tumors in adults. Exposure to solvents, pesticides, oil products, rubber, or vinyl chloride may increase the risk of developing a brain tumor, although no scientific evidence supports this.

STAY SAFE Avoid exposure to environmental and industrial chemicals.

SOURCES ASCO, cancer.net/cancer-types; CDC USCS, cdc.gov/cancer/dcpc/data/index.htm; NIH NCI, seer.cancer.gov

ASTROCYTOMA

ANNUAL US DEATHS Incidence too low for a reliable estimate

An astrocytoma is a tumor that begins in the brain or spinal cord in small, star-shaped cells called astrocytes. This is a rare cancer that mostly affects children and adolescents. In 2020, an estimated 930 children and adolescents younger than 19 were diagnosed with astrocytoma, or about 35 percent of childhood brain tumors. Because the 5-year survival rate is good—up to 97 percent for the most common type—accurate death rates aren't available.

Adults also get astrocytoma, but far less often. Less than 25 percent of all cases are in adults over age 20.

STAY SAFE There are no known risk factors for astrocytoma in children or adults.

SOURCES ASCO, cancer.net/cancer-types; CDC USCS, cdc.gov/cancer/dcpc/data/index.htm; NIH NCI, seer.cancer.gov

GLIOBLASTOMA MULTIFORME

ANNUAL US DEATHS No reliable statistics

Also called stage IV astrocytoma, glioblastoma multiforme (GBM) is rare, with an incidence of 2 to 3 per 100,000 adults per year. It is the most common, most aggressive, and most deadly form of brain cancer and is almost always lethal within 18 months of diagnosis. The median survival time is 14.6 months; the 5-year survival rate is less than 1.0 percent. About 16 percent of all primary brain tumors are GBM.

On August 1, 1966, "Texas Tower Sniper" Charles Whitman killed his mother and his wife with knives and then went to the University of Texas tower in Austin and used multiple firearms to shoot and kill 14 people and wound 31. He was killed by police and posthumously diagnosed with glioblastoma multiforme. The commission investigating the shooting concluded the tumor "conceivably could have contributed to his inability to control his emotions and actions."

Senator Ted Kennedy (D-MA) died of a malignant brain tumor (glioblastoma) on August 25, 2009. His friend and colleague Senator John McCain (R-AZ) died of GBM on August 25, 2018, exactly 9 years later. President Joe Biden's son Beau died of GBM in 2015 at the age of 46.

STAY SAFE There are no known risk factors for glioblastoma multiforme.

SOURCES ASCO, cancer.net/cancer-types; CDC USCS, cdc.gov/cancer/dcpc/data/index.htm; NIH NCI, seer.cancer.gov

Breast Cancer

ANNUAL US DEATHS 43,600

Female breast cancer is the most common cancer in the US. Approximately 281,550 cases, or 14.8 percent of all cancer cases, are diagnosed each year. Based on data from 2014 to 2018, the death rate was 20.1 per 100,000 women per age-adjusted year. An estimated 43,600 women die of breast cancer each year—7.2 percent of all cancer deaths. Female breast cancer is the 4th leading cause of cancer death in the US. Among women, breast cancer is the 2nd most common cause of death from cancer after lung cancer.

In 2020, breast cancer overtook lung cancer as the most commonly diagnosed form of cancer worldwide, with an estimated 2.3 million new cases and 685,000 deaths. In the US, approximately 1 in 8 women (13 percent) will be diagnosed with invasive breast cancer in their lifetime. One in 39 women (3 percent) will die from breast cancer. The age-adjusted death rate declined on average 1.4 percent each year over 2009 through 2018. Because of early detection and improved treatment, the number of women who have died of breast cancer decreased by 41 percent from 1989 to 2018.

BREAST CANCER SUBTYPES

Female breast cancer is classified into 4 main cancer subtypes. In order of prevalence, they are:

HR+/HER2– (Luminal A)
HR–/HER2– (Triple Negative)
HR+/HER2+ (Luminal B)
HR–/HER2+ (HER2-enriched)

The most common subtype is HR+/HER2–. Based on 2018 data, the age-adjusted incidence rate is 88.1 new cases per 100,000 women. The HR+/HER2+ rate is 13.4, the HR–/HER2– rate is 5.5 and the HR–/HER2+ rate is 5.5.

RACIAL DISPARITIES

For 2013 to 2017, the overall death rate for female breast cancer was 19.6 per 100,000 women. When broken out by race/ethnicity, disparities become clear:

US DEATH RATE	19.6
White	20.3
Black	28.4
Hispanic	14.0
Asian/PI	11.5

The incidence rate of all breast cancer types is highest among White women at 130.8 per 100,000 female population. Incidence rates for Black women are close, at 126.7 per 100,000. Black women have higher incidence rates than White women before age 40 and are more likely to die from breast cancer at every age. In the period 2013 to 2017, the breast cancer death rate in Black women was 40 percent higher than for White women.

One factor in the disparity is that about 21 percent of breast cancers in Black women are the triple negative subtype, about double the proportion of this subtype than in other racial/ethnic groups. For all women, the triple negative subtype is more aggressive and has a poorer prognosis than other types of breast cancer.

GEOGRAPHIC TRENDS

In 2018, the states with the worst death rates per 100,000 women for female breast cancer were:

District of Columbia	26
Louisiana	23
Mississippi	23
Oklahoma	23

The states with the best death rates were:

Hawaii	16
Connecticut	17
Massachusetts	17

HISTORIC TRENDS

The overall breast cancer death rate increased by 0.4 percent per year from 1975 to 1989. It then began decreasing rapidly due to improved treatment and better early detection, for a total decline of 40 percent through 2017. Approximately 375,900 breast cancer deaths were averted in US women from 1989 to 2017. In the period 2011 to 2017, the 5-year relative survival rate for breast cancer was 90.3 percent.

MALE BREAST CANCER

ANNUAL US DEATHS 520 in 2020

Male breast cancer is rare, accounting for less than 1 percent of breast cancer cases in the US. The death rate for male breast cancer from 2013 to 2017 was 0.3 per 100,000 men. In 2020, an estimated 2,620 men were diagnosed with breast cancer; approximately 520 died.

STAY SAFE Early detection reduces your risk of death. The USPSTF recommends biennial screening mammography for women aged 50 to 74 years. Screening before age 50 is appropriate for some women; discuss mammograms with your physician.

SOURCES ACS *Breast Cancer Facts & Figures 2019–2020;* ASCO, cancer.net/cancer-types; CDC USCS, cdc.gov/cancer/dcpc/data/index.htm; IARC, iarc.who.int/branches-csu; NIH NCI, seer.cancer.gov

Cervical Cancer

ANNUAL US DEATHS   Estimated 4,290 in 2020

In 2020, an estimated 13,800 women in the US were diagnosed with cervical cancer, for 0.8 percent of all new cancer cases. The rate of new cervical cancer cases is 7.4 per 100,000 women per year. Approximately 0.6 percent of women will be diagnosed with cervical cancer at some point during their lifetime. Human papillomavirus infection (HPV) causes more than 90 percent of all cervical cancer cases.

Cervical cancer is the 20th most common cause of cancer death in the US. Estimated deaths in 2020 were 4,290 women, or 0.7 percent of all cancer deaths, a rate of 2.2 per 100,000 women per age-adjusted year. Regular Pap test screening of women between the ages of 21 and 65 improves early detection and decreases the chance of dying from cervical cancer.

STAY SAFE   The USPSTF recommends screening for cervical cancer every 5 years for women aged 21 to 65. Practice safe sex and get vaccinated for HPV.

SOURCES   ASCO, cancer.net/cancer-types; CDC USCS, cdc.gov/cancer/dcpc/data/index.htm; J. Chor et al. Cervical Cancer Screening Guideline for Individuals at Average Risk. *JAMA*, July 2021; NIH NCI, seer.cancer.gov

Colorectal Cancer

ANNUAL US DEATHS   About 52,980 in 2020

Colorectal cancer (CRC) is the 2nd leading cause of cancer deaths in the US for men and women combined. In 2021, an estimated 149,500 new cases were diagnosed, for approximately 7.9 percent of all new cancer cases. For colorectal cancer, death rates increase with age. In 2020, approximately 52,980 people died from CRC, or about 8.7 percent of all cancer deaths. In 2018, the death rate was 13.7 per 100,000 men and women per year age-adjusted 100,000 population.

More men die from colorectal cancer than women. The age-adjusted mortality rate is 16.9 per 100,000 population per year in men and 11.9 per 100,000 per year in women.

The rate of new cases of colorectal cancer was 37.8 per 100,000 men and women per age-adjusted year. About 4.4 percent of Americans will develop CRC in their lifetime. The lifetime risk of dying from CRC is 1.8 percent.

HISTORIC TRENDS

From 2013 to 2017, CRC incidence declined by about 1 percent per year. From 2014 to 2018, mortality from CRC declined by about 2 percent per year. In 2020, about 10.5 percent of new colorectal cancer cases were in people younger than 50. Cases among adults between 40 and 49 years increased by almost 15 percent from 2000–2002 to 2014–2016.

RACIAL DISPARITIES

Compared to White people, Black people are more likely to develop CRC at a younger age and are more likely to die of the disease. Black people have a 20 percent higher incidence of CRC, and their mortality rate is 43 percent higher than Whites. In 2020 actor Chadwick

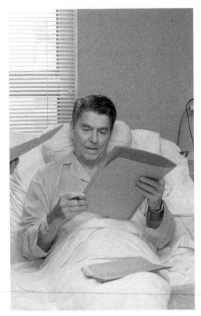

President Ronald Reagan at Bethesda Naval Hospital, about to undergo surgery to remove a suspicious polyp from his colon

Boseman died from colon cancer at age 43, illustrating the disproportionate impact of early-onset CRC among the Black community.

In 2018, the overall death rate from colorectal cancer was 13.7 per 100,000 men and women per year. For men, the rate was 16.3; for women, it was 11.5. For White men, it was 15.9, while for Black men it was 22.5, and for Hispanic men it was 14.0. For White women, the death rate was 11.3, while for Black women it was 14.8, and for Hispanic women it was 8.6.

STAY SAFE The USPSTF and American Cancer Society recommend screening for colorectal cancer starting at age 45 for people at average risk and continuing every 5 to 10 years until age 75 years. People at higher risk (family history of colon cancer or other factors) should discuss screening with their doctor.

SOURCES ASCO, cancer.net/cancer-types; CDC USCS, cdc.gov/cancer/dcpc/data/index.htm; NIH NCI, seer.cancer.gov

Endometrial Cancer

ANNUAL US DEATHS   Estimated 12,590 in 2020

Also called uterine cancer, endometrial cancer was diagnosed in 2020 in an estimated 65,620 women in the US, or 3.6 percent of all new cancer cases. The rate of new cases was 27.8 per 100,000 women per year. In the US, endometrial cancer is the most commonly diagnosed invasive cancer of the female reproductive system. Approximately 3.1 percent of women will be diagnosed with endometrial cancer at some point during their lifetime.

Endometrial cancer is the 9th most common cause of cancer death. In 2020, an estimated 12,590 women died, or 2.1 percent of all cancer deaths. The death rate was 4.9 per 100,000 women per age-adjusted year.

Racial disparities are apparent in endometrial cancer deaths. Although the incidence is roughly equal in White women (28.3 per 100,000) and Black women (27.9 per 100,000), the death rate for Black women (8.7 per 100,000) is almost double that for White women (4.5 per 100,000).

The most common risk factors for endometrial cancer are obesity, prediabetes, and type 2 diabetes. Taking tamoxifen to treat breast cancer is a risk factor.

STAY SAFE   Maintain a healthy weight. Discuss the risks of tamoxifen with your doctor.

SOURCES   ASCO, cancer.net/cancer-types; CDC USCS, cdc.gov/cancer/dcpc/data/index.htm; NIH NCI, seer.cancer.gov

Esophageal Cancer

ANNUAL US DEATHS   Estimated 16,170 in 2020

In 2020, an estimated 18,440 people in the US will be diagnosed with esophageal cancer, or 1.0 percent of all new cancer cases. The rate of new cases of esophageal cancer is 4.3 per 100,000 men and women per age-adjusted year. Approximately 0.5 percent of men and women will be diagnosed with esophageal cancer at some point during their lifetime. Esophageal cancer is the 18th most common cause of cancer death. In 2020, an estimated 16,170 people died of this cancer, or 2.7 percent of all cancer deaths. The death rate was 3.9 per 100,000 men and women per age-adjusted year.

Men are much more likely to get esophageal cancer and to die from it. Although esophageal cancer is overall the 18th most common cause of cancer death in the US, it is the 7th most common cause of cancer death among men. The incidence rate for men is 7.3 per 100,000 population; for women, it's 1.7. The death rate for men is 6.9 per 100,000 population; for women, it's 1.4.

Risk factors for esophageal cancer include smoking, heavy drinking, and gastroesophageal reflux disease (GERD) severe enough to cause the precancerous condition Barrett's esophagus.

Actor Humphrey Bogart died of esophageal cancer in 1957 at the age of 57. Author and journalist Christopher Hitchens died of esophageal cancer in 2011 at the age of 62.

STAY SAFE   Don't smoke. Avoid heavy drinking. If you have Barrett's esophagus, get regular endoscopic examinations.

SOURCES   ASCO, cancer.net/cancer-types; CDC USCS, cdc.gov/cancer/dcpc/data/index.htm; NIH NCI, seer.cancer.gov

Ewing Sarcoma

ANNUAL US DEATHS   Mortality statistics are unreliable, but some 200 US cases are diagnosed yearly.

Each year, about 200 children and young adults in the United States are diagnosed with Ewing tumors, most of which will be Ewing sarcoma. Ewing tumors make up 1 percent of all childhood cancers; Ewing sarcomas make up 2 percent of all childhood cancers. About half of all Ewing sarcoma diagnoses are in people between the ages of 10 and 20. These tumors can also affect younger children and young adults in their twenties and thirties. Almost all cases of Ewing sarcoma occur in White people. Because this cancer is rare, statistics are scattered. The incidence rate is 0.3 cases per million in children under age 3 and as high as 4.6 cases per million in adolescents aged 15 to 19.

When diagnosed and treated early, approximately 80 percent of patients survive for 5 years or longer.

STAY SAFE   Ongoing bone and joint pain and stiffness in children and adolescents should be discussed with a physician.

SOURCES   ASCO, cancer.net/cancer-types; CDC USCS, cdc.gov/cancer/dcpc/data/index.htm; NIH NCI, seer.cancer.gov

Hodgkin Lymphoma

ANNUAL US DEATHS   Estimated 960 in 2021

Hodgkin lymphoma is rare. In 2021, about 8,830 new cases of Hodgkin lymphoma were diagnosed in the US. It is the 27th most common cancer in the US.

The 5-year relative survival rate for Hodgkin lymphoma patients is about 87 percent. Hodgkin lymphoma is the most common cancer diagnosed in adolescents aged 15 to 19, comprising 6 percent of all childhood cancers. Ninety to 95 percent of children with Hodgkin lymphoma can be cured.

In 2021, about 960 people died of this cancer. In 2018 the death rate was 0.3 per 100,000 men and women per age-adjusted year. Better treatments mean the adult death rate for this cancer over the past 50 years has fallen faster than for any other malignancy.

The risk of Hodgkin lymphoma rises again in late adulthood (after age 55). The relative survival rate for older adults is 65 percent. Based on 2018 data, approximately 0.2 percent of men and women will be diagnosed with Hodgkin lymphoma at some point during their lifetime.

STAY SAFE   Epstein-Barr virus (EBV) has been implicated in some cases of Hodgkin lymphoma.

SOURCES   ASCO, cancer.net/cancer-types; CDC USCS, cdc.gov/cancer/dcpc/data/index.htm; NIH NCI, seer.cancer.gov

HPV-Associated Cancer

ANNUAL US DEATHS   See entries for the individual cancers mentioned in this entry.

Human papillomavirus (HPV) can cause cancer, although most people infected with HPV do not develop cancer. Because HPV is sexually transmitted, it can cause cervical cancer as well as some cancers of the vagina, vulva, penis, anus, and the oropharynx (the back of the throat, including the base of the tongue and the tonsils).

Statistics for HPV-associated cancer are not regularly reported to cancer registries, so the available data is extrapolated from other data for cancers that occur in parts of the body where HPV is often found. About 90 percent of cervical and anal cancers are attributable to HPV; about 70 percent of oropharyngeal, vaginal, and vulvar cancers are attributable to HPV; abut 60 percent of penile cancers are attributable to HPV.

About 45,300 new cases of HPV-associated cancer occur in the US each year. About 25,400 cases are among women; about 19,400 cases are among men.

In 2017, the most common HPV-associated cancer for women was cervical cancer (12,143 cases, or 48 percent), followed by anal (4,751 cases, or 19 percent), oropharyngeal (3,530 cases, or 14 percent), vulvar (4,114 cases, or 16 percent), and vaginal (867 cases, or 3 percent).

In 2017, the most common HPV-associated cancer for men was oropharyngeal (16,245 cases, or 81 percent), anal (2,332 cases, or 12 percent), and penile (1,348 cases, or 7 percent). Women should follow the USPSTF recommendations for Pap smears.

STAY SAFE Practice safe sex. Vaccination can prevent over 90 percent of cancers caused by HPV. Women should follow the USPSTF recommendations for Pap smears.

SOURCES ASCO, cancer.net/cancer-types; CDC USCS, cdc.gov/cancer/dcpc/data/index.htm; NIH NCI, seer.cancer.gov; USPSTF, uspreventiveservicestaskforce.org/uspstf/recommendation-topics

Kidney Cancer

ANNUAL US DEATHS Estimated 13,780 in 2021

In 2021, an estimated 13,780 people died of kidney or renal pelvis cancer, for 2.3 percent of all cancer deaths. The death rate was 3.6 per 100,000 men and women per year. Approximately 1.7 percent of men and women will be diagnosed with kidney and renal pelvis cancer at some point during their lifetime. Kidney and renal pelvis cancer is the 8th most common cause of cancer death in the US. New cases of kidney cancer have been rising slowly but steadily since 1992, but the age-adjusted death rate has been falling on average 1.4 percent each year over 2009 to 2018.

Men die of kidney cancer more often than women. In 2018, the age-adjusted death rate per 100,000 population was 5.3 for men but only 2.3 for women. American Indians/Alaska Natives have a very high death rate for this cancer: 8.3 per 100,000 for men and 3.2 for women.

STAY SAFE Smoking is a risk factor for kidney cancer.

SOURCES ASCO, cancer.net/cancer-types; CDC USCS, cdc.gov/cancer/dcpc/data/index.htm; NIH NCI, seer.cancer.gov

Laryngeal Cancer

ANNUAL US DEATHS Estimated 3,770 in 2021

In 2021, 3,770 people died of cancer of the larynx (voice box) in the US, for 0.6 percent of all cancer deaths. The death rate was 0.9 per 100,000 men and women per year. Approximately 0.3 percent of men and women will be diagnosed with laryngeal cancer at some point during their lifetime.

Laryngeal cancer is rare; it is the 22nd most common cause of cancer death. The downturn in smoking means that deaths from this cancer have dropped sharply since 1992, when the death rate was 1.6 per 100,000 population. Age-adjusted death rates have been falling on average 2.3 percent each year over 2009 to 2018.

Men die of laryngeal cancer more than women: 1.7 per 100,000 population for men, compared to 0.4 for women. Black men are at highest risk for death, with a death rate of 2.9 per 100,000 population. The disparities in death rates are due to differences in smoking rates.

STAY SAFE Don't smoke. Smoking causes almost all cases of laryngeal cancer. Heavy drinking is another risk factor.

SOURCES ASCO, cancer.net/cancer-types; CDC USCS, cdc.gov/cancer/dcpc/data/index.htm; NIH NCI, seer.cancer.gov

Leukemia

ANNUAL US DEATHS Estimated 23,660 in 2021

Leukemia is an overall term for cancer of the body's blood-forming tissues, including the bone marrow and the lymphatic system. The bone marrow produces excess numbers of abnormal blood cells, which crowd out normal infection-fighting White blood cells, oxygen-carrying red blood cells, and platelets, tiny particles needed to form blood clots. Leukemia has 4 main types:

Acute lymphocytic leukemia (ALL)

Acute myelogenous leukemia (AML)

Chronic lymphocytic leukemia (CLL)

Chronic myelogenous leukemia (CML)

These types are discussed in greater detail below.

Rare types of leukemia include hairy cell leukemia, acute promyelocytic leukemia, and a number of myeloproliferative syndromes that can become leukemia.

In 2021, estimated deaths from all types of leukemia totaled 23,660, for 3.9 percent of all cancer deaths. The death rate was 6.3 per 100,000 men and women per age-adjusted year. Leukemia is the 10th most common cause of cancer death. Approximately 1.6 percent of men and women will be diagnosed with leukemia at some point during their lifetime, based on 2018 data.

Men die from leukemia more often than women. In 2018, the death rate for men was 8.4 per 100,000 population and 4.7 for women. Although leukemia is one of the most common childhood cancers, most cases are among older adults. The median age at death for leukemia is 75.

Better treatment for leukemia has greatly improved survival and death rates since 1975. The 5-year survival rate for leukemia was 33 percent in 1975. In 2005, it was 59 percent; in 2021, it was 65 percent. Age-adjusted death rates have been falling on average 1.7 percent each year over 2009 to 2018.

Despite considerable research, the evidence for exposure to electromagnetic fields (e.g., power lines, microwaves, cell phones) as a risk factor for leukemia is inconclusive.

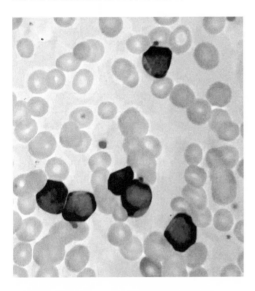

STAY SAFE Smoking is a risk factor for leukemia.

SOURCES ASCO, cancer.net/cancer-types; CDC USCS, cdc.gov/ cancer/dcpc/data/index. htm; Leukemia and Lymphoma Society, lls. org/disease-information; NIH NCI, seer.cancer.gov

Lymphoblasts (the dark areas in this photomicrograph) proliferate wildly in acute lymphoblastic leukemia.

ACUTE LYMPHOBLASTIC LEUKEMIA (ALL)

ANNUAL US DEATHS   Estimated 1,580 in 2021

Acute lymphocytic leukemia (ALL) is an aggressive type of leukemia characterized by the presence of too many lymphoblasts or lymphocytes in the bone marrow and blood. Without treatment, ALL usually progresses quickly. In 2021, an estimated 5,690 new cases were diagnosed in adults; 1,580 died.

ALL is the most common cancer diagnosed in children, representing approximately 25 percent of all cancer diagnoses among children younger than 15 years.

ALL is rare. Most children diagnosed with ALL are aged 2 to 3. In this age group, there are approximately 90 cases per 1 million per year. The incidence rate is approximately 41 cases per 1 million people aged 0 to 14 years and approximately 17 cases per 1 million people aged 15 to 19 years. In total, about 3,100 children and adolescents under age 20 are diagnosed each year. About 2 to 3 percent of childhood ALL cases occur in children with Down syndrome.

The survival rate for childhood ALL is very high, at 91.9 percent for children and adolescents younger than 15 years, and 94.1 percent for children younger than 5 years.

STAY SAFE   Exposure to high levels of radiation is a risk factor for ALL. Exposure to benzene is a possible risk factor.

SOURCES   ASCO, cancer.net/cancer-types; CDC USCS, cdc.gov/cancer/dcpc/data/index.htm; Leukemia and Lymphoma Society, lls.org/disease-information; NIH NCI, seer.cancer.gov

ACUTE MYELOGENOUS LEUKEMIA (AML)

ANNUAL US DEATHS   Estimated 11,400 in 2021

Acute myelogenous leukemia is an aggressive type of leukemia characterized by abnormal myeloblasts, a type of white blood cell. AML is rare, comprising about 1 percent of all cancers. In 2021, an estimated 11,400 people died from AML.

Previous treatment with chemotherapy or radiation is a risk factor for AML. Smoking is also a risk factor.

STAY SAFE   Avoid smoking. Exposure to benzene is a possible risk factor.

SOURCES ASCO, cancer.net/cancer-types; CDC USCS, cdc.gov/cancer/dcpc/data/index.htm; Leukemia and Lymphoma Society, lls.org/disease-information; NIH NCI, seer.cancer.gov

CHRONIC LYMPHOCYTIC LEUKEMIA (CLL)

ANNUAL US DEATHS Estimated 4,320 in 2021

Chronic lymphocytic leukemia (CLL) is a cancer of the lymphocytes, white blood cells important for the body's immune system. For some people with CLL, the disease progresses slowly, and treatment can be delayed or may even not be needed. In other patients, the disease progresses faster, and treatment is needed sooner.

In 2021, an estimated 4,320 people died of CLL. CLL is most common in older adults; about 90 percent of people diagnosed with CLL are older than 50. The average age at diagnosis is 71. The US Department of Veterans Affairs lists CLL as a disease associated with exposure to Agent Orange, a defoliating chemical used during the Vietnam War.

STAY SAFE Avoid smoking. Get regular medical checkups if you were exposed to Agent Orange.

SOURCES ASCO, cancer.net/cancer-types; CDC USCS, cdc.gov/cancer/dcpc/data/index.htm; Leukemia and Lymphoma Society, lls.org/disease-information; NIH NCI, seer.cancer.gov

CHRONIC MYELOGENOUS LEUKEMIA (CML)

ANNUAL US DEATHS Estimated 1,220 in 2021

CML is the most significant of the myeloproliferative disorders, a group of related blood diseases characterized by abnormal blood cells. The others include hairy cell leukemia, polycythemia vera, myelofibrosis, and essential thrombocythemia, which are all very rare and reliable statistics for incidence and death rates aren't available.

STAY SAFE Consider genetic testing.

SOURCES ASCO, cancer.net/cancer-types; CDC USCS, cdc.gov/cancer/dcpc/data/index.htm; Leukemia and Lymphoma Society, lls.org/diseaseinformation; NIH NCI, seer.cancer.gov

Liver Cancer

> **ANNUAL US DEATHS** Estimated 30,230 in 2021

Statistics for liver cancer and intrahepatic bile duct cancer are usually combined. Intrahepatic bile duct cancer is a subtype of liver cancer.

In 2021, an estimated 30,230 people died of liver cancer in the US, for 5.0 percent of all cancer deaths. The death rate was 6.6 per 100,000 men and women per age-adjusted year. Liver cancer is the 6th most common cause of cancer death. Approximately 1.0 percent of men and women will be diagnosed with liver and intrahepatic bile duct cancer at some point during their lifetime.

Liver cancer kills more men than women. The death rate per 100,000 population is 9.7 for men and 4.0 for women. Racial and ethnic minority men have a much higher risk of death from liver cancer. The death rate for White men is 9.0 per 100,000 population, but for Black men it is 13.1, for Asian Americans and Pacific Islanders it is 13.1, for Native Americans/Alaska Natives it is 14.8, and for Hispanics it is 13.3. The death rate for Native Americans/Alaska Natives is so high because of higher rates of exposure to risk factors for liver cancer, such as hepatitis B and hepatitis C infection, chronic liver disease, obesity, alcohol consumption, smoking, and diabetes.

Liver cancer death rates jumped 43 percent in the US between 2000 and 2016 and continue to rise. In 2000, liver cancer was the 9th leading cause of cancer death; it rose to 6th in 2016. The higher death rate is accompanied by an increase in liver cancer incidence. In 2021, an estimated 42,230 people were diagnosed with liver cancer, for 2.2 percent of all new cancer cases.

Liver cancer usually develops in people with existing liver damage, including cirrhosis, hepatitis B, hepatitis C, and fatty liver disease.

STAY SAFE Practice safe sex, avoid excess alcohol, and maintain a healthy weight.

SOURCES ASCO, cancer.net/cancer-types; CDC USCS, cdc.gov/cancer/dcpc/data/index.htm; NIH NCI, seer.cancer.gov

Lung Cancer

> **ANNUAL US DEATHS** Estimated 131,880 in 2021

In 2021, an estimated 131,880 people died of lung cancer, for 21.7 percent of all cancer deaths. The death rate was 38.5 per 100,000 men and women per age-adjusted year. Approximately 6.1 percent of men and women will be diagnosed with lung and bronchus cancer at some point during their lifetime. Lung cancer is the 3rd most common cancer diagnosis in the US and is the leading cause of cancer death. Lung cancer causes more deaths than the next 3 leading causes of cancer death combined (colorectal, breast, and pancreatic).

Smoking is a leading cause of lung cancer. Since the mid-2000s, lung cancer incidence has decreased steadily by about 2 percent per year overall. Lung cancer mortality has declined by 54 percent since 1990 in men and by 30 percent since 2002 in women due to reductions in smoking. As more people stop smoking or never smoke, mortality reduction has accelerated. Age-adjusted death rates from lung cancer fell on average 3.6 percent each year from 2009 to 2018. From 2014 to 2018, the rate decreased by more than 5 percent per year in men and 4 percent per year in women. In 2018, the death rate for all men was 46.9 per 100,000 population; for women, it was 32.0.

Cigarette smoking is by far the most important risk factor for lung cancer. Approximately 90 percent of lung cancer deaths in the US are caused by smoking. Almost all deaths from small cell lung cancer are caused by smoking. You are most at risk if you are aged 50 to 80, have a 20-pack-year or more smoking history, and currently smoke or quit within the past 15 years.

TYPES OF LUNG CANCER

Lung cancer has 2 main types: non-small cell lung cancer and small cell lung cancer.

Non-small cell lung cancer (NSCLC) has different types, including adenocarcinoma, squamous cell carcinoma, large cell carcinoma, and several less common types. NSCLC is the most common type of lung cancer, comprising 84 percent of all lung cancer diagnoses. The 5-year survival rate is 25 percent.

Small cell lung cancer (SCLC) has 2 main types: small cell carcinoma (oat cell cancer) and combined small cell carcinoma. About 10 to 15 percent of people with lung cancer have SCLC. The 5-year survival rate is 7 percent.

RACIAL DISPARITIES

Black men are about 15 percent more likely to get lung cancer than White men. Black women are 14 percent less likely to get lung cancer when compared with White women. In 2018, the death rate from lung cancer for White men was 47.1 per 100,000 population and 55.4 for Black men. The death rate for White women was 33.3 and 29.7 for Black women.

BEST AND WORST STATES

Lung cancer incidence and deaths vary considerably by state and region. In 2018, 21 states had a higher lung cancer death rate than the national rate, 15 states and DC had lower death rates, and 14 states had rates that were not statistically different from the national rate. Most states with higher death rates were in the Midwest or Southeast.

The 5 states with the highest age-adjusted lung cancer death rates were Kentucky (53.5), West Virginia (50.8), Mississippi (49.6), Arkansas (47.4), and Oklahoma (46.8). The 5 states with the lowest lung cancer death rates were Utah (16.4), New Mexico (22.5), Colorado (23.0), DC (24.6), and California (25.0).

LUNG CANCER IN NEVER SMOKERS

ANNUAL US DEATHS   16,000 to 24,000 nonsmoker deaths

An estimated 10 to 15 percent of lung cancer deaths, or about 16,000 to 24,000 deaths annually, are caused by factors other than smoking. Almost all these deaths are from non-small cell lung cancer, primarily adenocarcinoma. Lung cancer in never smokers is associated with exposure to radon, secondhand tobacco smoke, and asbestos. However, a large portion of lung cancer in never smokers has no known environmental cause. The Queen of Disco, Donna Summer, a never smoker, died of lung cancer in 2012. Dustin Diamond, the actor who played Samuel "Screech" Powers in *Saved by the Bell,* a never smoker, died in 2021 of lung cancer.

STAY SAFE   Don't smoke. If you are age 50 or up and currently smoke or quit within the past 15 years, the USPSTF recommends yearly screening for lung cancer using low-dose CT imaging.

SOURCES   ASCO, cancer.net/cancer-types; CDC USCS, cdc.gov/cancer/dcpc/data/index.htm; NIH NCI, seer.cancer.gov; USPSTF, uspreventiveservicestaskforce.org/uspstf/recommendation-topics

Lymphoma

ANNUAL US DEATHS   Estimated 20,720 in 2021

Lymphoma is a type of cancer that arises in the lymph system. Lymphomas fall into 2 categories: non-Hodgkin lymphoma and Hodgkin lymphoma. This entry is for non-Hodgkin lymphoma.

In 2021, an estimated 20,720 people died of non-Hodgkin lymphoma (NHL) in the US, for 3.4 percent of all cancer deaths. Non-Hodgkin lymphoma is the 8th leading cause of cancer death. In 2018, the death rate was 5.4 per 100,000 men and women per age-adjusted year. Better treatments have caused the age-adjusted death rate to fall on average 2.2 percent each year over 2009 to 2018.

Non-Hodgkin lymphoma has numerous subtypes, some very slow growing. Approximately 2.1 percent of men and women will be diagnosed with NHL at some point during their lifetime. Older age, being male, and having a weakened immune system are risk factors for non-Hodgkin lymphoma. People with HIV/AIDS have an increased risk of developing NHL. The percent of non-Hodgkin lymphoma deaths is highest among people aged 75 to 84. The median age at death is 76. Former first lady Jackie Kennedy died of non-Hodgkin lymphoma in 1994. Paul Allen, cofounder of Microsoft, died of NHL in 2018.

Although the weed killer glyphosate (Roundup) has been implicated in some cases of NHL, in 2020 the U.S. Environmental Protection Agency found that it is not a carcinogen.

STAY SAFE Your risk increases with age.

SOURCES ASCO, cancer.net/cancer-types; CDC USCS, cdc.gov/cancer/dcpc/data/index.htm; Leukemia and Lymphoma Society, lls.org/disease-information; NIH NCI, seer.cancer.gov

Melanoma of the Skin

ANNUAL US DEATHS Estimated 2,000 (basal-cell skin cancer) and 7,180 (invasive melanoma)

Skin cancer is the most common cancer in the US. An estimated 3.6 million new cases of basal cell carcinoma (BCC) are diagnosed every year. An estimated 1.8 million new cases of squamous cell carcinoma (SCC) are diagnosed each year. These cancers are highly treatable and are rarely fatal. In fact, because deaths from these cancers are so uncommon, they are not tracked by any federal agencies. The American Academy of Dermatology estimates that only about 2,000 people die in the US each year from these cancers.

Invasive melanoma of the skin is the 3rd most common skin cancer type. This cancer can be deadly. In 2021, 7,180 people died of melanoma, for 1.2 percent of all cancer deaths. The death rate was 2.3 per 100,000 men and women per age-adjusted year. Because of new treatments for advanced melanoma approved since 2011, the death rate decreased by nearly 5 percent per year on average from 2012 to 2016.

Melanoma is more common in men than women. It is more common among fair-skinned people and people who have been exposed to

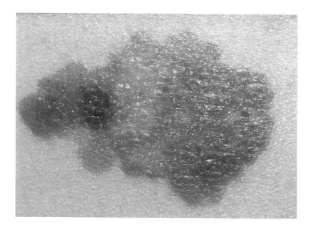

Melanoma, a form of skin cancer, develops from the skin's pigment-producing melanocyte cells.

natural or artificial sunlight (such as tanning beds) over long periods of time. There are more deaths among Whites than any other racial/ ethnic group. The melanoma death rate among Black men is 0.4 per 100,000 population but 3.9 for White men.

The World Health Organization says UV tanning devices cause cancer in humans. The FDA classified UV tanning devices as moderate to high risk in 2014. At all ages, the more women tan indoors, the higher their risk of developing melanoma.

STAY SAFE   Do not use UV tanning devices. Avoid exposure to UV light.

SOURCES   ASCO, cancer.net/cancer-types; CDC USCS, cdc.gov/ cancer/dcpc/data/index.htm; NIH NCI, seer.cancer.gov; Skin Cancer Foundation, skincancer.org/skin-cancer-information

Merkel Cell Carcinoma

ANNUAL US DEATHS   No data available

Merkel cell carcinoma (MCC) is a very rare, aggressive form of skin cancer that is very hard to treat. It has a high risk of recurring and metastasizing within 3 years of diagnosis. Approximately 3,000 new cases are diagnosed annually in the US. The death rate is unknown, but the 5-year survival rate for early stage MCC is about 50 percent; if the cancer has spread to distant sites, the 5-year survival rate is about 14 percent. Merkel cell carcinoma is 40 times rarer than melanoma, with an estimated rate of 1 case per 100,000 people.

STAY SAFE   Avoid excessive UV exposure.

SOURCE   Skin Cancer Foundation, skincancer.org/
skin-cancer-information

Mesothelioma

ANNUAL US DEATHS   No data available

Malignant mesothelioma is a cancer of the thin tissue (mesothelium) that lines the lung, chest wall, and abdomen. This cancer is rare in the US. About 3,000 new cases are diagnosed each year. The death rate is unknown, but the 5-year survival rate is approximately 10 percent.

The major risk factor for mesothelioma is asbestos exposure, usually in the workplace.

STAY SAFE   Avoid asbestos exposure.

SOURCES   ASCO, cancer.net/cancer-types; CDC USCS, cdc.gov/
cancer/dcpc/data/index.htm; NIH NCI, seer.cancer.gov

Myelodysplastic Syndromes

ANNUAL US DEATHS   6,611 in 2019

Myelodysplastic syndromes (MDS) are a group of blood malignancies that occur when the blood-forming stem cells in bone marrow become abnormal. This leads to low numbers and poor quality in red and white blood cells and platelets. Six different types of MDS are recognized, but most patients have MDS with multilineage dysplasia (MDS-MLD). Patients with the rarer type known as MDS with excess blasts (MDS-EB) are most likely to progress to acute myeloid leukemia (AML).

MDS is diagnosed in slightly more than 10,000 people in the United States yearly, for an annual age-adjusted incidence rate of approximately 4.4 to 4.6 cases per 100,000 population. About 40 percent of patients diagnosed with MDS have higher-risk disease and are likely to progress to acute myeloid leukemia and death, usually within 18 months to 2 years of diagnosis. Patients with lower-risk disease don't get leukemia, but these types of MDS can still cause death. In 2019, 6,611 people died from all myelodysplastic syndromes combined, for a mortality rate of 2.01 per 100,000 population.

Almost all cases of MDS occur in older adults in their 70s and 80s. Risk factors for MDS include previous treatment with chemotherapy and/or radiation for cancer.

STAY SAFE   If you are a cancer survivor who received chemotherapy or radiation therapy, be aware of the risk of MDS later in life.

SOURCES   ASCO, cancer.net/cancer-types; CDC USCS, cdc.gov/cancer/dcpc/data/index.htm; NIH NCI, seer.cancer.gov

Multiple Myeloma

ANNUAL US DEATHS   12,410 in 2021

Myeloma is a cancer of the infection-fighting plasma cells in the bone marrow. Because myeloma often causes multiple bone lesions, it is usually called multiple myeloma.

Multiple myeloma is rare. In 2021, an estimated 12,410 men and women died of this cancer, for 2.0 percent of all cancer deaths. The death rate was 3.2 per 100,000 men and women per age-adjusted year. Multiple myeloma is the 14th most common cause of cancer death in the US. Myeloma is more common in men than women. The death rate for men is 8.8 per 100,000 and 5.7 for women.

For unknown reasons, multiple myeloma occurs twice as frequently in Black people than in White people. The death rate for Black men in 16.3 per 100,000 but only 8.2 for White men. For Black women, the death rate is 12.1 but only 5.0 for White women.

Only 2 percent of myeloma cases occur in people under 40. Myeloma occurs most commonly in people over 60. The average age at diagnosis is 70.

STAY SAFE   Aging is the only known risk for multiple myeloma.

SOURCES   ASCO, cancer.net/cancer-types; CDC USCS, cdc.gov/cancer/dcpc/data/index.htm; NIH NCI, seer.cancer.gov

Nasal Cavity and Paranasal Sinus

ANNUAL US DEATHS   About 2,000 diagnosed but death data unavailable

The nasal cavity is the space just behind the nose where air passes on its way to the throat. The paranasal sinuses are air-filled areas that surround the nasal cavity. Cancer in either area is rare. About 2,000 people are diagnosed with nasal cavity or paranasal sinus cancer in the US every year. These cases make up about 3 to 5 percent of all head and neck cancers. The death rate is unknown, but the 5-year survival rate is between 84 and 42 percent, depending on the stage of the cancer and if it involves both the nasal cavity and the paranasal sinuses.

The incidence and death rate for this cancer are much higher for Asian Americans and Pacific Islanders compared to any other group. The incidence rate is 2.3 per 100,000 population for Asian Americans/ PIs and 0.3 for Whites, for a rate ratio of 7.7. The death rate for Asian Amercians/PIs is 0.7 and 0.1 for Whites, for a rate ratio of 7.0

Workplace exposure to high levels of dust is a risk factor for nasal cavity and paranasal sinus cancer.

STAY SAFE   Take appropriate safety measures to reduce exposure to dust in the workplace.

SOURCES   ASCO, cancer.net/cancer-types; CDC USCS, cdc.gov/ cancer/dcpc/data/index.htm; NIH NCI, seer.cancer.gov

Neuroendocrine Tumors

ANNUAL US DEATHS   Because these tumors take several forms and develop in different parts of the body, no death data is available.

A neuroendocrine tumor (NET) begins in the specialized cells of the body's neuroendocrine system. These cells are found throughout the body, but most NETs develop in the lungs, pancreas, or gastrointestinal tract. Every year in the US, about 12,000 people are diagnosed with a neuroendocrine tumor. Because these tumors can take a number of forms and can grow in several different parts of the body, death rates aren't known. The 5-year survival rate varies with the type of tumor and its location. The causes of NETs are unknown and no avoidable risk factors are known.

Steve Jobs, cofounder and CEO of Apple, died of a neuroendocrine tumor of the pancreas is 2011. Aretha Franklin, the Queen of Soul, died of a neuroendocrine tumor of the pancreas in 2018.

STAY SAFE   No avoidable risk factors are known.

SOURCES   ASCO, cancer.net/cancer-types; CDC USCS, cdc.gov/ cancer/dcpc/data/index.htm; NIH NCI, seer.cancer.gov

Oral Cavity and Pharynx (Oropharyngeal) Cancer

ANNUAL US DEATHS   Estimated 10,850 in 2021

Cancers of the oropharynx (back of the throat) include tongue cancer (the base of the tongue), throat (pharynx) cancer, and tonsil cancer. All types of oropharyngeal cancer are often grouped together for statistical purposes, so some data is not available for specific types.

In 2021, there were 54,010 estimated new cases of oral cavity and pharynx cancer in the US, or 2.8 percent of all new cancer cases. In 2018, the rate of new cases of oral cavity and pharynx cancer was 11.5 per 100,000 men and women per age-adjusted year. Based on data from 2016 to 2018, approximately 1.2 percent of men and women will be diagnosed with oral cavity and pharynx cancer at some point during their lifetime. About 70 percent of oropharyngeal cancer is attributable to HPV infection.

An estimated 10,850 people died of oral cavity and pharynx cancer in 2021, or 1.8 percent of all cancer deaths. In 2018, the death rate was 2.5 per 100,000 men and women per age-adjusted year.

Oral cavity and pharynx cancer is more common among men than women. The incidence rate for men is 17.3 per 100,000 population, but only 6.5 per 100,000 for women. The death rate is 3.9 per 100,000 population for men and 1.3 per 100,000 for women.

STAY SAFE Don't smoke. Practice safe sex and get vaccinated for HPV.

SOURCES ASCO, cancer.net/cancer-types; CDC USCS, cdc.gov/cancer/dcpc/data/index.htm; NIH NCI, seer.cancer.gov

Ovarian Cancer

ANNUAL US DEATHS Estimated 13,770 in 2021

Ovarian cancer describes cancers that begin in the cells of the ovary, fallopian tube, or peritoneum. The cancers are closely related and are generally treated the same way. In 2021, an estimated 13,770 women died from ovarian cancer, or about 2.3 percent of all cancer deaths. The death rate was 6.7 per 100,000 women per year. Ovarian cancer is the 13th leading cause of cancer death in the US, but it is the 5th leading cause of cancer deaths in women. It accounts for more deaths than any other cancer of the female reproductive system. The lifetime risk of dying from ovarian cancer is about 1 in 108.

Women who have a family history of ovarian cancer or who carry the *BRCA1* or *BRCA2* genes are at an increased risk of ovarian cancer. *Saturday Night Live* cast member Gilda Radner died of ovarian cancer in 1989. Civil rights leader and widow of Martin Luther King Jr., Coretta Scott King, died of ovarian cancer in 2006.

STAY SAFE   Discuss your genetic risk of ovarian cancer with your doctor.

SOURCES   ASCO, cancer.net/cancer-types; CDC USCS, cdc.gov/cancer/dcpc/data/index.htm; NIH NCI, seer.cancer.gov

Pancreatic Cancer

ANNUAL US DEATHS   Estimated 48,220 in 2021

In 2021, an estimated 48,220 people died of pancreatic cancer in the US, for 7.9 percent of all cancer deaths. The death rate was 11.0 per 100,000 men and women per age-adjusted year. Pancreatic cancer is the 11th most commonly diagnosed cancer, but it is the 3rd leading cause of cancer death. The 5-year survival rate is 10.8 percent.

Pancreatic cancer is more common with increasing age and slightly more common in men than women. Risk factors include smoking, severe obesity, and a personal history of diabetes or chronic pancreatitis. The biggest risk factor, however, is advancing age. Alex Trebek, host of *Jeopardy!*, died of pancreatic cancer in 2020. Supreme Court justice Ruth Bader Ginsburg died of pancreatic cancer that same year. Actor Patrick Swayze died of pancreatic cancer in 2009. Tenor Luciano Pavarotti died of the disease in 2007.

STAY SAFE   Don't smoke, and maintain a healthy weight.

SOURCES   ASCO, cancer.net/cancer-types; CDC USCS, cdc.gov/cancer/dcpc/data/index.htm; NIH NCI, seer.cancer.gov

Penile Cancer

ANNUAL US DEATHS   About 460

Penile cancer is very rare in the US, making up less than 1.0 percent of all cancer diagnosed in men. The annual incidence is approximately 1 in 100,000 men. An estimated 2,210 men in the United States are diagnosed with penile cancer every year and about 460 of them die.

The primary risk factor for penile cancer is infection with the human papillomavirus (HPV). About 60 percent of all cases of penile cancer are HPV-associated.

STAY SAFE   Practice safe sex; get vaccinated for HPV.

SOURCES   ASCO, cancer.net/cancer-types; CDC USCS, cdc.gov/cancer/dcpc/data/index.htm; NIH NCI, seer.cancer.gov

Pheochromocytoma

> **ANNUAL US DEATHS** 2 to 8 deaths per million

Pheochromocytoma is a rare type of neuroendocrine tumor that usually affects the adrenal glands. The incidence rate of pheochromocytoma is unknown but very low, probably between 2 and 8 people per million population. Most pheochromocytomas are benign; only about 10 percent spread to other parts of the body. Because these tumors are often detected early, the death rate is unknown but very low. The 5-year survival rate for small tumors is about 95 percent. The 5-year survival rate for recurrent tumors or tumors that have spread to other parts of the body is between 34 and 60 percent.

No known risk factors are linked to pheochromocytoma.

STAY SAFE There are no known risk factors.

SOURCES ASCO, cancer.net/cancer-types; CDC USCS, cdc.gov/cancer/dcpc/data/index.htm; NIH NCI, seer.cancer.gov

Prostate Cancer

> **ANNUAL US DEATHS** Estimated 34,130 in 2021

Prostate cancer is the most common cancer among men in the US. The lifetime risk of being diagnosed with prostate cancer is approximately 13 percent, but the lifetime risk of dying of prostate cancer is only 2.5 percent. Prostate cancer is the 2nd most commonly diagnosed cancer and the 5th leading cause of cancer death. The 5-year survival rate is 97.5 percent.

In 2021, approximately 248,530 new cases of prostate cancer were diagnosed, or about 13.1 percent of all new cancer cases. Deaths from prostate cancer were approximately 34,130, or 5.6 percent of all cancer deaths. The death rate was 19.0 per 100,000 men per age-adjusted year, or about 1 man in 41. Rock musician Frank Zappa died of prostate cancer in 1993.

Advancing age is the only known risk factor for prostate cancer.

RACIAL DISPARITIES

Most men diagnosed with prostate cancer do not die from it. More than 3.1 million men who have been diagnosed with prostate cancer at some point are still alive. For unknown reasons, Black men are about 1.6 times more likely than all other men to get prostate cancer and are twice as likely to die from it. Prostate cancer in Black men is more likely to be more aggressive and progress faster. The prostate cancer death rate for White men is 17.9 per 100,000 men, but for Black men, it is 37.4.

SCREENING

Screening for prostate cancer using the PSA blood marker is controversial. In 2018, the USPSTF issued new screening recommendations. For men aged 55 to 69 years, the decision to undergo periodic PSA-based screening for prostate cancer should be an individual one and should include discussion with their doctor of the potential benefits and harms of screening. The USPSTF recommends against PSA-based screening for prostate cancer in men 70 years and older.

STAY SAFE Discuss PSA-based screening with your doctor.

SOURCES ASCO, cancer.net/cancer-types; CDC USCS, cdc.gov/cancer/dcpc/data/index.htm; NIH NCI, seer.cancer.gov, USPSTF, uspreventiveservicestaskforce.org/uspstf/recommendation-topics

Salivary Gland Cancer

ANNUAL US DEATHS No data available

Malignant tumors of the salivary glands (parotid gland, submandibular glands, sublingual glands) are rare. In 2021, an estimated 1 out of 100,000 people in the US were diagnosed with salivary gland cancer. This type of cancer makes up less than 1 percent of all cancers and approximately 3 to 5 percent of all head and neck cancers each year. The death rate is unknown, but the 5-year survival rate is approximately 75 percent.

Advancing age is the biggest risk factor for salivary gland tumors. Most patients are over age 60. No other risk factors have been identified.

STAY SAFE The causes of most salivary gland cancers are unknown.

SOURCES ASCO, cancer.net/cancer-types; CDC USCS, cdc.gov/cancer/dcpc/data/index.htm; NIH NCI, seer.cancer.gov

Small Intestine Cancer

> **ANNUAL US DEATHS** Estimated 2,100 in 2021

The most common type of small intestine cancer is adenocarcinoma. Other small intestine cancers include sarcoma, carcinoid tumor, gastrointestinal stromal tumor (GIST), and lymphoma. Small intestine cancer is rare. In 2021 an estimated 11,390 new cases were diagnosed, or 0.6 percent of all new cancer cases in the US. It is the 23rd most common type of cancer. Based on 2016 to 2018 data, the lifetime risk of developing small intestine cancer is 0.3 percent.

Estimated deaths for 2021 were 2,100, or 0.3 percent of all cancer deaths. The death rate in 2018 was 0.4 per 100,000 men and women per age-adjusted year.

STAY SAFE Risk factors for small intestine cancer include eating a high-fat diet, Crohn disease, celiac disease, and familial adenomatous polyposis (FAP).

SOURCES ASCO, cancer.net/cancer-types; CDC USCS, cdc.gov/cancer/dcpc/data/index.htm; NIH NCI, seer.cancer.gov

Stomach Cancer

> **ANNUAL US DEATHS** Estimated 11,180 in 2021

In 2021, an estimated 11,180 people died of stomach cancer in the US, or 1.8 percent of all cancer deaths. The death rate was 3.0 per 100,000 men and women per age-adjusted year. Stomach cancer is relatively rare. It is the 15th most commonly diagnosed cancer in the US. Approximately 0.8 percent of men and women will be diagnosed with stomach cancer at some point during their lifetime.

Stomach cancer is more common in men than women. White women have the lowest death rate at 1.9. Black men are more likely to die of stomach cancer than White men. The death rate for Black men is 7.6 per 100,000 population compared to 3.5 per 100,000 population for White men. For Hispanic men, the death rate is 6.3 per 100,000; for Hispanic women it is 3.9 per 100,000.

The risk of stomach cancer increases with age. Diet and infection with the bacterium *Helicobacter pylori* can increase the risk of stomach cancer.

STAY SAFE Eat a healthy diet low in processed foods and get treated for stomach ulcers.

SOURCES ASCO, cancer.net/cancer-types; CDC USCS, cdc.gov/cancer/dcpc/data/index.htm; NIH NCI, seer.cancer.gov

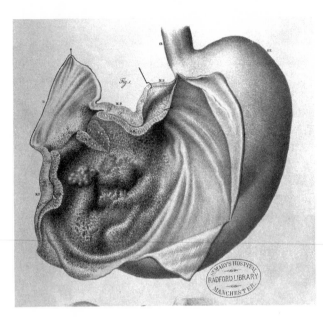

A nineteenth-century illustration of a stomach cancer tumor within a dissected stomach

Testicular Cancer

ANNUAL US DEATHS Estimated 440 in 2021

Testicular cancer is rare—it is the 24th most commonly diagnosed cancer in the US. In 2021, an estimated 9,470 new cases were diagnosed in the US. Because this cancer is highly treatable and even curable, there were only 440 deaths, or 0.1 percent of all cancer deaths. The death rate was 0.3 per 100,000 men per year.

Testicular cancer is most common in young men. Half of all cases occur in men aged 20 to 34. This cancer is rare in Black men. In 2018, the incidence rate was 7.1 per 100,000 men for White men but only 1.7 for Black men. The death rate for White men was 0.3; for Black men, it was 0.1. Olympic figure skater Scott Hamilton is a testicular cancer survivor. So is bicycle racer Lance Armstrong.

STAY SAFE There are no known risk factors for testicular cancer.

SOURCES ASCO, cancer.net/cancer-types; CDC USCS, cdc.gov/cancer/dcpc/data/index.htm; NIH NCI, seer.cancer.gov

Thyroid Cancer

> **ANNUAL US DEATHS**   Estimated 2,200 in 2021

Thyroid cancer is rare. In 2021, an estimated 44,280 new cases were diagnosed, or 2.3 percent of all new cancer cases, but only an estimated 2,200 men and women died from it, or 0.4 percent of all cancer deaths. The rate of new cases is 15.5 per 100,000 men and women per age-adjusted year; the death rate is 0.5 per 100,000 men and women. Thyroid cancer is the 12th most commonly diagnosed cancer.

Approximately 1.2 percent of men and women will be diagnosed with thyroid cancer at some point during their lifetime. Thyroid cancer is the most commonly diagnosed cancer in women aged 20 to 34, making women about 3 times more likely than men to have it. In 2018, 8.0 men per 100,000 population were diagnosed with thyroid cancer, compared to 22.8 per 100,000 for women. The death rate for men and women is approximately the same.

The number of people diagnosed with thyroid cancer has been rising since the 1980s, but the number of deaths has stayed about the same. This cancer is highly treatable and usually curable.

Risk factors for thyroid cancer include advancing age and being female.

STAY SAFE   There are no modifiable risk factors for thyroid cancer.

SOURCES   ASCO, cancer.net/cancer-types; CDC USCS, cdc.gov/cancer/dcpc/data/index.htm; NIH NCI, seer.cancer.gov

Tongue Cancer

> **ANNUAL US DEATHS**   Estimated 2,870 in 2021

Tongue cancer is rare—it is the 19th most common cancer in the US. In 2021, there were 17,960 estimated new cases, or 0.9 percent of all new cancer case. The rate of new cases was 3.6 per 100,000 men and women per age-adjusted year. Based on data from 2016 to 2018, approximately 0.4 percent of men and women will be diagnosed with tongue cancer during their lifetime.

Deaths from tongue cancer in 2021 were an estimated 2,870, or 0.5 percent of all cancer deaths. The death rate was 0.7 per 100,000 men and women per age-adjusted year.

For 2009 to 2018, age-adjusted rates for new tongue cancer cases rose on average by 2.2 percent each year. The age-adjusted death rate did not change over that period.

About 70 percent of tongue cancer cases are associated with HPV infection.

STAY SAFE   Don't smoke. Practice safe sex and get vaccinated for HPV.

SOURCES   ASCO, cancer.net/cancer-types; CDC USCS, cdc.gov/cancer/dcpc/data/index.htm; NIH NCI, seer.cancer.gov

Tonsil Cancer

See **Oral Cavity and Pharynx (Oropharyngeal) Cancer.**

Uterine Cancer

See **Endometrial Cancer.**

Vaginal and Vulvar Cancer

See **HPV-Associated Cancer.**

Rare Cancers

ANNUAL US DEATHS   No reliable data for this group of cancers

The definition of a rare cancer varies among authoritative sources. The National Cancer Institute defines a rare cancer as one that occurs in fewer than 15 out of 100,000 people each year. The American Cancer Society (ACS), however, follows the definition adopted in Europe by the RARECARE consortium in 2011 and defines rare cancers as those with fewer than 6 cases per 100,000 people per year, or fewer than 40,000 new cases per year. For perspective on this number, the incidence rate for both breast and prostate cancer, the most common cancers in women and men, respectively, is currently about 123 cases per 100,000.

Using the ACS definition, 1 in 5 cancers diagnosed in the US is a rare cancer. Among children and adolescents with cancer, 71 percent are rare cancers, compared with less than 20 percent of cancers diagnosed in patients aged 65 and older.

In 2017, approximately 208,000 rare cancers were diagnosed, not including those diagnosed with a rare subtype of a more common cancer. Some are so rare that under 10 cases are diagnosed every year. Cancerous primary cardiac tumors, for example, are extremely rare, with most major cancer centers reporting 0 to 2 cases a year. Fashion designer Virgil Abloh died in 2021 at age 41 of cardiac angiosarcoma.

Treatment options for rare cancers are often limited and less effective than for more common cancers. Because these cancers vary considerably in anatomical site and treatability, and because many are diagnosed only at an advanced stage, death statistics are difficult to track.

From 2009 to 2013, 181 rare cancers with overall incidence rates of fewer than 6 cases per 100,000 per year were diagnosed in the US. Together, these rare cancers represented 20 percent of all cancers diagnosed in that time frame. Of these 181 rare cancers, 119 cancers were very rare, with an incidence rate of 0.5 or less per 100,000. These very rare cancers represent only about 3 percent of all cancers diagnosed each year.

STAY SAFE Don't smoke. Maintain a healthy weight, eat a good diet, and get 30 minutes of exercise daily. Get cancer screenings as recommended by the U.S. Preventive Services Task Force for your age, sex, family history, and other characteristics.

SOURCES CDC USCS, cdc.gov/cancer/dcpc/data/index.htm; C.E. DeSantis et al. The burden of rare cancers in the United States. *CA: A Cancer Journal for Clinicians*, July 2017; NIH NCI, seer.cancer.gov

Cancer Death Disparities

In the US, cancer disparities—differences in cancer measures among groups—affect racial and ethnic minorities in every area: screening rates, incidence, stage at diagnosis, prevalence, deaths, survival, morbidity, and financial burden. Overall, Blacks have higher death rates than all other racial/ethnic groups for many, although not all, cancer types. For example, Black and White women have similar rates of breast cancer, but Black women are more likely to die of the disease. American Indians/Alaska Natives have higher death rates from kidney cancer than any other racial/ethnic group.

Other population groups also are affected by cancer disparities. For example, the incidence and death rates for colorectal, lung, and cervical cancer are much higher in rural Appalachia than in urban areas in the region.

In 2020, of the 606,520 people who died from cancer, about 75,030 were Black and about 42,700 were Hispanic. The overall death rate for cancer in the US is 146.2 per age-adjusted 100,000 population. For Whites the cancer death rate is 151.4. For Blacks, the cancer death rate is 171.0; for Hispanics it is 105.6; for Asian Americans/Pacific Islanders it is 92.1; and for American Indians/Alaska Natives, it is 123.6.

Much progress has been made in reducing disparities in cancer incidence and deaths. From 2006 to 2015, the overall cancer death rate declined faster among Black men and women than White men and women—the rate was 2.6 percent versus 1.6 percent per year for men and 1.5 percent versus 1.3 percent per year for women. Continuous declines since 1994 have meant more than 462,000 fewer cancer deaths in the Black population. The declines are largely driven by sharper drops in smoking rates since 1979 among Blacks than Whites.

Among men, the overall cancer death rate was 47 percent higher for Blacks than for Whites in 1990 versus 19 percent higher in 2016. Among women, the disparity decreased from 19 percent to 13 percent over the same period, with the gap nearly eliminated for some age groups. The disparity in the overall cancer death rate between Blacks and Whites has been nearly eliminated among men younger than 50 and women ages 70 or older. However, Black Americans still have the highest overall cancer death rate and lowest survival rate of any racial or ethnic group in the country.

The overall cancer death rates among Hispanics in the US are about 32 percent lower than for Whites for the most common cancers: breast cancer, lung cancer, prostate cancer, colorectal cancer, and melanoma. However, death rates for Hispanics for some other types of cancer are significantly higher than for non-Hispanic Whites, including stomach cancer, liver and intrahepatic bile duct cancer, thyroid cancer, and cervical cancer.

In 2018, the cancer mortality rates by race/ethnicity per 100,000 population were:

All Races	155.5
White	156.3
Black	177.5
Hispanic	110.8
Asian/PI	97.2
American Indian/Alaska Native	141.1

SOURCES CDC USCS, cdc.gov/cancer/dcpc/data/index.htm; NIH NCI, seer.cancer.gov

Diabetes

ANNUAL US DEATHS 87,647 in 2019

Diabetes is a chronic metabolic disease characterized by elevated levels of blood glucose caused by an impaired ability to produce or respond to the hormone insulin. Diabetes is the 7th leading cause of death in the

US, with 87,647 deaths in 2019. Worldwide, in 2019 diabetes entered the top 10 causes of death in the 9th position, with approximately 1.5 million deaths.

The 3 main types are type 1 diabetes, type 2 diabetes, and gestational diabetes. People with type 1 and type 2 diabetes are at increased risk of death from heart disease, stroke, and kidney disease. The majority of people with diabetes have type 2 diabetes, largely as a result of excess body weight and physical inactivity. Women with gestational diabetes are at increased risk of having a stillborn baby.

STAY SAFE Type 2 (not type 1) diabetes is closely linked to obesity and a sedentary lifestyle. Maintain a normal weight and get regular exercise. If you are at risk of diabetes, see your doctor regularly. Early diagnosis and careful management can help avoid many complications of diabetes such as vision loss and amputations.

SOURCES CDC National Diabetes Statistics Report 2020, cdc.gov/ diabetes/library/features/diabetes-stat-report.html; NIDDK, Diabetes in America, 3rd edition, niddk.nih.gov/about-niddk/strategic-plans-reports; WHO, who.int/health-topics/diabetes

Type 1 Diabetes

Type 1 diabetes is thought to be caused by an autoimmune reaction that destroys the beta cells in the pancreas that produce insulin. To survive, people with type 1 diabetes need to inject insulin daily. This disease is usually diagnosed in children, teens, and young adults, but about half of type 1 diabetes cases occur in adults over age 20. Epidemiologic data on youth-onset type 1 diabetes is scarce. Because epidemiological data is very scarce for adult-onset type 1 diabetes, this section will discuss youth-onset diabetes only.

Approximately 5 to 10 percent of people with diabetes in the US have type 1. In 2018, nearly 1.6 million Americans had type 1 diabetes, about 5.2 percent of all US adults with diagnosed diabetes. About 187,000 children and teens have type 1 diabetes. It affects 1 in every 518 young people under age 20, or 1.93 per 1,000 people. Worldwide, an estimated 79,100 children under age 15 are diagnosed with type 1 diabetes annually. In the US, approximately 18,000 new cases in people under age 20 are diagnosed annually.

People with type 1 diabetes usually have shortened life expectancy. The estimated life expectancy after age 20 for a man is about 46 years, compared to about 57 years for a man without the condition. For women, the estimated life expectancy after age 20 is about 48 years, compared to 61 years among women without the condition. Black Americans with type 1 diabetes are at increased risk of premature death, with a mortality rate that is 2.5 times higher compared to White Americans.

STAY SAFE The causes of type 1 diabetes are unknown, and no preventive measures can be taken. Careful management helps avoid life-threatening complications and prolongs life expectancy.

Type 2 Diabetes

In 2019, diabetes was the 7th leading cause of death in the US, based on 87,647 death certificates listing it as the underlying cause of death. But that number alone understates the danger of the disease: diabetes was listed as a contributing cause of death on an additional 195,154 death certificates. Moreover, some researchers believe diabetes is underreported as a cause of death overall, since it is not listed anywhere on the death certificates of about 60 percent of people with diabetes who die.

In 2018, 34.2 million people of all ages, or 10.5 percent of the US population, had diabetes. Most were adults with type 2 diabetes: 34.1 million, or 13.0 percent of all US adults. About 7.3 million of these were unaware they had the disease.

The prevalence of type 2 diabetes has increased substantially since 2000. In 2000, the estimated prevalence was 9.8 percent of the population. In 2018, it was 14.3 percent of the population.

STAY SAFE Type 2 diabetes is closely linked to obesity and a sedentary lifestyle. Maintain a normal weight and get regular exercise. If you are at risk of diabetes, see your doctor regularly. Early diagnosis and careful management can help avoid many complications of diabetes such as vision loss and amputations.

By Age

The percentage of adults with diabetes increases with age. In 2019, 4.0 percent of people aged 18 to 44 years had diabetes, 17.0 percent of those aged 45 to 64 years had it, and 25.2 percent of those aged over 65 had it. Adults over age 50 with type 2 diabetes develop disability 6 to 7 years earlier than adults without diabetes and die 4.6 years earlier.

By Race/Ethnicity

American Indian adults have the highest rates of diagnosed diabetes, at 14.7 percent per 1,000 persons. The rate among Hispanics is 12.5 percent; among Blacks it is 11.7 percent; and among Whites it is 5.0 percent.

Mortality rates for type 2 diabetes show sharp racial and ethnic disparities. In 2019, the overall death rate for diabetes was 21.6

per 100,000 population. Among White people, the rate was 19.0 per 100,000 population, but for Black people, the rate was double, at 38.2 per 100,000, and for Native Americans, the rate was 41.5 per 100,000. For Hispanic Americans, the rate was 25.6 per 100,000 and for Asians/ Pacific Islanders, it was 16.6 per 100,000.

Gestational Diabetes

> **ANNUAL US DEATHS** Rarely causes death to mother, but increases the risk of stillbirth and preeclampsia

Gestational diabetes is defined as a rise in blood glucose and other symptoms of diabetes in a pregnant woman not previously diagnosed with diabetes. About 7 percent of all pregnant women in the US develop gestational diabetes. It is more common in Black and ethnic minority women.

Gestational diabetes increases the risk of preeclampsia in the woman. The overall risk of stillbirth for women with gestational diabetes is 17.1 per 10,000 deliveries, compared to 12.7 for women without gestational diabetes.

STAY SAFE Pregnant women should be screened for gestational diabetes between 24 and 28 weeks of pregnancy.

Kidney Disease

> **ANNUAL US DEATHS** 51,386 deaths in 2020

The kidney diseases nephritis, nephrotic syndrome, and nephrosis are the 9th leading cause of death in the US. In 2020, 51,386 people died from these forms of kidney disease, or 1.8 percent of all deaths, for a death rate of 15.7 per 100,000 population. Another 35,835 people died of essential hypertension and hypertensive renal (kidney) disease, making that the 13th leading cause of death, or 1.3 percent of all deaths, for a death rate of 11.0 per 100,000 population. Each year, kidney disease kills more people than breast or prostate cancer.

About 37 million US adults—roughly 15 percent of the population, or more than 1 in 7—are estimated to have chronic kidney disease (CKD), where the kidneys gradually lose the ability to filter waste and extra fluid from the blood. Chronic kidney disease refers to all 5 stages of kidney damage, from very mild damage in stage 1 to complete kidney failure in stage 5. About 40 percent of people with severely reduced kidney function are not aware of having CKD.

Untreated CKD may progress to kidney failure, or end-stage renal disease (ESRD), where the kidneys no longer work and dialysis or a kidney transplant is needed. In 2019, of all the people who died of kidney disease, approximately 47,000 people had end-stage renal disease. In 2021, nearly 786,000 people in the US (2 per 1,000 population) were living with ESRD.

Major risk factors for kidney disease include diabetes, high blood pressure, and family history of kidney failure. In the US, diabetes and high blood pressure are the leading causes of kidney failure, representing about 3 out of 4 new cases. These 2 conditions were the primary diagnosis in 76 percent of kidney failure cases between 2015 and 2017. Diabetes causes nearly half of all cases of kidney failure in the US. High blood pressure causes about 1 out of 4 cases.

Cardiovascular disease is a leading cause of death for people with chronic kidney disease. Individuals with ESRD are an extremely high-risk population for cardiovascular death from heart attack or stroke. People with kidney disease are 5 to 10 times more likely to die prematurely than they are to progress to long-term dialysis or kidney transplantation. Cardiovascular disease contributes to more than half of all deaths among patients with ESRD.

STAY SAFE   Prevention and early detection are key in kidney disease. Don't smoke. Maintain a healthy weight, treat high blood pressure, and manage blood sugar levels if you have diabetes. Reduce salt intake, avoid aspirin and ibuprofen, exercise regularly, and stay hydrated. Know your family medical history.

SOURCES   American Kidney Fund, kidneyfund.org/kidney-disease; CDC, Chronic Kidney Disease Initiative, cdc.gov/kidneydisease/ publications-resources; NIH NIDKK, US Renal Data System, adr.usrds.org/2021; National Kidney Foundation, kidney.org/ kidney-basics

By Age

Because end-stage renal disease is overwhelmingly a disease of older adults, mortality statistics are based on Medicare data collected by the United States Renal Data System (USRDS), part of the NIH National Institute of Diabetes and Digestive and Kidney Diseases (NIDDK). The system tracks kidney disease in adults aged 65 and up. End-stage renal disease is progressive, and the length of time someone can live with it varies. To express death rates for progressive diseases accurately, epidemiologists use person-years, a measurement that takes into account both the number of people with the disease and the amount of time they live with the disease and are at risk of death.

Based on the 2020 USRDS Annual Data Report, the adjusted CKD mortality rate among Medicare beneficiaries aged 66 or older was 96.0 per 1,000 person-years—more than twice as high as the rate of 41.0 per 1,000 person-years among beneficiaries without CKD. Better treatment for underlying hypertension and diabetes means the adjusted mortality for CKD declined by 20 percent between 2009 and 2018, going from 121.6 deaths per 1,000 person-years in 2009 to 97.7 per 1,000 person-years in 2018.

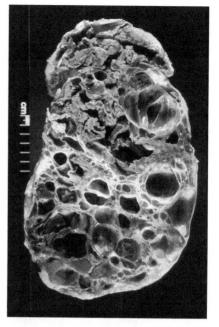

A polycystic kidney removed from a kidney cancer patient

By Sex

Mortality was higher among men (107.0 per 1,000 person-years) than women (88.1 per 1,000 person-years). Women with kidney failure are more likely to die prematurely of all causes than men with kidney failure, losing an average of 3.6 years more life than men. They experience 11 times the expected deaths, compared to 7 times in men.

By Race/Ethnicity

In 2018, 16.0 percent of all Black Americans had chronic kidney disease; 15.7 percent of White Americans had it, and 11.9 percent of Hispanic Americans had it. Among adults aged 65 and up, however, kidney disease is highest among Blacks at 33 percent, followed by American Indians or Alaska Natives at 30 percent, Hispanics at 28 percent, and Asian Americans or Pacific Islanders at 26 percent. White Medicare beneficiaries had the lowest percentages of kidney disease at 23 percent.

Black Americans are about 3 times more likely than Whites to develop kidney failure—they are about 13 percent of the US population but represent 35 percent of those with kidney failure. They are more at risk for kidney failure than any other race—more than 1 in 3 kidney failure patients are Black. The high numbers are because Black Americans get diabetes more often (1 in 9, or 11.7 percent) and high blood pressure (over 42 percent) compared to Whites (about 1 in 13, or 7.5 percent, have diabetes).

In 2020, mortality from CKD was similar for White and Black individuals at 97.3 versus 97.8 per 1,000 person-years. However, Black Americans developed ESRD sooner and at a younger age. Because Hispanic Americans have high rates of diabetes (about 1 in 8, or 12.5 percent) and high blood pressure (almost 1 in 4, or 22.5 percent), they are at high risk of CKD. Diabetes causes kidney failure more often in Hispanics than Whites. Native Americans are twice as likely to get kidney failure as Whites. This population has very high rates of diabetes and high blood pressure.

All racial and ethnic minorities are more likely to have diabetes and high blood pressure and are more likely to develop CKD that progresses faster to kidney failure than White Americans. The disparity happens in large part because of limited access to medical care, which mean these diseases are more advanced when they are diagnosed.

SOURCE NIH NIDKK, US Renal Data System, adr.usrds.org/2021

Geographic Variation

Deaths from chronic kidney disease vary considerably among the states. The best states are Vermont, at 3.6 deaths per 100,000 population, followed by Washington (4.5 per 100,000), South Dakota (6.2 per 100,000), Minnesota (7.4 per 100,000), and Oregon (8.0 per 100,000). The worst states are Mississippi, at 21.8 per 100,000 population, followed by Louisiana (19.4 per 100,000), Missouri (18.4 per 100,000), Kentucky (18.2 per 100,000), and Georgia (18.1 per 100,000).

SOURCE CDC, Chronic Kidney Disease Initiative, cdc.gov/kidneydisease/publications-resources

Dialysis

ANNUAL US DEATHS 100,000 in 2018 due to complications from dialysis, stroke, or heart attack

When kidney function drops to only about 10 to 15 percent of normal, they no longer remove enough wastes and fluid from the body. Dialysis can then be used to remove toxins, waste products, and excess fluids by filtering the blood through an artificial kidney. In 2017, 47 percent of new dialysis patients had a primary diagnosis of diabetes, the leading reason for needing dialysis, while 29 percent of new dialysis patients had a primary diagnosis of hypertension, the 2nd leading reason for needing dialysis. In 2018, about 130,000 people started dialysis treatment; more than 100,000 people on long-term dialysis died in 2018. The highest mortality rate for new dialysis patients is within the first 6 months. The 5-year survival rate for a patient undergoing long-term dialysis is approximately 35 percent, but only about 25 percent in patients with diabetes. Almost all of deaths are from complications from the dialysis process or from a heart attack or stroke.

Kidney Transplants

ANNUAL US DEATHS   On average, 4,380 people die annually waiting for a transplant.

In 2020, 22,817 people in the US received a kidney transplant. About one-third of these transplants came from living donors. The survival rate for kidney transplant recipients is about 80 percent after 5 years. The waiting list for a kidney transplant as of May 2021 had 90,694 people on it. Twelve people die every day while waiting for a kidney transplant.

For information on kidney donation (living and deceased) and to register as a donor, contact Donate Life America (donatelife.net).

STAY SAFE   Register as a kidney donor to help others.

SOURCES   American Kidney Fund, kidneyfund.org/kidney-disease; United Network for Organ Sharing (UNOS), unos.org/transplant

COVID-19 EXCESS MORTALITY

ANNUAL US DEATHS   In 2020, 6,953 to 10,316 excess deaths occurred among end-stage renal disease patients during the first 7 months of the pandemic.

On February 1, 2020, 798,611 patients were on chronic dialysis or had a kidney transplant. From February 1 through August 31, 2020, during the early months of the COVID-19 epidemic, significant excess mortality was recorded among these patients. Over this 7-month period, an estimated 6,953 to 10,316 deaths occurred among end-stage renal disease patients, or an estimated death rate of 10.8 to 16.6 per 1,000 patients. In this time, among patients undergoing dialysis, the risks of dying from any cause were 17 percent higher compared to death rates in 2017 through 2019.

SOURCES   R. Ziemba. Excess Death Estimates in Patients with End-Stage Renal Disease — United States, February–August 2020.

Morbidity and Mortality Weekly Report, 1 June 2021; E.D. Weinhandl et al. COVID-19-Associated Decline in the Size of the End-Stage Kidney Disease Population in the United States. *Kidney International Reports*, October 2021.

Polycystic Kidney Disease

ANNUAL US DEATHS   No statistics available

Polycystic kidney disease (PKD) is a genetic disorder that causes many fluid-filled cysts to grow in the kidneys and can reduce kidney function. PKD is one of the most common genetic disorders. In the US, about 600,000 people have PKD. It causes about 5 percent of all cases of kidney failure, making it the 4th leading cause. Half of all patients with PKD need dialysis or transplantation by age 60. The median age of death for people with PKD is 60.5 years. Nearly half will die of cardiovascular problems caused by their failing kidneys.

STAY SAFE   There is no cure for polycystic kidney disease, but a kidney transplant can extend life for some patients. Careful management of blood pressure, attention to diet, and treating symptoms and complications can slow the disease. Register as a kidney donor to help others.

SOURCES   American Kidney Fund, kidneyfund.org/kidney-disease; CDC, Chronic Kidney Disease Initiative, cdc.gov/kidneydisease/publications-resources; NIH NIDKK, US Renal Data System, adr.usrds.org/2021; United Network for Organ Sharing (UNOS), unos.org/transplant

Lung Disease

ANNUAL US DEATHS   Estimated 156,979 in 2019

Chronic lower respiratory disease (CLRD), the cause of most deaths from lung disease, is the 4th leading cause of death in the US, claiming 156,979 people in 2019. The age-adjusted death rate is 38.2 per 100,000 population. CLRD has resulted in 49.2 deaths per 100,000 adults. Lung diseases fall into 3 main types:

- Airway diseases. These diseases affect the airways, the tubes or bronchi that carry air into and out of the lungs. Airway diseases such as emphysema narrow or block the tubes, making it difficult to breathe.
- Lung tissue diseases (interstitial diseases). These diseases affect the structure of the spongy lung tissue. They cause scarring

or inflammation, making it difficult for the lungs to expand fully and limiting their ability to take in oxygen and expel carbon dioxide. People with lung tissue diseases such as pulmonary fibrosis or black lung disease can't breathe deeply.

- Lung circulation diseases. These diseases affect the blood vessels in the lungs, causing clotting, scarring, or inflammation and limiting the ability of the lungs to take in oxygen and expel carbon dioxide. Lung circulation diseases such as pulmonary hypertension often also affect the heart and cause shortness of breath.

Many lung diseases involve 2 or 3 of these types. Death from lung disease may be from the disease itself, from infections such as pneumonia that attack the weakened lungs, or from acute respiratory distress syndrome (ARDS) associated with an underlying disease.

Most lung disease is caused by exposure to airborne toxins from smoking, secondhand smoke, air pollution, and occupational exposure to particulates such as coal dust. Others are caused by bacteria, viruses, or fungi or have no known cause.

This section discusses chronic lung diseases such as emphysema and pulmonary fibrosis. Separate entries discuss other forms of lung disease such as asthma and infectious diseases that attack the lungs, such as Valley fever.

STAY SAFE Don't smoke. If you smoke, quit. Free resources to help you quit are available from BeTobaccoFree.gov, the National Cancer Institute quitline at 1-877-44U-QUIT, and the American Lung Association at lung.org/quit-smoking. Avoid exposure to secondhand smoke and industrial particulates. Wear appropriate protective gear if you work in a high-risk occupation.

SOURCES American Lung Association, lung.org/research/trends-in-lung-disease; CDC NCHS, cdc.gov/nchs/ndi/index.htm

Acute Respiratory Distress Syndrome (ARDS)

ANNUAL US DEATHS About 10,592 deaths are attributed to ARDS, which, however, is also associated with other diseases.

Acute respiratory distress syndrome (ARDS) is a life-threatening lung condition that causes a buildup of fluid in the organ's air sacs (alveoli). The fluid prevents enough oxygen from passing into the bloodstream and makes the lungs heavy and stiff so they cannot expand efficiently. Common causes include aspirating vomit or other substances, pneumonia, chemical inhalation, septic shock, and trauma. In 2013, influenza and pneumonia were listed as underlying causes of death in 35.1 percent of ARDS-related deaths. ARDS often occurs along with organ failure. Cigarette smoking and heavy alcohol use may be risk factors for its development.

Deaths from ARDS are usually attributed to the underlying cause and not tabulated separately. Almost all—94.9 percent—occur in hospital inpatients. From 2014 through 2018, there were 52,958 ARDS-related deaths (approximately 10,592 per year), for a mortality rate of 3.1 per 100,000 population.

STAY SAFE   Don't smoke. If you smoke, quit. Free resources to help you quit are available from BeTobaccoFree.gov, the National Cancer Institute quitline at 1-877-44U-QUIT, and the American Lung Association at lung.org/quit-smoking.

SOURCES   NIH NHLBI, nhlbi.nih.gov/health-topics; V. Parcha et al. Trends and geographic variation in acute respiratory failure and ARDS mortality in the United States. *Chest*, 2020.

Asthma

ANNUAL US DEATHS   3.524 in 2019

Deaths from asthma are rare and largely preventable. In 2019, 3,524 people died from asthma, including 178 children under age 18. The asthma death rate was 1.1 per 100,000 population and 9.7 per 1,000,000 population.

The rate of asthma deaths decreased from 4,269, or 15 per 1,000,000 population, in 2001 to 3,518, or 10 per 1,000,000 population, in 2016.

The worst state for asthma deaths in 2019 was Mississippi with 58 deaths, or 18.1 per 1,000,000 population.

Black adults and children were more likely to die of asthma. The total death rate for Blacks in 2019 was 22.3 per 1,000,000 population, while for Whites it was 8.2 and for Hispanics it was 7.8. The total death rate for women was 11.2 per 1,000,000 population; for men, it was 8.5.

The death rate for children was 2.8 per 1,000,000 population; for adults, it was 13.3. The highest death rate was for adults 65 and older: 29.2 per 1,000,000 population.

Nationally in 2019, 25,131,132 people had asthma. In the period 2016 to 2018, the average number of emergency room visits for asthma was 1,656,920.

STAY SAFE Work with your doctor to keep your asthma under control. Don't smoke and avoid secondhand smoke and air pollution.

SOURCES CDC NCHS, cdc.gov/nchs/ndi/index.htm; NIH NHLBI, nhlbi.nih.gov/health-topics/asthma

Chronic Obstructive Pulmonary Disease (COPD)

ANNUAL US DEATHS 156,045 in 2018

Chronic obstructive pulmonary disease (COPD) is a group of progressive lung diseases that affect the airways and air sacs of the lungs. COPD includes 2 main conditions: emphysema and chronic bronchitis. Emphysema damages the air sacs in the lungs and makes them less elastic. In 2013, 8,282 people died of emphysema, for a death rate of 2.6 per 100,000 population. Chronic bronchitis happens when the lining of the airways in the lungs is constantly irritated and inflamed, which makes them swell and fill with mucus. In 2016, 8.6 million people were diagnosed with chronic bronchitis. There were 518 deaths, for a death rate of 0.2 per 100,000 population.

In 2015, 3.2 million people died from COPD worldwide, or approximately 5 percent of all deaths. In 2018, 156,045 people in the US died from COPD, making it the 4th overall leading cause of death behind heart disease, cancer, and accidents and unintentional injuries, or the 3rd disease-related cause of death.

In 2018, more women than men died from COPD (82,158 versus 73,887), but the COPD death rate is higher among men than women (42.9 versus 35.8 per 100,000) because the older female population is larger than the older male population. In 2017, chronic lower respiratory disease, primarily COPD, was the 3rd leading cause of death among US women.

About 86 percent of deaths from COPD are among those aged 65 or older. The number of deaths is greatest for those age 75 to 84 years, but the death rate is greatest for those age 85 years or older. Older adults with COPD have a death rate that is 2.5 times higher than older adults without COPD. Younger adults (age 35 to 55 years) have a death rate that is 5 times higher than younger adults without COPD.

The best state for COPD in 2014 was Hawaii, with 15.2 deaths per 100,000; the worst was Kentucky, with 62.8 deaths per 100,000. Death rates from COPD are greater in rural areas (about 55 per 100,000 people) versus large metropolitan centers (32 per 100,000 people).

The primary risk factor for COPD is smoking. Other risk factors include secondhand smoke, air pollution, and environmental or occupational exposure to fumes or particulates. Compared to never smokers, COPD rates are 4.3 times higher among current smokers and 3.7 times higher among former smokers.

STAY SAFE Don't smoke. If you smoke, quit. Free resources to help you quit are available from BeTobaccoFree.gov, the National Cancer Institute quitline at 1-877-44U-QUIT, and the American Lung Association at lung.org/quit-smoking. Avoid secondhand smoke and exposure to particulates and fumes.

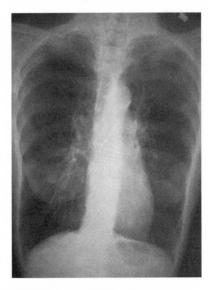

Chest X-ray of a COPD patient

SOURCES American Lung Association, lung.org/lung-health-diseases; CDC, cdc.gov/copd/index.html

EVALI

ANNUAL US DEATHS 68 from September 2019 to February 2020

EVALI (e-cigarette or vaping associated lung injury) is caused by vaping vitamin E acetate, an additive used in vape pen liquid. Between September 2019, when the condition was first reported, and February 2020, when the vitamin E was removed from e-cigarettes, 2,807 people in the US were hospitalized for EVALI. Of these, 68 died. The CDC stopped tracking cases in February 2020.

In 2019, a man in Texas was killed when his e-cigarette exploded and tore his carotid artery.

STAY SAFE Don't vape.

SOURCE CDC, cdc.gov/tobacco/basic_information/e-cigarettes/index.htm

Idiopathic Pulmonary Fibrosis (IPF)

> **ANNUAL US DEATHS**   No data available

Pulmonary fibrosis is an interstitial lung disease that causes inflammation and fibrosis (scarring) around the alveoli (air sacs) in the lungs. While inhaling hazardous chemicals can cause pulmonary fibrosis, in most cases the cause is unknown. Idiopathic pulmonary fibrosis (IPF) is quickly fatal. Most patients survive only 2 to 5 years after diagnosis. Approximately 50,000 new cases of IPF are diagnosed each year. The estimated mortality rate is 64.3 deaths per million in men and 58.4 deaths per million in women.

STAY SAFE   Avoid inhaling hazardous chemicals.

SOURCES   American Lung Association, lung.org/lung-health-diseases; CDC NCHS, cdc.gov/nchs/ndi/index.htm; NIH NHLBI, nhlbi.nih.gov/health-topics/idiopathic-pulmonary-fibrosis

Occupational Lung Diseases

> **ANNUAL US DEATHS**   See the individual diseases in this entry.

Occupational lung diseases are generally caused by repeated, long-term exposure to toxic fibers, dust, or vapor in the workplace. They are the primary cause of occupation-associated illness in the US. The 3 most common occupational lung diseases—asbestosis, black lung, silicosis—are all pneumoconiosis diseases, or lung disease caused by inhaling dust.

Smoking can increase both the severity of an occupational lung disease and the risk of lung cancer.

Asbestosis is a type of pulmonary fibrosis caused by inhaling asbestos fibers or dust. It killed 613 Americans and 3,495 people worldwide in 2016.

Coal worker's pneumoconiosis (CWP), also known as black lung disease, is caused by inhaling coal dust. About 2.8 percent of coal miners in the US have coal worker's pneumoconiosis. The Federal Coal Mine Health and Safety Act of 1969 introduced safety measures that brought the prevalence of black lung disease down sharply. Between 1970 and 1974, 32 percent of miners who had worked for 25 years or more had black lung. This fell to 9 percent by 2006.

In 1999, 409 workers died of CWP; in 2016, 112 died. In 2018, data from the CDC's Coal Workers' Health Surveillance Program showed an uptick in black lung disease, with 1 in 10 underground coal miners who had worked in mines for at least 25 years having it. Coal miners in central Appalachia are disproportionately affected; as many as 1 in 5 long-term coal miners now have CWP. Because black lung disease is entirely preventable, the increase is worrisome and indicates safety failures.

Silicosis is a type of pulmonary fibrosis lung disease caused by inhaling free crystalline silica, a mineral found in sand, quartz, and many other types of rock. Workers in mines, foundries, and stone, clay, and glass manufacturing are particularly at risk. Silicosis increases the risk of death from other lung diseases. Approximately 2.3 million workers per year are exposed to workplace silica.

In the late 1960s, deaths from silicosis were well over 1,000 annually. Because of improved safety regulations, from 2001 to 2010, a total of 1,437 people died from silicosis. The annual number of deaths declined from 164 in 2001 to 101 (0.39 per 1 million) in 2010.

In the early 1930s, about 750 workers (possibly as many as 2,000) died from silicosis contracted while working on the construction of the Hawks Nest Tunnel near Gauley Bridge, West Virginia. The tunnel was built to channel water to a hydropower plant owned by Union Carbide. The rock was high in silica, well known at the time to cause silicosis, but safety gear wasn't provided. The incident caused a national outcry and eventually led to improved safety regulations. Hawks Nest is considered one of America's worst industrial disasters.

Popcorn lung (obliterative bronchiolitis) is a type of airway disease that causes scarring of the bronchioles, the smallest airways in the lungs. The damage is cause by exposure to harmful vapors containing the flavoring chemical diacetyl and some other substances used as food flavorings. The disease became known as popcorn lung because it was first discovered in 2000 among workers at a microwave popcorn plant who had inhaled diacetyl. Death statistics for popcorn lung are unknown.

STAY SAFE Don't smoke. Wear appropriate protective devices when around airborne irritants and dusts.

SOURCES BLS, bls.gov/data/#injuries; CDC NIOSH, cdc.gov/ eWorld/Set/Work-Related_Respiratory_Diseases

Pulmonary Arterial Hypertension (PAH)

ANNUAL US DEATHS Statistics unavailable for this rare disease

Pulmonary arterial hypertension (PAH) is one form of pulmonary hypertension, or high blood pressure in the lungs. In PAH, the high pressure is caused by narrowing of the small arteries in the lung. In most cases the cause is unknown. PAH is rare and incurable. It usually affects women between the ages of 30 and 60. About 30 percent of people with PAH also have a connective tissue disease such as scleroderma; about 20 to 25 percent have a genetic form of PAH.

Approximately 500 to 1,000 new cases of PAH are diagnosed each year in the US. New cases are estimated to occur in 1 to 2 individuals per million population each year. The mortality rate is high. Only 57 percent of people with PAH are alive 5 years after diagnosis. Most people with PAH die of a condition or complication caused by the disease. Right-sided heart failure or sudden cardiac death kill somewhere between 44 and 73 percent of people with PAH. The singer Natalie Cole died of PAH in 2016.

STAY SAFE No preventive measures are known.

SOURCES CDC NCHS, cdc.gov/nchs/ndi/index.htm; NIH NHLBI, nih.gov/health-topics/pulmonary-hypertension

Infectious Diseases

Infectious diseases are disorders caused by bacteria, viruses, fungi, or parasites. An infectious disease makes you sick. Your body's immune system can fight off many infectious diseases by itself, but infectious diseases can also be severe enough to require medical attention. And sometimes infectious disease is deadly.

Some infectious diseases are contagious. They can spread from person to person through direct physical contact, through breathing in droplets of infected body fluids from someone nearby, and by touching an object an infected person has touched or used. Not all infectious diseases are contagious, however. Infections that spread to people from animals, insects, and ticks, such as malaria or Lyme disease, are not contagious.

This section discusses infectious diseases caused by bacteria and viruses, with the exception of diseases spread by insects and ticks—see the separate entries for these. Also see the separate entries for infectious diseases caused by fungi and parasites and the separate entry for food poisoning.

Deadliest Infectious Diseases

ANNUAL US DEATHS 113,650 in 2014

Every year between 1980 and 2014, 6 diseases in the US each caused 1 percent or more of deaths due to infectious disease:

- Lower respiratory infections
- Diarrheal diseases
- HIV/AIDS
- Meningitis
- Hepatitis
- Tuberculosis

All these diseases are discussed in detail later in this section. See also the separate discussion of COVID-19.

In 2014, 113,650 Americans died of infectious diseases in the above categories, for an overall mortality rate of 34.10 deaths per 100,000 population. Lower respiratory infections accounted for 78.80 percent of all infectious disease mortality. Between 1980 and 2014, overall mortality from infectious diseases decreased from 42.95 to 34.10 deaths per 100,000 persons, a drop of almost 19 percent over 35 years.

STAY SAFE Prevention is key for avoiding infectious disease. Get vaccinated. Wash your hands often. Wear a face mask when the risk of COVID is high in your community. Use insect repellent outdoors. Practice safe sex. Don't share needles. Don't share razors, combs, toothbrushes, and other personal hygiene items.

SOURCE CDC NCHS, cdc.gov/nchs/fastats/infectious-disease.htm

Anthrax

ANNUAL US DEATHS No deaths since 2001

Anthrax is a serious infectious disease caused by the *Bacillus anthracis* bacterium. Anthrax occurs naturally in the soil and primarily affects domestic and wild animals. It can cause severe illness and death in animals and humans. Today, however, routine vaccination of domestic animals prevents outbreaks, and anthrax is very rare in the US. Worldwide, at least 2,000 cases occur a year. In the US, about 2 cases are diagnosed each year; there have been no deaths since the anthrax outbreak of 2001 (see below). More than 95 percent of the cases are cutaneous anthrax, a skin infection that has a 23.7 percent risk of death without treatment but is cured by timely antibiotic treatment. Respiratory anthrax has a mortality rate of 50 to 80 percent, even with antibiotic treatment.

Death from anthrax in the US is extremely rare. In September 2001,

On October 9, 2001, US Senator Thomas Daschle was one of several persons who received, in the mail, an envelope containing anthrax-laced powder.

an outbreak of respiratory anthrax was caused by a suspected domestic terrorist who mailed letters containing anthrax spores to several news media offices and 2 US Senators. Numerous postal workers were exposed; 20 were infected, along with 2 other people. Of the 22 people infected, 11 developed cutaneous anthrax and 11 developed respiratory anthrax. Five people died. The prime suspect was never charged and killed himself in 2008.

STAY SAFE   Be cautious handling leather hides, which may carry anthrax spores.

SOURCE   CDC, cdc.gov/anthrax

Antibiotic-Resistant Infections

ANNUAL US DEATHS   35,000+

Infections caused by antibiotic-resistant germs are difficult, and sometimes impossible, to treat. More than 2.8 million antibiotic-resistant infections occur in the US each year, and more than 35,000 people die as a result. On average, an antibiotic-resistant infection occurs every 11 seconds and a death occurs every 15 minutes. When *Clostridioides difficile,* a deadly bacterium that is not antibiotic-resistant but is associated with antibiotic use, is considered, an estimated 223,900 infections and at least 12,800 deaths need to be added to the 35,000.

The CDC tracks 18 different antibiotic-resistant bacteria and fungi considered urgent, serious, and concerning threats. The deadliest are:

- In 2017, methicillin-resistant *Staphylococcus aureus* (MRSA) caused an estimated 323,700 infections and 10,600 deaths.
- In 2017, ESBL-producing *Enterobacteriaceae* caused an estimated 197,400 infections and 9,100 deaths.
- In 2017, vancomycin-resistant *Enterococcus* (VRE) caused an estimated 54,500 infections and 5,400 deaths.
- In 2014, drug-resistant *Streptococcus pneumoniae* caused an estimated 900,000 infections and 3,600 deaths.
- In 2017, multidrug-resistant *Pseudomonas aeruginosa* caused an estimated 32,600 infections and 2,700 deaths.
- In 2017, drug-resistant *Candida* species caused an estimated 34,800 infections and 1,700 deaths.
- In 2017, carbapenem-resistant *Enterobacterales* (CRE) caused an estimated 13,100 infections and 1,100 deaths.
- In 2016, clindamycin-resistant group B *Streptococcus* (GBS) caused an estimated 13,000 infections and 720 deaths.
- In 2017, carbapenem-resistant *Acinetobacter* caused an estimated

8,500 infections in hospitalized patients and 700 estimated deaths.

Each year, drug-resistant *Campylobacter* causes an estimated 448,400 infections and 70 deaths.

Each year, drug-resistant nontyphoidal *Salmonella* causes an estimated 212,500 infections and 70 deaths.

In 2017, multidrug-resistant tuberculosis caused 847 infections and 62 deaths.

STAY SAFE   Wash your hands often. Get vaccinated. Use antibiotics appropriately and only when your doctor tells you they are needed. Prepare food safely.

SOURCE   CDC, 2019 AR Threats Report, cdc.gov/drugresistance/biggest-threats.html

COVID-19

ANNUAL US DEATHS   From February 6, 2020 (first confirmed US death) to May 1, 2022: 993,813

COVID-19 (coronavirus disease 2019) is a contagious viral disease caused by severe acute respiratory syndrome coronavirus 2 (SARS-CoV-2). The first reported case was identified in Wuhan, China, in December 2019. Since then, the disease has become a global pandemic. As of May 1, 2022, 993,813 Americans, or almost in 1 in 300, were dead from COVID-19.

In 2020, COVID-19 was the 3rd leading cause of death in the US, behind heart disease and cancer and above unintentional injury. COVID-19 was reported as the underlying cause of death or a contributing cause of death for an estimated 385,343 deaths, or 11.3 percent of all deaths in 2020. The death rate was 91.5 per 100,000 population. From 2019 to 2020, the estimated age-adjusted death rate increased by 15.9 percent, from 715.2 to 828.7 deaths per 100,000 population. The per capita mortality rate from the disease is among the highest in the world. In mid-September 2021, US deaths from COVID-19 topped 675,000, surpassing the number of Americans killed by the 1918–1919 flu pandemic. By mid-November 2021, the number of US COVID-19 deaths was more than the total for 2020: 386,301 versus 385,343. At the start of November 2021, the global death toll surpassed 5.4 million. In mid-December 2021, the US death count passed 800,000, the highest in the world. The US population is 4 percent of the world population, but accounts for about 15 percent of the world's deaths from COVID.

The death toll from COVID-19 was so high that life expectancy at birth in the US dropped 1.8 years from 2019 to 2020, falling back to 77 years, its lowest point since 2000. The drop is the biggest since WWII. The impact on longevity was felt most profoundly by minority populations. Provisional figures from the CDC indicate that life expectancy for Black and Hispanic Americans fell at least 3 years from 2019 to 2020.

Another way to look at the toll COVID-19 has taken is by years of life lost, a number that is derived by subtracting age at death from remaining life expectancy. In 2020 approximately 380,000 Americans died from COVID-19, for an estimated 5.5 million years of life lost. That's more than all the years of life lost from accidental deaths in 2019 (5.320 million) and nearly triple the number of years of life lost to diabetes in 2019 (1.5 million).

A Note on Sources

At the start of the pandemic in late 2019, the US was unprepared for an event of this magnitude. The unprecedented speed and scope of the pandemic led to confusing, inconsistent, and outright wrong messaging, starting at the highest level. Unreliable and incorrect information was amplified, while objective authorities were sidelined or silenced. Statistical misinformation and misinterpretation, particularly on some media outlets and social media, continues to abound. Many aspects of the pandemic remain to be fully examined—for example, the effects of obesity and other underlying conditions on COVID-19 death rates, or the costs and benefits of lockdowns and other policy measures.

Overall Case Statistics

COVID-19 statistics as of May 1, 2022:
Global confirmed cases: 514,024,181
Global deaths: 6,237,390
US confirmed cases: 81,403,552
US deaths: 993,813

The global death toll from the COVID-19 pandemic is a significant undercount due to poor public health infrastructure in many parts of the world. WHO officials estimate that as of April 2022 the true number of direct and indirect deaths is 14.9 million. The total number of US cases is almost certainly higher because many mild cases went unnoticed or were never reported, especially early in the pandemic when testing was largely unavailable. The death number is probably slightly higher than reported because some deaths early in the pandemic, especially among the elderly, were misattributed to other causes.

As of May 1, 2022, the overall US case fatality rate was 1.2 percent. Due in part to the spread of the milder omicron variant, this was down from a fatality rate of 1.8 percent earlier in 2022.

SOURCE   Johns Hopkins Coronavirus Resource Center, coronavirus. jhu.edu

Age Disparities

Death rates from COVID-19 increase rapidly with age. Based on cumulative deaths through May 1, 2022, the risk of death for someone aged 30 to 39 was 18 times higher than that of someone aged 5 to 17; for someone aged 40 to 49, 44 times higher; for someone aged 50 to 64, 130 times higher; for someone aged 65 to 74, 310 times higher; for someone aged 75 to 84, 700 times higher; and for someone aged 85 or older, 1,800 times higher. Put another way, as of May 1, 2022, 1 in 30 Americans over 85 had died of COVID-19; 1 in 120 Americans aged 65 to 84 had died of COVID-19; and 1 in 415 Americans aged 50 to 64 had died of COVID-19. In other words, about 1 of every 150 Americans over 50 had died of the disease.

As of May 1, 2022, the CDC had age data for 842,903 COVID-19 deaths in the US:

AGE GROUP	PERCENT OF DEATHS	DEATHS	PERCENT OF POPULATION
0–49	6.8	57,199	65.5
50–64	18.1	152,774	19.2
65–74	22.7	190,995	9.6
75–84	26	219,096	4.9
85+	26.4	222,839	2.0

SOURCE   CDC COVID Data Tracker, covid.cdc.gov/ covid-data-tracker/#demographics

Sex Disparities

Men were more likely to die of COVID-19 than women. Of the 838,555 COVID-19 deaths for which the CDC had sex data as of May 1, 2022, females were 44.8 percent of deaths but 50.75 percent of the population. Men were 55.2 percent of deaths but 49.25 percent of the population.

SOURCE   CDC COVID Data Tracker, covid.cdc.gov/ covid-data-tracker/#demographics

Racial/Ethnic Disparities

Racial and ethnic minority communities were disproportionately affected by COVID-19 deaths. Many community members were unable to isolate during the worst of the pandemic and became infected. Unequal access to medical care and higher rates of chronic disease worsened the death toll. Statistics for deaths in these populations are probably undercounted. In addition, not all geographic areas reporting deaths to the CDC contribute data on race/ethnicity. The data below is based on available information for approximately 84 percent of reported deaths.

As of May 2, 2022, percent of deaths and percent of US population by race/ethnicity were as follows. At first glance, the only differences that stand out are a relatively higher rate of death for American Indians and a lower one for Asian Americans.

RACE/ETHNICITY	PERCENT OF DEATHS	DEATHS	PERCENT OF POPULATION
Hispanic/Latino	17.1	122,977	18.45
American Indian/ Alaska Native	1.1	8,023	0.74
Asian	3.3	23,460	5.76
Black	13.5	96,949	12.54
Hawaiian/Pacific Islander	0.2	1,709	0.18
White	62.5	447,924	60.11
Multiple/Other	2.2	15,993	2.22

But the racial disparities become clearer when we take the age distribution of different racial and ethnic groups into account. (The median age of the White population of the US is much higher than that of minority populations.) As of April 29, 2022, the age-adjusted risk of death from COVID-19 over the course of the pandemic was as follows:

American Indian/Alaska Native: 2.4× the White rate

Asian: 0.8×

Black: 1.7×

Hispanic/Latino: 1.8×

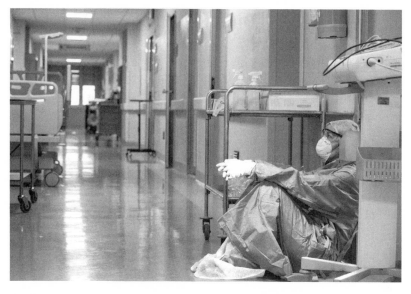

San Salvatore Hospital, Pesaro, Italy, during the COVID-19 pandemic

Earlier in the pandemic, these ratios were even higher for American Indian, Black, and Hispanic Americans.

Another consequence of the racial age distribution is that White Americans make up a large proportion of the elderly population most vulnerable to COVID. White Americans aged 65 and over are just 12.5 percent of the US population overall. However, they make up 76 percent of the population in their age bracket, and account for fully half of all US COVID deaths through May 2, 2022.

SOURCES CDC COVID Data Tracker, covid.cdc.gov/covid-data-tracker/#demographics; Johns Hopkins Corona Virus Resource Center, coronavirus.jhu.edu/data/disparity-explorer

Other Vulnerable Populations

INCARCERATED PERSONS

ANNUAL US DEATHS 2,879 as of May 2, 2022

COVID-19 disproportionately killed incarcerated persons and institution staff. As of May 2, 2022, 583,283 cases were reported among persons incarcerated in federal, state, and immigration institutions, with 2,879 deaths. Among prison staff, 199,320 cases were reported, with 277 deaths. As of April 1, 2021, at least 39 percent of the incarcerated population in the US had been infected, far higher than the overall population rate of 10 percent. Incarcerated persons had a 5.5 times greater risk of infection and a 3 times greater risk of death compared to the rest of the population. Visitors to correctional facilities, released prisoners and detainees, and staff were all significant vectors of infection into surrounding communities, making the corrections system a public health hazard.

SOURCE   COVID Prison Project, covidprisonproject.com

NURSING HOMES

ANNUAL US DEATHS   See data in entry.

COVID-19 swept through nursing homes across the country, causing deaths among both residents and staff. Exact numbers of deaths will probably never be known, because many residents of nursing homes were already in poor health, leading to underreporting of COVID-19 deaths, particularly in the early months of the pandemic when testing was largely unavailable. In addition, until May 2020 the Centers for Medicare and Medicaid Services (CMS) didn't require COVID deaths among nursing home residents and staff to be reported.

In some cases, nursing home deaths were significantly and sometimes deliberately underreported. In New York, for example, deaths in nursing homes and long-term care facilities were substantially underreported well into the first year of the pandemic. The undercount was a result of confused, inaccurate, and inconsistent reporting, aided by a political desire to keep the numbers low.

As of April 17, 2022, CMS reported that a total of 152,026 nursing home residents died from COVID-19. A total of 2,362 staff workers died.

SOURCES   AARP Nursing Home COVID-19 Dashboard, aarp.org/ppi/issues/caregiving/info-2020/nursing-home-covid-dashboard.html; CDC COVID Data Tracker, covid.cdc.gov/covid-data-tracker; CMS, data.cms.gov/covid-19/covid-19-nursing-home-data

VULNERABLE WORKERS

ANNUAL US DEATHS   See data in entry.

Workers in some industries were particularly hard-hit by the COVID-19 pandemic. Leading the list are healthcare personnel. In addition to the nursing home staff members discussed above, the CDC recorded 4,120 known deaths among doctors, nurses, and other healthcare personnel as of May 2, 2022. An ongoing analysis by Kaiser Health News and the *Guardian*, "Lost on the Frontline," reported 3,607 deaths among healthcare workers from March 2020 to March 2021. This analysis includes deaths among nursing home staff. The report found that almost a third of those who died were nurses, followed by support staff members (20 percent) and physicians (17 percent). As of this writing, no national registry for healthcare worker deaths has been established, making an accurate national count impossible.

The meatpacking industry was also very affected by COVID-19; at least 286 workers are known to have died.

SOURCES CDC COVID Data Tracker, covid.cdc.gov/covid-data-tracker; the *Guardian*, theguardian.com/us-news/series/lost-on-the-frontline

COVID-19 Variants

The original SARS-CoV-2 strain of the virus that causes COVID-19 was first identified by Chinese scientists in early January 2020. The virus quickly spread around the world. Like all other RNA viruses, the coronavirus that causes COVID-19 has mutated over time. In December 2020, a new variant scientists called B.1.1.7 was identified in the United Kingdom. Because this was the first variant identified, under standardized WHO nomenclature it was called alpha. Variants beta and gamma followed, but were quickly overtaken by the delta variant, first identified in India in December 2020. The delta variant is about twice as infectious as the original SARS-CoV-2 strain. It caused a huge surge in cases in mid-April 2021. In November 2021, another new variant, omicron, was identified in South Africa and quickly spread from there. The first confirmed case of omicron in the US was identified on December 1, 2021. This highly transmissible variant spread quickly, leading to the largest wave of COVID-19 infections yet, which peaked at more than 800,000 daily cases in the middle of January 2022. However, the omicron variant caused milder illness and fewer deaths than previous strains; despite the tremendous case numbers, daily deaths in the omicron wave did not exceed those during the previous peak in January 2021.

STAY SAFE The current vaccines plus a booster shot provide good protection against all variants to date. Get vaccinated and boosted; continue to take precautions against infection.

SOURCES CDC, cdc.gov/coronavirus/2019-ncov/variants; CDC Corona Virus Tracker, covid.cdc.gov/covid-data-tracker

Vaccines

Highly effective vaccines against COVID-19 became available for adults early in 2021. Booster shots 6 months after full vaccination were approved for adults in October 2021. Vaccines for children age 5 and up were approved in early November 2021. The federal government provides the COVID-19 vaccine free of charge to all people living in the US, regardless of their immigration or health insurance status.

The US COVID-19 Vaccination Program began on December 14, 2020. As of May 2, 2022, 220 million people were fully vaccinated with 2 doses of vaccine, for 70 percent of the population over age 5. More than 100 million people had received a booster dose, for 32 percent of the population over age 5.

Of the population aged 18 and older, 197 million, or 76 percent, were fully vaccinated with 2 shots. Among people aged 65 and older, 49 million, or 90 percent, were fully vaccinated.

There is strong evidence for the safety of the COVID-19 vaccines. The mRNA-based vaccines may cause rare serious side effects—such as myocarditis, or inflammation of the heart muscle, primarily in young males—but no deaths have been conclusively linked to them. A very small number of deaths from a rare blood clotting condition may be linked to the J&J/Janssen vaccine. A recent study in France of nearly 3.9 million persons 75 years or older who received COVID-19 vaccines found no increase in heart attacks, strokes, or pulmonary embolisms.

SOURCES CDC COVID Data Tracker, covid.cdc.gov/covid-data-tracker; M.J. Jabagi et al. Myocardial Infarction, Stroke, and Pulmonary Embolism After BNT162b2 mRNA COVID-19 Vaccine in People Aged 75 Years or Older. *JAMA*, November 2021; Johns Hopkins Coronavirus Resource Center (CRC), coronavirus.jhu.edu

Vaccination by State

Vaccination levels vary considerably by state. As of May 1, 2022, the average for all 50 states, the District of Columbia, and Puerto Rico was 66.6 percent. The 5 states with the highest percentage of people fully vaccinated (with 2 shots) were:

Rhode Island	82.7%
Vermont	81.1%
Maine	80.1%
Massachusetts	79.4%
Connecticut	79.0%

The 5 states with the lowest percentage of fully vaccinated people were:

Arkansas	54.8%
Louisiana	53.3%
Mississippi	51.7%
Alabama	51.4%
Wyoming	51.4%

There is a correlation between vaccination rates and death rates. The 5 states with the highest COVID-19 death rates in the 6 months from October 26, 2021, to April 26, 2022, had rates of vaccination (as of May 1, 2022) lower than the national average:

STATE	DEATHS PER 100,000 POPULATION	VACCINATION RATE
Tennessee	141	55.5%
West Virginia	140	56.8%
Missouri	129	57.4%
Michigan	125	60.3%
Kentucky	125	57.6%

And 4 of the 5 states with the lowest COVID-19 death rates in the same period had rates of vaccination higher than the national average:

STATE	DEATHS PER 100,000 POPULATION	VACCINATION RATE
Massachusetts	18	79.4%
Hawaii	36	77.3%
Vermont	42	81.1%
California	45	72.5%
Utah	48	66.5%

SOURCE CDC COVID Data Tracker, covid.cdc.gov/covid-data-tracker; Johns Hopkins Coronavirus Resource Center (CRC), coronavirus.jhu.edu; USAFacts, usafacts.org

Vaccine Effectiveness

The COVID-19 vaccines were highly effective against the initial variants of the virus. From the beginning of April through mid-July 2021, 92 percent of all COVID cases, 92 percent of hospitalizations, and 91 percent of deaths were reported among persons not fully vaccinated. In the same time frame, 8 percent of cases, 8 percent of hospitalizations, and 9 percent of deaths were reported among fully vaccinated persons (2 shots, no booster).

The rate of breakthrough infections among vaccinated people increased significantly with the spread of the omicron variant, but the vaccines remained effective against severe illness. According to the CDC, in February 2022 unvaccinated people aged 5 and older were 2.8 times more likely to test positive for COVID and 10 times more likely to die from COVID than fully vaccinated people. And in the same period, the unvaccinated were 3.1 times more likely to test positive and 20 times more likely to die than fully vaccinated people who had also received a booster.

A member of the California Army National Guard collects a specimen at a drive-through COVID-19 testing site in Long Beach, April 27, 2020

According to a study from the Commonwealth Fund, without a vaccination program, there would have been approximately 1.1 million additional COVID-19 deaths and more than 10.3 million additional COVID-19 hospitalizations in the US by November 2021. COVID-19 deaths would have been approximately 3.2 times higher and COVID-19 hospitalizations approximately 4.9 times higher than the actual toll during 2021. If no one had been vaccinated, daily deaths from COVID-19 could have jumped to as high as 21,000 per day—nearly 5.2 times the level of the record peak of more than 4,000 deaths per day recorded in January 2021.

STAY SAFE   Get fully vaccinated and get booster shots as recommended. Wear a face mask whenever and wherever it is mandated or suggested. Wash your hands often. Avoid contact with people who are ill with COVID-19. If you are exposed to COVID-19, get tested and follow current isolation instructions. If you are ill with COVID-19, isolate according to current medical advice.

SOURCES   APM Research Lab, apmresearchlab.org/covid; CDC COVID Data Tracker, covid.cdc.gov/covid-data-tracker; the Commonwealth Fund, Eric C. Schneider et al. *The U.S. COVID-19 Vaccination Program at One Year: How Many Deaths and Hospitalizations Were Averted?* December 2021, commonwealthfund. org/publications/issue-briefs/2021/dec/us-covid-19-vaccination-program-one-year-how-many-deaths-and; the COVID Prison Project, covidprisonproject.com; Food and Environmental Reporting Network, thefern.org/?s=covid; the *Guardian*, theguardian.com/us-news/series/lost-on-the-frontline; M.J. Jabagi et al. Myocardial Infarction, Stroke, and Pulmonary Embolism After BNT162b2 mRNA COVID-19 Vaccine in People Aged 75 Years or Older. *JAMA*, November 2021; Johns Hopkins Coronavirus Resource Center (CRC), coronavirus.jhu.edu; E. Reinhart et al. Carceral-community epidemiology, structural racism, and COVID-19 disparities. *Proceedings of the National Academy of Science USA*, May 2021; S.H. Woolf et al. Effect of the covid-19 pandemic in 2020 on life expectancy across populations in the USA and other high income countries: simulations of provisional mortality data. *BMJ*, June 2021; H.M. Scobie et al. Monitoring Incidence of COVID-19 Cases, Hospitalizations, and Deaths, by Vaccination Status — 13 U.S. Jurisdictions, April 4–July 17, 2021. MMWR Morbidity and Mortality Weekly Report, September 17, 2021; WHO, who.int/emergencies/diseases/novel-coronavirus-2019

Diarrheal Disease

ANNUAL US DEATHS   About 8,000 in 2014

Diarrheal diseases were the 2nd leading cause of death from infectious disease in 2014. About 8,000 people died, or 2.41 deaths per 100,000 population, or 7.07 percent of all infectious disease deaths. From 1980 to 2014, the death rate for diarrheal diseases increased from 0.41 to 2.41 deaths per 100,000 population.

Childhood deaths from diarrheal diseases declined substantially after the introduction of rotavirus vaccines in 2006. Before the vaccines, about 369 children aged 5 and younger died of diarrheal diseases each year, or 2.3 deaths per 100,000 children. Black children died at almost 4 times the rate of White children. The vaccines are highly effective at reducing the need for hospitalization and have cut deaths from all diarrheal diseases to about 300 a year.

Globally, 1 in 9 childhood deaths are due to diarrhea. These diseases kill 2,195 children a day, or 801,000 a year.

STAY SAFE   Wash your hands after using the toilet. Vaccinate infants for rotavirus.

SOURCES   CDC, cdc.gov/healthywater/global/diarrhea-burden.html; WHO, who.int/news-room/fact-sheets/detail/diarrhoeal-disease

Diphtheria

ANNUAL US DEATHS   Thanks to vaccination, 0.045

Diphtheria is a serious bacterial infection caused by the bacterium *Corynebacterium diphtheriae*. The bacterium produces a toxin that causes difficulty breathing, heart failure, paralysis, and even death. CDC recommends vaccines for infants, children, teens, and adults to prevent diphtheria. The estimated overall case fatality ratio for diphtheria is 5 to 10 percent.

During the 1920s, 100,000 to 200,000 cases of diphtheria (140 to 150 cases per 100,000 population) and 13,000 to 15,000 deaths from diphtheria were reported each year. After diphtheria vaccines became available in the 1940s, cases declined by 1945 to about 19,000 a year (15 cases per 100,000 population). Universal childhood vaccination soon reduced diphtheria cases and deaths to very low numbers. From 1996 through 2018, only 14 cases of diphtheria were reported in the United States, an average of less than 1 per year. Of those cases, 1 was fatal.

In 1925, an outbreak of diphtheria in Nome, Alaska, was stopped when 20 relays of sled dog teams carried antitoxin 674 miles in 127 hours. This heroic effort has been commemorated every year since 1973 as the Iditarod Trail Sled Dog Race. A statue of Balto, the lead dog of the team for the final leg, stands in New York City's Central Park.

STAY SAFE   Get all recommended childhood vaccines and adult boosters on schedule.

SOURCES   CDC, cdc.gov/diphtheria; Iditarod.com, iditarod.com/booms-and-busts-iditarod-trail-history

Encephalitis

ANNUAL US DEATHS   205 (herpes encephalitis) in 2019

Encephalitis, or inflammation of the brain, can be caused by a bacterium or fungus, but it is more usually viral. Several thousand cases of encephalitis are reported each year, but many more—up to 60 percent—may go undiagnosed because the symptoms are very mild or unnoticed.

Most diagnosed cases of encephalitis in the US are caused by herpes simplex virus types 1 and 2, arboviruses, or enteroviruses. Encephalitis from arboviruses is spread by mosquitos and is discussed separately under Insect-Borne Diseases. This section includes only herpes simplex encephalitis (HSE) because death from enterovirus encephalitis is very rare.

Herpes simplex encephalitis (HSE) causes about 10 percent of all encephalitis cases, with a frequency of about 2 cases per 1,000,000 persons per year. More than half of untreated cases are fatal. In 2019, 205 people died of herpes encephalitis, for a mortality rate of 0.06 per 100,000 population.

STAY SAFE   Wash your hands.

SOURCES   CDC NCHS, cdc.gov/nchs/ndi/index.htm; NIH NINDS, ninds.nih.gov/Disorders/All-Disorders/Meningitis-and-Encephalitis-Information-Page

Flesh-Eating Bacteria

ANNUAL US DEATHS   700 to 1,200 cases, of which about one-third prove fatal

Necrotizing fasciitis, also known as flesh-eating bacteria or flesh-eating disease, is usually caused by group A *Streptococcus* (group A strep) bacteria. "Necrotizing" means causing the death of tissues, in this case the fascia (the tissue under the skin that surrounds muscles, nerves, fat, and blood vessels). The bacteria infects the body through breaks in the skin. Necrotizing fasciitis can spread throughout the body very rapidly and aggressively and lead to sepsis, shock, and organ failure. Patients often need emergency surgery to remove dead tissue and amputate limbs that have become gangrenous.

Data for necrotizing fasciitis is incomplete. Since 2010, approximately 700 to 1,200 cases occur each year in the US, but this is likely an underestimate. Early diagnosis and rapid, aggressive treatment are critical for fighting this infection, but even with prompt treatment, up to 1 in 3 people with necrotizing fasciitis die from it.

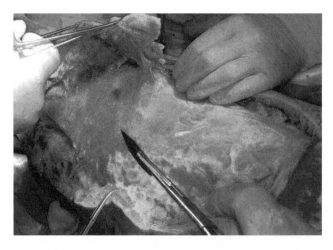

Surgeons operate to save the leg of a patient infected with flesh-eating bacteria.

STAY SAFE   Wash your hands. Prevent skin infections with good wound care. If you have an open wound or skin infection, avoid hot tubs, swimming pools, and natural bodies of water such as lakes and oceans.

SOURCE   CDC, cdc.gov/groupastrep/diseases-public/necrotizing-fasciitis.html

Hantavirus Disease

ANNUAL US DEATHS   265 deaths between May 1993 and January 2017

Hantavirus is a virus found in the droppings, urine, and saliva of infected rodents, including the deer mouse, white-footed mouse, red-backed vole, cotton rat, and rice rat. The virus causes a rare but serious lung disease called hantavirus pulmonary syndrome (HPS). Hantavirus was first identified in May 1993 when an outbreak of an unexplained pulmonary illness occurred in the Four Corners region of the American Southwest. As of January 2017, 728 cases of hantavirus disease have been reported. Of these people, about 40 percent, or an estimated 265 people, died from HPA.

STAY SAFE   Avoid contact with wild rodents and their nesting areas while hiking, camping, and working in barns. Remove wild rodents and clean up droppings before occupying vacant buildings.

SOURCE   CDC, cdc.gov/hantavirus

Hepatitis

ANNUAL US DEATHS   4,285 in 2018

Hepatitis is inflammation of the liver from a variety of causes, both infectious and noninfectious. Viral infection is by far the most common cause of hepatitis, but other causes include damage from drugs and toxins and autoimmune reactions. In 2018, 4,285 Americans died from viral hepatitis, for a death rate of 1.3 per 100,000 population.

Viral hepatitis has 3 main forms: hepatitis A, hepatitis B, and hepatitis C. All are reportable diseases tracked by the CDC's National Notifiable Diseases Surveillance System.

Hepatitis A is a vaccine-preventable liver infection spread by contact with contaminated food and water. In 2018, an outbreak of the disease led to 12,474 cases reported to the CDC. Because many cases go untreated and unreported, the actual number of infections was estimated to be 24,900. Because death from hepatitis A is rare, the death rate for this outbreak was estimated at 0.05 per 100,000 population. In 2017, a more typical year, there were 3,366 reported cases and an estimated 6,700 infections. The death rate was 0.02 per 100,000 population.

Hepatitis B is a vaccine-preventable liver infection spread by contact with blood, semen, or other body fluids from an infected person.

Infection can happen through sexual contact, sharing needles, or from mother to baby at birth. Chronic infection, usually present from birth, can lead to cirrhosis or liver cancer years later. In the US, routine vaccination at birth began in 1991, and few infections occur. In 2018, 3,322 cases were reported, resulting in an estimated 21,600 infections. In 2017, there were 1,727 deaths, for a death rate of 0.43 per 100,000 population.

Hepatitis C is a liver infection spread through contact with blood, semen, or other body fluids from an infected person. Infection can happen through sexual contact or sharing needles. In 2018, 3,621 cases were reported, resulting in an estimated 50,300 infections. In 2018, 15,713 deaths with hepatitis C as the underlying cause were reported. The death rate was 3.72 per 100,000 population.

The hepatitis C virus was discovered in 1989. Deaths in the US associated with hepatitis C reached an all-time high of 19,659 in 2014. Because death certificates often underreport hepatitis C, the death toll was likely much higher. In 1 analysis only 19 percent of hepatitis C patients who died had HCV listed on their death certificate.

There is no vaccine for hepatitis C, but the introduction of effective antiviral medications starting in 2013 means that it can be cured more than 95 percent of the time.

Hepatitis D, also known as delta hepatitis, occurs only in people who are also infected with the hepatitis B virus. Hepatitis D is spread when blood or other body fluids from an infected person enters the body of an uninfected person. Hepatitis D is usually an acute, short-term infection, but it can cause serious illness leading to lifelong liver damage and even death. Because hepatitis D is uncommon in the US and is not yet nationally notifiable, the number of cases and possible deaths is unknown. Most cases are among people who are in the US from other regions where the disease is endemic, including eastern Europe, southern Europe, the Mediterranean region, the Middle East, west and central Africa, east Asia, and the Amazon Basin in South America. There is no vaccine for hepatitis D.

Hepatitis E is a liver infection spread by contaminated drinking water. This disease is common in developing countries with inadequate water supply and poor environmental sanitation. Epidemics of hepatitis E occur among people living in crowded camps or temporary housing, including refugees and people who are internally displaced. In the US, hepatitis E is uncommon and occurs almost entirely among people returning from areas where the disease is endemic. There is no vaccine for hepatitis E.

STAY SAFE   Get vaccinated for hepatitis A and hepatitis B (if necessary). Practice safe sex and don't share needles. Avoid contaminated water and areas with poor sanitation.

SOURCE   CDC, cdc.gov/hepatitis

HIV/AIDS

> **ANNUAL US DEATHS**   15,820 (HIV) and 5,044 (AIDS) in 2018

HIV (human immunodeficiency virus) attacks the body's immune system, increasing the risk for serious infections. AIDS (acquired immunodeficiency syndrome) is the final stage of infection with HIV. Not everyone with HIV develops AIDS. Because HIV makes people very vulnerable to a range of fatal infections, deaths in HIV-infected people may be due to any number of causes. Many people with HIV/AIDS die of opportunistic infections such as pneumonia.

In 2019, an estimated 1.2 million people in the US and dependent areas were living with HIV. There were approximately 38,400 new diagnoses; gay, bisexual, and other men who have sex with men (MSM) accounted for 70 percent. Heterosexual contact accounted for 22 percent of new cases. About 7 percent of new cases were among people who inject drugs.

In 2018, 15,820 people with HIV died, for a death rate of 13.6 per 100,000 population. Of those deaths, 11,975 were among men; 8,049 were among gay, bisexual, and MSM; 4,905 were among people who inject drugs. In 2018, 5,044 people in the US died of AIDS, or 1.5 deaths per 100,000 population.

Since the AIDS epidemic began in the US in 1981, more than 700,000 people with AIDS have died. In the early 1980s, nearly half of all people infected with HIV died within 2 years. The introduction of combination antiretroviral therapy in 1996 has substantially reduced AIDS-related mortality and improved long-term outcomes for people with HIV. The overall death rate among persons diagnosed with HIV/AIDS in New York City, for example, decreased by 62 percent from 2001 to 2012. The age-adjusted HIV death rate has dropped by more than 80 percent since its peak in 1995.

By Age

HIV is a leading cause of death for certain age groups. In 2016, HIV was the 9th leading cause of death for those aged 25 to 34 and for those aged 35 to 44. In 2017, there were 149 deaths from HIV among youths aged 13 to 24. In 2018, there were 11,425 deaths among people aged 50 and older.

In 1993, AIDS was the leading cause of death among persons 25 to 44 years old and 8th overall among all causes of death, accounting for 2 percent of all deaths. In 1994 in the 25- to 44-year-old age group, AIDS accounted for 23 percent of all deaths among men and 32 percent of all deaths among Black men. It was 3rd overall among causes of death for women 25 to 44 years of age, accounting for 11 percent of deaths, but first among Black women, accounting for 22 percent of deaths.

Racial Disparities

Black people are disproportionately affected by HIV/AIDS. Of the more than 1.1 million people living with HIV/AIDS in the US, 476,100 are Black. Of people who die from HIV, 44 percent are Black. In 2017, HIV was the 6th leading cause of death for Black men ages 25 to 34 and 8th for Black women ages 35 to 44, higher than for any other racial or ethnic group. In 2018, there were 6,678 deaths among Black people with HIV in the US.

By Location

In 2018, HIV-related deaths in the US were mostly in the southern states. The geographic distribution was:
- 47 percent in the South
- 22 percent in the Northeast
- 17 percent in the West
- 12 percent in the Midwest
- 2 percent in US territories

The top 10 cities for HIV infection in 2020 ranked by cases per 100,000 population were:
1. Baton Rouge, LA, 44.7
2. Miami, FL, 42.8
3. New Orleans, LA, 36.9
4. Jackson, MS, 32.2
5. Orlando, FL, 28.8
6. Memphis, TN, 27.6
7. Atlanta, GA, 25.9
8. Columbia, SC, 25.6
9. Jacksonville, FL, 25.1
10. Baltimore, MD, 24.3

Worldwide HIV/AIDS

Approximately 37.7 million people across the globe were living with HIV/AIDS in 2020. Of these, 36 million were adults and 1.7 million were children under age 15.

There were an estimated 1.5 million new HIV infections worldwide in 2020, a 23 percent decline since 2010. Of these new infections, 1.5 million were among adults and 150,000 were among children under age 15.

In 2020, around 680,000 people worldwide died from an AIDS-related illness, a 37 percent decrease from 1.1 million in 2010 and a 59 percent decrease from the peak of 1.7 million in 2004. Since the start of the epidemic in the 1980s, approximately 79.3 million people have become infected with HIV and 36.3 million have died.

STAY SAFE Practice safe sex and don't share needles.

SOURCES CDC, cdc.gov/hiv/statistics/overview/index.html; KFF, kff.org/hivaids; UNAIDS Global HIV & AIDS statistics fact sheet, unaids.org/en/resources/fact-sheet

Influenza (Flu)

> **ANNUAL US DEATHS** 12,000 to 61,000, depending on the severity of variant

Influenza, or flu, is a contagious respiratory illness caused by human influenza A and B viruses. Variants on human influenza A and B viruses cause seasonal flu epidemics almost every winter in the US. Flu pandemics, where the disease spreads globally, are caused when a new and very different variant of the A virus emerges. The flu season varies in severity from year to year.

Influenza viruses infect the nose, throat, and sometimes the lungs, causing mild to severe illness that can sometimes lead to death. On average, about 8 percent of the US population gets sick with the flu each season. The incidence rate can range from 3 to 11 percent, depending on the season. Children under age 17 are most likely to become ill with flu. In a typical flu season, the median attack rate for children under 17 years is 9.3 percent; the rate is 8.8 percent for adults at 18 to 64, and 3.9 percent for adults over age 65. Children younger than 18 are more than twice as likely to get sick with flu than adults 65 and older.

People at high risk of flu complications and possible death include those 65 years and older, people with chronic medical conditions such as asthma, diabetes, or heart disease, pregnant women, and children younger than 5 years.

Death rates for flu vary yearly, depending on the severity of the season. In addition, many death certificates record flu as a contributory cause of death in people with chronic disease. Estimated annual deaths since 2010 range from a low of 12,000 in the very mild 2011–2012 season to a high of 61,000 in the 2017–2018 season. For the 2019–2020 season, the death toll was an estimated 22,000 out of an estimated 38 million cases and 400,000 hospitalizations.

Flu Vaccines

The annual flu vaccine is usually about 40 to 60 percent effective in preventing infection completely. It is highly effective in preventing severe infection that leads to hospitalization and death. During the 2019–2020 season, flu vaccination prevented an estimated 7.5 million cases of flu, 105,000 influenza-associated hospitalizations, and 6,300 deaths. Flu vaccination among adults can reduce the risk of being admitted to an intensive care unit (ICU) with flu by 82 percent. Flu vaccination can significantly reduce a child's risk of dying from flu.

The Flu Pandemic of 1918

The 1918 influenza pandemic, sometimes called the Spanish flu (although it probably originated in the US), was, until the COVID-19 pandemic began in late 2019, the most severe pandemic in recent history. An estimated 500 million people, or about one-third of the world's population at the time, got sick with flu. At least 50 million people died worldwide; about 675,000 people died in the US. Unusually for flu, this

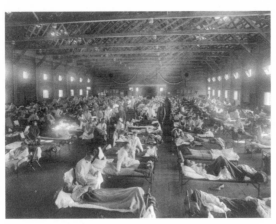

Camp Funston, at Fort Riley, Kansas, may have been the source of the devastating 1918 flu pandemic, which killed some 50 million worldwide–675,000 in the US.

virus caused a high death rate among healthy adults aged 15 to 34. The pandemic lowered the average life expectancy in the US by more than 12 years. Since 1918, 3 more flu pandemics, in 1957, 1968, and 2009, have killed many people, but the death toll was far below that of 1918. The 2009 pandemic had a worldwide estimated death toll of between 151,700 and 575,400. In the US, there were an estimated 60.8 million cases, 274,304 hospitalizations, and 12,469 deaths.

STAY SAFE   Get an annual flu shot early in the fall. Wash your hands. Cover coughs and sneezes. Stay home if you have the flu and avoid people who are sick with it.

SOURCE   CDC, www.cdc.gov/flu/

Legionnaires' Disease

ANNUAL US DEATHS   Only significant in outbreaks, as in 1976, when 29 died

Legionnaires' disease is a serious type of bacterial pneumonia caused by *Legionella* bacteria. The disease is spread through contaminated water droplets. *Legionella* also causes a less serious disease known as Pontiac fever. *Legionella* bacteria can grow and spread through building water systems, including showerheads, sink faucets, air-conditioning cooling towers, hot tubs, and hot water tanks. Outbreaks of Legionnaires' disease are usually building-associated and mostly occur in hotels, hospitals, and long-term care facilities.

Nearly 10,000 cases of Legionnaires' disease were reported in 2018 in the US. This number almost certainly undercounts the true incidence, which is probably 1.8 to 2.7 times higher. Legionnaires' disease is the cause of an estimated 2 to 9 percent of community-acquired pneumonia cases and is very dangerous. About 1 in 10 people who get it will die. It is particularly lethal for older adults, especially smokers, former smokers, and those with chronic diseases.

Legionella was first identified after an outbreak in 1976 among people who went to a convention of the American Legion in Philadelphia. Of the 4,000 attendees, 182 became ill with atypical pneumonia that was eventually found to be *Legionella*.

STAY SAFE   Avoid hot tubs and spas. Older adults with pneumonia symptoms (chest pain when breathing or coughing, low body temperature) should seek medical attention promptly.

SOURCE   CDC, cdc.gov/legionella/index.html

Leprosy (Hansen's Disease)

ANNUAL US DEATHS   Nominally zero

Once widely feared as a contagious source of crippling disfigurement, Hansen's disease today is very rare in the US. The disease is caused by a slow-growing bacterium called *Mycobacterium leprae*. When the disease is diagnosed early and treated promptly with antibiotics, it can be cured. Patients become noninfectious after only a few doses and don't need to be isolated. In 2019, 216 new cases were reported in the US. Worldwide, about 250,000 new cases are diagnosed each year.

The US government established the National Leprosarium in Carville, Louisiana, in 1917. The facility was closed and replaced with outpatient clinics in 1981. Today, the National Hansen's Disease Program is the center of care, research, and information in the US. Leprosy is very hard to catch—you need prolonged, close contact with someone with untreated leprosy; you can't catch it from casual contact.

STAY SAFE   Avoid contact with armadillos. Some armadillos in the Southwest are infected with the leprosy bacterium.

SOURCES   CDC, cdc.gov/leprosy; HSS HRSA, hrsa.gov/hansens-disease/index.html; WHO, who.int/news-room/fact-sheets/detail/leprosy

Leptospirosis

ANNUAL US DEATHS   Very low in the continental US; about 59,000 worldwide

Leptospirosis is a bacterial infection caused by the *Leptospira* family. It is fairly common in unvaccinated dogs, where it can cause illness and death in severe cases. In humans, leptospirosis is rare, with under 1,000 cases annually. Most infected people have only mild symptoms, but without treatment leptospirosis can lead to kidney damage, meningitis, liver failure, respiratory distress, and very rarely, death. Cases of leptospirosis can increase after hurricanes or floods when people may come into contact with contaminated water.

STAY SAFE   Wash your hands after contact with animal urine. Vaccinate pet dogs annually. Avoid drinking, wading in, or swimming in contaminated water.

SOURCE   CDC, cdc.gov/leptospirosis/index.html

Lower Respiratory Infection

> **ANNUAL US DEATHS**   89,880 in 2014; 43,881 pneumonia deaths in 2019

Lower respiratory tract infection (LRTI) broadly means pneumonia (infection in 1 or both of the lungs that makes the air sacs fill up with fluid or pus) or acute bronchitis (inflammation of the bronchial tubes, the airways that carry air to the lungs). Lower respiratory infections can be caused by a bacterium, a virus, or a fungus. Influenza, or flu, is a virus that affects both the upper and lower respiratory tracts. It can cause pneumonia and is discussed above in this section.

In 2014, lower respiratory infections of all types killed 89,880 people, causing 78.80 percent of all infectious disease deaths, for a death rate of 26.86 per 100,000 population. In 2019, 43,881 Americans died of pneumonia, for a death rate of 13.4 per 100,000 population.

STAY SAFE   Two vaccines are available to prevent infections from the pneumococcus bacteria, the most common type of bacteria that causes pneumonia. Get vaccinated if you are aged 65 or older, if you smoke, have impaired immunity, if you have a chronic disease such as cancer, HIV, asthma, or sickle cell disease, or if your spleen is damaged or has been removed. Children under age 2 should also be vaccinated.

SOURCES   CDC, cdc.gov/pneumonia; NIH NHLBI, nhlbi.nih.gov/health/pneumonia

Measles

> **ANNUAL US DEATHS**   Thanks to widespread vaccine usage, negligible

Measles is a dangerous and highly contagious disease of childhood caused by the rubeola virus. Before 1963, approximately 500,000 cases of measles and 500 deaths were reported annually in the US, with epidemic cycles every 2 to 3 years. Because most cases were unreported, the actual number was probably 3 to 4 million annually. After vaccines became available in 1963, the incidence of measles decreased by more than 95 percent and it no longer occurred in epidemic cycles.

From 1987 to 2000, about 30 percent of reported measles cases had 1 or more complications, including pneumonia, encephalitis, and death, mostly in children under age 5. Between 1 and 3 of every 1,000 children who become infected with measles will die from respiratory and neurologic complications.

The World Health Organization (WHO) defines measles elimination as "the absence of endemic measles virus transmission in a defined geographical area (e.g., region or country) for at least 12 months in the presence of a surveillance system that has been verified to be performing well." Measles was declared eliminated in the US in 2000. Outbreaks sparked by travelers bringing measles with them continue to occur among unvaccinated people.

In 2019, 22 outbreaks of measles were reported, accounting for 1,249 cases. Most cases (934, or 75 percent) occurred in undervaccinated, tight-knit communities of Orthodox Jews located in New York City and New York State. Overall, 119 patients (10 percent) were hospitalized with complications. None died. Before 2019, the highest number of measles cases following elimination in the United States occurred in 2014 when 667 cases were reported. None died. In 2020, 13 cases of measles and 0 deaths were reported.

Globally, measles vaccination has prevented 25.5 million deaths since 2000. Annual deaths from measles have declined 62 percent from 2000 to 2020. In 2019, however, cases reached a 23-year high of 869,770 cases; more than 207,000 people died.

STAY SAFE Vaccinate children on schedule with the MMR vaccine. Concerns that the measles vaccine causes autism are based on fraudulent research—the vaccine is very safe, far safer than having measles.

SOURCES CDC, cdc.gov/measles; WHO, who.int/news-room/fact-sheets/detail/measles

Meningitis

ANNUAL US DEATHS About 500

Meningitis is an inflammation of the meninges, the protective membranes covering the brain and spinal cord. The most common cause is a bacterial or viral infection of the fluid surrounding the brain and spinal cord, but infection with some types of fungus or parasites can also cause meningitis. Treatment varies depending on the cause.

Almost 4,100 cases of bacterial meningitis occur annually in the US, causing approximately 500 deaths. The annual incidence in the US is 1.33 cases per 100,000 population. Bacterial meningitis is a very serious illness. It is fatal in 1 in 10 cases, and 1 of every 7 survivors is left with a disability such as deafness or brain injury.

Viral meningitis is the most common type of meningitis. It is usually less severe than bacterial meningitis and most people recover. Viruses that can cause meningitis include the mumps virus, herpes viruses (including chicken pox), the measles virus, and the influenza virus.

Fungal meningitis can develop if a fungus such as *Cryptococcus*, *Histoplasma*, *Blastomyces*, *Coccidioides*, or *Candida* spreads from somewhere else in the body to the brain or spinal cord. Fungal meningitis is rare, but the death rate is high: Overall mortality for *Candida* meningitis is 10 to 20 percent, but rises to 31 percent for patients with HIV. Between 2012 and 2013, a multistate fungal meningitis outbreak from tainted steroid injections caused 753 cases; 64 people died.

STAY SAFE   Get vaccinated for bacterial meningitis. People with weakened immune systems, including people with HIV/AIDS, are at increased risk. Take appropriate precautions.

SOURCE   CDC, cdc.gov/meningococcal

Middle East Respiratory Syndrome (MERS)

ANNUAL US DEATHS   0 reported

Middle East respiratory syndrome (MERS) is a coronavirus infection first reported in the Arabian Peninsula in 2012. Since then, 27 countries have reported cases. As of April 1, 2021, a total of 2,574 laboratory-confirmed cases of MERS have been reported, with 885 associated deaths. The mortality rate for MERS is 34.4 percent. Almost all cases and deaths were in Saudi Arabia. Only 2 cases and 0 deaths have been reported in the US.

STAY SAFE   Avoid close contact with camels on the Arabian Peninsula.

SOURCE   WHO, who.int/health-topics/middle-east-respiratory-syndrome

Mumps

ANNUAL US DEATHS   None reported since childhood vaccination became universal

Mumps is a highly contagious viral illness caused by a member of the Rubulavirus family. Symptoms include pain, tenderness, and swelling in 1 or both parotid salivary glands. Complications can include testicular atrophy in males and deafness. Before 1967, when childhood mumps vaccination began, about 186,000 cases were reported each year in addition to many unreported cases. Death was rare. The incidence of mumps decreased 99 percent after routine childhood vaccination became universal in the US. Only a few hundred cases a year were reported by 2002. Annual cases have increased since then, ranging from about 200 to as many as several thousand in peak years. The majority of cases and outbreaks occur among people who are fully vaccinated and spend most of their time in close-contact or congregate settings. Death from mumps is exceedingly rare. There have been 0 mumps-related deaths reported in the United States during recent mumps outbreaks.

STAY SAFE   Vaccinate children with the MMR vaccine on schedule.

SOURCE   CDC, cdc.gov/mumps

Peritonitis

ANNUAL US DEATHS   No data available; fewer than 4 percent of treated peritonitis episodes result in death.

Peritonitis is an inflammation of the peritoneum, the tissue that lines the inner wall of the abdomen and covers and supports the abdominal organs. The disease is usually caused by a rupture within the abdominal wall resulting from a medical procedure, such as peritoneal dialysis, or by a ruptured appendix, perforated colon, pancreatitis, or diverticulitis—all conditions that allow bacteria or fungi to enter the peritoneum. Peritonitis is a medical emergency. If left untreated, the infection can spread to other organs and cause sepsis and death. The death rate for peritonitis is high because many patients also have other health problems. The death rate from peritonitis in those who also have cirrhosis can be as high as 40 percent. Peritonitis is the single most common cause of death from infectious disease in people using peritoneal dialysis, with rates estimated at 6 to 33 percent.

STAY SAFE   Use proper sterile techniques for peritoneal dialysis.

SOURCES   NIH NIDDKD, niddk.nih.gov/health-information/kidney-disease/kidney-failure/peritoneal-dialysis; N. Boudville et al. Recent peritonitis associates with mortality among patients treated with peritoneal dialysis. *Journal of the American Society for Nephrology*, August 2012.

Pertussis (Whooping Cough)

> **ANNUAL US DEATHS** For infants younger than 1 year, death rate is 1 to 2 percent.

Pertussis, also known as whooping cough, is a highly contagious respiratory disease caused by the bacterium *Bordetella pertussis*. Pertussis causes uncontrollable, violent coughing that makes it hard to breathe. The coughing makes people need to take deep breaths with a "whooping" sound. Pertussis can be very serious and even lethal for babies under age 1.

Before pertussis vaccination began in the 1940s in the US, cases frequently exceeded 100,000 per year. By 1965, cases had declined to fewer than 10,000 per year. Since the 1980s there has been a gradual increase in cases. In 2018, 15,609 cases were reported nationwide, down from a peak of 48,277 in 2012. The infection rate for children under 1 year is 52.8 per 100,000 population. Approximately 50 percent of babies under age 1 who get pertussis need hospitalization. Of those, about 1 or 2 out of 100 will die.

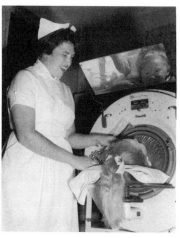

A nurse shares a smile with a child inside of an iron lung. Before a vaccine became available, polio killed or disabled millions, mostly children. When muscles involved in breathing were destroyed by the disease, the primitive "iron lung" respirator maintained life.

STAY SAFE Vaccinate children on schedule and get all booster shots. Pregnant women should get a booster shot. All other adults should discuss vaccination and booster shots with their doctor.

SOURCE CDC, cdc.gov/pertussis

Polio

> **ANNUAL US DEATHS** No cases reported since 1979

Massively successful vaccination programs in the US since the first polio vaccine was developed in 1955 mean that the US has been polio-free since 1979. Because poliovirus is still a threat in some countries, all children should continue to be vaccinated. Worldwide, about 400 to 500 cases of polio are reported every year.

STAY SAFE   Vaccinate children against polio on schedule. Get a polio booster shot if you plan on travel to a country where polio still occurs, including Afghanistan, Pakistan, Malaysia, the Philippines, Yemen, and most countries in Africa.

SOURCES   CDC, cdc.gov/polio/what-is-polio; Global Polio Eradication Initiative, polioeradication.org/polio-today; WHO, who.int/health-topics/poliomyelitis1

Q Fever

ANNUAL US DEATHS   None reported

The *Q* in Q fever stands for "query," because when this disease was first recognized in Australia in 1935 and in the United States in the early 1940s, the cause was unknown. We now know that Q fever is caused by infection with the bacterium *Coxiella burnetii*. Because cattle, sheep, and goats carry the bacteria, farmers, veterinarians, and others exposed to dust contaminated with infected animal body fluids are at risk.

The CDC started tracking Q fever as a notifiable disease in 1999. In 2017, 153 acute Q fever cases were reported, as well as 40 chronic Q fever cases. Acute Q fever is probably underreported, because only about 50 percent of all people infected with the bacteria show signs of illness and most people recover fully without any treatment. About 2 to 5 percent of adults develop severe illness and need to be hospitalized.

STAY SAFE   Avoid contact with animal fluids, particularly birthing fluids.

SOURCES   CDC, cdc.gov/qfever; NORD, rarediseases.org/rare-diseases/q-fever

Respiratory Syncytial Virus (RSV)

ANNUAL US DEATHS   Estimated 500

A common respiratory virus that usually causes only mild, coldlike symptoms, respiratory syncytial virus (RSV) can be serious in infants and older adults. In the US, RSV is the most common cause of bronchiolitis pneumonia in children under age 1. An estimated 58,000 children younger than 5 are hospitalized for RSV every year. About 500 people, mostly infants and young children, die every year. The mortality rate in children is 0.5 to 1.7 percent. In adults, RSV is more dangerous, with a mortality rate ranging from 11 to 78 percent (immunocompromised patients are more likely to die). In nursing homes, an estimated 5 to 27 percent of respiratory tract infections are caused by RSV. Of those patients, 10 percent will develop pneumonia and 1 to 5 percent will die.

STAY SAFE   Wash your hands. Cover coughs and sneezes. Seek medical attention for severe respiratory infection symptoms such as wheezing, shortness of breath, and high fever in babies, young children, and the elderly.

SOURCE   CDC, cdc.gov/rsv

Severe Acute Respiratory Syndrome (SARS)

ANNUAL US DEATHS   None reported

Severe acute respiratory syndrome (SARS) is a coronavirus infection first reported in Asia in February 2003. The illness spread to more than 2 dozen countries in North America, South America, Europe, and Asia before it was contained in July 2003. The global outbreak caused 8,098 cases and 774 deaths, for a mortality rate of 15 percent. There were no deaths in the US. Currently, there is no known SARS transmission anywhere in the world.

STAY SAFE   Wash your hands.

SOURCE   WHO, who.int/health-topics/severe-acute-respiratory-syndrome

Septicemia

ANNUAL US DEATHS   270,000

Septicemia, also known as blood poisoning, occurs when a bacterial infection spreads into the bloodstream, causing a condition called sepsis. In sepsis, the whole body becomes dangerously inflamed, causing blood clots, leaking blood vessels, and organ damage. Sepsis can lead to septic shock, a life-threatening condition that makes the blood pressure drop and organs shut down.

Sepsis is a medical emergency. Every year, approximately 1.7 million Americans develop sepsis, and nearly 270,000 die of it. One in 3 patients who dies in a hospital has sepsis. People with chronic health conditions such as diabetes, lung disease, kidney disease, and cancer are at higher risk of sepsis; so is anyone over age 65.

STAY SAFE   Wash your hands. Know the signs and symptoms of sepsis and act fast if you or a loved one has a bad infection that's not improving or is getting worse.

SOURCE   CDC NCHS, cdc.gov/nchs/pressroom/sosmap/septicemia_mortality/septicemia.htm

Shigellosis

> **ANNUAL US DEATHS**   40

Shigella bacteria cause an infection called shigellosis. Symptoms of *Shigella* infection include diarrhea (sometimes bloody), fever, and stomach cramps. *Shigella* causes approximately 450,000 illnesses, 6,000 hospitalizations, and 40 deaths in the US annually. Shigellosis is highly communicable. It's transmitted through fecal matter and food and water contaminated with fecal matter.

STAY SAFE   Wash your hands, especially after changing a diaper or helping to clean someone who used the toilet. Avoid swallowing water from ponds, lakes, and untreated swimming pools. Avoid sexual activity with those who have diarrhea or who have recently recovered from shigellosis.

SOURCE   CDC, cdc.gov/shigella

Strep and Staph Infections

> **ANNUAL US DEATHS**   1,200 to 1,900 from invasive group A strep infections

Streptococcus and *Staphylococcus* bacteria are the most common causes of deadly infections. Most group A streptococcal (GAS) infections cause relatively mild illnesses such as strep throat, pharyngitis, and impetigo. Sometimes, however, strep diseases such as scarlet fever, rheumatic fever, cellulitis, necrotizing fasciitis, and streptococcal toxic shock syndrome become invasive and life-threatening. Approximately 11,000 to 13,000 cases of invasive GAS infections occur each year in the US and kill between 1,200 and 1,900 people.

Staphylococcus aureus (staph) bacteria are naturally found on the skin. When this bacteria enters the body through a wound in the skin, it can cause serious infections. If staph gets into the blood, it can lead to sepsis and death. In 2017, nearly 119,247 *S. aureus* bloodstream infections occurred in the US and caused nearly 19,382 associated deaths. Infections from methicillin-resistant staph (MRSA) are very difficult to treat and can be deadly. Many MRSA infections are acquired in hospital settings, but better adherence to prevention measures has led to a steady, if slow, decline. At the same time, community-acquired MRSA infections are slowly rising. About 10 percent of serious staph infections occur in people who inject drugs such as opioids.

STAY SAFE   Wash your hands. Cover coughs and sneezes. Keep cuts, scrapes, and other skin wounds clean to prevent infections. Don't share personal items such as towels and razors.

SOURCE   CDC, cdc.gov/mrsa

Syphilis

ANNUAL US DEATHS   No recent adult deaths reported; globally, infection transmitted by pregnant women to their babies remains a significant danger.

Syphilis is a sexually transmitted infectious (STI) disease caused by the bacterium *Treponema pallidum*. In 2019, 129,813 cases of all stages of syphilis were reported, including 38,992 cases of primary and secondary (P&S) syphilis, the most infectious stages of the disease. Since reaching a historic low in 2000 and 2001, the rate of P&S syphilis has increased almost every year, increasing 11.2 percent from 2018 to 2019.

Men account for the most cases of syphilis, with the vast majority of those cases occurring among gay, bisexual, and other men who have sex with men (MSM). In 2019, 32,685 men were diagnosed with syphilis, compared to 6,545 women.

Syphilis is easily treated with antibiotics in the primary and secondary stages; no deaths among adults have been reported in recent years. Pregnant women with syphilis pass the disease to the fetus. This can cause stillbirth. If the baby is born alive, it has congenital syphilis. From 2018 to 2019, the number of syphilitic stillbirths increased from 79 to 94. In that same time frame, infant deaths from congenital syphilis increased from 15 to 34 deaths. In 2020, 2,022 cases of congenital syphilis were reported, including 139 infant deaths, for a rate of 49.5 cases per 100,000 live births.

Globally each year, an estimated 6 million new cases of syphilis are reported. Over 300,000 fetal and neonatal deaths are attributed to syphilis, with 215,000 additional infants placed at increased risk of early death.

STAY SAFE   Practice safe sex and always use a condom. All pregnant women should be screened for syphilis.

SOURCES   CDC, cdc.gov/std/syphilis; WHO, who.int/health-topics/sexually-transmitted-infections

Toxic Shock Syndrome

ANNUAL US DEATHS   Average of 3 cases per 100,000 population, with mortality between 30 and 70 percent

Toxic shock syndrome (TSS) is caused by severe infection with group A *Streptococcus* bacteria, which causes sudden onset of shock, organ failure, and frequently death.

TSS is a medical emergency. Prompt, aggressive treatment with antibiotics and surgery, including amputations if necessary, is needed. Even with aggressive treatment, the mortality rate for TSS ranges from 30 to 70 percent; death may occur within 2 days of onset. Fortunately, TSS is rare. On average, it occurs in about 3 people per 100,000 population each year, but among menstruating women, the incidence can be as high as 17 per 100,000 women.

An outbreak of TSS in women in the early 1980s was tied to the use of superabsorbent tampons. These products have been removed from the market, but TSS continues to be a concern for women who use tampons, menstrual sponges and cups, diaphragms, and cervical caps, and for women who have recently given birth. More than one-third of all cases of toxic shock syndrome involve women under 19, but people over age 65 are most at risk. Jim Henson, creator of the Muppets, died of TSS in 1990.

STAY SAFE   Wash your hands. Treat skin infections and wounds promptly. Don't use high-absorbency tampons and change tampons and other menstrual products frequently. Clean diaphragms and cervical caps thoroughly.

SOURCE   CDC, cdc.gov/groupastrep/diseases-public/streptococcal-toxic-shock-syndrome.html

Tuberculosis

> **ANNUAL US DEATHS**   542 in 2018

Tuberculosis (TB) is a contagious, infectious disease caused by the *Mycobacterium tuberculosis* bacterium. Historically known as consumption, TB killed millions each year around the world until the introduction of antibiotic therapy with isoniazid in 1952.

In the US, 8,916 TB cases were reported in 2019, for a rate of 2.7 per 100,000 persons—the lowest incidence rate since individual TB case reporting began in 1953. In 2018, 542 deaths were attributed to TB. For comparison, in 1953, 85,304 Americans had TB, for an incidence rate of 52.6 per 100,000 population. Today, some people have a type of TB that is drug-resistant to 1 or more antibiotics (MDR TB). In 2017, 847 out of 9,082 cases were drug-resistant; there were 62 deaths.

In 2019, 51 percent of TB cases were reported from just 4 states: California (23.7 percent), Texas (13.0 percent), New York (8.5 percent), and Florida (6.3 percent). A total of 71.4 percent of reported TB cases in the US occurred among people born outside the country. In 2019, Asian persons continued to represent the largest proportion of persons with TB (35 percent), followed by Hispanic persons (30.2 percent), Black persons (19.7 percent), and White persons (11.4 percent).

Worldwide in 2020, an estimated 10 million people had TB, for an estimated 127 cases per 100,000 population. In 2020, approximately 1.3 million HIV-negative people died of it; approximately 214,000 HIV-positive people died. TB is still one of the top global causes of death and is the leading cause of death from infectious disease.

STAY SAFE   In the US, most people with tuberculosis have compromised immunity or were close contacts of someone with the disease. In some countries, TB is much more common than in the US. Travelers should avoid close contact or prolonged time with known TB patients in crowded, enclosed environments such as clinics, hospitals, prisons, or homeless shelters. If you think you have been exposed to TB, see your doctor or your local health department to get tested. TB is easily cured if treatment starts soon after exposure.

SOURCES   CDC, cdc.gov/tb; WHO, Global Tuberculosis Report 2021, who.int/teams/global-tuberculosis-programme

Fungal Infections

ANNUAL US DEATHS   See individual diseases; data is provided as available.

Fungi are a group of living organisms that are separate from animals, plants, and bacteria. Fungal cells are more complex than bacteria. Like bacteria, fungi live in air, in soil, in water—and in and on us. Like bacteria, most fungi are harmless. Some reproduce through tiny spores, which travel through the air. Spores can be inhaled or land on the skin, which is why many fungal infections start in the lungs or on the skin. Fungal infections can be difficult to eliminate, but most, such as ringworm, athlete's foot, and yeast infections, aren't serious. However, some fungal infections, such as pneumocystis pneumonia, can cause serious illness and death. Dangerous fungal infections are more likely to happen in people with weakened immune systems, including people living with HIV/AIDS, organ transplant recipients, cancer patients, hospitalized patients, and people undergoing stem cell transplants. Some serious fungal infections have become resistant to the drugs used to treat them.

STAY SAFE   Practice good hygiene, especially if you are immunocompromised.

SOURCE   CDC, cdc.gov/fungal/index.html

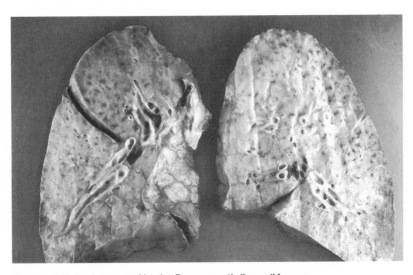

Pneumonia infection caused by the *Pneumocystis jirovecii* fungus

Aspergillosis

ANNUAL US DEATHS   No reliable data available

Aspergillosis is a disease caused by *Aspergillus*, a common mold (a type of fungus). There are a number of types of aspergillosis, all affecting the lungs. Most cause mild to moderate illness, but invasive aspergillosis can cause serious illness and death. Invasive aspergillosis is associated with a significant mortality rate of 30 to 95 percent. The 1-year survival rate for organ transplant recipients with aspergillosis is 59 percent; among stem cell transplant recipients with aspergillosis, the survival rate is 25 percent.

STAY SAFE   Practice good hygiene, especially if you are immuno-compromised.

SOURCE   CDC, cdc.gov/fungal/diseases/aspergillosis

Cryptococcosis

ANNUAL US DEATHS   330 in 2009

Cryptococcosis is caused by the fungi *Cryptococcus neoformans*. The infection starts in the lungs. If it stays there, symptoms are minimal or may even go unnoticed. If the infection spreads, it usually goes to the brain and can cause meningitis and death. *C. neoformans* is a major cause of illness in people living with HIV/AIDS; an estimated 220,000 cases of cryptococcal meningitis occur worldwide each year. Globally, cryptococcal meningitis is responsible for 15 percent of AIDS-related deaths. In the US, antiretroviral therapy has sharply reduced deaths from cryptococcal meningitis. In 2009, there were 330 deaths from cryptococcal meningitis.

STAY SAFE   Practice good hygiene, especially if you are immuno-compromised.

SOURCE   CDC, cdc.gov/fungal/diseases/cryptococcosis-neoformans

Histoplasmosis

ANNUAL US DEATHS   No data available; mortality rate for those hospitalized is 8 percent.

Histoplasmosis is an infection caused by a fungus called *Histoplasma*. The fungus is found in soil that contains large amounts of bird or bat droppings, mostly in the areas around the Ohio and Mississippi River valleys. Histoplasma spores are breathed in and can attack the lungs. The infection can be deadly if it spreads from the lungs to other organs. Among adults aged 65 and older, the incidence of histoplasmosis is 3.4 cases per 100,000 population. In the Midwest, where the fungus is more common, rates are estimated at 6.1 cases per 100,000 population. Among adult patients hospitalized for histoplasmosis, the mortality rate is 8 percent.

STAY SAFE   Practice good hygiene, especially if you are older or immunocompromised and live in areas where *Histoplasma* is common.

SOURCE   CDC, cdc.gov/fungal/diseases/histoplasmosis

Mucormycosis

ANNUAL US DEATHS   300+ deaths reported in a June 2021 outbreak in India

Mucormycosis is a serious fungal infection caused by a group of molds called mucormycetes. Mucormycosis is rare but deadly. Among stem cell transplant recipients, mucormycosis is the 3rd most invasive fungal infection and accounted for 8 percent of all invasive fungal infections. The overall cause case mortality rate for mucormycosis is 54 percent, rising to 76 percent for pulmonary infections and nearly 100 percent for disseminated (spread through the body by the bloodstream) murcomycosis. Because murcomycosis spreads through the respiratory tract and erodes facial structures, treatment sometimes involves surgically removing an eye to keep the infection from reaching the brain. In India, overuse of steroids and poor sanitation of respiratory equipment during the COVID-19 pandemic caused outbreaks of murcomycosis, or "black mold." In June 2021, in India, at least 11,700 cases were reported, with over 300 deaths.

STAY SAFE   Avoid breathing or contacting black mold and mitigate any black mold in homes or other buildings.

SOURCE   CDC, cdc.gov/fungal/diseases/mucormycosis

Pneumocystis pneumonia (PCP)

ANNUAL US DEATHS   No data available; last major outbreaks were associated with the AIDS epidemic in the early 1980s

Pneumocystis pneumonia (PCP) is a serious infection caused by the fungus *Pneumocystis jirovecii*. Clusters of PCP cases in the early 1980s were an early indication of the AIDS epidemic. Today, cases of PCP occur less often, but this disease is still very deadly. Overall, about 25 percent of patients die. Among those who need intensive care, the mortality rate is up to 58 percent.

STAY SAFE   Practice good hygiene.

SOURCE   CDC, cdc.gov/fungal/diseases/pneumocystis-pneumonia

Valley Fever

ANNUAL US DEATHS   About 200

Valley fever, also called coccidioidomycosis, is an infection caused by the fungus *Coccidioides*. The fungus is found in soil in the southwestern US. In 2018, 15,611 cases of Valley fever were reported to the CDC, mostly in Arizona or California. This number is an undercount, because thousands more cases probably occur but cause only mild or no symptoms. In Arizona, Valley fever causes an estimated 15 to nearly 30 percent of community-acquired pneumonia. On average, there are approximately 200 coccidioidomycosis-associated deaths each year.

STAY SAFE   If you are immunocompromised, be very cautious about activities that involve dusty areas, disturbed soil, or contact with birds and animals, especially in the American Southwest. Be aware that fungal diseases can resemble bacterial diseases. If you are being treated for a bacterial infection and it's not getting better, discuss the possibility of a fungal infection with your doctor.

SOURCE   CDC, cdc.gov/fungal/diseases/coccidioidomycosis

Insect-Borne Diseases

ANNUAL US DEATHS   US-specific statistics unavailable; 700,000+ deaths worldwide

Diseases carried by insects are a major cause of illness and death worldwide. Worldwide, vector-borne (carried by insects or arthropods)

diseases account for more than 17 percent of all infectious diseases. Many of these vectors are bloodsucking insects, which ingest disease-producing microorganisms during a blood meal from an infected host (human or animal) and later transmit it into a new host, after the pathogen has replicated. The burden of these diseases is highest in tropical and subtropical areas, and they disproportionately affect the world's poorest people, particularly children. More than 700,000 people worldwide die every year from infections caused by parasites, bacteria, or viruses transmitted by the bite of an insect or tick. (Tick-borne diseases are discussed in a separate section below.) Vector control is the primary means of prevention for insect-borne diseases. None have effective vaccines.

The incidence of mosquito-borne arboviral diseases such as eastern equine encephalitis has been increasing. In 2019, more cases of these diseases were reported to the CDC than in any previous year.

STAY SAFE   Protect yourself from mosquito bites and other insect bites. Avoid outdoor activities at dusk and dawn, the peak feeding time for mosquitoes, particularly in warmer months. Use insect repellents containing DEET according to label directions. Repellents containing up to 30 percent DEET are safe and effective for adults and children over 2 months of age. Alternative effective repellents include picaridin, IR3535, and oil of lemon eucalyptus. Pre-treat clothing, footwear, and outdoor gear with permethrin-based products (do not use permethrin on the skin). Wear loose-fitting, long-sleeved shirts and pants. Maintain screens on windows and doors. If you become ill after traveling outside of the US mainland, see your doctor and be sure to mention where you have been.

SOURCES   CDC NIOSH, cdc.gov/niosh/topics/outdoor/mosquito-borne; A. Freitas et al. Excess Mortality and Causes Associated with Chikungunya, Puerto Rico, 2014–2015. *Emerging Infectious Diseases*, December 2018; G.M. Vahey et al. West Nile Virus and other domestic nationally notifiable arboviral diseases - United States, 2019. *MMWR Morbidity and Mortality Weekly Report*, August 2021; WHO, who.int/news-room/fact-sheets/detail/vector-borne-diseases

Bubonic Plague

ANNUAL US DEATHS   12 US deaths since 2000

Bubonic plague, or just "plague," is caused by the bacterium *Yersinia pestis*. The disease is easily spread when humans are bitten by infected rodent fleas. Antibiotics are effective in treating plague, but without

Public health physicians examine rats for bubonic plague during a 1914 breakout in New Orleans.

prompt treatment, the disease can rapidly cause serious illness or death. If left untreated, the mortality rate is 30 to 100 percent.

The world has seen 3 massive epidemics of bubonic plague. The 1st was the Plague of Justinian, which ran from 541 to 549. It began in Constantinople and spread to the entire Mediterranean basin, Europe, Arabia, and Asia. The estimated death toll is between 30 and 50 million people, or about half the world population (excluding North and South America). The 2nd epidemic was the Black Death, which struck Europe and Asia in 1346 and lasted until 1353. The epidemic is the most fatal in world history, killing an estimated 25 million people in Europe, almost a third of the continent's population. In all, the Black Death probably killed 75 to 200 million people in Eurasia and North Africa and equal numbers in Asia. Outbreaks of bubonic plague continued for centuries—in the Great Plague of London in 1665 to 1666, 70,000 residents died. The 3rd global outbreak began in China in 1860. It spread through China and India and killed at least 10 million people and possibly more than 15 million. The pandemic was considered active until 1960 when deaths dropped to about 200 a year. Since then, deaths from plague have never again reached epidemic levels, but the disease is still with us. Since the 1990s, the 3 most endemic countries for plague are the Democratic Republic of the Congo, Madagascar, and Peru. From 2010 to 2015, 3,248 cases were reported worldwide, including 584 deaths.

The first plague epidemic in the US was in San Francisco from 1900 to 1908, when ships arriving from Asia brought infected rats. At least 280 people were infected and at least 172 died. An outbreak in New Orleans from 1919 to 1921 infected 25 and killed 11. Today, plague infections sometimes occur in rural areas in the western US, where rodents such as ground squirrels, prairie dogs, chipmunks, mice, voles, and rabbits are the reservoir. On average since 2000, 7 human plague cases are reported in the US each year. In 2006, cases reached a high of 17. There have been 12 deaths since 2000.

STAY SAFE   Infected fleas carried by rats and other rodents are the main vector for bubonic plague. Reduce rodent habitat around your home, workplace, and recreational areas. Use insect repellent to avoid rodent flea bites outdoors. Use flea control products on pets.

SOURCES   CDC, cdc.gov/plague; WHO. who.int/en/news-room/fact-sheets/detail/plague

Chagas Disease

ANNUAL US DEATHS   US data unavailable; 10,000 to 12,000 worldwide

Chagas disease is caused by the parasite *Trypanosoma cruzi*, which is transmitted to animals and people by contact with the droppings of an infected triatomine bug, also called the kissing bug because it often bites on the lips. Chagas disease is common in parts of Mexico, Central America, and South America, where triatomine bugs infest substandard housing. An estimated 8 million people globally are infected, and an estimated 300,000+ people with *T. cruzi* infection live in the US. Almost all these people contracted the disease elsewhere.

Chagas disease is named after the Brazilian physician Carlos Chagas, who identified the disease in 1909. Also called American trypanosomiasis, Chagas disease causes the highest burden of any parasitic disease in the western hemisphere. Untreated Chagas disease infection is lifelong and can lead to death years later from cardiomyopathy, heart arrhythmias, or gastrointestinal complications. The lifetime risk for fatal symptoms ranges from 10 to 30 percent. An estimated 10,000 to 12,000 people worldwide die each year from Chagas disease. The number of deaths annually in the US is unknown.

Because Chagas disease is largely a disease of poverty, it has been neglected. In 2019, WHO designated April 14 as World Chagas Disease Day.

STAY SAFE   Practice good hygiene and take care when traveling to areas where the disease is prevalent.

SOURCE   CDC, cdc.gov/parasites/chagas

Chikungunya

ANNUAL US DEATHS 31 reported in 2014–2015 Puerto Rican outbreak

Chikungunya is a viral disease transmitted to humans through the bites of *Aedes* mosquitoes. It was first described during an outbreak in southern Tanzania in 1952 and has now been identified in nearly 40 countries in Asia, Africa, and Europe. In 2013, local transmission of chikungunya was identified in Caribbean countries and spread throughout the Americas from there. Symptoms of chikungunya include fever and joint pain. The name comes from the African Makonde language and means "bent over in pain."

In 2014, an outbreak was identified in Puerto Rico and parts of the mainland US. A total of 2,811 chikungunya virus disease cases were reported in the US in 2014, almost all in travelers returning from affected areas. A total of 4,710 cases were reported from Puerto Rico, the US Virgin Islands, and American Samoa. Almost all were locally transmitted. The outbreak died away over the following years, and in 2020 only 29 chikungunya cases were reported, all but 1 in travelers returning from an affected area. No cases were reported in US territories.

There is no vaccine or antiviral drug treatment for chikungunya. Death from chikungunya is rare and usually occurs in older people with preexisting conditions. In 2014 to 2015 at the height of the outbreak, chikungunya caused 31 deaths in Puerto Rico, but as many as 1,310 deaths were possibly attributable to the disease in the territory.

STAY SAFE Practice good hygiene and take care when traveling to areas where the disease is prevalent.

SOURCE CDC, cdc.gov/chikungunya

Dengue Fever

ANNUAL US DEATHS 1 reported in 2020

Dengue, a dangerous mosquito-borne viral disease transmitted predominantly by *Aedes aegypti* mosquitoes, is found throughout the tropics in more than 100 countries. Dengue is the most common vector-borne viral disease in the world. It causes an estimated 50 to 100 million infections globally each year and 25,000 deaths. The incidence of dengue fever has increased rapidly since 2000. In 2019 to 2020, an epidemic of dengue fever hit countries in southeast Asia, including the Philippines, Malaysia, Vietnam, and Bangladesh. The epidemic caused an estimated 6.2 million cases and 3,930 deaths.

Global climate change causing higher temperatures and increased rainfall, along with unplanned rapid urbanization, favor increases in the *A. aegypti* mosquito population and widespread transmission of the virus. Today, 40 percent of the world's population, or about 3 billion people, lives in areas with a risk of dengue.

Dengue virus disease became a nationally notifiable condition in the US in 2010. In 2020, 332 dengue cases were reported in US states and 760 cases were reported in US territories, almost all in Puerto Rico. Only 1 death was reported.

There are no effective antiviral drugs for dengue. Vaccines for dengue fever have been developed but their effectiveness is controversial, and the cost is high.

STAY SAFE Take measures to protect yourself against mosquito infestation and mosquito bites.

SOURCE CDC, cdc.gov/dengue

Eastern Equine Encephalitis

ANNUAL US DEATHS 48 from 2010 to 2019

Eastern equine encephalitis virus (EEEV) is transmitted to humans through the bite of an infected mosquito. Human infection with EEEV is rare but can cause a deadly form of encephalitis. Almost all (96 percent) of people infected with EEEV have no symptoms. Of those who do have symptoms, however, about 33 percent die and most of the rest have ongoing and often severe neurological damage. Only between 5 and 10 cases are reported in the US each year, most in the northeastern and Gulf Coast states. From 2010 to 2019, a total of 107 cases were reported, with 48 deaths. There is no specific treatment for EEEV and there is no vaccine.

STAY SAFE Take measures to protect yourself against mosquito infestation and mosquito bites.

SOURCE CDC, cdc.gov/easternequineencephalitis

Malaria

ANNUAL US DEATHS 5

Malaria is a parasitic infection transmitted by *Anopheline* mosquitoes. Globally in 2020, an estimated 240 million cases of malaria occurred, causing more than 627,000 deaths; 96 percent were people living in sub-Saharan Africa. Most of the deaths were in children under age 5.

In the US, malaria has been almost entirely eliminated. About 2,000 cases of malaria occur each year, leading to 5 deaths a year on average. Almost all cases and deaths are in people returning from countries where malaria occurs, mostly sub-Saharan Africa and South Asia.

Malaria was endemic in much of the US in the nineteenth century, but by the early 1900s, malaria elimination programs had reduced or eliminated it in many parts of the country. By the 1930s, malaria was found mostly in the southeastern states. Malaria eradication in the region began in 1942 under the Office of Malaria Control in War Areas (the predecessor to the CDC) to limit the impact of malaria around military training bases. The National Malaria Eradication Program began in 1947 under the auspices of the CDC and state and local health agencies. Using methods that today would be frowned on (DDT spraying, for example), the program was highly successful and by 1951, malaria was eliminated in the US.

STAY SAFE   Take measures to protect yourself against mosquito infestation and mosquito bites, especially when traveling in sub-Saharan Africa and South Asia.

SOURCES   CDC, cdc.gov/malaria/about/index.html; WHO, who.int, *World Malaria Report 2021*

St. Louis Encephalitis

ANNUAL US DEATHS   6 from 2010 to 2019

St. Louis encephalitis (SLE) is a viral disease spread by the bite of an infected mosquito. Most people infected with SLE virus have no symptoms, but a small number become seriously ill with encephalitis or meningitis. About 5 to 20 percent of these people die. From 2010 to 2019, 97 cases of SLE were diagnosed in the US, with 6 deaths. Most cases of SLE occur in the eastern and central states.

The disease is named for St. Louis because in 1933, an outbreak of encephalitis in the area caused over 1,000 cases, with a mortality rate of about 14 percent. Researchers at the newly formed National Institutes of Health identified the previously unknown virus.

STAY SAFE   Take measures to protect yourself against mosquito infestation and mosquito bites.

SOURCE   CDC, cdc.gov/sle

West Nile Virus

ANNUAL US DEATHS   60 in 2019

West Nile virus (WNV) is the leading cause of mosquito-borne disease in the continental United States. Most people infected with WNV have no symptoms. About 1 in 5 infected people develop a fever and other symptoms. About 1 out of 150 infected people develop encephalitis, meningitis, or some other neuroinvasive disease. The CDC began collecting information on WNV in 1999. Between 1999 and 2019, a total of 51,801 cases were reported. Of these, 2,390, or 5 percent, were fatal. The worst year for total deaths from WNV was 2003, with 9,862 cases reported and 264 fatalities (3 percent of total cases). In 2019, 971 cases of WNV were reported to the CDC. Of these, 60, or 6 percent, were fatal. Reported cases of WNV have been declining. In 2019, the incidence was 53 percent lower than the median annual incidence from 2009 to 2018.

The most dangerous state for WNV is California, with 7,026 total cases reported between 1999 and 2019. The safest state is Hawaii, with only 1 case reported between 1999 and 2019. The safest mainland state is Maine, with only 4 cases reported between 1999 and 2019.

SOURCE   CDC, cdc.gov/westnile

Western Equine Encephalitis

ANNUAL US DEATHS   About 28

Western equine encephalitis (WEE) is a viral illness transmitted to humans by the bite of an infected mosquito. The virus is closely related to eastern equine encephalitis. Most cases of western equine encephalitis occur west of the Mississippi River, west of the Rocky Mountains,

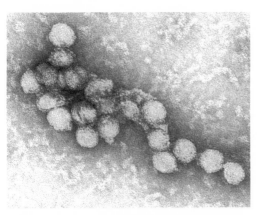

Digitally enhanced electron photomicrograph of the West Nile virus

and in California. The annual incidence of reported infections is highly variable. In 1941, a significant outbreak caused over 3,000 confirmed cases with an estimated 300 deaths. Since 1964, however, there have been fewer than 700 confirmed cases in the US. Most people infected with the virus have no or very mild symptoms. A small number will develop encephalitis. About 5 to 15 percent of these cases are fatal. Overall, mortality from WEE is around 4 percent.

STAY SAFE   Take measures to protect yourself against mosquito infestation and mosquito bites, especially in the western US.

SOURCES   CDC, cdc.gov/nchs/ndi/index.htm; Minnesota Department of Health, health.state.mn.us/diseases/weencephalitis

Zika Virus Disease

ANNUAL US DEATHS   No data available

Zika virus (ZIKV) disease is transmitted to humans by the bite of an infected *Aedes* mosquito. Most people infected with Zika virus have no or mild symptoms, including fever, rash, and conjunctivitis. Death from Zika virus is very rare. However, Zika can be passed from a pregnant woman to her fetus and can cause serious birth defects, including microcephaly (very small head) and eye abnormalities. The rate of birth defects among infected women is 5 to 10 percent and is 4 times higher than normal in areas with widespread local transmission of Zika virus.

A large outbreak of Zika virus in the Americas began in 2015. It caused an increase in travel-related cases in the US, widespread cases of local mosquito-borne transmission in Puerto Rico and the US Virgin Islands, and some cases of local transmission in Florida and Texas. In 2016, at the peak of the outbreak, 5,168 cases were reported in the US, with 4,897 cases in travelers returning from affected areas, 218 cases acquired through local transmission in Florida and 6 in Texas, and 47 cases acquired through other routes, including sexual transmission (the only vector-borne virus known to be transmitted this way). In the US territories, 36,512 cases were reported, with 36,367 cases acquired through local transmission. The outbreak began to decline in 2017. In 2020, only 4 Zika virus cases in travelers returning from affected areas were reported in the US; 57 cases of local transmission were reported in Puerto Rico and the US Virgin Islands.

STAY SAFE   Take measures to protect yourself against mosquito infestation and mosquito bites, especially during outbreaks.

SOURCE   CDC, cdc.gov/zika

Tick-Borne Diseases

> **ANNUAL US DEATHS**   See individual entries below.

Ticks are a major vector for transmitting disease, and they are a growing public health concern. As temperatures rise because of global climate change, tick habitat expands, their active period gets longer, and more people get sick. The number of annual reports of tick-borne bacterial and protozoan diseases to the CDC more than doubled between 2004 and 2016, from approximately 22,000 cases to over 48,000. These numbers are probably significant undercounts, because most cases of tick-borne disease are undiagnosed or unreported. Most tick-borne diseases are easily treated with antibiotics and deaths are rare.

STAY SAFE   Prevention is key to avoiding tick-borne diseases. Ticks are most active during warm weather but remain active as long as the temperature is above freezing. Ticks are most numerous in brushy, wooded, or grassy habitats. Stay on cleared trails when walking or hiking. Use a tick repellent with DEET, picaridin, or oil of lemon eucalyptus and apply permethrin to clothing, shoes, and camping gear (not skin). Wear light-colored clothing so ticks are visible. After spending time outdoors, check yourself, children, and pets carefully for ticks. Be aware of the symptoms of tick-borne illness, especially the bull's-eye rash of Lyme disease. If you have been someplace likely to have ticks and you develop symptoms of a tick-borne disease, seek medical attention right away.

SOURCES   CDC, cdc.gov/ticks; CDC NIOSH, cdc.gov/niosh/topics/tick-borne; E.J. Pace and M. O'Reilly. Tickborne diseases: diagnosis and management. *American Family Physician*, May 2020; R. Rosenberg et al. Vital signs: trends in reported vectorborne disease cases, United States and territories, 2004–2016. *Morbidity and Mortality Weekly Report*, May 4, 2018.

Anaplasmosis

> **ANNUAL US DEATHS**   Variable with a high of about 57

Anaplasmosis is a tick-borne disease caused by the bacterium *Anaplasma phagocytophilum*. The bacterium is transmitted to people by tick bites primarily from the blacklegged tick *(Ixodes scapularis)*.

Anaplasmosis symptoms may include fever, headache, chill, and muscle pain. Prompt treatment with antibiotics cures most cases.

Anaplasmosis was first recognized as a human disease in the mid-1990s in the US. The number of cases reported to the CDC varies from year to year but has been increasing steadily since 1999. In 2000, 348 cases were reported; in 2017, a peak of 5,762 cases were reported. The case fatality rate is low, less than 1 percent.

Anaplasmosis is most frequently reported in the northeastern states. Just 8 states (Maine, Massachusetts, Minnesota, New Hampshire, New York, Rhode Island, Vermont, and Wisconsin) account for nearly 9 in 10 of all reported cases.

STAY SAFE   See note for the overall **Tick-Borne Diseases** heading.

SOURCES   See note for the overall **Tick-Borne Diseases** heading.

Babesiosis

ANNUAL US DEATHS   1,126 cases reported in 2018, with mortality between 5 and 9 percent in nonimmunocompromised people

Babesiosis is a tick-borne disease caused by microscopic protozoa belonging to the *Babesia* family. The protozoa infect red blood cells and cause malaria-like symptoms. The disease is transmitted by black-legged ticks *(Ixodes scapularis),* also known as deer ticks. In the US, the disease mainly occurs in parts of the Northeast and upper Midwest, especially in parts of New England, New York, New Jersey, Wisconsin, and Minnesota. Babesiosis has been recognized as a cattle disease since the 1880s, but the first human case was reported on Nantucket Island in 1969. Cases have been increasing steadily since 2000, but babesiosis is still rare. In 2018, 1,126 cases were reported to the CDC. Babesiosis can be dangerous for the elderly and people with weak immune systems. The mortality rate for people with severe cases is estimated at between 5 and 9 percent but may be as high as 21 percent in immunocompromised people.

STAY SAFE   See note for the overall **Tick-Borne Diseases** heading.

SOURCES   See note for the overall **Tick-Borne Diseases** heading.

Ehrlichiosis

ANNUAL US DEATHS   About 18

Ehrlichiosis is the general name used to describe tick-borne diseases caused by the bacterium *Ehrlichia chaffeensis, E. ewingii,* or *E. muris eauclairensis* in the US. Most reported cases are from *E. chaffeensis.*

Ehrlichiosis symptoms may include fever, headache, chill, and muscle pain. Prompt treatment with antibiotics cures most cases.

Ehrlichiosis was first recognized as a human disease in the late 1980s. In 2018, 1,799 cases were reported to the CDC. The case fatality rate is about 1 percent.

Ehrlichiosis is most frequently reported in the southeastern and south-central United States. In 2018, 4 states (Arkansas, Missouri, New York, and Virginia) accounted for more than half of all reported cases.

STAY SAFE   See note for the overall **Tick-Borne Diseases** heading.

SOURCES   See note for the overall **Tick-Borne Diseases** heading.

Lyme Disease

ANNUAL US DEATHS   Between 1985 and 2019, 11 cases of fatal Lyme carditis were reported worldwide.

Lyme disease is the most common vector-borne disease in the US. Between 2004 and 2016, it accounted for 82 percent of all tick-borne disease reports. Lyme disease is caused by the spirochete bacterium *Borrelia burgdorferi.* It is transmitted to humans through the bite of infected blacklegged ticks *(Ixodes scapularis),* also known as deer ticks. Typical symptoms include fever, headache, muscle and joint pain, fatigue, and a characteristic bull's-eye skin rash called erythema migrans. If left untreated, infection can spread to joints, the heart, and the nervous system.

Each year, approximately 30,000 cases of Lyme disease in the US are reported to the CDC by state health departments. Most cases are never reported, however, and the CDC estimates that approximately 476,000 people get sick with Lyme disease every year. A large majority of the cases occur in New England and the mid-Atlantic states. Increasingly, cases are occurring in the Midwest and northern California.

Prompt treatment with antibiotics cures most cases. In about 1 out of 100 cases, the bacterium infects the heart and causes Lyme carditis. In very rare cases, Lyme carditis causes death from heart arrhythmias. Between 1985 and 2019, 11 cases of fatal Lyme carditis were reported worldwide.

Lyme disease gets its name from the town of Lyme in Connecticut, where the disease was first identified in 1975.

STAY SAFE   Seek prompt treatment if you are symptomatic; also see note for the overall **Tick-Borne Diseases** heading.

SOURCES   See note for the overall **Tick-Borne Diseases** heading.

Powassan Virus Disease

ANNUAL US DEATHS   21 between 2010 and 2019

The Powassan virus is transmitted by *Ixodes* ticks in the northeastern states and the Great Lakes region. In rare cases, it can cause encephalitis. The disease is named for Powassan, Ontario, where it was first identified in 1958 in a 5-year-old boy who died from encephalitis. The first human case in the US was identified in 1970 in New Jersey.

In 2010, 8 cases of Powassan virus disease were reported; in 2019, 39 cases were reported. Between 2010 and 2019, there were 181 cases of Powassan virus disease and 21 deaths. The fatality rate for cases involving encephalitis is 10 to 15 percent.

STAY SAFE   See note for the overall **Tick-Borne Diseases** heading.

SOURCES   See note for the overall **Tick-Borne Diseases** heading.

Spotted Fever Rickettsiosis (SFR)

ANNUAL US DEATHS   Variable, between 5 and 25

This term refers to a group of bacterial diseases, hard to distinguish clinically, that includes Rocky Mountain spotted fever, *Rickettsia parkeri rickettsiosis*, Pacific Coast tick fever, and rickettsialpox (the last transmitted by mouse mites rather than ticks).

The overall mortality rate of SFR is under 0.5 percent, but Rocky Mountain spotted fever is the most lethal tick-borne disease in the US, with a mortality rate of 5 to 10 percent in clinical reviews of confirmed cases.

The number of SFR cases has been rising steadily since 2000, when 495 cases were reported. In 2017, cases peaked at 6,248 and have declined slightly since then. SFR occurs throughout the contiguous US, but five states (Arkansas, Missouri, North Carolina, Tennessee, and Virginia) account for over 50 percent of cases.

STAY SAFE   See note for the overall **Tick-Borne Diseases** heading. Despite its name, the disease is found throughout the mainland US.

SOURCES   See note for the overall **Tick-Borne Diseases** heading.

The male adult dog tick transmits Rocky Mountain spotted fever.

Tick-Borne Relapsing Fever

ANNUAL US DEATHS About 2

Tick-borne relapsing fever (TBRF) is transmitted by multiple tick species, but the *Ornithodoros hermsii* tick is responsible for most cases in the US. The ticks transmit several different spirochetes, most commonly *Borrelia hermsii* and *B. turicata*. TBRF is characterized by episodes of fever that last for 3 to 5 days, with relapses 5 to 7 days after apparent recovery. Most cases are linked to sleeping in rodent-infested cabins in mountainous areas of California, Washington, and Colorado. In Texas, TBRF is linked to cave exposures. TBRF is rare, with 483 cases reported between 1990 and 2011. Death is very rare, with a mortality rate of less than 1 percent.

STAY SAFE See note for the overall **Tick-Borne Diseases** heading. Highest risk in the California mountains.

SOURCES See note for the overall **Tick-Borne Diseases** heading.

Tularemia

ANNUAL US DEATHS The disease is rare (126 cases annually) but has a mortality rate of 2 percent with some strains spiking to 24 percent.

Tularemia is caused by the bacterium *Francisella tularensis*. It can be transmitted to humans through tick bites and other insect bites, by eating undercooked meat, and in rare cases, by inhalation.

Tularemia is rare. Between 2001 and 2010, a total of 1,208 cases were reported, for a mean of 126 cases annually. Tularemia can be serious and potentially fatal, with a mortality rate of around 2 percent. However, for some strains, the mortality rate may be up to 24 percent.

STAY SAFE See note for the overall **Tick-Borne Diseases** heading.

SOURCES See note for the overall **Tick-Borne Diseases** heading.

Emerging Tick-Borne Diseases

ANNUAL US DEATHS Minimal data

New or previously obscure tick-borne diseases are a growing problem. The diseases listed below have all been recognized since 2004. Little is known about their incidence and lethal risk.

- *Borrelia miyamotoi* has recently been identified as a cause of relapsing fever. It is transmitted by *Ixodes* ticks and is found in areas where Lyme disease is prevalent.
- Bourbon virus disease has been identified since 2017 in a small number of people in the Midwest and southern United States. Some infected people died.
- Heartland virus. More than 30 cases of Heartland virus disease have been reported since 2017 in states in the Midwest and the South. It is probably transmitted by ticks.
- *Rickettsia parkeri* is transmitted by the Gulf Coast tick *(Amblyomma maculatum)* and causes an illness similar to Rocky Mountain spotted fever. It has been identified in many southern states, including Florida.
- Southern tick-associated rash illness (STARI) is a bacterial illness transmitted by the lone star tick *(Amblyomma americanum)*. It causes an illness similar to Lyme disease. The lone star tick is found in the southeastern and northeastern states.
- 364D Rickettsiosis is a new disease transmitted by the Pacific Coast tick *(Dermacentor occidentalis)*. It is found in California.

STAY SAFE See note for the overall **Tick-Borne Diseases** heading.

SOURCES See note for the overall **Tick-Borne Diseases** heading.

Parasitic Diseases

A parasite is an organism that lives on or in a host organism and gets its food from or at the expense of its host. Human parasites that can cause disease fall into 3 main classes: protozoa (one-celled organisms), helminths (worms), and ectoparasites (organisms that burrow into the skin). This section deals only with protozoa; ectoparasites are discussed under the sections Insect-Borne Diseases and Tick-Borne Diseases. Helminths cause illness in humans, but because fatalities are very rare, they are not discussed in this book.

STAY SAFE   Avoid contaminated water.

SOURCES   See individual entries.

Brain-Eating Amoeba

ANNUAL US DEATHS   Of 145 known cases between 1962 and 2018, only 4 people survived.

Primary amoebic meningoencephalitis (PAM) is caused by infection with *Naegleria fowleri*, better known as the brain-eating amoeba. This microscopic organism is found naturally in warm freshwater lakes, rivers, and hot springs and may infect swimming pools, splash pads, and similar facilities if water quality testing is neglected. Infection occurs when the amoeba enters the body through the nose (swallowing contaminated water doesn't lead to infection). From the nose, the amoeba travels to the brain and causes PAM, which has a fatality rate of nearly 100 percent. Of the 145 known cases between 1962 and 2018, only 4 people survived.

Your risk of being killed by brain-eating amoeba is negligible. From 2010 to 2019, 34 infections were reported. In 2021, a boy died of PAM in Texas. By comparison, in that same time frame, more than 34,000 people drowned.

STAY SAFE   Wear a nose plug when swimming in possibly contaminated warm lakes, rivers, and hot springs.

SOURCE   CDC, cdc.gov/parasites/naegleria/

Cryptosporidiosis

ANNUAL US DEATHS   0 to 19

Cryptosporidium is a microscopic parasite that causes the diarrheal disease cryptosporidiosis. Both the parasite and the disease are called crypto for short. Crypto is spread through contaminated drinking water and recreational water. It's a leading cause of waterborne disease in the US. The number of cases and deaths varies considerably from year to year. The range for hospitalization is between 58 and 518 people a year. The range for death from crypto is between 0 and 19 people a year.

In 1993, a contaminated water plant in Milwaukee caused an outbreak of crypto. About 403,000 people, or about 25 percent of the city's population, became ill with fever, stomach cramps, and diarrhea. Of those, 69 died, making the Milwaukee outbreak the largest waterborne disease outbreak in US history. Almost all—93 percent—of the deaths occurred in persons with AIDS.

STAY SAFE Don't drink contaminated water. Obey boil water advisories from local authorities. Maintain swimming pools, hot tubs, and spas safely.

SOURCES CDC, cdc.gov/parasites/crypto; Water Quality and Health Council, waterandhealth.org/healthy-pools/pool-maintenance/

Giardiasis

ANNUAL US DEATHS 0 to 3

Giardia duodenalis is a microscopic parasite that causes the diarrheal disease giardiasis. The *Giardia* parasite is widespread but is found most commonly in contaminated drinking water, freshwater lakes, streams, and rivers, and swimming pools. Swallowing *Giardia* can cause giardiasis, also known as beaver fever, because cases usually come from drinking contaminated fresh water in a natural setting.

Giardiasis is the most common intestinal parasitic disease in the US. It is easily spread through contact with the feces of an infected person. In 2018, there were 15,579 reported giardiasis cases in the US. The incidence of giardiasis is under 7 cases per 100,000 population. Only 0 to 3 deaths are reported each year.

STAY SAFE Avoid drinking untreated fresh water. Use a portable water filter or water purification tablets when camping and backpacking.

SOURCE CDC, cdc.gov/parasites/giardia

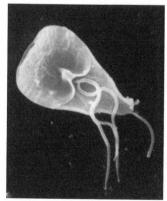

Photomicrograph of *Giardia duodenalis* [lamblia was its former name], the parasite that causes giardiasis

Toxoplasmosis

ANNUAL US DEATHS 200 to 482; high risk in pregnant women, both to mother and fetus

Toxoplasmosis is an infection caused by a single-celled parasite called *Toxoplasma gondii*. In the US, toxoplasmosis is the 3rd leading cause of death attributed to foodborne illness. Between 200 and 482 people die of toxoplasmosis in the US each year.

House cats are the primary host of *Toxoplasma gondii*. The parasite completes its life cycle in the intestines of cats. Oocysts (eggs) then pass into cat feces and can contaminate soil, drinking water, and food. Undercooked meat is another common route of infection.

The parasite is very common—more than 40 million people, or about 11 percent of the population, in the US carry it. Very few people have symptoms, but people with compromised immune systems and women who get infected with *Toxoplasma* before or during pregnancy can develop severe illness. People with weak immune systems who are infected with *Toxoplasma* may have symptoms that include fever, confusion, headache, seizures, nausea, and poor coordination. Pregnant women who are infected with *Toxoplasma* may have a miscarriage or stillborn child, or the child might be born with signs of congenital toxoplasmosis or develop vision loss, mental disability, and seizures later in life.

Toxoplasmosis has been suggested as a cause of reckless behavior that can lead to accidental death, but the evidence for this is inconclusive.

STAY SAFE Avoid contaminated food and water. Avoid close contact with animal feces, particularly cat feces and litter boxes. Cook all food thoroughly. Pregnant women and immunocompromised people should be particularly cautious.

SOURCE CDC, cdc.gov/parasites/toxoplasmosis

Medical Mishaps, Malpractice, and Murder

> **ANNUAL US DEATHS**   22,165 to 250,000

Medical knowledge has, unquestionably, advanced steadily. Still, errors are made, and even the best intentions can produce lethal results. Unfortunately, too, the practice of medicine is not immune to the occasional homicidal personality.

According to a 2016 report by Johns Hopkins Medicine, more than 250,000 annual US deaths are caused by medical error, making this the 3rd leading cause of death in the United States. A 2020 Yale study put the number much lower, at 22,165.

STAY SAFE   Be your own advocate and practice healthy skepticism. Ask questions. Get a second opinion. Research the caregivers and institutions you patronize.

SOURCE   M.A. Makary and M. Daniel. Medical error—the third leading cause of death in the US. *BMJ*, May 2016; B.A. Rodwin et al. Rate of Preventable Mortality in Hospitalized Patients: a Systematic Review and Meta-analysis. *Journal of General Internal Medicine*, July 2020.

Angels of Death

> **ANNUAL US DEATHS**   See data in entry.

So-called angels of death—healthcare workers, almost always nurses—use their position of trust as a caregiver to kill patients. These serial killers often drift from one hospital or nursing home to another, killing multiple patients before they are finally caught. Although they are prolific killers, angels of death are also very rare. Modern hospital record-keeping and statistical analysis quickly flag anomalies in death rates in hospital and clinic settings, but nursing homes and other care settings are far less vigilant. The biggest red flag for an angel of death is a history of frequent job changes, often for vaguely explained reasons.

Angels of death are usually suspected of killing many more patients than they are convicted for. The most prolific serial killer nurse in US history was Charles Cullen, who confessed to killing 40 patients between 1988 and 2003 at various hospitals and healthcare facilities in New Jersey. He later confessed to killing many more, possibly as many as 400 in total, which would make him the most prolific serial killer in American history. In 2006 he was sentenced to 11 consecutive life terms.

Cullen was able to continue as a serial killer for so long because the hospitals where he worked, fearing lawsuits, preferred to quietly fire him instead of investigating his suspicious actions. The Cullen case was an impetus for the federal Patient Safety and Quality Improvement Act of 2005, which improved patient safety by encouraging voluntary and confidential reporting of events that adversely affect patients.

Other prolific angels of death in the US include:

- Orville Lynn Majors, convicted in 1999 of 6 patient murders in Indiana but suspected of between 100 and 130
- Donald Harvey, convicted in 1987 of 36 patient murders in Cincinnati but suspected of 87
- Genene Jones, a pediatric nurse in Texas, convicted in 1984 for 1 patient murder but suspected of 60
- Kristen Gilbert, convicted in 2001 of 4 patient murders at a VA hospital in Massachusetts but suspected of at least 25
- Reta Mays, convicted in 2021 of 7 patient murders at a VA hospital in West Virginia
- Richard Angelo, convicted in 1989 of 4 patient murders on Long Island, NY, but suspected of 25

STAY SAFE Report any suspicious behavior by nurses or other staff in healthcare settings. Demand immediate follow-up for any unexpected life-threatening event or death of a hospitalized loved one, especially if it involved resuscitation. Angels of death often kill in the guise of being heroes for attempting to save a life.

SOURCES Wikipedia, Angel of Death; B.C. Yorker et al. Serial murder by healthcare professionals. *Journal of Forensic Sciences*, November 2006.

Hospital-Acquired Conditions

ANNUAL US DEATHS 3,219 in 2016

The Centers for Medicare and Medicaid Services (CMS) defines hospital-acquired conditions (HACs) as avoidable complications of care that could reasonably have been prevented through the application of evidence-based guidelines. Mortality from HACs is difficult to determine. Hospitalized patients who get an HAC tend to be elderly and very ill, and the cause of death is usually an underlying condition, not the HAC itself. The CMS tracks 14 HACs, including infection with *Clostridioides difficile*, pressure sores, catheter-associated infections, and surgical site infections. In 2016, 48,771 HACs resulted in 3,219 potentially avoidable deaths. These HACs added an average of 8.17 days per patient to the average length of stay, and increased mortality risk per patient by 72.32 percent over patients without an HAC. Because CMS determines Medicaid and Medicare reimbursement rates for hospitals and nursing homes and can impose financial penalties, its statistics are important baselines.

The Agency for Healthcare Quality and Research (AHQR), the lead federal agency charged with improving the safety and quality of America's healthcare system, tracks a somewhat different and larger set of HACs and uses a different metric: HACs per 1,000 hospital discharges per year. By this metric, in 2010 there were 145 HACs per 1,000 hospital discharges, in 2014 there were 99 HACs per 1,000 hospital discharges, and in 2017 there were 86 HACs per 1,000 discharges. The improvement from 2014 to 2017 led to an estimated 20,700 deaths averted.

STAY SAFE   Any hospital admission carries the risk of an HAC, and the longer the stay, the greater the risk. Work with your doctor to avoid or minimize hospital stays.

SOURCES   Centers for Medicare & Medicaid Services (CMS), cms.gov/Medicare/Medicare-Fee-for-Service-Payment/HospitalAcqCond/Hospital-Acquired_Conditions; Agency for Healthcare Research and Quality (AHRQ), ahrq.gov/patient-safety/index

Maternal and Pregnancy-Related Mortality

ANNUAL US DEATHS   Approximately 660 in 2018

Maternal mortality is defined by the CDC as the death of a woman during pregnancy, at delivery, soon after delivery, and up to 42 days after delivery from a cause related to pregnancy. Pregnancy-related mortality is defined as death during pregnancy and up to 1 year after giving birth. In 2018, there were approximately 660 maternal deaths, for a death rate of 17.4 per 100,000 pregnancies. About 31 percent of maternal deaths occur during pregnancy; 17 percent occur around the time of delivery; 52 percent occur up to 1 year after delivery.

Most pregnancy-related mortality is related to cardiovascular conditions, including cardiomyopathy (11 percent), blood clots (9 percent), high blood pressure (8 percent), stroke (7 percent), and a combination of those conditions (15 percent). Infection (13 percent) and severe postpartum bleeding (11 percent) are also leading causes of death. Most maternal deaths are preventable; missed or delayed diagnosis is a key factor.

Historically, maternal mortality in the US dropped for most of the twentieth century. In 1933, the death rate was 619 deaths per 100,000 live births. By 1960, the rate was 37 deaths per 100,000 live births, and by 1997 it was 9. It has been gradually rising since then.

The maternal death rate for Black women in 2018 was 37.1 per 100,000 pregnancies. This was 2.5 times the rate for White women (14.7) and 3 times the rate for Hispanic women (11.8). The disparity between Black and White deaths today is roughly the same as in the 1940s. Education levels are often seen as mitigating maternal deaths, but in fact, maternal death is more common among Black women with a college education than White women with less than a high school diploma (40.2 versus 25.0). The causes of this startling disparity are many, but structural racism plays an important role.

Accurate data for maternal mortality by state is unavailable for about half the states. In addition, small numbers in each state make the data statistically unreliable. Based on available information for 2018, Alabama, Arkansas, Kentucky, and Oklahoma have maternal death rates greater than 30 per 100,000 live births. California, Illinois, Ohio, and Pennsylvania have rates of under 15 deaths per 100,000 live births.

Compared to all other developed countries in the world, the US has the highest rate of maternal mortality, even when the data is limited to White women. At 17.4 deaths per 100,000 live births, the US rate is nearly double the rate in other wealthy developed nations. In France and Canada, the maternal death rate is 8.7 and 8.6, respectively. In Germany the rate is 3.2 and in Norway it is 1.8.

Worldwide, in 2017, about 810 women died every day from preventable causes related to pregnancy and childbirth. Between 2000 and 2017, however, the maternal mortality ratio (the number of maternal deaths per 100,000 live births) dropped by about 38 percent, largely due to improvements in healthcare delivery. About 94 percent of all maternal deaths occur in low- and lower-middle-income countries.

STAY SAFE Seek prenatal medical care as soon as you realize you are pregnant and keep all appointments during pregnancy and after delivery. Gain the right amount of weight during pregnancy and follow medical advice about blood pressure and gestational diabetes. Don't smoke, drink alcohol, or use drugs while you are pregnant. Speak up during pregnancy, labor, and after delivery if you feel healthcare personnel aren't listening to your concerns. Work in your community for better and more equitable maternity care and maternity leave.

SOURCES　CDC, cdc.gov/nchs/maternal-mortality/index.htm; the Commonwealth Fund, *Maternal Mortality in the United States: A Primer,* data brief, December 2020, commonwealthfund. org/publications; WHO, who.int/news-room/fact-sheets/detail/ maternal-mortality

Medication Nonadherence

ANNUAL US DEATHS　100,000 to 125,000

Half of all patients in the US with chronic diseases, such as hypertension and type 2 diabetes, fail to take medications as prescribed. More than 1 in 5 new prescriptions are never filled, and half of all patients with a chronic disease stop taking their medicine within 1 year. Medication nonadherence underlies about 10 percent of all hospitalizations each year and is associated with 100,000 to 125,000 deaths annually.

STAY SAFE　Take your medicine as prescribed by your doctor. If you are having trouble paying for your drugs, don't understand why you need them or how to take them, or want to stop because of side effects, talk to your doctor first. Solutions are available.

SOURCE　F. Kleinsinger. The unmet challenge of medication nonadherence. The *Permanente Journal,* July 2018.

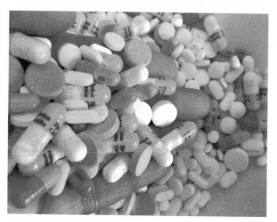

Modern pharmacology can save lives—if the patient takes the medication as prescribed.

Nursing Homes

ANNUAL US DEATHS   535,000 in 2017

In 2017 in the US, 535,000 people died in nursing homes. Most deaths in US nursing homes are unavoidable due to the advanced age and serious health issues of the residents, yet even in these cases, some deaths could have been delayed and the resident's nursing home stay made more comfortable by better quality care. Some deaths are due to neglect or abuse of nursing home patients. Abuse can be by staff, by other patients, or by relatives, and it can be physical, emotional, or sexual.

In 2014, over 14,000 abuse and neglect complaints were filed just with nursing homes' own ombudsmen. Data is not readily available about how many nursing home abuse/neglect cases might go undetected or unreported, but surveys suggest that about 25 percent of the roughly 1.4 million patients staying in any 1 year in the nation's 15,600 licensed nursing homes have suffered at least 1 incident of abuse there. Data is not readily available on how many nursing home deaths might have been premature or altogether preventable because they were brought on by abuse, neglect, or poor quality care.

STAY SAFE   For information about how to select a nursing home, see "How to Choose a Nursing Home" at nia.nih.gov, or "Finding a Nursing Home: Don't Wait Until You Need One to Do the Research" at aarp.org. The advice includes to "drop by and spy" to observe the state of grooming of patients and the attitude of staff during your reconnaissance visit.

SOURCES   AARP, aarp.org/caregiving/basics/info-2019/finding-a-nursing-home.html; CDC, cdc.gov/nchs/fastats/nursing-home-care.htm; National Association of Nursing Home Attorneys, nanha.org/resources/facts-statistics; NIA NIH, nia.nih.gov/health/how-choose-nursing-home

Outpatient Surgery Centers

ANNUAL US DEATHS   About 260 between 2013 and 2017

Around 5,700 Medicare-certified outpatient surgery centers, also known as ambulatory surgery centers (ASCs), provide services in the US. Outpatient surgery centers are licensed, freestanding outpatient facilities, often physician-owned. They usually specialize in same-day, scheduled procedures such as knee arthroplasty, colonoscopy, or cataract surgery. Today, nearly 75 percent of all surgery is performed on an outpatient basis. Of those, slightly more than half are done in an ambulatory surgery center.

While the safety record for outpatient surgery is excellent, emergencies and deaths do occur. A 2018 investigation by Kaiser Health News and the USA Today network found that between 2013 and 2017, at least 260 patients died at an outpatient surgery center. The number is only an estimate, because no national authority tracks these deaths. A 2017 report from the CDC estimated that only 2 percent of patients discharged from an outpatient surgery center were later admitted to the hospital as an inpatient.

Comedian Joan Rivers died in 2014 at age 81 after she stopped breathing following a routine endoscopy at an outpatient clinic in Manhattan.

STAY SAFE   Choose an outpatient surgery center that is Medicare-certified.

SOURCES   Kaiser Health News, khn.org/news/medicare-certified-surgery-centers-are-expanding-but-deaths-question-safety; M.J. Hall et al. *Ambulatory surgery data from hospitals and ambulatory surgery centers: United States, 2010.* National health statistics reports, no. 102. National Center for Health Statistics, 2017.

Polypharmacy

ANNUAL US DEATHS   No data available

Polypharmacy is the regular use of at least 5 prescription medications to treat 1 or more chronic health problems. It is common in older adults and increases the risk of adverse drug reactions, harmful drug interactions, functional and cognitive decline, frailty, and death from all causes. An estimated 44 percent of men and 57 percent of women older than age 65 take 5 or more medications. While these drugs may be needed to manage medical conditions, they are sometimes inappropriately prescribed or continued past their usefulness.

The risk of death for someone who regularly takes 5 prescription drugs is 1.31 times higher than for someone the same age who takes none. When the drug count rises to 6 or more, the risk of death rises to 1.59 times higher. Excessive polypharmacy (10 or more drugs) raises the risk of death by 1.96 times.

STAY SAFE   Discuss your medications with your doctor to be sure they are all needed.

SOURCE   N. Leelakanok et al. Association between polypharmacy and death: A systematic review and meta-analysis. *Journal of the American Pharmaceutical Association*, November 2017.

Postsurgical Death

ANNUAL US DEATHS   US statistics unavailable; estimated 4.2 million worldwide deaths within 30 days of surgery

Perioperative mortality is defined as death from any cause occurring within 30 days after surgery, in or out of the hospital. Causes can include postoperative infection, sepsis, blood clots, pulmonary complications, kidney failure, and liver failure. Globally, 4.2 million people are estimated to die within 30 days of surgery each year. Exact statistics for the US are surprisingly difficult to find. In 2014, a total of 14.2 million operative procedures were performed on inpatients at US hospitals. In general, mortality from surgery decreases as volume at the hospital for that procedure increases.

One study estimated that the incidence of postoperative hospital mortality was 0.57 percent within 48 hours and 2.1 percent within 30 days of an inpatient operation. Overall, postoperative mortality is estimated to range from 5 to 10 percent, depending on variables such as patient age, preexisting morbidities, and medical center quality. A study of VA patients found that any postsurgical infection in the first 30 days after surgery was associated with a 1.9 times increase in all-cause mortality within 1 year.

A 2020 study found that patients who underwent surgery on a surgeon's birthday had higher mortality compared with patients who underwent surgery on other days: 6.9 percent adjusted mortality rate for the birthday surgery, compared to 5.6 percent for surgery on other days.

STAY SAFE   Try to have surgery at a large hospital with a high volume of surgeries for your condition. Ask your surgeon their birthday and schedule surgery for a different day.

SOURCES   K. Fecho et al. Postoperative mortality after inpatient surgery: Incidence and risk factors. *Therapeutic and Clinical Risk Management*, August 2008; H. Kato. Patient mortality after surgery on the surgeon's birthday: observational study. *BMJ*, December 2020; W.J. O'Brien et al. Association of postoperative infection with risk of long-term infection and mortality. *JAMA Surgery*, January 2020.

Preventable Medical Error

ANNUAL US DEATHS   As many as 250,000

How many hospitalized people die every year from a preventable medical error is controversial. In 2000, the Institute of Medicine issued a report estimating the death rate for hospitalized patients from medical error at between 44,000 and 96,000 patients each year. A study by a Johns Hopkins researcher in 2016 suggested that more that 250,000 patient deaths per year were due to preventable medical errors, which would make them the 3rd leading cause of death. In 2020, a new study from researchers at Yale estimated the number of preventable deaths in hospitals at approximately 22,165 annually. Of these, 7,150 were in patients who had a life expectancy of greater than 3 months. In other words, a very large majority of patients who die from preventable medical error were already extremely ill with complex medical problems.

STAY SAFE   The main causes of preventable death are clinical monitoring or management, diagnostic error, and complications from surgery. To minimize these risks, if you need hospitalization, choose a large hospital that is designated as a center of excellence for whatever is wrong with you.

SOURCE   B.A. Rodwin et al. Rate of Preventable Mortality in Hospitalized Patients: A Systematic Review and Meta-analysis. *Journal of General Internal Medicine*, July 2020.

Pressure Injuries

ANNUAL US DEATHS   60,000, mostly among the elderly with comorbidities

Pressure injuries, also known as pressure ulcers, bedsores, or decubitus ulcers, are injuries to the skin and underlying tissue resulting from continuous pressure on the skin. Bedsores often occur when patients are immobile, bedridden, or have impaired sensation, particularly from spinal cord injuries. They develop on parts of the body that make the most contact with surfaces, including the tailbone, ankles, heels, and hips. When bedsores progress to the point of open wounds, they can lead to fatal complications such as osteomyelitis, sepsis, and gangrene.

Each year in the US, approximately 60,000 people, mostly elderly with underlying morbidities, die of complications from pressure ulcers. Individuals with pressure ulcers have a risk of death that is 4.5 times greater than persons with the same risk factors but without pressure injuries. Pressure ulcers are a direct cause of death for 7 to 8 percent of all patients with paraplegia.

Pressure ulcers are a very common problem in hospitals and nursing homes. As many as one-third of hospitalized patients with pressure ulcers will die during hospitalization. Nursing home residents with pressure ulcers experience a 6-month mortality rate of 77.3 percent.

STAY SAFE Malnutrition is an important risk factor for pressure injuries, especially in the elderly. Monitor the nutritional status of elderly loved ones, particularly those in nursing homes or hospitalized.

SOURCE G. Brown. Long-term outcomes of full-thickness pressure ulcers: healing and mortality. *Ostomy and Wound Management*, October 2003.

Reye Syndrome

ANNUAL US DEATHS Vanishingly low, with about 2 cases yearly and rarely any reported deaths

Reye syndrome is a rare, potentially life-threatening illness that affects the brain and liver. It is most common in children and teens who take aspirin (salicylates) or aspirin products during a viral illness or while recovering from one. Symptoms include frequent vomiting, sleepiness, muscle weakness, and seizures.

Reye syndrome today is very rare because doctors warn against aspirin use in children and teens. The FDA has required warning labels on aspirin since 1986. In 1980, about 555 cases of Reye syndrome were reported in the US. By 1994, the number had dropped to about 2 a year. Death is very rare.

STAY SAFE Reye syndrome is both rare and easily prevented. Never give aspirin to children or teens except on the advice of a doctor. Do not give children or teens nonprescription medications containing salicylates, the active ingredient in aspirin, including pink bismuth (Pepto-Bismol).

SOURCE J. Chapman and J.K. Arnold. Reye Syndrome (Updated July 2021). In: StatPearls, StatPearls Publishing, 2021.

Vaccines

> **ANNUAL US DEATHS**   See data in entry.

Death as a direct result of being vaccinated is extremely rare. Because of the high volume of use, however, coincidental adverse events, including deaths, do occur. Almost all deaths have only a temporal link to vaccination and can't be attributed to the vaccine itself.

Hundreds of millions of vaccinations are given in the US each year, in addition to well over 546.6 million doses of COVID-19 vaccine administered between December 2020 and mid-February 2022. Adverse reactions and deaths from vaccines are reported to the federal Vaccine Adverse Event Reporting System (VAERS). Serious adverse events are very uncommon and very few people have died from vaccines. Reports of death to VAERS following vaccination do not necessarily mean that a vaccine caused the death.

Between July 1997 and the end of 2013, VAERS received 2,149 death reports. After review of available medical records, autopsy reports, and death certificates, almost all the deaths were found to be only temporally associated with the vaccines. A 2013 study using VAERS data from 2005 to 2008 concluded that the age-adjusted death rate within 60 days of vaccination was 442.5 deaths per 100,000 population, lower than the overall death rate of 758.3 per 100,000 population.

The National Vaccine Injury Compensation Program (VICP) is a federal program that was created in 1986 to compensate people who may have been injured by certain vaccines. Between 2006 and 2019, 4,092,757,049 vaccine doses were administered. Of those, 5,755 were considered compensable, almost entirely for injuries. This means for every 1 million doses of vaccine that were distributed, approximately 1 individual was compensated. Between 1988 and 2021, a total of 1,350 death claims were submitted to VICP; almost all were not considered compensable.

STAY SAFE   Vaccines save lives. Get all recommended vaccines and boosters on a timely basis.

SOURCES   HSS VAERS, vaers.hhs.gov/data.html; HRSA VICP, hrsa.gov/vaccine-compensation/data/index.html; P.L. Moro et al. Deaths reported to the Vaccine Adverse Event Reporting System, United States, 1997–2013. *Clinical and Infectious Diseases*, September 2015; N.L. McCarthy et al. Mortality rates and cause-of-death patterns in a vaccinated population. *American Journal of Preventive Medicine*, July 2013.

COVID-19 Vaccines

ANNUAL US DEATHS   See data in entry.

More than 442 million doses of COVID-19 vaccines were administered in the US from December 14, 2020, through November 15, 2021. During this time, the CDC's Vaccine Adverse Event Reporting System (VAERS) received 2,794 reports of death (0.00167 percent) among people who received a COVID-19 vaccine. All had only a temporal, not causal, association with the vaccine. No deaths in the US from any version of the COVID-19 vaccine had been confirmed as of June 2021. It is possible that 5 deaths caused by thrombosis with thrombocytopenia syndrome (TTS), a rare and serious adverse event that causes blood clots with low platelets, may be linked to the J&J/Janssen COVID-19 vaccine.

STAY SAFE   Vaccines save lives. Get all recommended vaccines and boosters on a timely basis.

SOURCES   HSS VAERS, vaers.hhs.gov/data.html; HRSA VICP, hrsa. gov/vaccine-compensation/data/index.html; P.L. Moro et al. Deaths reported to the Vaccine Adverse Event Reporting System, United States, 1997-2013. *Clinical and Infectious Diseases*, September 2015; N.L. McCarthy et al. Mortality rates and cause-of-death patterns in a vaccinated population. *American Journal of Preventive Medicine*, July 2013.

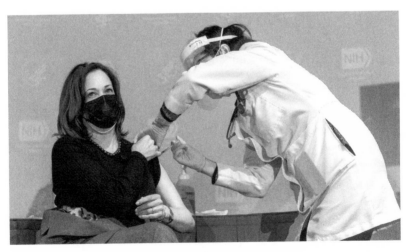

Vice President Kamala Harris being vaccinated against COVID-19

Waiting for Organ Transplant

ANNUAL US DEATHS   Estimated 6,205

As of February 2021, over 107,000 men, women, and children were on the national organ transplant waiting list. Another person is added to the list every 9 minutes. Over 40,000 transplants were performed in 2020; of those, about 6,500 were from living donors. Each day, 17 people die waiting for an organ transplant.

Information about becoming an organ donor is available from United Network for Organ Sharing (UNOS), the private, nonprofit organization that manages the nation's organ transplant system under contract with the federal government. Additional information is available from the Division of Transplantation (DoT) within the Department of Health and Human Services at organdonor.gov.

STAY SAFE   Sign up to be an organ donor at your state motor vehicle department, through UNOS at unos.org, or at organdonor.gov.

SOURCES   Organdonor.gov/learn/organ-donation-statistics; UNOS, unos.org/data

Not Dead Yet

Brain Death

> **ANNUAL US DEATHS**   In effect, all deaths: 3,358,814 in 2020

The invention of the mechanical ventilator in the 1960s, and improved techniques for resuscitating people with no respiration, heartbeat, or other external signs of life, meant that the definition of death had to change. The 2 traditional criteria—failure to respond to painful stimuli or lack of a heartbeat and respiration—no longer necessarily meant someone was dead. Instead, death increasingly came to be seen as the cessation of brain activity, but an exact definition of brain death was needed. In 1968, an ad hoc committee at Harvard Medical School defined brain death as unresponsiveness and lack of receptivity, the absence of movement and breathing, the absence of brain stem reflexes, and coma whose cause has been identified. Brain death today is medically defined as the irreversible loss of all functions of the brain, including the brain stem. The 3 essential findings in brain death are coma, absence of brain stem reflexes, and cessation of breathing. A patient determined to be brain dead is legally and clinically dead even if their heart is still beating. Organs may be removed for donation from someone who is brain dead.

In 1981, a presidential commission proposed the Uniform Determination of Death Act as a model statute for states to adopt. The act defines legal death as when the breathing and circulation have irreversibly stopped or when the entire brain has irreversibly stopped functioning. Most but not all states have adopted the statute and every state today recognizes brain death as legal and clinical death. However, variation among the states, including religious exemptions, means you can be legally dead in some states but still alive in others. In New Jersey, for instance, someone who is clinically brain dead cannot be declared legally dead without the family's consent. In 2013, 13-year-old Jahi McMath was declared brain dead in California following complications from a tonsillectomy. Her family refused to accept this, based on their strong Christian beliefs, and went to court to keep Jahi alive on a ventilator. She was moved to New Jersey, where state law allowed her to remain on the ventilator, and finally died there in 2018.

STAY SAFE   As Benjamin Franklin said, "Nothing is certain except death and taxes."

Cotard Syndrome

> **ANNUAL US DEATHS**   0

Named for the Parisian neurologist who first described it in 1880, Cotard syndrome is the delusion that your body has lost organs, blood, body parts, or, in extreme versions, that you are dead. Cotard syndrome is exceedingly rare, with about 200 known cases worldwide. From what little is known of the disease, it is usually a mental illness, but some cases may have a physiological origin.

STAY SAFE   If you think you're dead, you're not.

SOURCE   G.E. Berrios and R. Luque. Cotard's syndrome: analysis of 100 cases. *Acta Psychiatrica Scandinavica*, March 1995.

Declared Legally Dead: Missing Persons

> **ANNUAL US DEATHS**   Some 20,000 unresolved missing person cases remain open in the US at any one time.

Sometimes a person goes missing, just vanishing without a trace, and nobody ever finds out what happened to them. They might actually be alive and well, having run off to change their identity and start a whole new life in a different place. They might have died, perhaps by natural causes or in an accident, or perhaps by murder or suicide, and their body is simply never found, or is eventually found but never identified. Nobody knows.

When someone disappears, their next of kin can make a motion in court to have them declared Legally Dead. This typically requires that a period of several years has gone by with no sign of life from the missing person. The waiting period varies by state; in Georgia it is 4 years, but in other states it can be 7 or even 10 years. One exception, which allows prompt declaration of Legal Death, is when the person is known to have been in mortal peril, as was the case with the victims of the sinking of the *Titanic*.

Of course, many people who are reported missing to the police do sooner or later turn up safe and sound. But over 600,000 people go missing in the US annually, of which a few tens of thousands are still unaccounted for after 1 year. About 4,400 dead bodies are recovered every year and are never identified. One estimate is that about 20,000 unresolved missing person cases are still open in the US at any one time. Data on the total number declared Legally Dead each year is not available.

STAY SAFE   The National Institute of Justice runs NamUs, a national clearinghouse that helps families and law enforcement try to resolve cases of missing persons and of unidentified bodies.

SOURCE   Namus.gov

Determination of Death Act

The Uniform Determination of Death Act was formulated in 1980 by the National Conference of Commissioners on Uniform State Laws. It reads:

> An individual who has sustained either (1) irreversible cessation of circulatory and respiratory functions, or (2) irreversible cessation of all functions of the entire brain, including the brain stem, is dead. A determination of death must be made in accordance with accepted medical standards.

This definition was approved by the American Medical Association in 1980 and by the American Bar Association in 1981. The Uniform Determination of Death Act has been adopted by 39 states.

Under current ethical rules, organs can be removed from your body if you are brain dead but your heart is still beating.

SOURCE   National Conference of Commissioners on Uniform State Laws, uniformlaws.org/committees/community-home

Faked Deaths

ANNUAL US DEATHS   Not applicable

Faking your own death is a time-honored method of disappearing, but it rarely works. Pseudocide, as it is technically known, may be attempted for life insurance fraud, to avoid prison, to escape debt, or to escape an undesirable personal situation. If done for fraudulent purposes, faking your own death is a crime.

In April 2009, Samuel Israel III, a hedge fund manager facing a 20-year prison sentence for fraud, faked his suicide by leaving his car on the Bear Mountain Bridge in New York. When authorities couldn't find his body, they grew suspicious. After a month in hiding and seeing himself on *America's Most Wanted,* Israel turned himself in and was sentenced to 22 years (2 extra years for bail jumping).

SOURCE   Association of Certified Fraud Examiners, acfe.com/fraud-examiner.aspx?id=4294994951

Hospice Care

ANNUAL US DEATHS   1.55 million Medicare beneficiaries received hospice care in 2018.

In 2018, 1.55 million Medicare beneficiaries received hospice care, either at home, in a nursing home or assisted living facility, or in one of 4,639 Medicare-certified hospices. Because 16 percent of hospice patients are under age 65 and aren't covered by Medicare, this number is somewhat lower than the actual number.

The average length of stay was 89.6 days. However, 53.8 percent of patients were enrolled in hospice for 30 days or fewer; only 20.8 percent were enrolled for 180 days or longer. About 1.3 million hospice patients died, but 17 percent of them were discharged alive.

Deaths by diagnosis were:

Cancer: 336,307, or 29.6 percent

Heart disease: 196,971, or 17.4 percent

Dementia: 177,490, or 15.6 percent

Other: 166,848, or 14.7 percent

Respiratory: 124,407, or 11.0 percent

Stroke: 107,439, or 9.5 percent

Chronic kidney disease: 25,221, or 2.2 percent

More than half of all hospice patients (55.1 percent) were female; 46.1 percent were male.

SOURCE   National Hospice and Palliative Care Organization (NHPCO) *Facts and Figures 2020,* nhpco.org/hospice-care-overview/hospice-facts-figures

Lazarus Syndrome

ANNUAL US DEATHS   US data unavailable; between 1982 and 2018, 65 patients worldwide were pronounced dead but revived; 18 of these fully recovered.

In the New Testament, Lazarus of Bethany was brought back to life by Jesus Christ 4 days after his death. In modern medicine, Lazarus syndrome is the delayed return of spontaneous circulation (ROSC) after CPR has ceased. In other words, patients who were pronounced dead return to life through spontaneous autoresuscitation. Lazarus syndrome is very rare, although concerns about malpractice mean cases are probably underreported. A medical review article from 2020 identified 65 patients worldwide between 1982 and 2018. Of these, 18, or 28 percent, made a full recovery.

In 2020, Timesha Beauchamp of Detroit was mistakenly declared dead after a heart attack. Funeral home workers discovered she was still alive when they began preparations for embalming. She was transported to a hospital and placed on a respirator; she died 8 weeks later.

STAY SAFE   Current guidelines for medical personnel recommend 2 to 5 minutes of observation after the heart has stopped beating before declaring death.

SOURCE   L. Gordon et al. Autoresuscitation (Lazarus phenomenon) after termination of cardiopulmonary resuscitation – a scoping review. *Scandinavian Journal of Trauma, Resuscitation, and Emergency Medicine*, January 2020.

The Gospel of John relates how the stricken Lazarus of Bethany, namesake of the Lazarus syndrome, was restored to life through a miracle of Jesus.

Physician-Assisted Suicide

> **ANNUAL US DEATHS**   Unknown, but see text regarding prescriptions associated with physician-assisted suicide.

Physician-assisted suicide, also known as medical aid in dying or death with dignity, is an end-of-life option that allows certain eligible individuals to legally request and obtain medications from their physician to end their life. As of 2021, physician-assisted suicide was legal in California, Colorado, the District of Columbia, Hawaii, Maine, New Jersey, New Mexico, Oregon, Vermont, and Washington.

In states where it is legal, between 0.3 and 4.6 percent of all deaths are reported as physician-assisted suicide. Very few people use death with dignity laws, and about one-third of those who obtain the prescribed medication never take it. Two out of 3 patients are aged 65 years or older; the median age at death is 72 years. Most patients have cancer (77.9 percent in Oregon) or ALS (7.8 percent). The 3 most frequently cited end-of-life concerns are loss of autonomy (90.9 percent), decreasing ability to participate in activities that made life enjoyable (89.5 percent), and loss of dignity (75.7 percent). Pain was a concern for 25.8 percent.

Exact numbers for physician-assisted deaths are not known. However, the number of prescriptions for the drugs in 2019 is as follows:

California	618
Colorado	170
DC	4
Hawaii	30
Maine	1
New Jersey	12
New Mexico	No data
Oregon	290
Vermont	34
Washington	251

SOURCES   Death with Dignity National Center, deathwithdignity. org; E.J. Emanuel et al. Attitudes and practices of euthanasia and physician-assisted suicide in the United States, Canada, and Europe. *JAMA*, July 2016.

Undead

ANNUAL US UNDEAD   Unknown

"Undead" refers to a person who is no longer alive but is animated by a supernatural force, as in vampires and zombies, or, like Dr. Frankenstein's monster, a fictional corpse reanimated by fictional electricity. The term can also be used as a collective noun for undead beings.

Cryonics

Cryogenically frozen individuals can be considered as undead awaiting reanimation. The bodies are cooled down with liquid nitrogen to $-196°$ degrees C (-321 degrees F), which cryopreserves them in a solid state. As of 2021, nearly 400 individuals worldwide are now stored in cryonic facilities in the US. The Alcor Life Extension Foundation (alcor.org), founded in 1972 in Scottsdale, Arizona, houses 183 cryonic patients. One of them is famed Red Sox baseball player Ted Williams, who died in 2002. His head and body are preserved separately. The Cryonics Institute (cryonics.org) in Michigan, founded in 1976, houses 206 patients.

Despite persistent rumors, the body of Walt Disney is not cryopreserved in a vault under the Pirates of the Caribbean ride at Disneyland. He died of lung cancer on December 16, 1966. His body was cremated and the ashes interred at Forest Lawn Memorial Park in Glendale, California, the next day.

Vampires

Supernatural undead beings who live on blood are found in the folklore of many cultures (e.g., chupacabras in Puerto Rico), but the modern legend of the vampire as an undead human originated in southeastern Europe in the early eighteenth century. The modern vampire as a debonaire and immortal fictional character began in 1819 with the publication of *The Vampyre* by the English writer John Polidori. Bram Stoker's definitive vampire novel, *Dracula*, was published in 1897. Vampires ever since have been major figures in the horror genre.

Zombies

In Haitian folklore, a zombie is an undead corpse reanimated through voodoo. Zombies are primitive, very violent, and vengeful. Zombies became part of popular culture in the twentieth century through sci-

Bela Lugosi in the 1932 film *White Zombie*

ence fiction authors and comic books such as *Tales from the Crypt* in the 1950s. George A. Romero's low-budget 1968 film *Night of the Living Dead* created the concept of zombies who eat human flesh. That they prefer brains was confirmed by Romero's 1985 sequel *Return of the Living Dead*. A zombie apocalypse occurs when hordes of zombies overwhelm civilization, leaving only a small number of survivors. The CDC Zombie Task Force was disbanded in 2012.

STAY SAFE If you wish to be cryogenically preserved, make arrangements (and substantial payments) well in advance. Cryopreservation works best if the procedure begins as soon as possible after death. Vampires can be warded off with garlic and with Christian symbols; they can be killed by a stake through the heart or by exposure to sunlight. Zombies can be killed by destroying their brains with any means handy.

SOURCES CDC Zombie Preparedness, cdc.gov/cpr/zombie/index. htm; C. Nugent et al. The undead in culture and science. *Proceedings* (Baylor University Medical Center), April 2018; P. Gwynne. Preserving Bodies in a Deep Freeze: 50 Years Later. *InsideScience*, January 12, 2017.

Other (More or Less) Natural Causes

Autoerotic Asphyxiation

ANNUAL US DEATHS   250 to 1,000

Death by autoerotic asphyxiation (the practice of intentionally reducing blood flow to the brain as a way to increase solo sexual pleasure) is very hard to track. Also known as breath play and AEA, this dangerous practice is estimated to cause anywhere from 250 to 1,000 deaths a year in the US. Most deaths are White males; most deaths involve hanging or ligatures. Deaths from autoerotic asphyxiation are often reported as suicides to avoid family embarrassment.

In 2009, actor David Carradine, star of the 1970s TV drama *Kung Fu,* was found dead in the closet of his Bangkok hotel room. He had a cord wrapped around his neck and his genitals. The death was ruled accidental autoerotic asphyxiation.

STAY SAFE   Avoid solo breath play; have a designated rescuer nearby.

SOURCES   A. Sauvageau. Autoerotic deaths in the literature from 1954 to 2004. *Journal of Forensic Science,* 2006.

On May 5, 1968, veteran character actor Albert Dekker was found dead in his Hollywood home, kneeling in the bathtub, a noose tightly wrapped around his neck and looped around the shower curtain rod. He was also blindfolded, handcuffed, and ball-gagged. Two hypodermic needles were inserted in one arm, and his naked body bore "explicit words and drawings" in red lipstick. The coroner ruled his death the result of accidental autoerotic asphyxiation.

Dying to Win

ANNUAL US DEATHS   See data in entry.

Die-hard fans of a sports team may indeed die during stressful times for their team, such as during a playoff game. An increase in heart-related events and deaths has been noted during important sporting events such as the World Cup. In the US, a study that looked at death rates when the LA Rams were in the Super Bowl in 1980 and 1984 found differences in the death rate for 14 days following the events. In 1980, when LA lost to Pittsburgh, the death rate in Los Angeles County rose for the 14 days following the Super Bowl. The all-cause death rate per 100,000 population rose to 2.45 from 2.1; the death rate from acute heart attacks rose from 0.27 from 0.23; while the death rate for ischemic heart disease rose to 0.85 from 0.71. In 1984, when LA beat the Washington Redskins, the death rate in Los Angeles County fell for the 14 days following the Super Bowl. The all-cause death rate fell to 2.18 per 100,000 population from 2.32. In Munich during the World Cup matches from June 9 to July 9, 2006, the incidence of cardiac emergencies on days of matches involving the German team was 2.66 times that of the control period from the previous year. The incidence was 3.26 times higher for men and 1.82 times higher for women.

STAY SAFE   If you are a die-hard fan, reduce your risk of death by heart attack during important games by avoiding tobacco and not drinking to excess.

SOURCES   R.A. Kloner et al. Comparison of total and cardiovascular death rates in the same city during a losing versus winning super bowl championship. *American Journal of Cardiology*, June 2009; U. Wilbert-Lampen et al. Cardiovascular events during World Cup soccer. *New England Journal of Medicine*, January 2008; MedicareAdvantage.com, medicareadvantage.com/resources/super-bowl-fan-health-study

Energy Drink Abuse

ANNUAL US DEATHS   About 3

Energy drinks, which combine high doses of caffeine and sugar, are popular for gaining more energy and alertness, as a mixer with alcohol, and as a hangover treatment, especially among young people. Drinking too much in too short a period can be harmful. These products contain high levels of caffeine, which elevates blood pressure and heart rate, and they can impair the heart's electrical system, causing fatal arrhythmia.

Since 2005, at least 35 cases in the US have been reported where death was attributed, at least in part, to abusing energy drinks. The implied annual US deaths average about 3 people.

STAY SAFE   Consume energy drinks in moderation.

SOURCES   American College of Cardiology, acc.org/latest-in-cardiology/articles/2018/02/28/10/46/stimulant-containing-energy-drinks; FDA, fda.gov/food/dietary-supplement-products-ingredients/pure-and-highly-concentrated-caffeine

Scared to Death

ANNUAL US DEATHS   See data in entry.

In very rare instances, it is possible to be literally scared to death. Extreme fright in the face of a situation perceived to be life-threatening triggers a sudden flood of the stress hormone adrenaline. In large amounts, adrenaline is toxic and can cause a fatal heart arrhythmia. Researchers have noted increases in heart attacks following large-scale very stressful events. In the aftermath of the Northridge earthquake that struck the Los Angeles area on January 17, 1994, the coroner noted a sharp increase in the number of deaths from sudden cardiac arrest. The average daily death rate from heart disease was 4.6, but on the day of the earthquake, 24 deaths were reported. In the 14 days following the 9/11 terrorist attacks in New York City, the number of people treated for heart attacks at a hospital in nearby Brooklyn increased by 35 percent.

STAY SAFE   Avoid frightening situations if possible, especially if you have heart disease.

SOURCES   J. Feng et al. Cardiac sequelae in Brooklyn after the September 11 terrorist attacks. *Clinical Cardiology*, January 2006; J. Leor et al. Sudden cardiac death triggered by an earthquake. *New England Journal of Medicine*, February 1996.

Sexual Activity

> **ANNUAL US DEATHS**   Possibly average of 0.25 deaths

Dying in the saddle is a death many hope for but very few achieve. A 2011 study in *JAMA* found that each additional hour of sexual activity per week resulted in an increased risk of 2 to 3 myocardial infarctions and 1 sudden cardiac death per 10,000 person-years. A 2017 research letter published in the *Journal of the American College of Cardiology* reported that of more than 4,500 cases of cardiac arrest in Portland, Oregon, over a 10-year period, only 34 deaths were related to having sex in the previous hour. In a very detailed 2019 study published in *Journal of Sexual Medicine*, German researchers reviewed 16,437 autopsies over a 25-year period (1993 to 2017). They found that only 74 (43 males and 31 females), or about one-half of 1 percent, died while engaged in a range of consensual sexual practices. In 2018, a research letter in the journal *Circulation* reported that sudden cardia death in males during or within 1 hour of sexual activity occurs in approximately 0.2 percent of natural deaths, predominantly in middle-aged men.

People who die during consensual sex may become posthumously famous for it. In 1964, the Condor Club in San Francisco became famous as the first topless bar. To begin their act, scantily clad go-go dancers would descend from the ceiling while sitting on a white baby grand piano that was lowered by a hydraulic motor. In 1983, bouncer Jimmy Ferrozzo was having sex with his girlfriend on top of the piano when the motor was accidentally activated. Jimmy was crushed against the ceiling and asphyxiated; his girlfriend survived. The house specialty drink at the Condor Club is called Sex on the Piano.

Nelson Rockefeller, heir to the Rockefeller family fortune, governor of New York from 1959 to 1973, and the 41st vice president of the US under Gerald Ford, died in 1979 in New York City of a heart attack at the age of 70. He was in the apartment of his secretary; the circumstances strongly suggested he had died of a heart attack during sexual intercourse.

In 2007, a young couple fell to their deaths from an office building in Columbia, South Carolina. Their nude bodies were found by a taxi driver that morning; their clothing was later discovered on the steeply sloped roof 4 stories up.

Death by sexual activity can pay off for the victim's family. In 2009, a police officer in Georgia died of a heart attack while engaged in 3-way sex with a woman (not his wife) and another man. His widow successfully sued his doctor for malpractice because he didn't correctly diagnose the officer with atherosclerotic heart disease and warn him against strenuous sex. In 2019, the family of a French businessman, who died soon after having sex with a woman while on a business trip, won a lawsuit claiming his death was a workplace accident.

STAY SAFE   Be very careful about where you have sex.

SOURCE   L. Bunzel et al. Non-Natural Death Associated with Sexual Activity: Results of a 25-Year Medicolegal Postmortem Study. *Journal of Sexual Medicine*, October 2019.

Substance Abuse

Alcohol Abuse

> **ANNUAL US DEATHS**   95,000

Alcohol abuse includes chronic excessive consumption (alcoholism) and/or episodic excessive drunkenness (binge drinking, high-intensity drinking, acute alcohol poisoning). All significantly raise the risk of death, and available data is known to underreport the problem.

An estimated 95,000 people in the US die each year from all causes directly or indirectly related to excessive alcohol consumption. Almost half are due either to liver failure or to a prescription or street/illegal drug overdose while drunk. This makes alcohol abuse the 3rd worst preventable lifestyle-related cause of death, after excessive tobacco use, bad diet, and lack of exercise. Alcohol abuse truncates the life expectancy of each victim by an average of 29 years.

Drunk driving killed 10,600 Americans in 2020, 28.0 percent of all driving-related fatalities. Alcohol abuse also heightens the risk of murder and suicide, of deadly firearm and power tool mishaps, of drowning, and of lethal slips and falls.

Alcohol abuse elevates the chance of eventually dying of cancer, heart attack, stroke, or liver disease. Alcohol increases the potency of some medications, such as opioids, in a dangerous way; alcohol is involved in 22.1 percent of drug overdose deaths. Alcohol abuse also causes about 2,200 deaths in the US each year through acute alcohol poisoning.

Alcohol-related deaths increased in the US between 1999 and 2017. In that timespan, nearly 1 million people died of alcohol-related causes. In 1999, 35,914 death certificates mentioned alcohol; in 2017, 72,558 did, meaning the number of alcohol-related deaths doubled over an 18-year span. In 2017, alcohol played a role in 2.6 percent of all deaths in the US. Alcohol-related liver disease caused 22,245 deaths, or 31 percent; overdoses of alcohol alone or with other drugs caused 12,954 deaths, or 18 percent.

STAY SAFE   The SAMHSA National Helpline is 1-800-662-4357. Help is available 24/7/365. All calls are strictly confidential.

SOURCES   CDC, cdc.gov/alcohol; CDC, Alcohol-Related Disease Impact (ARDI) Application, nccd.cdc.gov/DPH_ARDI; Governors Highway Safety Association, ghsa.org/issues/alcohol-impaired-driving; NIH NIAAA, niaaa.nih.gov/alcohols-effects-health

By Age

The harmful effects of alcohol abuse can begin before birth. During 2019, 9.5 percent of pregnant women surveyed in the US had used alcohol in the past month. Fetal alcohol syndrome was estimated to occur in between 0.5 and 3.0 cases per 1,000 live births; an estimated 1 percent to 5 percent of first graders have fetal alcohol spectrum disorders.

Alcohol-related deaths have the highest mortality rates for ages between 45 and 74. The biggest increase in recent years has been among people ages 25 through 34. Alcohol-related deaths among younger people are also a serious problem. Alcohol kills about 3,000 Americans under age 21 each year overall and about 1,500 college students ages 18 through 24.

STAY SAFE The SAMHSA National Helpline is 1-800-662-4357. Help is available 24/7/365. All calls are strictly confidential.

SOURCE NIH NIAAA, niaaa.nih.gov/alcohols-effects-health.

By Race/Ethnicity

Alcohol-related death rates vary significantly by race/ethnicity. Age-adjusted deaths per 100,000 during 2016 gave these comparisons:

Among men, the highest rate was 113.2 among Native Americans and Alaska Natives. It was 21.9 among Hispanic men, 18.2 among White men, 13.8 among Black men, and 4.4 for Asians and Pacific Islanders.

Among women, the highest age-adjusted death rate per 100,000 was 58.8 among Native Americans and Alaska Natives. It was 7.6 among White women, 4.7 among Hispanic women, 4.6 among Black women, and 1.0 among Asians and Pacific Islanders.

STAY SAFE The SAMHSA National Helpline is 1-800-662-4357. Help is available 24/7/365. All calls are strictly confidential.

SOURCE P.A. Vaeth et al. Drinking, Alcohol Use Disorder, and Treatment Access and Utilization Among U.S. Racial/Ethnic Groups. *Alcoholism: Clinical and Experimental Research*, January 2017.

By Sex

Overall, men are more likely than women to drink excessively. In 2019, 7 percent of men had an alcohol use disorder compared with 4 percent of women. Almost 75 percent of deaths from excessive drinking are among men, or about 68,000 deaths each year.

Among drivers in fatal motor vehicle traffic crashes, men are 50 percent more likely to have been drunk compared with women.

More men than women die from alcohol abuse in the US, but this gender disparity is decreasing as heavy drinking rises faster among women. Between 1999 and 2017, alcohol-involved deaths increased 85 percent for females, compared to 35 percent for males.

Of the estimated 95,000 Americans who died from the direct and indirect effects of alcohol in 2020, 68,000, or 71.6 percent, were male, and 27,000, or 28.4 percent, were female. Women have higher risk than men for getting potentially fatal alcohol-related diseases such as heart attacks and strokes, liver failure, and cancer.

STAY SAFE   The SAMHSA National Helpline is 1-800-662-4357. Help is available 24/7/365. All calls are strictly confidential.

SOURCE   Drugabuse.gov, Sex and Gender Differences in Substance Use report, drugabuse.gov/publications/research-reports

Drunk Driving

ANNUAL US DEATHS   10,600

Drunk driving killed 10,600 Americans in 2020, 28.0 percent of all driving-related fatalities. Male drunk drivers accounted for 80 percent of all such deaths. The figures were very similar for 2017, for which more data breakdowns are available: The distribution of drunk driving fatalities by vehicle type in 2017 was 44 percent passenger cars, 19 percent pickup trucks, 18 percent SUVs, 3 percent vans, 1 percent large trucks, and 15 percent motorcycles.

Based on 2017 age-adjusted drunk driving deaths per 100,000, the 5 worst states were Wyoming at 7.60, then South Carolina at 6.23, North Dakota at 6.09, New Mexico at 5.75, and Alabama at 5.50. The 5 best states were Texas at 0.63, then New Jersey at 1.39, New York at 1.49, Minnesota at 1.52, and Utah at 1.71.

STAY SAFE   Don't drink and drive. The SAMHSA National Helpline is 1-800-662-4357. Help is available 24/7/365. All calls are strictly confidential.

SOURCE   Governors Highway Safety Association, ghsa.org/issues/alcohol-impaired-driving

Worst and Best Cities

Alcohol-related death statistics broken down by US city are incomplete, but an informative proxy measurement ranks the overall severity of residents' exposure to alcohol's various dangers. A survey published in early 2021 used a composite score weighting recent binge/heavy drinking, underage drinking, DUI driving deaths, and deaths from liver failure and chronic alcoholism. It ranked 100 major US cities.

The worst city was Reno, Nevada. The other worst 5 were Denver, Colorado, then Billings, Montana, Milwaukee, Wisconsin, and Fargo, North Dakota.

The best city was Memphis, Tennessee. Next best was Miami, Florida, followed by Jackson, Mississippi, then Salt Lake City, Utah, and Newark, New Jersey.

STAY SAFE   The SAMHSA National Helpline is 1-800-662-4357. Help is available 24/7/365. All calls are strictly confidential.

SOURCE   *Men's Health*

Worst and Best States

ANNUAL US DEATHS   95,000

The 50 states, though very diverse in population size and demographic mix, can be compared meaningfully using alcohol-related age- and sex-adjusted deaths per 100,000.

The top 5 worst states for alcohol-related deaths as of 2020 were:

1. New Mexico, 12.4 per 100,00 population
2. Montana, 7.6
3. Alaska, 7.2
4. Wyoming, 6.7
5. North Dakota, 6.3

The top 5 best states for alcohol-related deaths as of 2020 were:

1. Maryland, 1.7 per 100,000 population
2. Hawaii, 1.7
3. Alabama, 1.8
4. Louisiana, 1.8
5. Texas, 2.0

STAY SAFE   The SAMHSA National Helpline is 1-800-662-4357. Help is available 24/7/365. All calls are strictly confidential.

Texas DWI highway sign

SOURCES   CDC, cdc.gov/alcohol; CDC, Alcohol-Related Disease Impact (ARDI) Application, nccd.cdc.gov/DPH_ARDI

Urban versus Suburban/Rural

> **ANNUAL US DEATHS**   See data in entry.

US alcohol-related death rates for both males and females have been increasing faster in suburban and rural areas than in urban areas. This trend has reversed the previous relative ranking: In 2000, large metropolitan areas had the highest deaths, but by 2018, suburban and rural areas did.

In 2018, for all males over age 25, the age-adjusted death rate per 100,000 from alcohol was 26.7 in suburbs and 25.3 in rural areas, compared to 22.8 in central big cities and 17.5 on the fringes of big cities. For all females over age 25, these figures were 10.5 and 9.9 for suburbs and rural areas, compared to 8.0 and 6.9 for central big cities and the fringes of big cities.

STAY SAFE   The SAMHSA National Helpline is 1-800-662-4357. Help is available 24/7/365. All calls are strictly confidential.

SOURCE   CDC, cdc.gov/alcohol

Alcoholism/Alcohol Use Disorder

ANNUAL US DEATHS   95,000

Alcoholism, also known as alcohol use disorder (AUD), is defined by the National Institute on Alcohol Abuse and Alcoholism (NIAAA) as problem drinking that becomes severe. AUD is a chronic disease that affects about 14.1 million adults ages 18 and older, or about 5.6 percent of this age group. Among youth, an estimated 414,000 adolescents ages 12 to 17, or 1.7 percent of this age group, is affected by alcoholism.

Alcohol use disorder is deadly. An estimated 95,000 people (approximately 68,000 men and 27,000 women) die from alcohol-related causes each year, or 261 deaths per day, making alcohol the 3rd-leading cause of preventable death in the US. Alcohol abuse has a causal relationship with more than 200 diseases and related health problems listed in the bible of medical diagnosis, International Statistical Classification of Diseases (ICD-10). About 40 of them are fully (100 percent) attributable to alcohol. An example is alcoholic cardiomyopathy, a form of heart failure that kills 2 in 1 million Americans every year.

In 2018, 37,198 Americans died from alcohol-induced diseases, or 15.3 deaths per 100,000 population. The sex breakdown was 26,727 men, or 22.6 per 100,000 population, and 10,471 women, or 8.6 per 100,000 population. In 2000, the death rate for males was 3.6 times the rate for females (17.5 and 4.9, respectively); in 2018, the rate for males was 2.6 times the rate for females (22.6 and 8.6, respectively). Of all alcohol-attributable deaths in 2018, more than 70 percent involved men and more than 80 percent involved adults aged 35 or older.

SOURCE   CDC, cdc.gov/alcohol/features/excessive-alcohol-deaths. html

Alcoholic Liver Disease

ANNUAL US DEATHS   23,172 in 2018

More than half of the deaths attributable to alcohol are from the long-term health impacts, such as cancer, heart disease, and liver disease. From 2015 to 2017, there were 65,089 deaths from alcohol-related cirrhosis. In 2018, 23,172 Americans died of alcoholic liver disease, for a death rate of 7.1 per 100,000 population. Most were male: 16,099 men and 7,0773 women. The majority were White (16,438); 3,626 Hispanics died, 1,711 Blacks died, and 817 American Indians died.

In 2017, the rate of alcohol-related cirrhosis deaths was highest among those aged 55 to 64 years, with around 19.4 deaths per 100,000 population.

SOURCE   CDC, cdc.gov/alcohol

Binge Drinking

ANNUAL US DEATHS   2,200 (from alcohol poisoning)

Binge drinking is a pattern of drinking that brings a person's blood alcohol concentration (BAC) to 0.08 g/dl or above. To raise the BAC that high usually means consuming 5 or more drinks in about 2 hours for men and 4 or more drinks for women. Binge drinking is the deadliest form of excess alcohol use in the US. Although 1 in 6 US adults binge drinks about 4 times a month, consuming about 7 drinks per occasion, most people who binge drink don't have severe alcohol use disorder. However, over 90 percent of people who drink excessively report binge drinking in the past 30 days. In 2019, 25.8 percent of people ages 18 and older engaged in binge drinking in the past 30 days. Men are almost twice as likely to binge drink than women. Approximately 22 percent of men report binge drinking and on average do so 5 times a month, consuming 8 drinks per binge.

Binge drinking causes about 2,200 deaths from alcohol poisoning each year, or about 6 a day. Three out of 4 alcohol poisoning deaths are among adults ages 35 to 64. About 76 percent of those who die are White men. Binge drinking is strongly associated with risky behavior that leads to fatalities, especially drunk driving. Alcohol at very high levels can suppress the gag reflex. People who pass out from binge drinking are at risk of choking death from aspirating their own vomit.

A worrying new trend in alcohol abuse is high-intensity drinking, defined as consumption of 2 or more times the gender-specific thresholds for binge drinking, meaning 10 or more standard drinks for men and 8 or more for women per occasion. High-intensity drinking is most common among college students, often in conjunction with special events,

One result of binge drinking is unconsciousness—which may lead to asphyxia or coma.

such as holidays, spring break, and 21st birthday celebrations. High-intensity drinking fatalities are tallied with binge drinking fatalities.

STAY SAFE The more you drink, the greater your risk of death. Limit alcohol consumption to 1 drink daily for women and 2 for men. Don't drive drunk. For immediate help with alcohol use disorder, call the SAMHSA National Helpline at 1-800-662-4357. Help is available 24/7/365. All calls are strictly confidential. To find AUD treatment, use the NIAAA Alcohol Treatment Navigator at alcoholtreatment.niaaa.nih.gov.

SOURCES CDC, cdc.gov/alcohol; NIH NIAAA, niaaa.nih.gov/publications/brochures-and-fact-sheets/binge-drinking; WHO, Global Status Report on Alcohol and Health 2018, who.int/publications/i/item/9789241565639

Trends

Alcohol-related deaths more than doubled between 1999 and 2017. In 1999, 35,914 deaths were alcohol-related, for a death rate of 16.9 per 100,000 population. In 2017, 72,558 deaths were alcohol-related, for a death rate of 25.5 per 100,000 population. In 2017, 2.6 percent of all deaths involved alcohol.

STAY SAFE The more you drink, the greater your risk of death. Limit alcohol consumption to 1 drink daily for women and 2 for men. Don't drive drunk. For immediate help with alcohol use disorder, call the SAMHSA National Helpline at 1-800-662-4357. Help is available 24/7/365. All calls are strictly confidential. To find AUD treatment, use the NIAAA Alcohol Treatment Navigator at alcoholtreatment.niaaa.nih.gov.

SOURCES CDC, cdc.gov/alcohol; NIH NIAAA, niaaa.nih.gov/publications/brochures-and-fact-sheets/alcohol-facts-and-statistics; WHO, Global Status Report on Alcohol and Health 2018, who.int/publications/i/item/9789241565639

Drug Abuse and Overdose

ANNUAL US DEATHS 93,331 in 2020

In *You Bet Your Life*, the drug overdose cause of death means unintentional (accidental) death. Intentional overdoses are included under murder and suicide entries.

Most fatal drug overdoses in the US (about 70 percent) involve natural and synthetic opioid drugs such as heroin, oxycontin, and fentanyl. Stimulants such as methamphetamine and cocaine cause most other fatal overdoses. Many overdose deaths involve a combination of drugs and/or alcohol.

Fentanyl, a fast-acting synthetic opioid that is 100 times more powerful than morphine, is a significant underlying cause for the increase in overdose fatalities since 2016. The drug is used deliberately by itself and is also often added surreptitiously to other drugs to increase the effect. In 2020, there were more overdose deaths from fentanyl than overdose deaths from all drugs in 2016.

Drug abuse and overdose deaths are serious and worsening public health crises in the US. In 2019, reported drug-involved overdose deaths totaled 70,630, almost double the total of 38,329 for 2010. In 2020, overdose deaths jumped to 93,331. And for the 12-month period between April 2020 and April 2021, more than 100,000 people died of drug overdoses, up almost 30 percent from the same period a year earlier. That works out to about 275 deaths a day.

Reasons for the sharp increase include the increase in drugs tainted with fentanyl, lack of overdose prevention centers, lack of access to drug treatment during the COVID-19 pandemic, and an increase in mental health disorders such as anxiety and depression during the pandemic.

STAY SAFE   In case of a drug overdose, call 911 immediately. People at risk of opioid overdose should keep the drug naloxone (Narcan) nearby to reverse overdoses. Use fentanyl test strips to check for tainted drugs. For help with drug abuse or addiction, call the SAMHSA National Helpline at 1-800-662-4357. Help is available 24/7/365. All calls are confidential.

SOURCES   CDC, cdc.gov/drugoverdose; Drug Policy Alliance, drugpolicy.org; Governors Highway Safety Association, ghsa. org/issues/drug-impaired-driving; Kaiser Family Foundation, kff. org/statedata/collection/opioid-epidemic; NIH NIDA, drugabuse.gov/ drug-topics

By Age

Drug overdose deaths in the US are a serious and rising problem for all age groups.

In 2019, drug overdoses killed 4,777 Americans age 15 through

24; 16,375 in the age range 25 to 34; 16,859 in the age range 35 to 44; 15,083 in the age range 45 to 54; 12,896 in the age range 55 to 64; and 4,469 at ages 65 and older.

STAY SAFE   In case of a drug overdose, call 911 immediately. People at risk of opioid overdose should keep the drug naloxone (Narcan) nearby to reverse overdoses. Use fentanyl test strips to check for tainted drugs. For help with drug abuse or addiction, call the SAMHSA National Helpline at 1-800-662-4357. Help is available 24/7/365. All calls are confidential.

SOURCES   CDC, cdc.gov/drugoverdose; Drug Policy Alliance, drugpolicy.org; Governors Highway Safety Association, ghsa. org/issues/drug-impaired-driving; Kaiser Family Foundation, kff. org/statedata/collection/opioid-epidemic; NIH NIDA, drugabuse.gov/ drug-topics

By Race/Ethnicity

In 2019, drug overdoses in the US killed 35,779 White individuals; 7,331 Black individuals; 5,264 Hispanic individuals; and 1,184 people of other races and ethnicities, including Native American and Asian.

Age-adjusted death rates per 100,000 by race/ethnicity for the total US population are not available, but a study for 2019 across the state of Minnesota is representative. This mortality rate due to drug overdoses was 11.6 for White individuals, compared to 20.2 for Black individuals and 80.7 for Native Americans.

STAY SAFE   In case of a drug overdose, call 911 immediately. People at risk of opioid overdose should keep the drug naloxone (Narcan) nearby to reverse overdoses. Use fentanyl test strips to check for tainted drugs. For help with drug abuse or addiction, call the SAMHSA National Helpline at 1-800-662-4357. Help is available 24/7/365. All calls are confidential.

SOURCES   CDC, cdc.gov/drugoverdose; Drug Policy Alliance, drugpolicy.org; Governors Highway Safety Association, ghsa.org/ issues/drug-impaired-driving; Kaiser Family Foundation, kff.org/ statedata/collection/opioid-epidemic; K.M. Lippold et al. Racial/Ethnic and Age Group Differences in Opioid and Synthetic Opioid–Involved Overdose Deaths Among Adults Aged ≥18 Years in Metropolitan Areas — United States, 2015–2017. MMWR *Morbidity and Mortality Weekly Report*, November 2019; NIH NIDA, drugabuse.gov/drug-topics

By Sex

ANNUAL US DEATHS In 2019, 44,941 overdose deaths for males; 22,426 for females

The number of drug overdose deaths in the US has been rising for both males and females. It doubled for males from 2009 to 2019 and almost doubled for females. In 2019, drug overdoses caused the deaths of 44,941 males and 22,426 females.

The data is also stark when adjusted for changes in total US population over time. For males, the age-adjusted death rate per 100,000 was 29.6 in 2019. For females, it was 13.7. The comparable statistics 20 years earlier, in 1999, had been only 8.2 and 3.9, more than tripling over those 2 decades.

STAY SAFE In case of a drug overdose, call 911 immediately. People at risk of opioid overdose should keep the drug naloxone (Narcan) nearby to reverse overdoses. Use fentanyl test strips to check for tainted drugs. For help with drug abuse or addiction, call the SAMHSA National Helpline at 1-800-662-4357. Help is available 24/7/365. All calls are confidential.

SOURCES CDC, cdc.gov/drugoverdose/deaths/index.html; Drug Policy Alliance, drugpolicy.org; Governors Highway Safety Association, ghsa.org/issues/drug-impaired-driving; Kaiser Family Foundation, kff.org/statedata/collection/opioid-epidemic; NIH NIDA, drugabuse. gov/publications/research-reports/substance-use-in-women/ sex-gender-differences-in-substance-use

Worst and Best States

Based on age- and sex-adjusted death rates per 100,000 of population in 2018, West Virginia ranked deadliest for drug overdoses at 52. Then came Delaware at 44, followed by Maryland at 38, Pennsylvania at 36, and then Ohio at 35.5.

The state with the lowest age-adjusted drug overdose death rate per 100,000 in 2018 was South Dakota, at 7. Second best was Nebraska at 7.5, then Iowa at 9, North Dakota at 10, and Texas at 10.5.

In 2020, 28 states had more than a 30 percent increase in overdose deaths compared to 2019, including 10 that increased by more than 40 percent. California, Colorado, Washington, and Wyoming all had increases above 35 percent. Nine of the top 15 increases occurred in southern or Appalachian states.

STAY SAFE   In case of a drug overdose, call 911 immediately. People at risk of opioid overdose should keep the drug naloxone (Narcan) nearby to reverse overdoses. Use fentanyl test strips to check for tainted drugs. For help with drug abuse or addiction, call the SAMHSA National Helpline at 1-800-662-4357. Help is available 24/7/365. All calls are confidential.

SOURCES   CDC, cdc.gov/nchs/fastats/drug-use-illicit.htm; Kaiser Family Foundation, kff.org/statedata/collection/opioid-epidemic; NIH NIDA, drugabuse.gov/drug-topics/opioids/opioid-summaries-by-state

Worst Cities

Data on age-adjusted death rates, or on total number of deaths, is generally not available on a comparable basis for all drug overdoses combined for all big US cities. Locales can be ranked by the severity of their overall drug abuse problem, as a proxy for per capita risk of death, using data on drug overdose emergency room visits, calls to drug addiction help lines, and drug arrests.

Cincinnati, Ohio, has the worst drug abuse problem of any US city. It is a city of 300,000 people, 29 percent of whom live below the poverty level. Drug overdose deaths increased more than 10-fold in a recent 5-year period. Users mixing fentanyl and cocaine account for many of the fatalities.

Wilmington, North Carolina, has the 2nd worst drug problem. Its population is 125,000, of whom 23 percent live in poverty. Abuse of prescription opioids is a particularly severe problem there.

Louisville, Kentucky, is 3rd worst. Seventeen percent of Louisville's 770,000 people live in poverty. Methamphetamine abuse is a serious, surging problem there.

Detroit, Michigan, comes in 4th. Its population is 670,000, of whom 35 percent live below the poverty level. Fentanyl, and heroin spiked with fentanyl, account for many of Detroit's drug abuse deaths.

Dayton, Ohio, comes in 5th worst of US cities for drug-related fatalities. It has 140,000 people, with a poverty rate of 35 percent. The deadliest problems in Dayton come from fentanyl, cocaine, heroin, and prescription opioid abuse.

STAY SAFE   In case of a drug overdose, call 911 immediately. People at risk of opioid overdose should keep the drug naloxone (Narcan) nearby to reverse overdoses. Use fentanyl test strips to check for tainted drugs. For help with drug abuse or addiction, call the SAMHSA National Helpline at 1-800-662-4357. Help is available 24/7/365. All calls are confidential.

SOURCES   CDC, cdc.gov/nchs/nvss/vsrr/prov-county-drug-overdose. htm; Drug Policy Alliance, drugpolicy.org; Kaiser Family Foundation, kff.org/statedata/collection/opioid-epidemic

COVID Pandemic Overdose Death Surge

During the COVID-19 pandemic, deaths from drug overdoses, mostly opioids and synthetic opioids tainted by fentanyl but also meth, rose sharply. From April 2020 to April 2021, there were an estimated 100,306 drug overdose deaths in the US. This was an increase of 28.5 percent over the 78,056 deaths during the same period the year before. For all of 2020, drug deaths spiked by 30 percent over 2019, reaching an unprecedented high of 93,331 people. Overdose deaths rose in every state but New Hampshire and South Dakota. The estimated years of life lost come to about 3.5 million.

The COVID-19 pandemic exacerbated an existing epidemic of drug overdose deaths. The death spike is driven by social isolation, increased mental health distress, and disrupted access to support groups and treatment programs. Drugs tainted with fentanyl also contribute to the problem.

STAY SAFE   In case of a drug overdose, call 911 immediately. People at risk of opioid overdose should keep the drug naloxone (Narcan) nearby to reverse overdoses. Use fentanyl test strips to check for tainted drugs. For help with drug abuse or addiction, call the SAMHSA National Helpline at 1-800-662-4357. Help is available 24/7/365. All calls are confidential.

SOURCE   CDC, cdc.gov/nchs/nvss/vsrr/drug-overdose-data.htm

Drugged Driving

ANNUAL US DEATHS   Estimated 16,333 in 2016

Data is scarce on deaths from driving under the influence of drugs. Drunk drivers can be detected with alcohol breathalyzers and reliable field sobriety tests, but detecting drugged drivers is more difficult. Blood and urine tests performed later may produce ambiguous results, because traces of some drugs stay in the bloodstream for days—a test result may come back positive even though the person had sobered up from the drug by the time they drove. In addition, many impaired drivers are under the influence of both drugs and alcohol, or they use more than 1 drug while behind the wheel.

A study for 2016 estimated that, of the total of 37,461 US deaths on the roads from all causes, drugged driving was a factor in 43.6 percent, or 16,333 deaths that year.

STAY SAFE Don't drive drugged. In case of a drug overdose, call 911 immediately. For help with drug abuse or addiction, call the SAMHSA National Helpline at 1-800-662-4357. Help is available 24/7/365. All calls are confidential.

SOURCES CDC, cdc.gov/transportationsafety/impaired_ driving; Governors Highway Safety Association, ghsa.org/issues/drug-impaired-driving

Anabolic Steroids

ANNUAL US DEATHS No data available

Anabolic steroids are abused, usually by athletes, for faster muscle building and greater athletic strength. Hard data on their use is scarce, and death appears to be rare. One study showed that anabolic steroid abuse results in a 3 times higher rate of premature death, often due to a heart attack.

STAY SAFE In case of a drug overdose, call 911 immediately. For help with drug abuse or addiction, call the SAMHSA National Helpline at 1-800-662-4357. Help is available 24/7/365. All calls are confidential.

SOURCE H. Horwitz et al. Health consequences of androgenic anabolic steroid use. *Journal of Internal Medicine*, November 2018.

Anabolic steroid tablets

Antidepressants

ANNUAL US DEATHS 5,175 in 2019

In 2018, 7.2 percent of US adults surveyed reported having an episode of major depression within the past 12 months. Antidepressants (Zoloft, Paxil, Prozac, Lexapro) are commonly prescribed to treat this, and when taken as directed they are generally safe and effective. Because of their psychoactive effects, elevating mood and inducing euphoria, they are sometimes abused.

Antidepressant overdoses killed 5,175 people in the US in 2019. These fatalities mainly resulted from the victims simultaneously abusing opioids and/or alcohol.

STAY SAFE   In case of a drug overdose, call 911 immediately. For help with drug abuse or addiction, call the SAMHSA National Helpline at 1-800-662-4357. Help is available 24/7/365. All calls are confidential.

SOURCES   CDC, CDC, cdc.gov/drugoverdose/deaths; NIH NIDA, drugabuse.gov/drug-topics/trends-statistics/overdose-death-rates

Barbiturates

ANNUAL US DEATHS   3,000+

Abuse of barbiturates (phenobarbital, Seconal, Nembutal) is estimated to kill over 3,000 people per year in the US.

STAY SAFE   In case of a drug overdose, call 911 immediately. For help with drug abuse or addiction, call the SAMHSA National Helpline at 1-800-662-4357. Help is available 24/7/365. All calls are confidential.

SOURCES   CDC, cdc.gov/drugoverdose/deaths; NIH NIDA, drugabuse.gov/drug-topics/trends-statistics/overdose-death-rates

Bath Salts

ANNUAL US DEATHS   No cumulative data available but known to be a drug of potentially lethal potency

Synthetic cathinone, known by the street name bath salts, have no relation to Epsom salts. They are chemically similar to the cathinone present in khat, a shrub found in parts of Africa and the Middle East. Chewing khat leaves is mildly stimulating and suppresses appetite. The synthetics have potency so high that they can kill people, but no reliable data on deaths from bath salts is available. Isolated incidents of suicide and of family annihilation have been reported.

STAY SAFE   In case of a drug overdose, call 911 immediately. For help with drug abuse or addiction, call the SAMHSA National Helpline at 1-800-662-4357. Help is available 24/7/365. All calls are confidential.

SOURCES   CDC, cdc.gov/drugoverdose/deaths; NIH NIDA, drug-abuse.gov/drug-topics/trends-statistics/overdose-death-rates

Benzodiazepines

ANNUAL US DEATHS   9,711 in 2019

Benzodiazepines (Valium, Xanax, Ativan) killed 9,711 people in the US in 2019. These fatalities occurred mainly because the victims simultaneously abused opioids.

STAY SAFE   In case of a drug overdose, call 911 immediately. For help with drug abuse or addiction, call the SAMHSA National Helpline at 1-800-662-4357. Help is available 24/7/365. All calls are confidential.

SOURCES   CDC, cdc.gov/drugoverdose/deaths; NIH NIDA, drugabuse.gov/drug-topics/trends-statistics/overdose-death-rates

Club Drugs

ANNUAL US DEATHS   No data available

Club drugs (MDMA/Ecstasy, GHB) are typically used to enhance sensory experiences and reduce interpersonal inhibitions at social gatherings involving music and dancing. Some, such as GHB (rohypnol), can be misused as date rape drugs. They are more potent if taken with alcohol, as is common at raves and nightclubs. Information on deaths caused by club drugs is very scarce. Although it's possible to overdose and experience severe toxic effects, most deaths from club drugs involve multiple drugs.

STAY SAFE   In case of a drug overdose, call 911 immediately. For help with drug abuse or addiction, call the SAMHSA National Helpline at 1-800-662-4357. Help is available 24/7/365. All calls are confidential.

SOURCES   CDC, cdc.gov/drugoverdose/deaths; NIH NIDA, drugabuse.gov/drug-topics/trends-statistics/overdose-death-rates

Cocaine and Crack

ANNUAL US DEATHS   15,883 in 2019

Cocaine in various forms killed 15,883 people in the US in 2019.

STAY SAFE   In case of a drug overdose, call 911 immediately. For help with drug abuse or addiction, call the SAMHSA National Helpline at 1-800-662-4357. Help is available 24/7/365. All calls are confidential.

SOURCES   CDC, cdc.gov/drugoverdose/deaths; NIH NIDA, drugabuse.gov/drug-topics/trends-statistics/overdose-death-rates

Fentanyl

ANNUAL US DEATHS   36,359 in 2019

Fentanyl and related drugs are synthetic opioids (other than methadone) that are extremely potent and deadly even in very small amounts. They killed more than 36,359 people in 2019. Many street drugs are now contaminated with added fentanyl. In 2020, overdose deaths at least partially attributable to fentanyl and other synthetic opioids increased by around 20,000 (54 percent). Deaths from overdoses of cocaine increased 21 percent, and deaths from methamphetamine and similar drugs rose by 46 percent. By comparison, in 2015, synthetic opioids were involved in only 18 percent of all overdose deaths.

STAY SAFE   In case of a drug overdose, call 911 immediately. People at risk of opioid overdose should keep the drug naloxone (Narcan) nearby to reverse overdoses. Use fentanyl test strips to check for tainted drugs. For help with drug abuse or addiction, call the SAMHSA National Helpline at 1-800-662-4357. Help is available 24/7/365. All calls are confidential.

SOURCES   CDC, cdc.gov/drugoverdose/deaths; NIH NIDA, drugabuse.gov/drug-topics/trends-statistics/overdose-death-rates

Hallucinogens

ANNUAL US DEATHS   No data available

LSD is effective in such low doses (micrograms) that death from overdose is exceedingly rare. Despite sensationalized media reports, LSD use has over decades been the primary cause of only a few suicides or accidental fatalities. The potential for untimely death is due to the warped perceptions and consequent aberrant behaviors that any hallucinogen can trigger.

Mescaline comes from the peyote cactus, and some mushrooms found in hot/humid climates produce psilocybin. These natural hallucinogens can also be made synthetically. Their effects and risks are similar to those for LSD, including the danger of toxic contaminants introduced along the illicit supply chain.

STAY SAFE In case of a drug overdose, call 911 immediately. For help with drug abuse or addiction, call the SAMHSA National Helpline at 1-800-662-4357. Help is available 24/7/365. All calls are confidential.

SOURCES CDC, cdc.gov/drugoverdose/deaths; NIH NIDA, drug-abuse.gov/drug-topics/trends-statistics/overdose-death-rates

Heroin

ANNUAL US DEATHS 14,019 in 2019

Heroin is a form of refined opium, which is itself made from poppy plants. Sometimes heroin is taken in combination with other opioids such as fentanyl or is used instead of a prescription opioid. Heroin killed 14,019 people in the US in 2019.

STAY SAFE In case of a drug overdose, call 911 immediately. For help with drug abuse or addiction, call the SAMHSA National Helpline at 1-800-662-4357. Help is available 24/7/365. All calls are confidential. Naloxone can help an opioid overdose victim if given promptly.

SOURCES CDC, cdc.gov/drugoverdose/deaths; NIH NIDA, drugabuse.gov/drug-topics/trends-statistics/overdose-death-rates

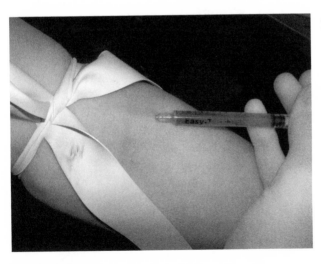

An addict injecting heroin

Inhalants

ANNUAL US DEATHS   Estimated 100 to 200

Inhalants include glues, paint thinner, gasoline, cleaning products, flammable gases such as butane or propane, chloroform, and laughing gas (nitrous oxide). Hard data is scarce, but studies estimate that about 100 to 200 people per year are killed by inhalant abuse in the US. Death is usually from heart arrythmias followed by cardiac arrest (sudden sniffing death syndrome).

STAY SAFE   In case of a drug overdose, call 911 immediately. For help with drug abuse or addiction, call the SAMHSA National Helpline at 1-800-662-4357. Help is available 24/7/365. All calls are confidential.

SOURCES   CDC, cdc.gov/drugoverdose/deaths; NIH NIDA, drugabuse.gov/drug-topics/trends-statistics/overdose-death-rates

Ketamine, PCP, and Other Dissociative Drugs

ANNUAL US DEATHS   Insufficient data available

In addition to changed sensory perceptions like those produced by naturally occurring hallucinogens, dissociative drugs such as ketamine, PCP (angel dust), salvia, and others also cause the user to feel detached, dissociated, or floating away from their body and their environment. Ketamine is approved by the FDA to treat severe depression when used under medical supervision.

Dissociative drugs in high doses can cause suicidal thoughts or trigger respiratory distress, seizures, or comas leading to death. They can also trigger delirium, psychosis, and violent behavior, leading to injury or death of the user and/or bystanders and police. Studies in Asia showed that drivers under the influence of dissociative drugs caused a disproportionately high number of fatal car accidents, compared to DUIs that instead involved alcohol or other controlled substances. Hard data for the US is scarce.

STAY SAFE   In case of a drug overdose, call 911 immediately. For help with drug abuse or addiction, call the SAMHSA National Helpline at 1-800-662-4357. Help is available 24/7/365. All calls are confidential.

SOURCE   CDC, cdc.gov/drugoverdose/deaths

Methamphetamine

ANNUAL US DEATHS 16,167 in 2019

Prescription methamphetamines such as Ritalin or Concerta, and amphetamines such as Aderall or Dexedrine, are stimulants used to treat attention deficit hyperactivity disorder (ADHD) and narcolepsy (a sleeping disorder). Although these drugs are sometimes abused or diverted, illegal methamphetamine (meth, speed, ice, crank) is the more serious concern. Methamphetamine abuse and other psychostimulants with abuse potential killed 16,167 people in 2019.

STAY SAFE In case of a drug overdose, call 911 immediately. For help with drug abuse or addiction, call the SAMHSA National Helpline at 1-800-662-4357. Help is available 24/7/365. All calls are confidential.

SOURCES CDC, cdc.gov/drugoverdose/deaths; NIH NIDA, drugabuse.gov/drug-topics/trends-statistics/overdose-death-rates

Opioids

ANNUAL US DEATHS 14,139 in 2019

Prescription opioids (oxycontin, oxycodone, morphine, methadone) are legitimately used for relief of severe pain, but can lead to addiction, abuse, and overdose death. These drugs are often illegally diverted from legitimate distribution and sold as street drugs. Prescription opioid overdoses killed 14,139 people in the US in 2019.

STAY SAFE In case of a drug overdose, call 911 immediately. For help with drug abuse or addiction, call the SAMHSA National Helpline at 1-800-662-4357. Help is available 24/7/365. All calls are confidential. Naloxone can help an opioid overdose victim if given promptly.

SOURCES CDC, cdc.gov/drugoverdose/deaths; NIH NIDA, drugabuse.gov/drug-topics/trends-statistics/overdose-death-rates

Synthetic Cannabinoids

ANNUAL US DEATHS Sporadic lethal outbreaks in the low double digits

Although natural marijuana and its derivative, hashish, are generally not direct causes of mortality, synthetic cannabinoids (spice, K2) have been known to kill people. Synthetic cannabinoids contain brodifacoum, the same blood-thinning agent present in rat poison. They can cause severe bleeding, requiring hospitalization. In 2015, a spate of abuse in Mississippi caused 17 deaths over 2 months. In 2018, in an outbreak of several hundred spice-related hospitalizations across 11 states, there were 8 more deaths. In 2021 in Florida, at least 2 people died and at least 41 were hospitalized after using synthetic cannabinoids.

STAY SAFE In case of a drug overdose, call 911 immediately. For help with drug abuse or addiction, call the SAMHSA National Helpline at 1-800-662-4357. Help is available 24/7/365. All calls are confidential.

SOURCES CDC, cdc.gov/drugoverdose/deaths; NIH NIDA, drug-abuse.gov/drug-topics/trends-statistics/overdose-death-rates

Contributory Factors

Air Pollution

> **ANNUAL US DEATHS** 60,200 to 200,000 excess deaths

Air pollution, defined as increased ambient exposure to particulate matter 2.5 nanometers (PM2.5) or smaller is associated with numerous health problems, such as lung disease, asthma, and heart disease and contributes to all-cause mortality from these and other diseases. Particulate matter is the cause of most deaths related to air pollution, but ground-level ozone (found in smog) is another important contributor.

In the US, deaths attributable to air pollution have dropped significantly since 1991, when PM2.5 began decreasing due to improved regulation under the Clean Air Act. In 1990, an estimated 103,700 people died prematurely from exposure to air pollution; in 2019, 60,200 people died. By some estimates, that number should be much higher, somewhere between 85,000 to 200,000 excess deaths each year. Worldwide, more than 8 million people, including 1.7 million children, died in 2018 from fossil fuel air pollution—a little less than 1 out of every 5 deaths, or about 18 percent of total global deaths.

Factory smoke and other sources of smoky particulates are highly visible sources of air pollution.

Compared with White Americans, Black Americans are disproportionately affected by exposure to air pollution from every type of source. People of color in general are exposed to more pollution from almost every type of source,

Air pollution from farms leads to approximately 17,900 deaths each year. Almost all are from fine particulate pollution related to animal agriculture.

Assuming no improvement in climate change mitigation effort, combined ozone and particle health effects will cause an estimated 1,000 to 4,300 additional premature deaths nationally per year by 2050.

STAY SAFE   Stay indoors when air pollution levels are high. Check the EPA Air Quality Index (AQI) at airnow.gov. Use home air filters with HEPA filters.

SOURCES   CDC, cdc.gov/air/pollutants.htm; EPA, epa.gov/environmental-topics/air-topics; N.G.G. Domingo et al. Air quality-related health damages of food. *Proceedings of the National Academy of Sciences USA*, May 2021; C.W. Tessum et al. PM2.5 polluters disproportionately and systemically affect people of color in the United States. *Science Advances*, April 2021.

Homelessness

ANNUAL US DEATHS   Contributed to between 17,500 and 46,500 deaths in 2018

In 2020, an estimated 580,466 people experienced homelessness in the US on a single night, or about 18 per 100,000 population. This number is an increase of 12,751 people, or 2.2 percent, from 2019. People experiencing homelessness (PEH) in the US die at higher rates than the general population. Because the number of PEH in the US is unknown, and the housing status of someone who dies is rarely recorded, statistics in this area are inconsistent and very approximate. About 553,000 people (probably a significant undercount) were homeless in 2018. Among these people, somewhere between 17,500 and 46,500 died.

Homeless deaths have increased in recent years. In New York City, for example, 177 deaths were recorded in 2008, rising to 290 in 2018, for an increase of nearly 50 percent. In Los Angeles, 518 deaths were recorded in 2008, rising to 1,038 in 2019. The state of California has the highest homeless population, with about 151,278 homeless people in 2020, or about 20 percent of the total homeless population in the US.

Causes of death among PEH are inconsistently reported. Accidental death rates are high; in Denver in 2019, 47 percent of deaths were accidents. In Los Angeles, 24 percent of deaths were due to trauma or violence. In New York City, 32 percent of deaths were due to substance abuse. Homeless people also die disproportionately from COVID-19. As of December 19, 2021, for example, LA County reported 267 COVID-19 deaths among people experiencing homelessness. COVID-19 deaths among this population are very difficult to track accurately, however, and the number of deaths is almost certainly an undercount, with some deaths probably attributed to other causes.

From March to August 2020, 104 homeless individuals died in New York City, 95 of them in shelters. The mortality rate among PEH for this period was 4 per 1,000, or 78 percent higher than the city's overall mortality rate.

Military veterans compose 11 percent of the US homeless adult population, or about 37,252 individuals in 2020. They have a 32.3 percent increase in all-cause mortality compared to nonhomeless veterans.

SOURCES   National Health Care for the Homeless Council, nhchc. org/understanding-homelessness/fact-sheets; Department of Housing and Urban Development (HUD), 2020 Annual Homeless Assessment Report Part 1, huduser.gov; J.A. Schinka et al. Mortality and Cause of Death in Younger Homeless Veterans. *Public Health Reports*, March/April 2018; US Interagency Council on Homelessness, usich.gov/tools-for-action

EVERYTHING
ELSE

ACCIDENTAL HARM AND INJURY

Acts of God and Nature

Asteroid/Comet Impacts

ANNUAL US DEATHS Very near 0 in recorded history, but extinction events are possible

Unlike a meteor, which is the smallest space object that could do any real damage by impacting Earth, asteroids and comets are much more substantial in size and mass. One entering the atmosphere would cause severe regional or even planetwide devastation. Modern human civilization, and even all life on our planet, could be wiped out. Such impacts are known by scientists to have occurred several times in the planet's 4.5-billion-year history.

The most recent such near-extinction event was the Chicxulub, Mexico, asteroid impact, 66 million years ago, which wiped out the dinosaurs (except for birds, which are a type of dinosaur). This impact is estimated to have had the force of almost 1 trillion Hiroshima-sized atomic bombs. More recently, the "Tunguska event," an explosion in Tunguska, Siberia, on June 30, 1908, is usually attributed to an asteroid. Some scientists, however, believe it was actually an icy comet that completely vaporized in midair (no impact crater was ever found) with the force of about 1,000 Hiroshima A-bombs. Because the area was a sparsely populated wilderness at the time, only 2 or 3 people were killed, but many animals died, and some 80 million trees over 830 square miles were knocked down. It is the largest impact event to hit Earth in recorded history. About 3,600 years ago, a Tunguska-sized airburst may have destroyed Tall el-Hammam in the Jordan Valley near the Dead Sea. This may be the historic event behind the Biblical story of the destruction of Sodom and Gomorrah.

Asteroids large enough to be potentially dangerous do speed by

Nineteenth-century newspaper illustration of the discovery of a comet at Greenwich Observatory, outside of London

Earth, at a safe distance, with some regularity. For instance, on June 1, 2021, an asteroid called 2021 KT1, described by NASA's Center for Near-Earth Object (NEO) Studies as "the size of a skyscraper," passed by at a closest distance of about 4.5 million miles—about 18 times the distance to the moon and about one-twentieth the distance from the Earth to the sun. The asteroid was only 1 of 7 NEOs that came near Earth that same week. NASA researchers have calculated that Bennu, a 1,700-foot-wide asteroid discovered in 1999 and visited by the spacecraft ORISIS-REx mission in 2018, has a 1 in 1,750 chance of hitting Earth between now and the year 2300. In 2021, NASA launched the Double Asteroid Redirection Test (DART) to learn if a spacecraft can hit a space rock and change its trajectory.

STAY SAFE Urge the US government and other world authorities to continue funding and urgently developing solar system near-Earth object-tracking systems and effective planetary defense systems to protect life on Earth against threatening space object impactors.

SOURCES T.E. Bunch et al. A Tunguska sized airburst destroyed Tall el-Hammam a Middle Bronze Age city in the Jordan Valley near the Dead Sea. *Science Reports*, September 2021; NASA Center for Near Earth Object Studies, cneos.jpl.nasa.gov/ca/intro.html

Avalanches

ANNUAL US DEATHS 9.8 to 27

A 1999 study of US snow avalanche deaths over the previous 45-year period found 440 deaths in 324 avalanches, each with at least 1 death. The average annual death toll is 9.8 people per year. It also indicated a death toll averaging 1.36 people per avalanche. Of the victims, 87.7 percent were fully buried by the avalanche that killed them, 4.7 percent were partially buried, and 7.6 percent were found not buried. A more recent study of the 10-year period extending through the winter of 2010 to 2020 indicated that the average annual avalanche death toll in the US had risen significantly, to about 27 people per year. Causes include more people going further into backcountry areas with greater risk and climate warming processes that are causing more unstable snow covers. Globally, about 150 people are killed in avalanches each year.

Of the victims in the earlier study, 87.3 percent were male and 12.7 percent were female. The victims included rock and mountain climbers (25.5 percent), backcountry skiers (22.7 percent), out-of-bounds skiers (10.0 percent), snowmobilers (6.8 percent), in-bounds skiers (5.2 percent), ski patrol and various workers (7.2 percent), local residents (4.5 percent), and motorists (3.0 percent).

By state, the most deaths occurred in Colorado, with 33.0 percent of all deaths, followed by Washington, with 13.2 percent, and Alaska, with 12.0 percent. Almost all fatalities happened in the months of January through April.

STAY SAFE   Most avalanches occur on slopes greater than 30 degrees. Stay on designated trails to benefit from avalanche prevention strategies at parks, trails, and ski resorts. Wear an avalanche locator beacon to aid rescuers in finding you. Heed local forecasts and warning flags for high avalanche risk conditions and go somewhere safer instead. Take an avalanche safety and awareness training course.

SOURCES   Avalanche.org/avalanche-accidents; Colorado Avalanche Information Center, avalanche.state.co.us/accidents/statistics-and-reporting; NWS, weather.gov/safety/winter-avalanche

Bee, Wasp, and Hornet Stings

ANNUAL US DEATHS   Average of 62

Stings from bees, wasps, and hornets killed 1,109 people in the US between 2000 and 2017, for an average of 62 deaths per year. The annual figures fluctuate, but there has been a steady upward trend from 2012 (about 60 deaths) to 2017 (about 90 deaths). About 80 percent of all deaths are in men. The lifetime odds of dying from a hornet, wasp, or bee sting are 1 in 59,507. Death from these flying insects is usually the result of a severe allergic reaction (anaphylactic shock) to the venom.

Bees, wasps, and hornets mostly sting to protect their hives if disturbed; some species will simply repel you from near their home, while others will aggressively chase after you for considerable distances. A bee's stinger has a large barb, which gets yanked out of the bee's body when it stings; it can only sting once and then dies. Hornets, which can be over 2 inches long, are actually large wasps. Wasp and hornet stingers do not break off, so a single insect can sting you repeatedly.

STAY SAFE   Be alert for stinging insects when you're outdoors. Call a professional pest removal service to remove nests that are on homes and other structures or near areas with high foot traffic. If you know you or your child is allergic to insect stings, carry a fresh epinephrin auto-injector (EpiPen) and know how to use it. If you are stung multiple times or know you are allergic, seek emergency medical help immediately!

SOURCES   CDC NIOSH, cdc.gov/niosh/topics/insects/beeswasphornets.html; *QuickStats*: Number of Deaths from Hornet, Wasp, and Bee Stings, Among Males and Females—National Vital Statistics System, United States, 2000–2017. MMWR *Morbidity and Mortality Weekly Report*, July 2019.

Droughts

ANNUAL US DEATHS   No data available; estimated 7,000 drought-related deaths during the Great Depression–era Dust Bowl (1934, 1936, 1939–1940)

Serious droughts (lack of rainfall) are often accompanied by prolonged, very hot weather. Droughts can devastate crops and impair water supplies needed by humans and animals.

In the US, deaths brought on in recent years by drought conditions have usually resulted from dehydration, hyperthermia (heat stroke), or wildfires. Individual records of causes of death are generally kept only about these specific diagnoses. National data on US deaths attributable specifically to droughts are not readily available and need more study.

America's worst-ever drought was the 1930s Dust Bowl dry spell and heat wave in the Great Plains, which overlapped with the very tough era of the Great Depression. Historians estimate that over a period of several years, about 7,000 people died from Dust Bowl–related causes such as starvation, respiratory disease ("dust pneumonia"), and transportation accidents while migrating out of the worst affected areas.

Drought is a very serious problem around the globe, both in economic devastation and in loss of life. The World Meteorological Organization (WMO) estimates that, globally, about 16,000 people die because of regional drought conditions every year.

STAY SAFE   See entry on **Heat-Related Deaths**

SOURCES   National Integrated Drought Information System, drought.gov; World Meteorological Organization (WMO), public.wmo.int/en/our-mandate/water/drought

Earthquakes

> **ANNUAL US DEATHS**   3

From 1990 through 2012 in the US, 70 people were killed by earthquakes. This is an average annual death toll of 3 people per year. Earthquakes occur continually, but most are so weak that they are detected only by sensitive seismographs, not by people.

The deadliest earthquake in US history was the Great San Francisco Earthquake of April 1906. It had a magnitude of 7.9 and caused massive fires, leading to a total death toll of over 3,000. The fires caused most of the deaths. The most powerful earthquake ever recorded in the US occurred in Alaska in March 1964. It had a magnitude of 9.2, causing widespread massive destruction of buildings and utilities, plus earth fissures, landslides, soil liquefaction, and a devastating tsunami in Prince William Sound. In some areas, the terrain was permanently raised by 30 feet, while in others it dropped by 8 feet. About 140 people were killed.

STAY SAFE   If you live in an earthquake-prone area, have emergency plans and practice them periodically with family and with co-workers. In case of an earthquake while indoors, stay inside, take shelter in a doorway or under a strong table or desk, and hold on, away from the

Aftermath of the Great San Francisco Earthquake of 1906. Fire killed more people than the quake itself.

kitchen, fireplaces, windows, appliances, and other furniture. Be wary of aftershocks and listen for emergency information from local authorities. If in a vehicle, get out of traffic, stop in the open on solid ground away from power lines, and stay in the vehicle until the tremors subside. Homeowners, business owners, and building owners should do an engineer's assessment and take earthquake risk-mitigation measures, such as retrofitting structural reinforcement if needed. Go to fema.gov for more info, such as their "Designing for Earthquakes" manual.

SOURCES FEMA, ready.gov/earthquakes; PolicyAdvice, policyadvice.net/insurance/insights/earthquake-statistics; USGS, usgs.gov/programs/earthquake-hazards/earthquakes

Flooding

ANNUAL US DEATHS About 250

Flooding can occur from coastal seawater flooding from a king tide (highest predicted high tide of the year in a coastal location) or storm surge; flash flooding or river flooding from a distant and/or nearby heavy rainstorm or hurricane; or from the sudden collapse of an ice dam, a man-made dam, a retaining pond or storage pool, a water main break, or an aqueduct failure.

Flooding from various causes kills about 250 people a year in the US. There is some uncertainty in this figure because (for instance) some deaths caused by hurricanes, which were actually caused by drowning in resulting floodwaters, might not be reported separately from hurricane deaths by other causes. In some cases, a death might be double-counted as due to both flooding and hurricane. In addition, some death certificates might record the death as drowning and not note that the cause was flooding. People die in floods from other causes, including blunt force trauma, electrocution, exposure to toxins in floodwater, and hypothermia.

STAY SAFE Safety from flash floods and river floods begins with situational awareness. Know in advance if you live or work on the floodplain of a major river or in an area subject to flash floods. Be prepared with a flooding safety plan and survival supplies. (See the CDC's bulletin "Floodwater After a Disaster or Emergency.") If water is flowing strongly, stay away from banks that might crumble underfoot and from makeshift bridges that might collapse; don't go swimming if the current is strong enough to carry you away or drown you. Do not enter floodwaters higher than your ankles; keep children and pets out of danger; be especially careful at night when hazardous conditions are hardest to see. Do not drive a vehicle into roadway flooding, unless you are very familiar with the area and are sure the floodwaters are shallow and flowing weakly. Be aware that floodwaters can be contaminated by raw sewage, toxic pollutants, and/or rotting carcasses and corpses. Floodwaters can also carry dangerous moving debris such as uprooted trees and can also be deadly due to live voltages from downed power lines. Tap water may become unsafe to drink. Obey public health warnings to boil water or use bottled water. If you come into contact with floodwater, thoroughly decontaminate yourself afterward using hand sanitizer, diluted bleach, or soap with water you know to be germ-free.

SOURCES CDC, cdc.gov/disasters/floods; NOAA, nssl.noaa.gov/education/svrwx101/floods; NWS, weather.gov/arx/usflood

Dam Failures

ANNUAL US DEATHS 14 annually from 1966 to 1985 but rising in recent years

There are on average about 10 dam failures a year across the US, varying greatly in size and lethality. Dams can be either constructed or natural, and either type can vary from small to massive both in size and in how much water the dam holds back under normal circumstances. Constructed dams are usually built of earth and/or reinforced concrete, and they can fail due to overtopping (about one-third of failures), foundation collapse (about one-third of failures), cracking, clogging of spillways, or progressive seepage and piping (about one-fifth of failures). Natural dams can result from clogging by ice on a river (ice dam), beaver activity, or a seismic event such as a landslide, earthquake, or volcano. Natural dams can occur and then fail very unpredictably, especially during periods of prolonged, heavy rainfall and/or sudden warming that melts built-up ice. Constructed dams usually give some advance warning of an impending failure, which can be detected by regular engineering inspection.

A study for the years 1966 through 1985 found an average of 14 people per year were killed by dam failures in the US. However, due to a combination of increased development, climate change, and infrastructure neglect, the risk of dam collapse catastrophes in America may be increasing.

From 2005 to mid-2013, 173 dam failures were reported in the US. Death tolls are not readily available for all of these but appear to be few to none in each case. During the 1970s, 4 different dam failures around the US killed between about 30 and 250 people each.

The worst-ever dam failure in US history was the Johnstown Flood, due to a dam collapse in western Pennsylvania in 1889. An old earthen dam had been restored by a private club to create an exclusive sailing and fishing lake, but the heightened dam was badly designed and poorly maintained. Lacking proper spillways and working emergency drainage gates, the earth barrier gave way due to overtopping by heavy rainfall. Suddenly, some 20 million tons of water plowed downstream to and then through the town, picking up more and more debris. This debris then piled up against a great stone bridge on the downstream side of Johnstown. The massive debris pile, including uprooted trees, railroad cars, and entire buildings, caught fire. More than 2,200 people were killed.

STAY SAFE An estimated 1,600 to more than 15,000 dams around the US are reportedly in poor condition and could kill people if they failed. Urge your state and congressional representatives to fund and implement infrastructure renewal programs so that these dangers can be averted.

SOURCES Association of State Dam Safety Officials, damsafety. org/dam-failures; National Performance of Dams Program, Stanford University, npdp.stanford.edu/dam_failures_us

Flash Floods and River Floods

ANNUAL US DEATHS 118 but rising in recent years

From 2015 through 2019 in the US, flash floods and river floods killed an average of about 118 people per year. There has been a rising trend in the death toll in recent decades, in part due to increased habitation in areas that are at risk, and also due to changes in weather and rainfall patterns.

STAY SAFE Before venturing into a dry arroyo or gully, check the weather forecast. Be aware of rain at higher elevations that drain into the channel; flash floods can originate at considerable distances, and then arrive suddenly as a deadly wall of water.

SOURCES NOAA, nssl.noaa.gov/education/svrwx101/floods; NWS, weather.gov/arx/usflood

King Tides

ANNUAL US DEATHS No available data

A king tide is an unusually high tide that occurs 2 to 4 times per year, depending on the geographic area, and is caused by astronomical conditions rather than Earth weather. The sun and moon line up so that their gravitational pull is especially strong. The height of a king tide above the annual average of high tides in an area also varies geographically; the greater the normal tidal range, as a rule, the higher above normal is a local king tide. (There are also unusually low tides twice a year, when the gravitational pull of the sun and moon is especially weak.)

In the US, deaths due to king tides are rare when not accompanied and exacerbated by windstorm-driven coastal surge floods. King tide flooding onto coastlines and up estuaries can be dangerous to people and damaging to property. In January 2021, 2 young children were killed when they and their father were pulled out to sea by a "massive swell" caused when a king tide overlapped with a major storm. (See also **Storm Surges and Storm Tides**.)

King tides cause what is called, in low-lying areas of the South (such as Savannah, Georgia, Norfolk, Virginia, and southern Florida), "sunny day flooding," in which parks, freeway underpasses, and other properties at low elevations near the coast are temporarily inundated. This global problem is expected to get worse over time, as sea levels continue to rise due to climate change.

STAY SAFE Check the calendar for king tides in your area, and check the weather forecast for coastal storms near you. Be extra careful visiting an ocean beach or shoreline, especially if a king tide coincides with a major storm, as the combination can cause extralarge waves and/or severe surge flooding driven by onshore winds. Do not drive your vehicle into road flooding, unless you are very familiar with that area, and you are sure the flooding is only of minor depth. For more information on king tide safety, see the bulletin "King Tides" from Palm Beach County's Office of Resilience at discover.pbcgov.org.

SOURCES NOAA, oceanservice.noaa.gov/facts/kingtide.html; Palm Beach County, Office of Resilience, discover.pbcgov.org/resilience/Pages/King-Tides.aspx

Storm Surges and Storm Tides

ANNUAL US DEATHS No available data; individual occurrences can be deadly

A storm surge is an abnormally large rise in the level of the water along a shoreline, caused by sustained high onshore winds from a major storm such as a hurricane. These winds, via their constant friction with the water's surface, cause water to bunch up against the shore and surge inland. A storm tide is the coincidental, mutually strengthening combination of a storm surge, due to weather events, and a king tide, due to astronomical events. (See **King Tides**.) In the US, storm tides can reach 20 feet or more above regular high tides, causing massive flooding.

The shape of the shoreline is a factor in storm-surge and storm-tide severity. A shallow slope to the inshore seafloor, and/or a funnel-like inward narrowing of the local shoreline, can channel water into very dangerous flooding. Large, powerful waves may add further destructive power to a storm surge or storm tide. Extrastrong sea currents can cause severe coastline erosion, and salt intrusion from seawater flooding can cause costly economic and environmental damage.

One of the worst storm surges in US history was caused by Hurricane Katrina in 2005, in New Orleans and elsewhere along the Gulf of Mexico coast. A storm surge estimated at 28 feet above normal tide level was a major contributing factor in over 1,600 deaths. Another severe storm surge, about 14 feet above normal, hit the New York City area during Hurricane Sandy in 2016, causing $42 billion in damage there and killing 147 people in the US (53 of them in NYC). The entire East Coast from Florida to Maine was hit by this unusually wide cyclonic storm system.

A storm tide hit the east coast of Florida in 2019 when Hurricane Dorian (by then reduced to a Category 2 storm) made landfall there during a coincidental king tide. At least 6 people in the US were killed by Dorian, although 4 of these deaths were due to accidents such as falls or electrocutions during storm preparation and cleanup, not flooding. Dorian hit the Bahamas earlier, while still a Category 5 hurricane, killing at least 75 people there.

STAY SAFE Check the calendar for king tides in your area, and check the weather forecast for coastal storms near you. Be extra careful visiting an ocean beach or shoreline, especially if a king tide coincides with a major storm, as the combination can cause extralarge waves and/or severe surge flooding driven by onshore winds. Do *not* drive your vehicle into road flooding, unless you are very familiar with that area, and you are sure the flooding is only of minor depth.

SOURCE   NOAA, National Hurricane Center, nhc.noaa.gov/surge

Tsunamis

ANNUAL US DEATHS   Average of 2

A tsunami is a massive tidal wave caused by an undersea or coastal earthquake, volcano, and/or landslide. Since 1737, 72 tsunamis have hit some part of the US, with a total of 548 deaths reported. The annual death toll over this 274-year period averages exactly 2 people per year.

The single most deadly tsunami to hit the US, with a wave height of almost 130 feet, resulted from a magnitude 8.6 seismic event in March 1946, at Unimak Island, Alaska, which killed 164 Americans. This tsunami also hit 6 other countries. The next most deadly tsunami in the US, with a wave height of over 220 feet, occurred in Prince William Sound, Alaska, and killed 124 Americans. It was caused by the Great Alaskan Earthquake of March 1964, a magnitude 9.2 megathrust quake that was the strongest ever in the US and the 2nd strongest ever recorded on Earth. This tsunami also hit 5 other countries.

STAY SAFE   If you are in an area that receives a tsunami warning, proceed immediately to high ground.

SOURCES   NOAA Tsunami Program, tsunami.noaa.gov; WorldData. info, worlddata.info/natural-disasters.php#tsunamis

Aftermath of the 2018 Sunda Strait tsunami in Indonesia

Water Main Breaks and Aqueduct Accidents

ANNUAL US DEATHS   No available data

Water main breaks and aqueduct accidents have only ever killed a handful of people in the US, in a few different ways. Flooding of low-lying areas from a water main break can kill people by drowning or by trench collapse, but such deaths appear to happen rarely and only among utility repair workers. Two workers drowned together this way in Boston in 2016. One worker was killed by a trench collapse while repairing a water main break in Eastpointe, Michigan, in 2020.

Aqueducts can (at least in theory) also fail and cause deadly flooding, but because they are open, aboveground, and sometimes located next to roadways and walkways, or are used for swimming, they can also kill people who fall, drive, or swim in them and then drown. Such events may be accidents or suicides.

Fourteen people were killed in 12 different accidents between early 2014 and early 2018 at the California Aqueduct in San Bernardino County, when they or their car fell in, and they drowned. A mother and her 3 children also died there in 2005.

STAY SAFE   If you are a water utility or trench worker or supervisor, follow all OSHA rules about safety equipment and precautions against trench collapse and flooding. If a water main break or aqueduct failure contaminates your drinking water, follow public health notices to boil water or use only bottled water until tap water becomes safe again. Be careful walking, jogging, playing, or driving near any aqueduct, canal, or river, and keep children and pets away from the water.

SOURCES   *Boston Herald*, bostonherald.com/2016/10/22/ruptured-water-main-floods-south-end-trench-killing-two-trapped-workers; ClickOnDetroit.com clickondetroit.com/news/local/2020/12/31/eastpointe-city-worker-dies-after-trench-collapses; *San Bernardino Sun*, sbsun.com/2018/04/27/deadly-waters-california-aqueduct-drownings-renew-safety-concerns-in-san-bernardino-countys-high-desert

Heat Waves

ANNUAL US DEATHS   650

An average of about 650 people per year in the US are killed by exposure to extremely hot weather. The CDC reported 505 heat-related deaths in 2019, but this doesn't capture all deaths because death certificates in general do not record weather conditions or indicate if excess heat was an underlying cause of death. There is no standard definition, applying nationally, as to exactly what temperature level, over how long a duration, constitutes a heat wave at any given location. In 1995, a heat wave in Chicago killed 739 people, leading many municipalities to create cooling centers.

The unprecedented heat wave that struck the Pacific Northwest in June 2021 killed at least 96 people in Oregon. For comparison, in all of Oregon between 2017 and 2019, 12 people died of hyperthermia. At least 112 died in Washington, for the state's deadliest weather disaster.

STAY SAFE   See entry on **Heat-Related Deaths**. Many municipalities and community organizations set up free cooling centers during heat waves. Check in on family, neighbors, and vulnerable individuals.

SOURCES   CDC, cdc.gov/disasters/extremeheat/index.html; NWS, weather.gov/mkx/heatwaves

Landslides, Mudslides, and Rockfalls

ANNUAL US DEATHS   25 to 50

Landslides

A landslide (one which isn't of the mudslide subtype) happens when a large amount of (relatively dry) earth, rocks and boulders, and debris falls down a slope due to gravity. Landslides can occur when the normal stability of a hillside is disrupted, for instance by a prolonged drought lowering the water table, or by the shaking of an earthquake or a volcano eruption. Technically, a landslide can combine 1 or more of 5 types of what geologists call mass wasting: falls, topples, slides, spreads, and flows.

The largest landslide in US history, by volume of the material involved, occurred during the eruption of Mount St. Helens in 1980, in which 2.9 cubic kilometers of the mountainside collapsed. The eruption killed 57 people, due not just to the landslide but also to exposure to devastating blast forces, poisonous fumes, superheated gases, burial under tons of volcanic ash, and other trauma.

STAY SAFE For extensive information on avoiding, preventing, and surviving landslides, see the USGS's detailed checklist "Landslide Preparedness" at usgs.gov.

SOURCES CDC, cdc.gov/disasters/landslides.html; USGS, www.usgs.gov/faqs/how-many-deaths-result-landslides-each-year

Mudslides

A mudslide, properly called a debris flow, is a specialized type of landslide, one which is triggered, exacerbated, and accelerated by large quantities of water or other liquid, moving down a slope due to gravity. This fluid component can come from a heavy rainstorm and/or river runoff, or from the collapse of a dam at a reservoir, retention pond, or storage pool. In some cases, an earthquake or a volcano eruption has triggered mudslides.

The deadliest individual mudslide in US history occurred in 2014 at Oso, Washington, where 43 people were killed by a massive debris flow that resulted from weeks of heavy rain. In 2018, in Montecito, California, 23 people were killed by a series of mudslides caused by heavy rainfall on steep hillsides after wildfires had depleted the plant cover that prevented erosion. In 1969, in Nelson County, Virginia, Hurricane Camille dropped 27 feet of rain that triggered a widespread series of mudslides in which 153 people were killed.

STAY SAFE Be aware of mudslide risk if you live in an area prone to wildfires, heavy rainfall, and previous mudslides. Obey any mandatory evacuation orders from authorities. Have an emergency plan for yourself, your family, and your business. During severe or prolonged rainstorms, watch for tilting trees and telephone poles, and listen for rumbling sounds. In an emergency, move to high ground immediately, away from the path of any potential debris flows. Beware of broken gas mains and downed electrical lines.

SOURCES CDC, cdc.gov/disasters/landslides.html; USGS, usgs.gov/faqs/how-many-deaths-result-landslides-each-year

Rockfalls

Rockfalls and rockslides are specific types of landslides, which can vary in size from an individual rock or boulder tumbling onto a climber's head, up to an entire rock-laden mountainside giving way under gravity. Sometimes a small rockfall is only the beginning of what expands rapidly into a massive rockslide.

Possibly the single largest rockslide in US history was the Gros Ventre slide in Wyoming in 1925. Although no one was killed, the mountainside is dramatically scarred and bare of trees even today, almost a century later.

In 2013, in Nathrop, Colorado, 5 of 6 members of a family were killed and 1 was trapped but rescued when a rockslide created a debris field half the size of a football field, including boulders the size of cars. Geologists believe the fall was triggered when rocks near a scenic waterfall were loosened by a hard freeze following heavy rains, possibly also lubricated by rainwater permeating the fractures between the rocks.

STAY SAFE   For detailed information on safety from rockfalls while climbing, see The Mountaineers bulletin "'Rock, Rock, Rock, Rock, Rock!' How to Avoid Rockfall" at mountaineers.org/blog/rock-rock-rock-rock-rock. Always wear a helmet when near a crag, even if you're not climbing, because other climbers could knock loose dangerous rocks above you.

SOURCES   CDC, cdc.gov/disasters/landslides.html; The Mountaineers, mountaineers.org; USGS, usgs.gov/faqs/how-many-deaths-result-landslides-each-year

Meteorite Strikes

ANNUAL US DEATHS   Near 0

Meteoroids (solar system objects ranging in size from sand grain to boulder) that plunge into Earth's atmosphere are called meteors (or shooting stars). Meteors that don't entirely burn up on the way down and hit the ground are called meteorites. Meteorites often land in showers after a meteor explodes in midair from intense thermal shock and aerodynamic stress.

Between ancient times and the Tunguska event (now believed to be an asteroid or comet impact) in Siberia in 1908, there are 8 known recorded instances of space objects killing humans. The death tolls in most of these different events were only 1 or 2 people, hit individually by a small stone or killed by air burst. But tens of thousands died in the cases of Tall el-Hamman (Jordan) around 1700 BCE and Ch'ing-yang (China) in 1490 CE, due to a massive air burst, as powerful as a nuclear detonation.

In 1954, in Alabama, a woman sleeping in bed was hit but not killed by a meteorite that penetrated the roof of her house. In 2013, in Chelyabinsk, Russia, about 1,600 people were injured by a meteorite's air burst, mostly by falling bricks or flying glass, and several fragments of the object were later recovered. There were no fatalities.

Based on 3 millennia of experience, the odds of someone born today being killed by a meteorite at sometime during their lifespan are very roughly estimated at about 1 in 200,000 to 1 in 250,000.

STAY SAFE Urge the US government and other world authorities to continue funding and urgently developing solar system object-tracking systems and effective planetary defense systems, to protect life on Earth against large space object impactors. Nothing can be done to prevent small meteorite impacts, which hit the upper atmosphere at random all the time. The odds of being killed by one are remote.

SOURCE NASA Center for Near Earth Object Studies, cneos.jpl. nasa.gov

Quicksand

ANNUAL US DEATHS Near 0

Quicksand exists in a few places in the US, but being killed by falling into it is extremely rare. Quicksand results when a rising underground source of water, such as a rising tide in a coastal area or a surging underground stream, separates and suspends the particles of a granular soil such as sand. The result is ground that appears solid but gives no support. While it is a myth that quicksand can suck down and bury a person without a trace, it can be so sticky that it is very difficult to escape from without outside help. Deaths in quicksand result from simple drowning, hypothermia in cold water, dehydration from prolonged entrapment, a heart attack from overexertion, or other cause.

In 2015, a Texas man drowned when he ventured into an area of quicksand adjoining the San Antonio River. He was said to be fond of fishing on the river but was alone at the time of his death, so the exact circumstances are not known.

In 1997, 2 Illinois youths died of drowning and hypothermia in quicksand that had formed in an excavation area at a construction site, where they apparently had been playing.

STAY SAFE If you fall into quicksand, do not struggle in a panic. It is not hard to float on the surface of quicksand due to its considerable buoyancy. To escape unaided, wave your arms and legs gently to separate the soil particles from the water and create an area of clear water, then extricate yourself. Walk in quicksand areas only in groups, so others can help you get out. Use a walking stick or cane to test the ground ahead of you and to help you lever yourself out or to reach toward a rescuer back on firm ground.

SOURCES N. Bakalar, Quicksand Science: Why It Traps, How to Escape, *National Geographic*, September 28, 2005; *Chicago Tribune*, chicagotribune.com/news/ct-xpm-1997-05-05-9705050054-story.html; A. Khaldoun et al. Rheology: liquefaction of quicksand under stress. *Nature*, September 2005.

Severe Storms

Falling Trees and Branches

ANNUAL US DEATHS 31.3

From 1995 through 2007 in the US, 407 people were killed by falling trees or branches. The average annual toll is 31.3 deaths per year.

Most deaths occurred when trees were uprooted or broken by high winds during bad weather. In some cases, heavy buildup of ice and/or snow during blizzards and ice storms (freezing rain) added to the other mechanical forces that caused the tree or branch to fail, fall, and kill someone. In a few cases, prolonged heavy rain had softened the ground to the point that a leaning tree was uprooted by its own weight, even in the absence of high winds.

STAY SAFE Property owners should check that their trees and branches are healthy and sound. Remove trees or branches at risk of falling, especially if they might land on a building or vehicle. The risk of a falling tree or branch is greater after a windstorm, ice storm, blizzard, or heavy rain or snowfall. Be careful while walking outside, camping, or driving underneath big trees.

SOURCE T.W. Schmidlin. Human fatalities from wind-related tree failures in the United States, 1995–2007. *Natural Hazards*, 2009.

Hailstorms

ANNUAL US DEATHS   About 0.2

Hail is a dangerous form of precipitation, produced during warmer weather by some thunderstorms. Frozen balls of ice build up in layers within the clouds, due to powerful updrafts, and then hit the ground below at high speed. Hailstones range in size from 0.2 inches up to 4 or even 6 inches, and a single very big hailstone can weigh over 1 pound.

Since 2000, only 4 people in the US have been killed by hail, the most recent in 2008. However, hail can cause severe damage to vehicles, structures, livestock, and other property, costing the American economy over $1 billion per year. About 24 people per year are hospitalized due to injuries from falling hailstones.

The worst hailstorm in modern US history occurred in 1995 during the Mayfest outdoor festival in Fort Worth, Texas, in which a sudden severe thunderstorm system caught over 10,000 attendees in the open. More than 400 were hurt by large hailstones. The same storm killed 13 people through flash floods or lightning strikes.

The deadliest hailstorm in modern world history occurred in India in 1888, where hail "as large as goose eggs, oranges, and cricket balls" fell with great speed and killed 246 people. Some of the deaths were at a racecourse and at a wedding. About 1,600 head of livestock were also killed.

STAY SAFE   Check weather forecasts. If a hail warning is issued, seek sturdy overhead shelter (away from trees) immediately. If in a vehicle, stop your car and pull over, or try to drive into a garage, under an overpass, or into some other covered area. The vehicle's roof can shield you from the hail, but the windshield could be shattered by very large hailstones.

SOURCES   NOAA, nssl.noaa.gov/education/svrwx101/hail; NWS, weather.gov/fwd/mayfest15; World Meteorological Organization, World Weather & Climate Extremes Archive, wmo.asu.edu

Hurricanes

ANNUAL US DEATHS   Average of 130; 24 in 2020

From 2006 through 2020, hurricanes making landfall in the continental US (sometimes after they had declined in wind strength to tropical or subtropical storms) killed an average of about 130 people per year. The annual death toll varied from 0 in 2006, up to about 1,800 in 2005 (the year of very deadly Hurricane Katrina). The number of such storms hitting the US varied during this 16-year period from 2 (in 2013) up to 15 (in 2005). These statistics do not include the severe death toll of over 2,900 people on the island of Puerto Rico (a US territory) in 2017 from Hurricane Maria.

Satellite photograph of Hurricane Florence, 2006

The 2020 hurricane season was the most active on record. It had 30 named storms (top winds of 39 mph or greater), the most in 170 years of record-keeping. (The previous record was 28 in 2005.) Of these, 14 became hurricanes (top winds of 74 mph or greater), including 7 major hurricanes (top winds of 111 mph or greater). Of all these storms, 11 made landfall in the continental US, surpassing the previous record of 9 set in 1916. The 2020 season caused 46 direct fatalities and 51 indirect fatalities, with at least 19 from carbon monoxide poisoning.

The 2020 season quickly exhausted the 21-name Atlantic list when Tropical Storm Wilfred formed on September 18. For only the 2nd time in history, the Greek alphabet was used for the remainder of the season, extending through the 9th name in the list, Iota. For context, an average hurricane season has 12 named storms, 6 hurricanes, and 3 major hurricanes.

The single deadliest hurricane ever hit Galveston, Texas, in 1900. The US death toll is estimated at between 8,000 and 12,000 people.

A hurricane is a tropical cyclonic (rotating) storm in the Atlantic Ocean that has sustained winds above 74 mph. (In the Pacific, such storms are called typhoons.) A hurricane's wind speed is rated by category, with Category 1 having the least intense winds and Category 5 the most intense. Hurricanes are typically about 300 miles wide and have a clear, quiet eye at the center that is usually about 20 to

40 miles wide. Continuous, extremely high winds, and the hurtling debris they pick up, cause destruction and death. Hurricanes also contain powerful thunderstorms, which can cause lightning strikes and throw off tornadoes. Hurricanes can deliver catastrophic rainfalls, causing severe flooding over wide areas, and may cause severe coastal seawater flooding via very high ocean storm surges. (See **Storm Surges and Storm Tides**, and **Flooding**.) The increased hurricane activity in recent years is attributed to the warm phase of the Atlantic Multi-Decadal Oscillation (AMO), which began in 1995 and has favored more frequent, stronger, and longer-lasting storms since that time. Such active eras for Atlantic hurricanes have historically lasted about 25 to 40 years.

STAY SAFE For information on the extensive preparations needed before a hurricane hits, and about how to stay safe during one and in the aftermath of one, see the CDC bulletin "Preparing for a Hurricane or Other Tropical Storm" (cdc.gov/disasters/hurricanes/before.html) and the page on "Floodwater After a Disaster or Emergency" (cdc.gov/disasters/floods/floodsafety.html). Especially if you live in a hurricane-prone area, it is important to develop a detailed, practical hurricane safety plan well in advance; obtain the preparation and survival supplies you need (such as bottled water, batteries, plywood, medicine, gas for your car) *before* a hurricane's imminent approach causes local shortages; own a battery powered or hand-cranked emergency radio, fire extinguishers, and a first aid kit; locate an animal shelter, pet-friendly motel, or willing friend or relative to accommodate your pets in safety, away from the hurricane's path.

SOURCES CDC, cdc.gov/disasters/hurricanes; CDC NIOSH, cdc.gov/niosh/topics/emres/flood.html; III, iii.org/fact-statistic/facts-statistics-hurricanes; NOAA National Hurricane Center, nhc.noaa.gov/dcmi.shtml

Sandstorms and Dust Storms

ANNUAL US DEATHS No available comprehensive data

Blowing sand and dust kill people in the US primarily from acute or chronic lung diseases and respiratory failures caused by inhaling sand and dust, and fatalities from motor vehicle accidents caused by obscured visibility. Comprehensive national statistics are not available. A causal link between windblown dust and Valley fever, a sometimes lethal respiratory disease caused by subsoil fungus, is suspected but has not been proven. (See the entry on **Valley Fever**.)

A study in Arizona found that over the 57-year period from 1955 through 2011, an average of 3 people per year were killed in highway crashes that occurred when dust storms obscured visibility.

In July 2021 in Utah, 8 people, including 5 members of a single family, were killed when a sandstorm caused a 22-vehicle crash on an interstate. The worst-ever dust storm in US history, known as Black Sunday, occurred in April 1935, when "a mountain of blackness swept across the High Plains . . . resembling a land-based tsunami." The suffocating Black Sunday dust storm killed at least 20 people in the Panhandles of Texas and Oklahoma during the infamous years-long Dust Bowl era environmental crisis. (See **Droughts**.)

STAY SAFE   If you live in an area prone to windblown sand and dust, use air-conditioning and keep the windows closed; if you need to go outside, wear a mask and goggles to protect your mouth, nose, and eyes. If visibility becomes obscured while you are driving, pull over and stop until the dust storm passes.

SOURCES   The *New York Times*, nytimes.com/2021/07/26/us/sandstorm-crash-millard-county-dead.html; NOAA, NOAA Technical Memorandum NWS-WR 290, Blowing Dust and Dust Storms: One of Arizona's Most Underrated Weather Hazards; NWS, weather.gov/safety/wind-dust-storm; World Meteorological Organization (WMO), public.wmo.int/en/our-mandate/focus-areas/environment/SDS

Sleet Storms

> **ANNUAL US DEATHS**   900 vehicle deaths during sleet and snow conditions

Sleet is a dangerous form of precipitation, usually occurring during cold weather, consisting of falling raindrops that freeze into ice pellets on their way down to the ground. Sleet is sometimes mixed with snow or freezing rain. Typically, sleet conditions are just 1 part of a broader winter storm system that can also cause icing (including treacherous black ice on roadways), along with buildups of drifting snow and reduced visibility.

While reliable statistics are scarce on exactly how many US deaths occur annually specifically and only due to sleet, each year, 24 percent of weather-related vehicle crashes occur on snowy, slushy, or icy pavement and 15 percent happen during snowfall or sleet. Over 1,300 people are killed and more than 116,800 people are injured in vehicle crashes on snowy, slushy, or icy pavement annually. Every year, nearly

900 people are killed and nearly 76,000 people are injured in vehicle crashes during snowfall or sleet.

In February 2015, a severe winter storm cut a wide swath from Missouri to Maine, which included heavy rain, snow, and sleet, during cold temperatures that set new record lows. At least 30 people were killed by hypothermia (freezing to death) or by motor vehicle accidents.

STAY SAFE   See **On the Road: Weather Conditions** entries.

SOURCES   CDC, cdc.gov/injury/features/global-road-safety; FHWA, ops.fhwa.dot.gov/weather/weather_events/snow_ice.htm; NWS, www.weather.gov/lot/2015_Feb01_Snow

Snowstorms and Freezes

ANNUAL US DEATHS   About 400

Some 1,330 people die in the US every year from very cold weather. About 700 of these victims die directly from hypothermia. About 400 are killed in various ways by blizzards, which are snowstorms with sustained wind speeds of 35 mph or more. The blizzard deaths are caused by hypothermia, heart attacks while clearing snow or struggling through deep snow, other preexisting health problems worsened by the extreme cold, motor vehicle accidents on treacherous roads in poor visibility, and carbon monoxide poisoning from faulty heating.

One study found that 75 percent of hypothermia deaths occurred outdoors, while 25 percent occurred indoors due to loss of heat. The risk of dying from all causes during snowstorms and other freezing weather is significantly exacerbated by homelessness. (See **Homelessness**.)

In Texas in February 2021, Winter Storm Uri and an unprecedented period of very cold weather caused extensive power failures and resulted in the deaths of at least 246 people in the state, most from hypothermia.

STAY SAFE   See **Hypothermia**. Be sure to have a properly working, properly installed carbon monoxide detector (or more than 1) in your home. See the EPA's bulletin "Where should I place a carbon monoxide detector?" at epa.gov.

SOURCES   J. Berko et al. Deaths attributed to heat, cold, and other weather events in the United States, 2006–2010. National health statistics reports, number 76, National Center for Health Statistics, 2014; CDC NCHS, cdc.gov/nchs/ndi/index.htm; EPA, epa.gov/indoor-air-quality-iaq; NOAA National Severe Storms Laboratory, ssl.noaa.

Killer Tornadoes

ANNUAL US DEATHS   78 in 2020

The worst killer tornado in US history was the Tri-State Tornado of March 18, 1925. It killed 695 people in Missouri (11), Illinois (613), and Indiana (71). The worst year since then for tornado deaths was 1953, when 515 people died in a number of events.

The Fujita scale rates the intensity of damage caused by a tornado, ranging from F0 (low) to F6 (high). In 2019, 12 different severe tornadoes (Fujita scale F2 or higher) killed a total of 41 people across 8 states (Alabama, Iowa, Ohio, Oklahoma, Louisiana, Mississippi, Missouri, and Texas). The single most dangerous tornado in 2019 killed 23 people on March 3 in Lee, Alabama. Of all tornado fatalities in 2019, 28 were residents of mobile homes.

On December 10 through 11, 2021, a deadly outbreak of tornadoes across the Southeast and Ohio Valley caused extensive damage and at least 89 deaths, including 77 in Kentucky. Known as the Quad-State outbreak, the storm system spawned 62 confirmed tornadoes across Arkansas, Kentucky, Missouri, and Tennessee, making it the deadliest December tornado event ever confirmed in the US.

STAY SAFE   Be alert to tornado warnings and sirens. Take cover quickly in a sheltered place, especially if you live in a mobile home.

SOURCE   NWS, weather.gov/lmk/tornado_climatology

Weather-Related Power Outages

ANNUAL US DEATHS   Annual data not available; death toll has exceeded 100 in some outages

Although a complete national tally of the number of people in the US killed by power outages each year is not available, the death toll from individual power outages has sometimes exceeded 100.

In February 2021, a severe winter storm hit 6 southern states, causing widespread, prolonged power outages that killed at least 246 people in Texas alone, mostly from hypothermia when their homes lost heat. It's possible that the death toll is actually as high as about 700 due to unacknowledged deaths as a direct consequence of the power outages.

Studies suggest that climate change, which can cause more frequent extreme weather such as heat waves and deep freezes, puts added strains on the national power grid that increase both the number and severity of regional power outages and the number of deaths caused by each outage.

Power outages can cause deaths in other ways than hypothermia, depending on the weather, such as by hyperthermia from lack of air-conditioning and cooling fans, dehydration from loss of water pressure for drinking water, and loss of power for essential medical devices such as respirators and kidney dialysis machines. In addition, lethal accidents such as fires, falls, and motor vehicle mishaps are more likely if/when candles and lighters are substituted for electric lighting, and when traffic signals fail. Loss of refrigeration can lead to risk of food poisoning and food shortages.

STAY SAFE   See the entries for **Hypothermia, Heat-Related Deaths, Fires, Slips, Trips, and Falls,** and **On the Road: Weather-Related Accidents.** A backup/emergency generator, with a safely stored fuel supply to keep it running until mainline power is restored, can help keep a home or business going during a prolonged power outage, as can a solar power system if the weather is sunny enough. If you lose refrigeration for any length of time, dispose of any food that may have spoiled.

SOURCES   BuzzFeed News, buzzfeednews.com/article/peteraldhous/texas-winter-storm-power-outage-death-toll; the *Texas Tribune,* texastribune.org/2022/01/02/texas-winter-storm-final-death-toll-246

Windstorms and Rainstorms

ANNUAL US DEATHS   Estimated 300

Windstorms and rainstorms (other than hurricanes and tornadoes), such as nor'easters, thunderstorms, tropical storms, gales, and blizzards, kill a few hundred people in the US per year in various ways. This is a very approximate figure, since some death certificates omit weather conditions that could be the underlying causal mechanism of mortality.

In March 2018, a powerful nor'easter killed at least 9 people along the East Coast between Virginia and Massachusetts. Several victims were hit by uprooted trees, and 1 was electrocuted by a downed power line. In March 2010, a bad nor'easter killed 7 people between West Virginia and New Hampshire. Most of the victims were killed by falling trees.

The National Weather Service recorded 55 wind-related fatalities in 2020, mostly people caught in high winds while boating, camping, or outside. Six people were killed when high winds toppled mobile homes. An additional 44 people were killed during thunderstorms with high winds.

STAY SAFE   See the entries for **Lightning, Flooding,** and **Hurricanes.**

SOURCES   J. Berko et al. Deaths attributed to heat, cold, and other weather events in the United States, 2006–2010. National health statistics reports, number 76, National Center for Health Statistics, 2014; NOAA, noaa.gov/explainers/severe-storms; NWS, weather.gov/aly/2Mar18Noreaster and www.weather.gov/phi/03122010wss

Volcanoes

ANNUAL US DEATHS   Very roughly 2.7

In the past 2,000 years, at 17 different volcanoes in the US, there have been a total of 29 substantial eruptions, killing more than 5,400 people. The average annual death toll from volcanoes in the US is at least 2.7 people per year. In January 2022 on the Big Island of Hawaii, a 75-year-old man fell to his death while attempting to get close to the lava lake from an eruption of the Kilauea volcano. His body was found 100 feet below the crater rim.

There are 70 volcanoes in the US considered by geologists to be active: 69 on land and 1 off the coast; there are also 15 inactive (but not extinct) volcanoes. The active volcanoes in Hawaii and in Yellowstone National Park, Wyoming, are popular tourist attractions. Most US volcanoes are located in Alaska and are part of the Pacific Ocean Ring of Fire. An additional 91 volcanoes are considered potentially active.

The deadliest single eruption occurred at Kilauea, Hawaii, in 1790 (before Hawaii became part of the US), killing 5,405 people. The next deadliest eruption was Mount St. Helens, Washington, in 1980, in which 57 people were killed.

STAY SAFE   If you are not a trained geologist doing research, only visit an active volcano that has a preestablished safe viewing area, and never venture too close to lava or hot springs. In case of an eruption alert, follow instructions about evacuation.

SOURCES   CDC, cdc.gov/disasters/volcanoes/during.html; USGS, usgs.gov/programs/VHP and usgs.gov/volcanoes/mount-st.-helens/1980-cataclysmic-eruption

One of many hundreds of wildfires that plague California forests

Wildfires

ANNUAL US DEATHS 80 (nonfirefighters) in 2019

In 2019, fires that occurred outside, rather than in structures or vehicles, killed 80 civilians (nonfirefighters) in the US. This statistic does not separate wildfires (fires in undeveloped brush, grasslands, and forests) from fires that occur in out-of-doors businesses: rubbish piles, materials storage areas, crops (farmland), or timber (commercial forestry operations). This grouping, together under "Outside Fires," is used by the national fire reporting system because many large outside fires are hybrid/combination events, burning both undeveloped areas and residential and/or commercial properties.

For line-of-duty deaths of firefighters, both professional and volunteer, including in wildfires, see **Service Occupations, First Responders: Firefighters**.

In 2020, California suffered a particularly devastating complex of wildfires, which burned a record 4.3 million acres in that state, also damaging over 10,000 homes or businesses and killing 47 people. Oregon and Washington also suffered terribly in 2020 from wildfires, and tens of thousands of people were forced to evacuate. At least 10 people were killed in Oregon, and at least 1, a young child, died in

Washington. These fires inflicted billions of dollars of damage. In 2021, 6 people died in wildfires in California and the Pacific Northwest.

Wildfire causes can include lightning strikes, electrical utility sparks, careless campfires or discarded cigarettes, and arson. In 2020, many experts believe the massive West Coast wildfires were exacerbated by prolonged drought, climate warming trends, and errors in forest management.

STAY SAFE To help prevent wildfires, be sure all campfires and cigarette butts are completely extinguished after use. When wildfire risk is high, respect official warnings and closures. If you are in an active wildfire area, obey mandatory evacuation instructions and other instructions from authorities. If you are sheltering in place, choose a room you can close off from outside air. If you have a preexisting respiratory or cardiovascular condition or you are pregnant, get medical help as soon as possible. For more information, including about how to stay healthy if the wildfire causes a power outage, and how to protect pets and livestock from the smoke, see the CDC bulletin "Stay Safe During a Wildfire" at cdc.gov/disasters/wildfires/duringfire.html.

SOURCES AccuWeather, accuweather.com/en/severe-weather/2021-wildfire-season-economic-damages-estimate-70-billion-to-90-billion/1024414l; CDC, cdc.gov/disasters/wildfires; III, iii.org/article/background-on-wildfires; and iii.org/fact-statistic/facts-statistics-wildfires; NFPA, nfpa.org/Public-Education/Fire-causes-and-risks/Wildfire; National Interagency Fire Center (NIFC), nifc.gov/fire-information/statistics

Animal Attacks

ANNUAL US DEATHS   About 200

Encounters between animals and people can sometimes lead to the humans getting killed. The death can be due to a severe allergic reaction to a venomous bite or sting from an insect, spider, snake, or jellyfish. Death might result from being mauled or trampled by a larger animal or from a motor vehicle accident caused by hitting an animal.

Over 47,000 people annually in the US seek medical attention after being attacked or bitten by animals, including both domestic animals and wildlife. Between 2008 and 2015, the total number of humans killed by animals in the US was fairly constant, at just over 200 per year, or a death rate of about 6 per 10 million population. Venomous animals accounted for 43 percent of the total, or about 85 deaths per year, of which about 60 deaths per year were due to bee, wasp, and hornet stings that triggered lethal anaphylactic shock. In 2021, an Arizona man died after hundreds of stings from a bee swarm; a hive weighing an estimated 100 pounds was found in a nearby tree.

In addition, on average each year, wildlife–vehicle collisions cause more than 59,000 human injuries and around 440 human deaths. On average each year, wildlife–aircraft collisions (bird strikes) cause 16 injuries and 10 fatalities.

For the 20-year period from 1999 to 2019, Texas led the nation with 520 deaths caused by animals; California was next with 299 deaths. Florida was third with 247 deaths, followed by North Carolina with 180 and Tennessee with 170. Georgia and Ohio tied with 161 each, followed by Pennsylvania with 148, Michigan with 138, and New York with 124.

Most attacks were bites by snakes, birds, rodents, and raccoons. The most human deaths every year are caused by livestock and farm animals, followed by bees/wasps/hornets, followed by dogs, then the other creatures discussed below.

STAY SAFE   Be careful around animals. Never harass or provoke them. If you're allergic to stings, be sure to carry a fresh epinephrine auto-injector (EpiPen) or call 911 immediately.

SOURCES   CDC NCHS, data.cdc.gov; M.R. Conover. Numbers of human fatalities, injuries, and illnesses in the United States due to wildlife. *Human–Wildlife Interactions*, Fall 2019; the *Guardian*, theguardian.com/us-news/2021/jul/31/arizona-bee-swarm-man-died-stings; Outforia, outforia.com/animal-attacks

Alligators and Crocodiles

ANNUAL US DEATHS   Average of 1

The American alligator *(Alligator mississippiensis)* is found in the marshes, swamps, rivers, ponds, and lakes of 10 southern states in the US. The American crocodile *(Crocodylus acutus)* is found in Mexico, the Caribbean, and southern Florida. South Florida is the only place in the US where both species are found. Alligators have a more U-shaped snout, while crocodiles have a more pointed, or V-shaped, one. Alligators are black; crocodiles are grayish.

Alligators and crocodiles kill on average 1 person per year in the US. In the aftermath of Hurricane Ida in 2021, a man in Louisiana was eaten by a 12-foot, 504-pound alligator when floodwaters brought the animal close to his suburban home. The alligator bit off his arm, then carried him away. The water hazards on golf courses in Florida and other Southeast states attract alligators. If your golf ball lands on an alligator, you may take free relief according to Rule 16 of the U.S. Golf Association rule book.

STAY SAFE   Keep your distance. Do not harass or feed alligators. Be cautious when hunting for lost balls near or in the water. Do not take refuge on the roof of your cart, as an alligator can reach that high. These reptiles are fast but not very maneuverable. If one charges in your direction, zigzag while you run for your life until they lose interest and give up.

SOURCES   M.R. Conover, Numbers of human fatalities, injuries, and illnesses in the United States due to wildlife. *Human–Wildlife Interactions*, Fall 2019; Florida Fish and Wildlife Conservation Commission, myfwc.com/wildlifehabitats/wildlife/alligator/data; Reuters, reuters.com/business/environment/alligator-hurricane-ida-attack-found-with-human-remains-stomach-2021-09-14; U.S. Golf Association, usga.org/content/usga/home-page/rules-hub.html

Animals in the Wild

ANNUAL US DEATHS   See data in entry.

Animals in the wild can kill, but such incidents are rare. The death tolls vary both by animal species and by the type of interaction.

The single most deadly species-and-incident combination is a collision between a deer and a motor vehicle. Almost 60,000 of these are reported in the US every year, and they kill around 440 people. All other road collisions between drivers and animals kill on average, per year, 6 people via moose crashes and 1 via bird impacts. Bird strikes by aircraft kill on average 10 people in the US per year. This splits as 9 mishaps with military aircraft versus 1 with civilian aircraft.

Wild animal bites and attacks kill about 8 people per year. Someone is killed by a cougar about once every 3 years, by a coyote about once every 10 years, and by a wolf about once every 10 years. Alligators kill 1 person per year.

STAY SAFE   Be careful in the woods, and don't approach wild animals unless you're hunting them. If you're bitten or scratched by an animal, seek prompt medical attention. Drive carefully, especially at night in areas where deer crossing signs are posted.

SOURCES   M.R. Conover, Numbers of human fatalities, injuries, and illnesses in the United States due to wildlife. *Human–Wildlife Interactions*, Fall 2019; IIHS-HLDI, iihs.org/topics/fatality-statistics/detail/collisions-with-fixed-objects-and-animals; Pew Charitable Trusts, pewtrusts.org/en/research-and-analysis/articles/2021/05/10/wildlife-vehicle-collisions-are-a-big-and-costly-problem-and-congress-can-help

Bears

ANNUAL US DEATHS   Average of 1 death every 3 years (black bears)

Black bears kill someone on average once every 3 years. Despite their justifiably fierce reputation, polar bears have killed only 6 people in 20 years, and grizzly bears have killed only 10 in the same period.

Bear lover Timothy Treadwell, subject of the 2005 documentary *Grizzly Man*, was killed in 2003, along with his girlfriend, by a bear in Alaska's Katmai Peninsula. In June 2019, a hiker was killed by a black bear in New Jersey, the only such fatality ever recorded in the state. The hiker and 4 other members of the group scattered when the bear attacked; he lost a shoe and was overtaken and killed. The others escaped unharmed. The hiker's phone, found later with tooth marks on it, contained pictures of the bear approaching.

In April 2021, a backcountry guide was killed by a grizzly bear in southwestern Montana; in July 2021, another person was killed by a grizzly at a campsite in western Montana. In May 2021, a woman was killed by a black bear in Colorado. Authorities euthanized an adult female bear and 2 yearling bears and found human remains in the grown bear and 1 of the yearlings.

STAY SAFE Bears are generally fearful of humans. In bear country, make noise on the trail, keep your distance, and avoid getting between a mother bear and her cub(s). Carry bear spray in grizzly country. Hike with people who run slower than you.

SOURCES G. Bombieri et al. Brown bear attacks on humans: a worldwide perspective. *Scientific Reports*, June 2019; *Field & Stream*, fieldandstream.com/conservation/fatal-bear-attacks-2021; NJ.com, nj.com/passaic-county/2014/11/hiker_snapped_pictures_of_bear_before_fatal_attack_in_west_milford.html; North American Bear Center, bear.org/bear-facts/black-bears/basic-bear-facts

Dogs

ANNUAL US DEATHS Average of 34

Pet and stray dogs on rare occasions can be dangerous, attacking or even killing their owners or strangers. Considering that there were 77 million dogs in the US in 2020, the likelihood of being killed by your own or any dog is extremely low. Dogs can also spread rabies, which is usually fatal in humans if not treated early.

In the US, 272 people were killed by dog attacks between 2008 and 2015, for an average of 34 people killed by dogs per year. In 2001 in San Francisco, college lacrosse player and coach Diane Whipple was mauled and killed by 2 Presa Canarios dogs being cared for by her neighbor, Marjorie Knoller. Knoller was found guilty of involuntary manslaughter and owning a mischievous animal that caused the death of a human being. She was sentenced to 15 years to life and is eligible for parole in 2022. The probability of your dog killing someone is 0.00004 percent per year.

STAY SAFE Don't provoke, harass, or otherwise mistreat dogs. Always keep dogs on a leash, or otherwise restrained, when walking in public areas or if they might bite someone. Make sure your dogs get rabies vaccines and boosters as needed.

SOURCES American Veterinary Medical Association, avma. org/resources-tools/pet-owners/dog-bite-prevention; DogBiteLaw. com, dogbitelaw.com/diane-whipple/the-diane-whipple-case; National Canine Research Council, nationalcanineresearchcouncil.com, Final Report on Dog Bite-Related Fatalities, 2019

Dolphins

ANNUAL US DEATHS   0

Although dolphins in captivity have been known to injure participants in dolphin encounters by biting, butting, or dragging them so as to inflict cuts, bruises, and broken bones, no one in the US has ever been killed. Dolphins in the wild have never killed someone in US waters. One reported death of a swimmer caused by a dolphin, in Brazil, resulted from him harassing it and trying to restrain it.

STAY SAFE   Do not harass dolphins in the sea. During commercial dolphin encounters, obey all handler safety instructions.

SOURCES   Humane Society of the United States, humanesociety. org/resources/swim-dolphins-attractions; NOAA National Maritime Sanctuaries, sanctuaries.noaa.gov/dolphinsmart

Exotic Pets

ANNUAL US DEATHS   Average of 2.3

Owning and trafficking in exotic pets such as lions and tigers, wolves, chimpanzees, bears, and even elephants, along with nonnative snakes and other reptiles, is illegal or needs a permit in most but not all states. Between 1990 and 2011, exotic pets have killed 75 people in the US, for an average death toll of 2.3 people per year. Of the deaths, 21 were by big cats, 18 by reptiles (mostly snakes), and 14 by elephants.

STAY SAFE   Do not keep wild animals as pets.

SOURCE   Born Free USA, bornfreeusa.org/campaigns/animals-in-captivity

Horses

ANNUAL US DEATHS   Average of 20

Horses sometimes kill people by kicking or trampling them, or by rearing and throwing them, or by falling and squashing them. Human deaths have occurred during recreational horse riding, rodeos, equestrian competitions, and horse races.

Worldwide, over 100 deaths per year are estimated to result from equestrian-related activities. An average of 20 people are killed by horses in the US every year. For every fatality, there are probably 10 to 20 times more serious head injuries. The injury rate for horseback riding is estimated at 3.7 per 1,000 hours of participation. Approximately 10 to 30 percent of horse-related injuries are head injuries.

STAY SAFE   Wear a helmet while riding. Do not startle or harass horses.

SOURCES   L. Meredith et al. Epidemiology of equestrian accidents: A literature review. *Internet Journal of Allied Health Sciences and Practice*, 2018; Ohio State University Extension, ohioline.osu.edu/ factsheet/19

Jellyfish Stings

ANNUAL US DEATHS   0

Although the Australian box jellyfish, the most venomous marine animal in the world, kills between 50 and 100 people a year worldwide (mostly in the Indo-Pacific region), this lethal box jellyfish isn't found in US coastal waters. No one in the US has ever been reported killed by a jellyfish, although thousands of people have been stung by less venomous jellyfish in warm coastal waters. More than 800 people were stung over a single weekend in June 2018 at beaches in Florida. Contrary to popular belief, urinating on a jellyfish sting does not relieve the pain.

The box jellyfish's sting carries a potent toxin, killing between 4 and 40 people per year across its range around Australia, the Philippines, and Thailand.

STAY SAFE   Jellyfish season in Florida is from June to October. Lifeguards display a purple flag to warn swimmers of dangerous marine life. Avoid contact with jellyfish and see the lifeguard or seek medical attention if you are stung.

SOURCES   Florida Health, floridahealth.gov/diseases-and-conditions/food-and-waterborne-disease/_documents/stings.pdf; NOAA, oceanservice.noaa.gov/facts/box-jellyfish.html; University of Florida Health, ufhealth.org/jellyfish-stings

Killer Whales

ANNUAL US DEATHS   0 (in the wild)

Killer whales (orcas) in captivity at aquariums and sea parks have killed 3 people in the US, all at SeaWorld in Orlando, Florida. All 3—2 trainers and 1 trespasser—were killed by an orca named Tilikum, the largest in captivity (he died in 2017). Of the reported killer whale attacks in the wild, none have been fatal.

STAY SAFE   Boat crews and passengers should use care during whale-watching trips. Employees at sea parks and aquariums should follow all safety protocols.

SOURCE   Dolphin Project, dolphinproject.com

Livestock

ANNUAL US DEATHS   About 20

Cattle can kill people by kicking, trampling, crushing, or goring them. Human deaths have occurred at dairy farms, cattle ranches, and meat processors. About 20 humans a year in the US are killed by cattle, mostly while working with cattle in enclosed spaces such as pens and chutes. Almost all deaths were from blunt force trauma to the chest or head. In 2021, a volunteer at an animal therapy farm was killed when she was repeatedly rammed by a sheep.

Other livestock can also kill. Feral or domesticated hogs and stampeding herds of sheep, goats, and bison can also kill, but hard data is very scarce. In 1996 in Texas, a hunter was killed by a wild hog; in 2019 in Texas, a woman was killed by a herd of wild hogs outside her house.

STAY SAFE   Farm, ranch, feed lot, and slaughterhouse workers should exercise caution around livestock, and visitors should always exercise caution around livestock. Hunters and hikers should exercise care in areas frequented by feral hogs. Never harass a bull or a bison.

SOURCES   BLS Census of Fatal Occupational Injuries, bls.gov/iif/oshcfoi1.htm, the *New York Times*, nytimes.com/2021/12/06/us/massachusetts-sheep-woman-killed.html and nytimes.com/2019/11/26/us/texas-woman-killed-feral-hogs

Mosquitoes

> **ANNUAL US DEATHS**   167 in 2018 (West Nile virus vectored by mosquito)

The most dangerous animal in the world is also one of the smallest: the mosquito. Globally, mosquito-borne diseases kill more than 1 million people a year. In the US, illness and death from mosquito-borne diseases are rare. West Nile virus is the disease most commonly communicated this way, and in 2018 it killed 167 Americans. Other mosquito-borne diseases found in the US include Zika virus disease and eastern equine encephalitis. Deaths from these diseases are rare.

STAY SAFE   Precautions against mosquito bites include staying indoors in the evening during warm weather, wearing long-sleeved tops, and using mosquito repellent and/or citronella candles.

SOURCES   American Mosquito Control Association, mosquito.org/page/diseases; CDC NIOSH, cdc.gov/niosh/topics/outdoor/mosquito-borne/

Rabies

> **ANNUAL US DEATHS**   Average 1.5

Rabies is a fatal viral disease of the central nervous system and brain, found mostly in some wild mammals. In the US these include raccoons, foxes, skunks, and bats. Rabies can also occur in unvaccinated cats and dogs. It is spread among animals and to humans by blood contact through bites or scratches. If a person gets rabies and doesn't get medical treatment before overt symptoms start (which usually takes 1 to 3 months), they will almost certainly die.

Approximately 5,000 animal rabies cases are reported annually to the CDC. More than 90 percent of those cases occur in wildlife. Human fatalities are rare and typically occur in people who do not seek prompt medical care, probably because they didn't realize they had been exposed.

From 1960 to 2018, 127 human rabies cases were reported in the US, with roughly a quarter resulting from dog bites received during international travel. Of the infections acquired in the United States, 70 percent were attributed to bat exposures. Bats have small teeth and leave bite marks that are only the size of a pencil tip.

Because rabies can be treated, in the past 50 years it has killed only

1 or 2 people per year in the US. These cases were almost exclusively from being bitten by an infected dog. In 2021, however, rabies deaths increased to 4, all from exposure to rabid bats.

Worldwide, rabies remains a serious health problem, causing approximately 59,000 deaths each year. Exposure to rabid dogs is the cause of over 90 percent of human exposures to rabies and of 99 percent of human rabies deaths worldwide.

STAY SAFE Rabies in humans is curable if treated promptly. If you are bitten or scratched by a potentially rabid animal, seek medical attention quickly. Dog owners are required to vaccinate their pets against rabies in almost every jurisdiction in the US. Keep the vaccines up to date. Many animal welfare organizations offer free rabies shots.

SOURCES CDC, cdc.gov/rabies/location/usa/surveillance/human_rabies.html; *USA Today*, usatoday.com/story/news/nation/2022/01/06/cdc-warns-increase-rabies-deaths-bat-contact/9115089002/

Safari Parks and Petting Zoos

ANNUAL US DEATHS Extremely rare

On very rare occasions, Americans on safaris in Africa have been killed by a wild animal; the last reported incident was in 2015, when a woman from Rye, New York, was mauled through her car window by a lion.

In the US, safari park rides and petting zoos are more a concern because they can lead to transmission of diseases from the animals to the people who touch them. *E. coli* is usually the culprit. Outbreaks of this kind have occasionally sickened as many as 82 adults and children.

In 2005 in Kansas, a 17-year-old high school student had her senior picture taken with a 700-pound Siberian tiger at a farm that was licensed to rescue and shelter exotic animals. The tiger attacked her, and she later died from her injuries.

STAY SAFE Wild animals are not pets—stay at a safe distance. At safari parks, remain in your vehicle. Always use hand sanitizer after touching the animals at petting zoos or safari park rides.

SOURCES Big Cat Rescue, bigcatrescue.org and cubpet.com

Scorpions

ANNUAL US DEATHS   About 0.36 (bark scorpion)

Scorpions are found in warm, dry climates, such as in Arizona. Their sting carries a neurotoxin that can be deadly to humans. The young and the elderly are the most vulnerable; a sting can kill in as little as 15 minutes.

In the US, stings from the bark scorpion *(Centruroides exilicauda)* caused the deaths of 4 people over a recent 11-year period. In Mexico, about 1,000 people are killed each year by scorpions, partly due to less availability of medical care in rural areas.

STAY SAFE   Wear high, thick leather boots for protection when walking in areas where scorpions are found. Seek medical care immediately if stung. Be especially careful while traveling in Mexico.

SOURCE   Medscape, emedicine.medscape.com/ article/168230-overview

Shark Attacks

ANNUAL US DEATHS   3 in 2020

From 2011 to 2020, 7 fatal shark attacks occurred in the US; there were 448 nonfatal attacks. Three people were killed in the US by shark attacks in 2020, which marine biologists called "an unusually deadly year." One each occurred in Maine, California, and Hawaii. In 2021, a boogie boarder was killed in a shark attack in Morro Bay, California.

Between July 1 and July 12, 1916, 4 people were killed and 1 person was injured in a series of shark attacks along the Jersey Shore.

STAY SAFE   Heed beach patrol warnings of sharks in the area. Do not venture too far from shore.

SOURCE   Global Shark Attack File, sharkattackfile.net/ incidentlog.htm

Snakes

ANNUAL US DEATHS   About 5

Venomous snakes found in the US include rattlesnakes, copperheads, cottonmouths/water moccasins, and coral snakes. Each year, an estimated 7,000 to 8,000 people are bitten by venomous snakes; about 5 of those people die. Most bites occur from handling or accidentally touching snakes. A California study found that 80 percent of rattlesnake victims were young men.

Rattlesnakes and other venomous snakes are shy of humans and will slither away harmlessly unless threatened. According to the California Poison Control System, approximately 300 rattlesnake bites are reported annually in a state with 39 million residents. Of those who survive a rattlesnake bite, 10 to 44 percent will have permanent injury, such as losing part of a finger. US Poison Control Center data from 1983 through 2007 suggests approximately 1 death per 736 people with rattlesnake bites.

Constrictor snakes (anacondas, boa constrictors, pythons) squeezed 17 people to death over the last 4 decades; 7 were infants or young children. Despite the numerous invasive Burmese pythons in the Everglades, no deaths have been reported.

STAY SAFE   Avoid venomous or constrictor snakes. Do not keep them as pets. Use caution when outdoors in areas where venomous snakes are found. Look before you step or place your hands; stay on the trail and avoid tall grass and underbrush; wear long pants and sturdy shoes. Do not attempt to suck the venom from a snake bite; you will only add bacteria to the wound and possibly get venom into your mouth and esophagus.

SOURCES   California Poison Control System, calpoison.org/ topics/rattlesnakes; CDC NIOSH, cdc.gov/niosh/topics/snakes/ default.html; USGS, Venomous Snakes of the United States and Treatment of Their Bites, pubs.usgs.gov/publications/5230010; Utah State University Extension, extension.usu.edu/news_sections/ agriculture_and_natural_resources/snake-bite-risk

Spiders

ANNUAL US DEATHS   Average of 6

The US has several species of venomous spiders, either native or introduced. Their bites can be painful, permanently disfiguring, and even fatal. These include the black widow, brown recluse, tarantula, hobo, and yellow sac.

On average, 6 people per year die from spider bites in the US. Most are very young or very elderly.

STAY SAFE   Avoid spiders. If you are bitten and develop any serious symptoms, seek medical help immediately.

SOURCE   CDC NIOSH, cdc.gov/niosh/topics/spiders

Zoos

ANNUAL US DEATHS   Average of 1.3

Zoos teach the public about the animal world and play a vital role in conserving endangered species and researching animal health and medicine. Millions of people visit zoos every year without mishaps.

Over a recent 26-year period in the US, 33 humans died in zoos, for an average of fewer than 1.3 persons killed per year. On rare occasions, a zoo worker is inattentive and is killed by a dangerous animal. More often though still very infrequently, a visitor intentionally trespasses into an animal enclosure, or falls in accidentally, and is killed.

In 1982 in New York City's Central Park Zoo, a 29-year-old man climbed a series of fences to enter the polar bear enclosure and was killed by a 1,200-pound bear. In 1987 in Brooklyn's Prospect Park Zoo, an 11-year-old boy scaled a fence with 2 friends and entered the polar bear enclosure. He was killed by the 2 bears in the enclosure; his friends escaped. His last words were, "Get help! He's biting me hard."

In 2012 at the Pittsburgh Zoo, a mother lifted her 2-year-old son to the top of a 4-foot wall for a better view of the African painted dogs. The child fell backward into the enclosure and was killed by the dogs.

Numerous other cases of foolish, drunken, or illegal behavior by zoo visitors could be cited.

STAY SAFE   Respect the animals at the zoo. Never enter their enclosures.

SOURCES   J.P. Barreiros, Zoo animals and humans killed because of human negligent behavior. *Journal of Coastal Life Medicine*, October 2016; Leviathan Project, leviathanproject.us; the *New York Times*, nytimes.com/1987/05/20/nyregion/polar-bears-kill-a-child-at-prospect-park-zoo.html

At Home

A home can be anything from a log cabin in the woods, to a mobile home, an RV, or a houseboat, to a house in the suburbs or an apartment in a city high-rise. The type and size of the structure, the number of rooms, what's put into each, and what goes on in each, and how much yard or land area (if any) goes with the property are all almost infinitely variable.

People do much more than just sleep while at home. For the minutes and hours they're awake and not away from the house, they engage in personal care and grooming, parenting and child-rearing, food prep and mealtimes; they enjoy hobby and leisure/entertainment time, play sports, and do fitness exercise; they do household chores and home maintenance tasks. And, especially due to the COVID-19 pandemic and its aftermath, many people also now work from home some or all of the time.

Using a single-family house as an example to illustrate the anatomy of the home, the interior part of the house typically includes a garage, a basement or cellar, and an attic, along with 1 or more floors of various rooms for living, all enclosed between roof, foundation, and outer walls. The different rooms inside vary greatly in purpose, function, and contents: kitchen and pantry, dining, living, bathing and toilet, sleeping and wardrobe, and maybe also a mudroom, playroom, and nursery, a home gym, a media room, or a hobby/crafts room, and a laundry area (maybe part of the basement). The outdoor part of the property typically includes a front and/or back yard, a driveway, a front sidewalk, and perhaps a pool, a patio or deck, and a garden.

The various home areas, and their contents and functions, involve varying risks of death. This is especially true for the home's utilities: its fuel, hot water and heating systems, plumbing, electrical power, and air-conditioning equipment, along with kitchen stove; the sink, toilet, and shower or tub in the bathrooms; and many different electronic devices (plug-in or battery powered) in virtually every room. Then there are other appliances large and small, for cooking and for cleaning, plus power tools, automatic garage doors, paints, cleaning products, solvents, and adhesives, drugs in the medicine cabinet, motor vehicles in the garage or driveway, trucks that stop at the curb, plus rain gutters to clean, trees to prune, walks that need shoveling, and lawns that need mowing.

This section inventories the chief lethal risks that arise, where they occur around the house based on how the physical home space is subdivided, what functions are performed in different spaces, and the likely contents of each room. Human activities at home, their safety (or lack thereof), and their resultant mortality, are not reported and regulated in the same rigorous detail as are occupational (**On the Job**) or transportation (**On the Go**) deaths, acts of violence (**Homicide, Suicide**), or even most recreational (**At Play**) deaths.

On the Job or **On the Go**, numerous government regulatory bodies and agencies, such as OSHA, the NTSB, state motor vehicle departments, police, and many others issue safety guidelines, inspect for compliance, and usually investigate fatalities in considerable detail. This allows reliable tallies of deaths by specific cause. It also yields accurate death rates, based on good estimates for total units in the rate calculation's denominator, such as 100 million Vehicle Miles Traveled (VMT), or 100,000 Full-Time Equivalent (FTE) persons employed.

Not so at home. Virtually everyone in the US is at home some of the time, yet virtually no one keeps detailed tallies of cumulative time elapsed during our various indoor and outdoor residential risk exposures (such as cooking, bathing, mowing the lawn, or tinkering with the furnace). The best statistics to look for in comparing and evaluating different household risks are the total numbers of people killed in a year while performing different household activities, chores, tasks, or DIY jobs. An additional way to evaluate the risks is by each different tool, appliance, utility, chemical, piece of furniture, or other item used in the home.

This section considers death risks of residents while they are at home and includes some accident categories that can happen in various environments but most often occur in a home setting. The death risks for employees such as nannies, and workers who provide services such as garbage collection, landscaping, roofing, and others, are all covered separately in **On the Job**.

RVs and camper vans are included in the **On the Go** section, as is various motorized home equipment, such as rider mowers, lawn tractors, ATVs, and UTVs. Houseboats are included in the **At Play** section on **Recreational Boating**. Residents killed when hit by vehicles "at or just beyond the curb of their house," such as cars, ice cream trucks, mail trucks, garbage trucks, or school buses, are also included in the death statistics in the **On the Go** entries.

Most outdoor leisure activities, such as sports and picnicking, are included in the **At Play** section because most often take place away from home. Household chores, hobbies, crafts, and DIY home improvements, maintenance, and repairs are included here in **At Home.**

Some causes of death that *can* occur while at home but also elsewhere are covered in other sections to provide more detailed and accurate data. For example, exercise equipment, home playground equipment, and swimming pools are covered in **At Play**; lethal mishaps involving home lawn mowing, groundskeeping, and landscaping are covered in **On the Go** or **On the Job**.

Leading Causes of Accidental Death

ANNUAL US DEATHS   Roughly 120,000 in US homes

The leading causes among the approximately 120,000 accidental deaths in US homes were:

- *Poisoning*. Gases and fumes, medicines, toxins, and chemicals killed 64,795 people at home (2017).
- *Slips, trips, and falls*. These mishaps killed 36,338 people at home (2017).
- *Airway obstruction*. Choking, suffocating, strangling, and postural asphyxiation kills about 6,620 people per year. Many of these deaths occur at home.
- *Drowning*. This is the cause of death for some 4,000 people per year. Many drownings, especially of young children, occur at home.
- *Fires*. In American homes, almost 3,000 people die each year from smoke inhalation, heat, and/or flame.

Males are almost twice as likely to die from a home accident as females. On average, for every person killed in an accident, 13 others are admitted to a hospital and a further 129 need emergency room treatment.

Data is not always readily available to break down each of the broad causes of death by the 3 dimensions that would more precisely characterize any particular accidental death at home:

1. Where in the home the mishap occurred (e.g., kitchen, bedroom)
2. What item or content triggered the death (e.g., stove, cigarette)
3. What activity the deceased was engaged in at the time (e.g., cooking dinner, smoking in bed)

STAY SAFE   See "Home Accident Statistics: Is Your Home as Safe as You Think?" at asecurelife.com/home-accident-statistics for a valuable compendium of safety tips.

SOURCES asecurelife.com, asecurelife.com/home-accident-statistics; NFPA, nfpa.org/News-and-Research/Data-research-and-tools; NSC, injuryfacts.nsc.org/home-and-community/home-and-community-overview/introduction; Stop Drowning Now, stopdrowningnow.org/drowning-statistics

Worst States for Home Accidental Death

ANNUAL US DEATHS See data below.

The 5 states in 2017 with the highest rates of accidental death at home per 100,000 of population were:

West Virginia	38.4 per 100,000
New Mexico	38.2
Wisconsin	30.5
Rhode Island	30.4
Vermont	29.6

This data does not include all causes of unintentional injury or accidental death, but it is still informative. On the basis of what was included, the national death rate was 24.3 per 100,000 population.

STAY SAFE See "Fear vs. Reality: Which Household Accidents Should You Worry About?" at Safewise.com (safewise.com/blog/household-accidents) for another valuable compendium of safety tips.

SOURCES NSC, injuryfacts.nsc.org/home-and-community/home-and-community-overview/introduction; Safewise.com, safewise.com/blog/household-accidents

Safest States for Home Accidental Death

ANNUAL US DEATHS   See data below.

The 5 states in 2017 with the lowest rates of accidental death at home per 100,000 of population were:

Maryland	13.8 per 100,000
California	16.5
New York	17.2
Texas	17.3
Utah	17.8

This data does not include all causes of unintentional injury or accidental death, but it is still informative. On the basis of what was included, the national death rate was 24.3 per 100,000 population, which equates to an annual national death toll from home accidents of about 80,000.

STAY SAFE   See "Fear vs. Reality: Which Household Accidents Should You Worry About?" at Safewise.com (safewise.com/blog/household-accidents) for another valuable compendium of safety tips.

SOURCES   CDC, cdc.gov/nchs/fastats/accidental-injury.htm; NSC, injuryfacts.nsc.org/home-and-community/home-and-community-overview/introduction; Safewise, safewise.com/resources/home-safety

The Matrix of Death

ANNUAL US DEATHS   120,000

About 120,000 people in the US die each year in accidents at home. To help grasp the many ways this can happen, this 2-D Matrix of Death shows how different rooms in a house can kill you in different ways. An × indicates a particular room where the mishap occurs—for instance, drowning in the bathtub. An × in every horizontal box indicates the accident can happen anywhere in the house—for instance, a slip-and-fall.

AT HOME

CAUSE OF DEATH	BR	LR	K	BATH	CELLAR	ATTIC	GARAGE	DR	PR	CHR	NURSERY	MR	GYM
Fire, explosion, collapse	x	x	x	x	x	x	x	x	x	x	x	x	x
Carbon monoxide or other gas			x		x		x						
Slip/Fall	x	x	x	x	x	x	x	x	x	x	x	x	x
Drowning				x							x		
Electrocution	x	x	x	x	x	x	x	x	x	x	x	x	x
Crib/Bassinet											x		
Toys	x	x	x	x						x	x		
Appliance, furniture tip-over	x	x	x					x	x		x	x	x
Poisoning				x	x	x	x	x		x			
Home invasion	x	x	x								x		
Suffocation	x		x			x	x			x	x	x	
DIY/repairs	x	x	x	x	x	x	x	x	x	x	x	x	x
Cleaning	x	x	x	x	x	x	x	x	x	x	x	x	x
Tools				x			x				x		
Gun/knife	x	x	x	x	x	x	x	x	x	x	x	x	x

Key: BR = bedrooms; LR = living room; K = kitchen/pantry; Cellar = cellar, basement, utilities, laundry; DR = dining room; PR = playroom; CHR = crafts/hobby room, MR = media room.

STAY SAFE See "Home Accident Statistics: Is Your Home as Safe as You Think?" at Asecurelife.com, asecurelife.com/home-accident-statistics, plus STAY SAFE in the individual entries that follow.

SOURCES Asecurelife.com. asecurelife.com/home-accident-statistics; CDC, cdc.gov/nchs/fastats/accidental-injury.htm; NSC, injuryfacts.nsc.org/home-and-community/home-and-community-overview/introduction

Home Not Childproofed

> **ANNUAL US DEATHS** At least 10,000 children (age 17 and younger) die in home accidents each year.

Although estimates from different sources vary, at least 10,000 young people—under age 18—die in accidents at home every year. Many of these deaths, whether of infants, toddlers, children, or teenagers, could have been prevented by installing safety measures in and around the homes.

STAY SAFE Childproofing a home is a key responsibility of parenthood. For good information on this complicated topic, see CPSC, "Childproofing Your Home—12 Safety Devices to Protect Your Children," at cpsc.gov/safety-education/safety-guides/kids-and-babies/childproofing-your-home-12-safety-devices-protect and NSC, "Childproofing Your Home," at nsc.org/community-safety/safety-topics/child-safety/childproofing-your-home.

SOURCES CDC, cdc.gov/nchs/fastats/accidental-injury.htm; CPSC, cpsc.gov/Research-Statistics/NEISS-Injury-Data; NSC, injuryfacts. nsc.org/home-and-community/home-and-community-overview

Home Not Elder-Proofed

> **ANNUAL US DEATHS** From 2007 to 2016 the annual US deaths from falls for seniors, aged 65 and older, rose from 18,334 to 29,668. (The CDC did not record what portion of these mishaps occurred in or around the home.)

Elderly people (age 65+) are especially vulnerable to death or serious injury from accidents in and around their own homes. People over age 65 are about 3 times as likely to die in a fire as younger persons; more than 75 percent of crippling or lethal falls by elderly victims happen in or near their house or apartment.

Safety measures for older adults, especially to prevent falls, are vital even for those who are mentally alert and physically fit. With age, balance, vision, physical strength, bone strength, and vulnerability to severe brain injury from concussion all worsen.

STAY SAFE Elder-proofing a residence is an important and complex topic. For more information, see "How to Make Your Home Safe for Aging Parents" at aarp.org and "15 Top Tips for Elder-Proofing a Home" (especially for Alzheimer's patients) at alzlive.com.

SOURCES   AARP, aarp.org/home-family/your-home/info-2021/ fall-prevention-safety-tips.html; CDC, cdc.gov/homeandrecreational safety/falls/adultfalls.html; B. Moreland et al. Trends in Nonfatal Falls and Fall-Related Injuries Among Adults Aged ≥65 Years — United States, 2012–2018. MMWR *Morbidity and Mortality Weekly Report*, July 2020; B. Moreland et al. A Descriptive Analysis of Location of Older Adult Falls that Resulted in Emergency Department Visits in the United States, 2015. *American Journal of Lifestyle Medicine* (AJLM), August 2020.

Appliances, Belongings, and Furnishings

ANNUAL US DEATHS   120,000

The activities in which we engage at home are so commonplace that we easily fail to realize the potentially lethal risks. For example, the safety risks of a taking a bath including drowning, scalding, falls, and electrocution. That puts a different perspective on taking a bath.

Care and skill are needed to perform our various household activities and tasks. The many items we make use of must all be sensibly selected, then properly handled, operated, stored, and even (eventually) disposed of, to protect everyone in the home from accidental injury or untimely death.

This section covers common activities associated with the contents of a typical home, including furniture, fixtures, built-ins, appliances, tools, devices, equipment, supplies, and materials. It discusses everything from beds to staircases to bathtubs, from bleach to food processors to power tools, and more. Some duplication/repetition and cross-referencing to other sections of this book are unavoidable—and desirable. Redundancy is one of the most fundamental principles of safety engineering best practices.

Appliances

ANNUAL US DEATHS   10,000+

Over a recent 10-year period in the US, an average of 200 people a year were fatally electrocuted by home appliances. Major appliances accounted for 67 of these deaths (33.5 percent), small appliances for 61 (30.5 percent), power tools for 39 (19.5 percent), and lamps and light fixtures for 33 (16.5 percent). About 90 percent of the victims were

Ad for an Acorn Gas
Range, 1932

male. Most of the fatalities occurred while the victim was trying to repair the appliance.

Appliances can also kill by tip-overs and fall-ontos, fires or burns, gas or fume asphyxiations, explosions, entrapments, and so on. While comprehensive data is hard to come by specifically for these various appliance-related causes of death, that there are a total of some 120,000 accidental deaths at home in the US every year suggests that appliances, however people use or misuse them, might contribute to upwards of 10,000 home fatalities a year, if not significantly more.

STAY SAFE   Always unplug an appliance before moving it, cleaning it, or maintaining or repairing it. Make sure all power cords are not frayed, have 3-prong plugs, and are plugged into properly grounded 3-prong outlets. Never plug too many appliances into the same household electrical circuit.

SOURCES   Electrical Safety Foundation International (ESFI), esfi. org/home-safety; Safewise, safewise.com/resources/home-safety

Firearms Accidents

ANNUAL US DEATHS   486 in 2019

In 2019 in the US, 486 people were killed in firearm accidents. In addition, an unknown number of gun suicides are recorded as accidental each year. Sixteen states reported accidental fatal shootings to the CDC's National Violent Death Reporting System (NVDRS). Based on data from NVDRS, 1,260 unintentional firearms deaths occurred in the US between 2005 and 2015. Based on additional CDC data, however, 6,885 people were unintentionally killed by firearms between 2006 and 2016. In 2016, 495 incidents were recorded. In 2018, 458 people were killed in accidental gun deaths, for approximately 1 percent of all gun-related deaths (39,740) in the US.

About 77 percent of unintentional gun deaths occur in the home. Eighty percent of victims are male; most are under 25 years old. Common circumstances include playing with guns (28.3 percent of incidents), thinking the gun was unloaded (17.2 percent of incidents), and hunting (13.8 percent—see entry for **Hunting**). Other accidental gun deaths occurred while the owner was cleaning the gun, or during target shooting, or while transporting a gun in an unsafe manner. Alcohol consumption by the victim is suspected in nearly 25 percent of cases.

Another 27,000 people per year in the US are injured nonfatally by gun accidents.

STAY SAFE   For information about gun safety and preventing firearm accidents, see NY SAFE Act, safeact.ny.gov/gun-safety. Always handle firearms as if they are loaded. Always keep them pointed in a safe direction. Keep them secured, either in a gun safe or protected by a trigger lock. Take a course in firearms safety and marksmanship. If you have children at home, consider keeping all your firearms in a remote lock-up, such as those provided at many shooting ranges.

SOURCES   CDC NCHS, cdc.gov/nchs/ndi/index.htm; Nationwide Children's Hospital, nationwidechildrens.org/research; NY SAFE Act, safeact.ny.gov/gun-safety; S.J. Solnick and J.D. Hemenway, Unintentional firearm deaths in the United States 2005-2015. *Injury Epidemiology*, October 2019; Stanford Children's Health, stanfordchildrens.org/en/topic/default?id=firearms--injury-statistics-and-incidence-rates-90-P02982

Hoarding

ANNUAL US DEATHS   No US data, but Australia recorded 10 hoarder deaths (due to fire) over 10 years.

While data is not readily available on how many people in the US are killed every year due to hoarding of old papers, trash, and/or animals in their home (think "crazy cat lady"), the American Psychiatric Association estimates that between 2 and 6 percent of the adult population suffers from this disorder. Once considered a form of obsessive-compulsive disorder (OCD), hoarding is now seen as a separate mental illness. The affliction tends to strike older people; 3 times as many Americans over age 55 suffer from it than do those under age 44.

Hoarding can be life-threatening. Accumulated animal waste, garbage, and mold are health hazards. The hoarder's home has an extreme risk of deadly fire or material pile collapse, endangering not only the residents but neighbors. A 2009 study in Australia found that hoarding-related fires were very deadly. Over a 10-year period, 48 such fires were responsible for 10 fatalities. While these 48 incidents comprised only 0.25 percent of all residential fires in Melbourne during the study period, those 10 hoarder deaths were fully 24 percent of all residential fire fatalities: the death rate for hoarders in fires was almost 100 times the average for the whole city.

One of the most infamous hoarder residential fires occurred in an apartment building in Toronto, Canada, in 2010. Though no one was killed, 3 firefighters and 14 residents of the building were hospitalized overnight. A fire that ordinarily would have taken about an hour to extinguish by a single company of firefighters instead became a 6-alarm blaze needing 300 firefighters with 27 fire engines working for 8 hours. More than 1,000 residents were endangered. The fire was allegedly caused by a cigarette that someone flicked onto the hoarder's balcony, where he kept piles of papers and books; ironically, the hoarder himself never smoked.

In 2019, an 80-year-old woman in Detroit died in a house crammed with hoarded objects. She was found dead, eaten by her dog.

STAY SAFE Compulsive hoarders often respond with hostility to any attempt by relatives, friends, neighbors, and health/safety authorities to intervene. Experts say that hoarders themselves must want to change. For information on the symptoms of hoarding, and how you can try to help someone who has this problem, see the International OCD Foundation's "What is Compulsive Hoarding?" at iocdf.org. If you become aware of an imminent threat to life or health, such as a fire safety violation or a collapse hazard, call 911 right away.

SOURCES American Psychiatric Association, psychiatry.org/patients-families/hoarding-disorder/what-is-hoarding-disorder; R. Iyer and M. Ball. The Vulnerability of Compulsive Hoarders to Residential Fire Fatality: A Qualitative Enquiry. Presentation at 8th Asia-Oceania Symposium for Fire Science and Technology, 2010; NFPA, nfpa.org/Public-Education/Fire-causes-and-risks/Behavioral-risks/Hoarding; *USA Today*, usatoday.com/story/news/nation/2019/08/01/hoarder-living-squalor-found-dead-eaten-her-dog-detroit/

Hot Tubs and Spas

ANNUAL US DEATHS 40+

While hard data is scarce on how many of the 335 Americans who drown every year specifically do so in hot tubs or spas, circumstantial evidence hints at the potentially lethal dangers. The rate of death from drowning in states with high hot tub ownership, such as California, is about 3 times higher than it is in states with low hot tub ownership, such as New York. A study for 1990 through 2009 identified 800 hot tub deaths in the US, for an annual average of 40, which is probably an undercount.

The CPSC reported more than 800 deaths associated with hot tubs between 1990 and 2009, nearly 90 percent of them in children under age 3.

Hot tubs can cause fatal accidents in ways besides drowning, such as fatal slip-trip-and-fall accidents while getting in and out. Hot tubs can also kill if their nozzles and drains malfunction or if their safety grills come loose. People have had their hair or clothing sucked in so hard that they are pulled under the water, where they drown. Spending too much time in a hot tub can also eventually raise core body temperature sufficiently to induce fatal hyperthermia; this happened to a 23-year-old women in Los Angeles in 1996. The heat (and maybe too much physical exertion) can also induce heart attacks in people with preexisting health conditions. In 2015, a couple from Canada who were at a hotel in Mexico for their daughter's wedding were having sex in a hot tub when the man had a heart attack and died. He fell on top of the woman, drowning her.

STAY SAFE Never leave kids unattended in a hot tub. Never drink or use drugs while hot tubbing. Never get into a hot tub if the drain cover is missing. Always be careful entering and leaving the tub and know

The pleasures of a hot tub can turn lethal.

where the safety off switch for the pump system is located. The hot water of a hot tub can be a vector for infection.

SOURCE CPSC report, cpsc.gov, Pool or Spa Submersion: Estimated Nonfatal Drowning Injuries and Reported Drownings, 2021 Report

Nursery Furniture

ANNUAL US DEATHS Estimated 107

Over the 3-year period 2014 through 2016 in the US, 320 deaths were reported of children younger than 5, in which nursery furniture and/ or consumer products were in use at the time of death. (They might or might not have been a factor in the cause or causes of death.) This is an annual average of about 107 deaths a year.

A federal safety agency report in mid-2021 found at least 90 accidental deaths over the past several years associated with baby sleep products, some of which were designed to be portable and used for travel and/or outside the home. (See **Accidental Postural Asphyxiation**.)

The most deaths, including sudden unexplained infant death syndrome (SUIDS), occurred in cribs and on mattresses, accounting for 36 (33.6 percent) of the 107 deaths per year. Playpens and play yards were associated with 21 deaths (19.6 percent), bassinets and cradles with 20 deaths (11.8 percent), and infant carriers (excluding motor vehicle incidents) with 10 deaths (9.3 percent). Other products and furniture categories that each caused between 1 and 4 deaths were portable baby swings, inclined sleepers, bouncer seats, strollers and carriages, baby baths or bath seats or bathinettes, baby gates and barriers, high chairs, changing tables, and other items.

Furnishings or nursery arrangements contributing to infant fatalities were blankets, pillows, crib bumpers, and other soft bedding placed inside a crib or playpen, nearby window covering or baby monitor cords, or ill-fitting, nonoriginal mattresses or cushions. In addition, having more than 1 infant sleep or play in the same crib/playpen/yard was associated with several deaths.

STAY SAFE Always read the instructions carefully and be sure to properly assemble baby furniture and products. When using older baby furniture, check the CPSC website for safety information. Follow the guidelines at Safe to Sleep from the NIH (safetosleep.nichd.nih.gov).

SOURCES CPSC report, cpsc.gov, Injuries and Deaths Associated with Nursery Products Among Children Younger than Age Five; NIH NICHD, safetosleep.nichd.nih.gov/safesleepbasics/SIDS/fastfacts

Tip-Overs and Fall-Ontos

> **ANNUAL US DEATHS** 30.1

A tip-over occurs when a large piece of furniture, electronics, machinery, bookcases, shelving, or an appliance becomes imbalanced and falls over. A fall-onto occurs when any heavy object stored on a shelf or in an overhead cabinet drops down from a height. In the US, between 2000 and 2017, 450 of these accidents killed a child aged up to 14 years, 19 killed an adult aged up 59 years, and 73 killed seniors aged 60 and over. This gives an average annual US deaths from tip-overs and fall-ontos of 30.1 for people of all ages. Most of these fatalities result from head injuries and/or postural asphyxiation.

In addition, between 2015 and 2017, an average of 9,400 people of all ages were injured badly enough by tip-overs and fall-ontos to require an emergency room visit.

Except for rare events such as earthquakes or explosions, these mishaps are usually caused by the victim or another person exerting an upsetting force on the object, which then falls down or tips over because its center of gravity has shifted out of a stable position. The force can result from climbing on the object, reaching up to grab it but losing control of it, placing an excessive weight on or in it, tugging on it or lifting it or trying to roll it improperly, or pulling open too many drawers, doors, lids, or other parts.

Broken down by type of item that killed, 34 percent were TVs, 29 percent involved TVs and the furniture holding the TV, 30 percent involved chests, bureaus, or dressers, and 6 percent were appliances, mainly stoves and ovens. Ninety-two percent of these deaths occurred in homes; 50 percent of all fatal tip-overs occur in a bedroom.

STAY SAFE "Anchor It" (anchorit.gov) is a US government program that educates the public about tip-over risks and advocates properly anchoring home furnishings so that they cannot tip over. Heavy items, and all items of interest to kids such as toys and games, should be stored/displayed low to the floor so they can be easily reached without fall-onto or tip-over risk. Secure all cords and cables so they do not create a combined trip/snag/tug *and* tip-over/fall-onto risk. Be careful reaching up to get anything heavy or breakable stored on a high shelf, whether in a closet, a cabinet, a bookcase, or other shelving. Be careful whenever using a stool or ladder to reach things high up. Be equally careful not to let old stuff and clutter pile up higher and higher, so that it might all eventually fall onto you or someone you love (see the entry for **Hoarding**).

SOURCES Anchorit.gov, anchorit.gov/about-us; CPSC report, cpsc. gov, Product Instability or Tip-Over Injuries and Fatalities Associated with Televisions, Furniture, and Appliances: 2020 Report

Toys that Kill

ANNUAL US DEATHS 14 in 2019

In 2019 in the US, 14 toy-related deaths of kids ages 14 or younger were reported. Kick scooters caused 5 of these deaths, another 4 involved a child at play getting hit by a car, 1 death was due to a fall, and 4 children were killed by small plastic balls that obstructed airways.

In 2019, there were 224,200 nonfatal toy-related injuries that needed emergency room visits. Males comprised 59 percent of this group and females 41 percent; 73 percent of these accidental injuries were in kids 14 or younger; the remaining 27 percent were in older teenagers or adults. Cuts, scrapes, and bruises made up 41 percent of the injuries. The head and face were affected 46 percent of the time. Kick scooters caused the most injuries, at about 21 percent of the total.

Data for 2017 and 2018 gave other specific mechanisms for child deaths due to toys: choking on a rubber ball, a pacifier, or a toy dart; swallowing a small ball; suffocation by a foil balloon; strangulation by a kite string; postural asphyxiation when buried by toys and a blanket; drowning related to a toy; suffocation in a toy chest; airway obstruction from a rock or rock-shaped toy.

Over the years, several toys have proved dangerous, requiring recalls and discontinuance. One of these involved small magnets, which could be ingested by children and then clump together in the digestive system, causing blocked intestines and/or internal gangrene that required major surgery and led to at least 1 death.

One of the most dangerous toys ever was the U-238 Atomic Energy Lab, sold for a year in 1951 by A. C. Gilbert Company, the same firm that made the Erector Set construction toy. This U-238 "toy" included samples of 3 different types of uranium ore, with a Geiger counter and an electroscope that kids could use to observe the radioactivity given off by the uranium. The kit was discontinued—not because of safety concerns (consumer safety laws didn't exist at the time) but because it was so expensive ($49.50, or about $500 today) that sales were poor. Today, dangerous toys are regulated under the 1966 Child Protection Act, the 1969 Child Protection and Toy Safety Act, and the 1976 Toxic Substances Control Act. The Consumer Product Safety Commission was established in 1972.

STAY SAFE   Select toys, games, and sports equipment carefully. See "For Kids Sake—Think Toy Safety" at cpsc.gov, "Toy Safety Checklist" at cchp.ucsf.edu, and "Toy Safety Checklist" at safekid.org.

SOURCES   California Childcare Health Program, cchp.ucsf. edu/content/toy-safety-checklist; CPSC report, cpsc.gov, Toy-Related Deaths and Injuries Calendar Year 2019; IEE *Spectrum*, spectrum.ieee.org/fun-and-uranium-for-the-whole-family-in-this-1950s-science-kit; safekids.org, https://www.safekids.org/child-product-and-safety-news-2019

Chores and Pastimes

Chores

ANNUAL US DEATHS   Estimated 33,700

Household chores, which can be more formally defined as "unpaid work in and around the home," kill somewhere between hundreds and thousands of people in the US every year, depending on how broadly the activity is defined, but data is scarce. Comprehensive hard data on the overall annual death toll can be roughly estimated from studies in other countries, and from the magnitude of accidental deaths at home, which add up to about 120,000 per year globally.

A study in Australia from 1989 through 1992 found an average of 75 deaths per year from housework. The most dangerous chores were DIY home improvements/repairs/maintenance, gardening/groundskeeping, and working on your car. The most common fatal injuries were being hit by an inadequately braced car while working on it, falls from ladders, burns (including while cooking), and electrocution. About 83 percent of these deaths were males. If the deaths per million of population were applied to the US in 2020, they would imply death tolls of 22,200 from DIY home improvements/repairs/maintenance, 6,620 from working on cars, and 5,300 from gardening and groundskeeping/landscaping.

For other miscellaneous cautionary statistics, in the US every year about 5 people are killed in fires caused by failing to remove clothes dryer lint buildup, about 50 per year are killed by fires caused by frayed or overloaded extension cords, up to about 12,000 die from falling down stairs, and pesticide poisoning kills 23 Americans a year.

STAY SAFE   See the stay safe advice in other entries.

SOURCE   T.R. Driscoll et al. Unintentional fatal injuries arising from unpaid work at home. *Injury Prevention*, March 2003.

DIY Improvements, Repairs, and Maintenance

> **ANNUAL US DEATHS** Estimated 55,000

DIY activities around the house, yard, or involving personal motor vehicles can lead to potentially deadly accident risk exposures. Estimates of the annual US death toll while doing DIY projects run as high as 55,000, with another roughly 13 million being injured.

Data on annual US DIY deaths by specific cause of death tends to be incomplete or anecdotal, but the available hard data does provide important safety warnings. About 300 people a year are killed falling off ladders, and about 400 are electrocuted fixing things around the house. Other causes of DIY death are slips, trips, and falls, especially off roofs or scaffolds or over cords and materials; asphyxiation on fumes from portable generators, heaters, or air compressors; electrocutions by power tools, appliances, high-tension lines, or home electrical infrastructure; poisonings by toxic paints, solvents, lubricants, stains, adhesives, and similar materials.; fires and burns from welding, soldering, electrical shorts and arcs; tip-overs and fall-ontos of appliances or furniture being worked on, piles of materials and supplies being used, and especially of motor vehicles not properly secured on jacks, lifts, and maintenance stands.

Below is an inventory of the most dangerous DIY tools, equipment, and supplies, rated by the estimated number of emergency room visits they caused US householders in 2017, in descending order:

Ladders	194,000 hurt per year
Nails/screws/tacks	125,000
Lawn mowers	90,000
Power saws	81,000
Cleaning equipment	57,000
Manual gardening tools	56,000
Razor knives	46,000
Cleaning agents	45,000
Grinders/buffers/polishers	37,000
Chain saws	31,000
Hammers	30,000

About 70 to 80 percent of DIY injuries happen to males. The age group most prone to DIY injuries are people in their 50s. The months with the most injuries are June (136,000) and July (137,000).

Serious danger can arise when homeowners rent heavy construction equipment (such as Bobcats) for DIY projects that need big earthmoving and/or heavy lifting/loading machines. See the entry for **On the Road: Construction and Earthmoving Vehicles**.

STAY SAFE Ladders must be free of loose or broken parts and must always be planted solidly on the ground, placed 1 foot way from the wall for every 4 feet of height; never stand on the top rung. Always turn off all electrical circuits before working on them. Read all power tool instructions when you buy them; never operate in cramped or cluttered quarters and never wear loose clothing; wear eye and hearing protection; turn off and unplug the tool before changing blades or drill bits; be very careful to protect your fingers and your face, the body parts most prone to injury. Lay down nonslip mats and wear nonslip shoes; clean up the work area as you go and clean up all spills right away. Avoid walking on a slanted roof; use a ladder or scaffold to reach gutters, shingles, chimneys, antennas, etc. Know your limitations, and try not to work alone. If a project needs heavy equipment such as bulldozers or front-end loaders, hire a contractor.

Ladders are both indispensable and a source of injury—and sometimes death.

SOURCES CDC, cdc.gov/nchs/nvss/deaths.htm; NSC, injuryfacts.nsc.org/home-and-community/home-and-community-overview/introduction; SafetySkills.com, safetyskills.com/hand-power-tool-safety

Hobbies and Crafts

ANNUAL US DEATHS No available data

Sometimes the processes, equipment, and materials used in home hobbies and crafts can lead to unintentional injury and even death. Hard data is scarce on how many deaths occur every year in the US specifically from indoor hobbies. Some hobbies, such as pottery making (pottery wheels and kilns), woodworking (power tools), metalworking (lathes, drill presses), or glass blowing (furnaces) involve using heavy equipment and cross over into the DIY (Do It Yourself) home improvements, maintenance, and repairs category.

As an illustrative example of typical hobbies and crafts, consider model railroading. Beginning with nothing but raw materials, tools, and supplies, enthusiasts can construct over a period of months or years an entire realistic little world of tracks with bridges over rivers and tunnels under mountains, plus moving trains, greenery and crop fields, buildings, motor vehicles, and little people and animals. Tools such as Dremel multitools and static-grass applicators, not to mention power saws and razor knives, can cause trauma via electrocution, abrasions, lacerations, or traumatic amputations. The various paints, adhesives, and lubricants used in this hobby, if not used and stored safely, can cause poisonings, chemical burns, lung damage, or fires. The many small, colorful parts and accessories can appeal to children, who might swallow them and suffer airway obstructions. The complicated electrical systems for powering and controlling the model trains, layout lighting, sound effects (train noises, even thunderstorms), and animated accessories (little log loaders, tiny neon signs, etc.) present their own electrocution hazards, and also fire hazards from electrical shorts— accelerated by the flammable liquids and solids that abound on a typical model railroad and accompanying workbench. Display and storage shelving, tall stacks of heavy supplies, and bulky layout backdrops all present tip-over and fall-onto hazards.

STAY SAFE   Each home hobby will have its own inventory of tools, supplies, and materials that add to the home's contents.

SOURCE   NSC, injuryfacts.nsc.org/home-and-community/ home-and-community-overview/introduction

Snow Shoveling and Snowblowers

ANNUAL US DEATHS   100

Shoveling snow kills about 100 people a year in the US and causes injuries or health emergencies in about another 11,500. Many of the deaths result from overexertion, or from slip-and-fall accidents on ice. Males over age 55 are especially prone to heart-related symptoms while shoveling snow. Children under 18 are 15 times as likely as adults to get injured by being hit with a snow shovel, whether by accident or on purpose.

A study of snowblower mishaps between 2003 and 2018 found 648 deaths during that 16-year period, for average annual US deaths of 40.5. All but 1 of these deaths were from heart attacks due to overexertion while operating a snowblower. Nonfatal snowblower mishaps caused about 5,700 injuries per year. Half of these were amputations, lacerations, or fractures, mainly from reaching into the mechanism to clear clogs, or from slipping and falling on icy pavement.

STAY SAFE For information on snow shoveling safety, see "Snow Shoveling" at Nationwide Children's Hospital, nationwidechildrens. org/research/areas-of-research/center-for-injury-research-and-policy/ injury-topics/sports-recreation/snow-shoveling. For information on snowblower safety, see "Snowblower Safety: Tips for Keeping You Safe When Dealing with Snow" at Society Insurance, societyinsurance.com/ blog/snow-blower-safety-tips-keep-safe-dealing-snow. Wear sturdy shoes or boots that give good traction. If you have a preexisting medical condition or are not accustomed to vigorous exercise, arrange for someone else to do your snow clearing for you. When using a snowblower, avoid wearing loose clothing that can snag moving parts. If the snowblower clogs, turn it off and wait for the blades to stop moving, then use an unclogging tool or a stick; *never* put your hand into the chute. Wear eye goggles and hearing protection. Dress warmly, including a hat and gloves.

SOURCES NSC, nsc.org/community-safety/safety-topics/seasonal-safety/winter-safety/snow-shoveling; D.S Watson et al. Snow shovel-related injuries and medical emergencies treated in US EDs, 1990 to 2006. *American Journal of Emergency Medicine*, January 2011

YouTube Challenges

ANNUAL US DEATHS Likely under 100

Hundreds of young people have died from imitating deadly stunts seen in videos posted to YouTube, TikTok, and similar online video sites. These stunts include "choking games," "cinnamon challenge," "chubby bunny," "extreme fighting," "dry scoop challenge," and "the salt and ice challenge."

STAY SAFE Never try to duplicate a crazy stunt seen in an online video post; most websites have policies against them but some slip through and gain viewership.

SOURCES Children's Hospital of Philadelphia Center for Injury Research and Prevention, injury.research.chop.edu/blog/posts/

tiktok-challenges-latest-dangerous-trend; HealthyChildren.org,
healthychildren.org/English/family-life/Media/Pages/Dangerous-
Internet-Challenges.aspx

Criminal and Negligent Acts

Babysitters and Nannies

> **ANNUAL US DEATHS** No current statistics exist, but data from the
> late 1990s reveals 1,900 cases of criminal offenses against children
> committed by nannies or babysitters, with about 95 fatalities.

Many American families use nannies (full-time) or babysitters (occasionally) to help take care of their kids. Almost all these for-hire, in-your-home childcare providers are professional, capable, and law-abiding. In fact, only 4.2 percent of all crimes against young children are perpetrated by their paid caregiver. Many more are committed by the child's own family members or by complete strangers. Data is not readily available on how many child injuries or deaths are caused by caregivers who are law-abiding but lack the skills and attention to prevent accidents.

A study in the late 1990s of a sample of about 1,500 cases in the US of criminal offenses against children by their nannies or babysitters found that only 0.5 percent of these crimes—sexual assaults, physical attacks, kidnappings, etc.—resulted in a child's death.

In New York City in 2012, a nanny stabbed a 6-year-old girl and a 2-year-old boy repeatedly in a bathtub, leaving them both to bleed to death. The woman had previously seen a therapist due to depression and panic attacks; her lawyer at her trial said she felt overworked and underpaid. She was sentenced to life in prison. But crime and abuse are not a parent's only concern. It is also vital that the nanny or babysitter you hire is aware of, and puts into practice, thorough safety-conscious measures to protect your kids from unintentional or accidental injuries.

STAY SAFE Unless you know the babysitter well, run a background check on any candidates, looking at criminal records and sex-offender databases. Ask for and check references. Be aware that the candidate's friends and relatives may gain access to your children; make sure that the candidate's social circle is reputable and law-abiding. Go through a childcare safety checklist with the candidate, such as "Nannies, and Caregivers" from Kidpower.org. When hiring a local teenager, make sure they have taken a Red Cross babysitting class and been certified.

SOURCES American Red Cross, redcross.org/take-a-class/babysitting/babysitting-child-care-training/babysitting-classes; Child Welfare Information Gateway, childwelfare.gov/more; Kidpower International, kidpower.org/library/article/kidpower-safety-tips-for-babysitters-and-caregivers; *People*, people.com/crime/krim-killer-nanny-yoselyn-ortega-sentenced-prison; US Department of Justice, D. Finkelhor and R. Ormrod. Crimes Against Children by Babysitters. *Juvenile Justice Bulletin*, September 2001.

Daycare Centers

ANNUAL US DEATHS 24 to 36

For many parents, daycare centers are an alternative to nannies and babysitters for out-of-home preschool and/or after-school care. These facilities are not without their risks, however. Accurate, comprehensive data on numbers and causes of deaths of kids at US daycare centers is hard to come by. The available statistics suggest that, overall, due to licensing requirements, daycare centers are safer places for children than their own homes. Informal and unlicensed daycare centers may be in violation of local safety ordnances and building codes and may have untrained or inexperienced staff.

One federal government report for 2016 put the number of kids who died in childcare facilities nationwide due to neglect or abuse at 24, which probably undercounts such deaths and understates the prevalence of bad daycare center care in general. Another study, for 1985 through 2003, found 332 child deaths at daycare centers that were *not* due to sudden unexplained infant death syndrome (SUIDS). This implies an annual average non-SUIDS death of about 8.5 individuals. Data from around 2012 from just 4 states (Indiana, Minnesota, Missouri, and Texas) recorded about 36 kids a year dying from all causes in daycare centers. Again, this is probably an undercount.

STAY SAFE For information sources about selecting a daycare center, see "8 Tips for Choosing Child Care" at *Parents*, parents.com/baby/childcare/basics/8-tips-for-choosing-child-care.

SOURCES Administration for Children and Families, report, acf.hhs.gov, Contemporary Issues In Licensing: Reporting, Tracking, and Responding to Serious Injuries and Fatalities in Child Care; Child Care Aware of America, childcareaware.org/families/choosing-quality-child-care; https://www.acf.hhs.gov; J. Wrigley and J. Dreby. Fatalities and the organization of child care in the United States, 1985–2003. *American Sociological Review*, October 2005.

Home Invasions

ANNUAL US DEATHS   Estimated 100

About 100 people a year in the US are murdered during home invasions. This is out of an average of about 1.03 million home invasions per year over the period 1994 through 2010, during which about 267,000 people per year were victims of other violent crimes, such as assault, rape, or attempted murder. In 2018, about 2 percent of American households suffered at least 1 instance of trespassing or burglary; about one-third of these were home invasions because they happened while the residents were present.

STAY SAFE   For tips on safety from home invasions, see "10 Ways to Stop Criminals from Choosing Your Home" at Safe Home, safehome. org/blog/stop-criminals-from-choosing-your-home. Home security measures include good interior and exterior lighting (including motion sensitive activation), locks and/or bars on all windows and doors, an alarm system connected to a central station, and video doorbells and security cameras. Many security experts suggest getting a large dog as an effective deterrent. Panic rooms are usually unnecessary.

SOURCES   DOJ Bureau of Justice Statistics, bjs.ojp.gov/data/topic; FBI.gov, ucr.fbi.gov/crime-in-the-u.s/2019/crime-in-the-u.s.-2019/topic-pages/robbery

Fires, Explosions, and Other Disasters

Collapses

ANNUAL US DEATHS   No data available except for firefighters—2.2 deaths

Structure collapses of houses and apartment buildings are very often associated with a fire and/or an explosion. It can be unclear which came first and exactly which caused any deaths—eyewitness reports are often confused and contradictory. Some residential structure collapses can be caused by natural disasters, such as earthquakes, floods, or landslides and mudslides (see these entries in **Acts of God and Nature**). Structure collapses can in turn create other lethal conditions, including electrocution hazards, gas leaks, and toxic dust.

Complete annual national data for US deaths (excluding firefight-

A building on the verge of collapse

ers and other first responders) specifically in residential structure collapses is not readily available. Between 1997 and 2008, 26 firefighters were killed while responding to residential structure collapses. This is an occupational death toll of about 2.2 firefighters per year for this type of fire-and-rescue callout.

Structure collapses are low-incidence but high-consequence events. On June 24, 2021, the 13-story Champlain Towers South apartment building in Surfside, Florida, collapsed, killing 98 residents—the deadliest residential collapse in US history.

STAY SAFE Be alert for signs that a residential structure has become unsound and might be about to collapse: creaking and groaning noises, sticking doors and windows, leaks, and cracks in walls are all indicators of structural problems. Get a risk assessment from a structural engineer. If collapse seems imminent, evacuate the building and call 911.

SOURCES CNN, cnn.com/2021/07/15/us/miami-dade-building-collapse-thursday/index.html; NFPA, nfpa.org/News-and-Research/Data-research-and-tools/Building-and-Life-Safety

Fires

ANNUAL US DEATHS 2,870 in 2019

Residential building fires are the 2nd most common type of fire in the US, after fires outdoors (trash and dumpsters, storage yards, brush-fires, forest fires, etc.). Residential fires accounted for 29.9 percent of all fires in 2019. Residential fires are by far the deadliest type of fire, killing by smoke inhalation and burns fully 77 percent of the total of 3,704 people (excluding firefighters) killed in fires of all types in 2019. This is a death toll of 2,870 people killed by fires in a home, giving a death rate in 2019 of 8.4 per million of population. (Firefighter deaths are covered in the **On the Job: Firefighters** section.) A further 12,700 people (again, excluding firefighters) were injured in the US in 2019 in residential structure fires.

Fires in 1- and 2-family houses caused 2,390 deaths (83.3 percent), fires in apartments caused 380 deaths (13.2 percent), and 100 people (3.5 percent) died in other residential fires. Two out of 3 people who died in fires at home were asleep when the fire started, showing the tremendous importance of having properly working and installed smoke detectors—they double to triple the odds of surviving a house fire. People living below the poverty line, people living in rural areas, and people who have disabilities are especially vulnerable to residential fires.

The causes of the roughly 354,000 residential building fires in 2019 in the US were:

Cooking	50.2 percent
Permanent/portable heating systems	9.3 percent
Accidents and carelessness	7.7 percent
Frayed wires/electrical malfunctions	6.8 percent
Arson	4.5 percent
Open flames (candles, fireplaces, BBQ)	4.3 percent
Other sources of heat	3.4 percent
Appliances	2.9 percent
Other equipment failure	2.4 percent
Smoking	2.0 percent
Acts of nature	1.8 percent
Playing with fire	0.3 percent
Other	4.4 percent

The causes of the 1,900 *fatal* fires broke down as:

Accidents and carelessness	19.7 percent
Smoking	12.3 percent
Arson	9.3 percent
Frayed wires/electrical malfunctions	7.5 percent
All other known causes	33.8 percent
Under investigation	17.4 percent

SOURCE NFPA, nfpa.org/News-and-Research/Data-research-and-tools/US-Fire-Problem

CHRISTMAS TREE FIRES

ANNUAL US DEATHS 2

Between 2014 and 2018, an average of 160 home fires each year started with Christmas trees. These fires caused an average of 2 deaths, 14 injuries, and $10 million in direct property damage annually. On Christmas Day in 2021, a Christmas tree fire killed a father and 2 sons in Pennsylvania. In January 2022 in Philadelphia, a 5-year-old boy playing with a lighter set a Christmas tree on fire, causing a fire that killed 12 people.

STAY SAFE Use fire-resistant artificial Christmas trees. Remember to water live Christmas trees and remove the tree from the home when it is dry. Place trees at least 3 feet away from fireplaces, radiators, space heaters, and other heat sources.

SOURCES CNN, cnn.com/2021/12/25/us/pennsylvania-christmas-tree-fire/index.html; FEMA, usfa.fema.gov/prevention/outreach/holiday.html; the *New York Times*, nytimes.com/2022/01/11/us/philadelphia-fire-christmas-tree.html; NFPA, nfpa.org/News-and-Research/Data-research-and-tools/US-Fire-Problem/Christmas-tree-fires

CLOTHING FIRES

ANNUAL US DEATHS 60

An average of 60 senior citizens a year are killed in clothing fires in the US. Another 1,100 seniors are injured every year. The loose-fitting clothes that many seniors prefer wearing can catch fire very easily; cooking, burning trash, and use of space heaters are the main culprits. Data for the general population is not available.

STAY SAFE Never wear loose or billowing clothing around sources of heat or open flames. If your clothing catches fire, stop, drop, and roll to put out the flames. All-cotton and all-wool clothing is more flame resistant that polyester or blends. Make sure all kids' sleepwear and bedclothes contain flame retardants; check the labels before buying. Never *ever* smoke in bed.

SOURCES EMSA, emsaonline.com/resource-library/summer-safety-tips/fire-safety-tips-fire-hazards-and-clothing; FEMA US Fire Administration, usfa.fema.gov/data/statistics; NFPA, nfpa.org/News-and-Research/Data-research-and-tools/US-Fire-Problem

HOME HEATING FIRES

ANNUAL US DEATHS 500

Heating equipment is a leading cause of home fires. Annually, municipal fire departments respond to an average of 48,530 home structure fires caused by heating equipment. An average of 500 people die in these fires every year, for about one-fifth of all home fire fatalities. Space heaters account for about 2 in 5 home fires but are responsible for 81 percent of deaths.

In January 2022 in an apartment in the Bronx, New York, a fire started by a space heater killed at least 17 people, all by smoke inhalation.

SOURCES The *New York Times*, nytimes.com/2022/01/09/nyregion/cause-bronx-apartment-fire.html; NFPA report, nfpa.org/News-and-Research/Data-research-and-tools/US-Fire-Problem/Heating-equipment, Home Heating Fires

NATURAL GAS FIRES AND EXPLOSIONS

ANNUAL US DEATHS 40 (excluding firefighters)

An estimated average of 4,200 home structure fires per year start with the ignition of natural gas. These fires cause an average of 40 deaths per year, excluding firefighters. (Firefighter deaths are covered in the **On the Job: Firefighters** entry.) Cooking equipment was involved in 54 percent of all the natural gas fires; heating equipment, including water heaters, was involved in 25 percent.

The main cause of accidental explosions within homes are natural gas distribution line (local or in-structure) leaks, which allow gas to build up. The gas then detonates from a spark or flame.

In the 20-year period leading up to mid-2018, in the US, all natural gas distribution line explosions, both residential and commercial/industrial, killed about 221 people, excluding firefighters. This is an average annual death toll of roughly 11 people per year, either at home, in a restaurant, or on the job. About 177 million US households use natural gas for their heating or cooking.

In 2014, in New York City, 8 people were killed when a natural gas leak detonated, causing a tremendous explosion that leveled 2 small apartment buildings. Residents smelled gas and called the utility company, but repair workers did not arrive in time.

In 2018, in the Massachusetts towns of Lawrence, Andover, and North Andover, excessive pressure in natural gas lines caused a series of explosions and fires in some 40 homes, leading to more than 80 separate fires. Although the event caused the emergency evacuation of 30,000 people, only 1 person was killed.

On very rare occasions, such as during prolonged exposure to a fire, a propane tank explosion can occur from the pressure of the propane tank reaching higher than the pressure that the tank can safely vent, causing the tank to burst open. This kind of explosion is called a Boiling Liquid Expanding Vapor Explosion (BLEVE). Propane is a very safe fuel. OSHA data found only 1 propane-related death in 2018 among 13 incidents.

STAY SAFE   If you smell a gas leak, evacuate the residence to a safe distance and immediately call (first) 911 and (next) your local utility company.

SOURCES   American Gas Association, aga.org/research/data; the *New York Times*, nytimes.com/2018/09/13/us/lawrence-massachusetts-explosion-gas-fire.html and nytimes.com/2014/03/18/nyregion/final-victim-of-blast-is-identified.html; OSHA, osha.gov/data/work; NFPA report, nfpa.org/News-and-Research/Data-research-and-tools/US-Fire-Problem, Natural Gas and Propane Fires, Explosions and Leaks; Estimates and Incident Descriptions

HOME FIRE DEATHS BY AGE

ANNUAL US DEATHS   See breakdown in entry.

Annual home fire death rates per 1 million people in each age group, averaged over the period 2011 through 2015, were as follows:

Under 5	7.5
5 to 9	4.9
10 to 14	2.9
15 to 19	2.4
20 to 24	2.7
25 to 34	4.3
35 to 44	5.2
45 to 54	8.8
55 to 64	12.3
65 to 74	15.2
75 to 84	21.2
85+	26.0
All Ages	7.9

Those most vulnerable to being killed in a fire in a home are young children and the elderly.

STAY SAFE   The San Francisco Fire Department's "Home Fire Facts" (sf-fire.org/home-fire-facts) is a good source on home fire safety for individuals and families.

SOURCE   NFPA report, nfpa.org/News-and-Research/Data-research-and-tools/US-Fire-Problem, Fire Loss in the United States During 2020

HOME FIRE DEATHS BY SEX

ANNUAL US DEATHS   See breakdown in entry.

Annual home fire death rates per 1 million people in each age group, averaged over the period 2011 through 2015, were as follows:

DEATHS

Males: 1,440 deaths, 57 percent of all deaths

Females: 1,070, 43 percent

INJURIES

Males: 6,630 injuries, 54 percent of all injuries

Females: 5,670, 46 percent

SOURCE   NFPA report, nfpa.org/News-and-Research/Data-research-and-tools/US-Fire-Problem, Home Fire Victims by Age and Gender

Nonresidential Structure Fires

ANNUAL US DEATHS   110 civilians (nonfirefighters) in 2019

In 2019 in the US, 110 civilians (nonfirefighters) were killed in fires in nonresidential buildings and other structures. A further 1,200 civilians were injured. Note that this data does not separate those victims who were on the job at the time from those who were patrons/customers, guests, or visitors to the building at the time of the fire. (Data on occupational fire deaths is given for different major occupation/job groups in the **On the Job** section.)

STAY SAFE   In case of fire or if a fire or smoke alarm sounds, evacuate immediately and call 911. Do not use the elevators; use the designated fire stairs. Many local building codes require fire safety public address systems; comply with any announcements.

SOURCES   FEMA, usfa.fema.gov/data/statistics/#who; NFPA, nfpa. org/News-and-Research/Data-research-and-tools/US-Fire-Problem/ Fire-loss-in-the-United-States

Lightning Strikes at Home Indoors

ANNUAL US DEATHS   Average of 49 lethal strikes, a fraction of which occur indoors

Death by electrocution via a lightning strike while someone is home indoors is very rare but does occur. On average, 49 people a year in the US are killed by lightning strikes, usually when the person is outdoors and struck directly. Death generally occurs from a combination of severe electrical burns, interference with heart rhythms, and neurological damage. The same fatal injuries can occur when the victim is inside a structure if the electrical current from the lightning strike gets conducted into their body through metal or by arcing across air gaps. Indoor lightning shocks, when they do occur, usually result from talking on a landline telephone at the time the bolt hits the house. The shock is usually not lethal, however.

Ground strike

Lightning bolts possess immense power compared to normal, in-residence electrical circuits. On average, lightning carries about 300 million volts and 30,000 amperes. For comparison, in the US, house current is usually only 120 volts and 15 amperes.

Lightning can also kill people at home by starting a fire. An average of 4,400 lightning-induced residential structure fires in the US occur every year. Lightning-induced fires kill an average of 9 people in the US every year, but not all of these deaths occur in residences.

In 2017, at an undisclosed location, a man was indoors renovating his cottage, which had exposed steel beams, when the cottage was struck by lightning. His body was found severely burned, surrounded by metal tools and lying between 2 metal sawhorses. The current from the lightning strike passed down inside the house, arced from a metal sawhorse to his left foot, passed through his torso, stopping his heart and burning him, and then passed out of his body via his right thumb and arced to the other sawhorse.

STAY SAFE   See the NWS checklist "Lightning Tips" at weather.gov/safety/lightning-tips. Make sure your house has an effective lightning rod and that the electrical and plumbing infrastructures are both thoroughly grounded per the residential building code for your area. Older structures should upgrade to today's safety standard. Whenever you can hear thunder, the lighting is close enough to strike your location. Stay away from all electrical and electronic equipment, appliances, outlets, and switches, and all metal plumbing such as pipes, faucets, and showerheads until the storm has passed, and you can no longer hear the thunder.

SOURCES   CNBC, cnbc.com/2017/03/03/how-a-man-was-struck-by-lightning-in-his-own-house.html; III, iii.org/fact-statistic/facts-statistics-mortality-risk; NWS, weather.gov/hazstat

Home Hazards

Accidental Airway Obstruction

ANNUAL US DEATHS   6,620

Accidental/unintentional airway obstruction kills about 6,620 people in the US per year. Many of these fatalities occur at home, but they can occur anywhere. Annual national statistics on such deaths are not available broken down by location, and airway obstruction is a broad category that can be separated into 4 subgroups:

1. **Suffocation, also known as asphyxiation or smothering:** Something blocks, covers, fills, or smothers the mouth and nose, preventing breathing.
2. **Strangling:** Something presses against or around the throat, cutting off air flow.
3. **Postural asphyxiation:** Something, including the person's own awkward physical position, compresses, crushes, or restricts their chest or abdomen enough to prevent breathing.
4. **Choking:** Something goes down the throat, lodges in or above the trachea, and cannot be removed before causing death via anoxia (lack of oxygen).

In 2018, suffocation, including hanging, asphyxiation, smothering, and other mechanical and nonmechanical threats to oxygenation (e.g., being trapped in a low-oxygen environment), caused 18,924 deaths. Approximately 44 percent of these deaths occurred in the home. An exact breakdown of the national death toll of accidental airway obstructions into each of these 4 separate categories is not readily available, for 2 main reasons. Often the victim is moved from the original scene of the accident as part of emergency transport and medical treatment, leaving a coroner to determine that the cause of death was anoxia but unable to specify the exact circumstances. Sometimes an airway obstruction is a hybrid of more than 1 of the 4 subgroups, such as when a victim falls within a grain silo and inhales grain while also being both smothered and crushed by it, or when a faulty baby playpen/play yard collapses, both smothering the infant in baby blankets and compressing its chest lethally under debris.

Both the youngest and the oldest members of the US population are most vulnerable to such fatalities. Several major consumer product recalls over the years, especially for some baby carriers, sleepers, and toys, were due to regulatory loopholes and insufficient testing that allowed sales of dangerous items that led to tragic deaths.

STAY SAFE   See entries below.

SOURCES   American Academy of Pediatrics, healthychildren. org/english/safety-prevention/Pages/default.aspx; CDC, cdc.gov/ injury; CPSC, cpsc.gov/Safety-Education; NSC, injuryfacts.nsc.org/ home-and-community/deaths-in-the-home

ACCIDENTAL POSTURAL ASPHYXIATION

ANNUAL US DEATHS No data available; however, by 2021, inclined sleepers were implicated in at least 94 US deaths.

Postural asphyxiation occurs when a person's bodily position itself, and/or external constraints on moving and expanding the chest, interfere with being able to draw air into the lungs, resulting in anoxia. The bodily position may interfere with breathing if it is very awkward or contorted. Postural asphyxiation, like the other subgroups of accidental airway obstruction, is a special risk for children, the elderly, the ill or disabled, anyone taking prescribed medication that can cause drowsiness or grogginess, and anyone who is impaired by alcohol or drug abuse.

Postural asphyxiation has been a problem with some badly designed baby sleeping furniture or portable infant carriers. Inclined sleepers allow babies to sleep at an angle that can restrict their airway if they fall asleep in a chin-to-chest position. These sleepers have been tied to at least 94 deaths by 2021. Many have been recalled—check the CPSC website to see which.

STAY SAFE Baby sleep products that incline more than 10 degrees are not safe.

SOURCES American Academy of Pediatrics, healthychildren. org/english/safety-prevention/Pages/default.aspx; CPSC, cpsc.gov/ Research--Statistics/NEISS-Injury-Data

ACCIDENTAL STRANGULATION

ANNUAL US DEATHS 960 in 2019

Strangulation occurs when something presses against or around the throat, preventing air from passing into the lungs. (Intentional strangulation is covered in the **Homicide** and **Suicide** sections.) Unintentional/accidental strangulation usually occurs when a cord, rope, wire, item of clothing, or a bedsheet or similar item catches a person around the neck/throat, and also gets tangled or snagged on something else, and gets pulled too tight. Accidental strangulation can occur on play equipment, for example. Infants and young children are especially vulnerable to such mishaps. Their clothing, furniture, bedding, and play equipment should always be chosen with care.

STAY SAFE See the entries for **Accidental Strangulation: Window Cords and Sashes** and **Clothing Drawstrings and Scarves**. See also

"Choking and Suffocation Prevention: Children Ages Birth to 19 Years" at health.ny.gov/prevention/injury_prevention/children.

SOURCES   CPSC, https://www.cpsc.gov/Safety-Education/Safety-Education-Centers/Window-Covering; New York State Department of Health, health.ny.gov/prevention/injury_prevention/children

ACCIDENTAL STRANGULATION: CLOTHING DRAWSTRINGS AND SCARVES

ANNUAL US DEATHS   Estimated 0.75

Between 1984 and 2019 in the US, clothing drawstrings on children's outerwear caused at least 26 deaths and 73 other injuries. This is an estimated 0.75 deaths per year.

The most common lethal mishap is for a jacket or hoodie neck drawstring, especially its knot or toggle, to get snagged on a playground slide or other home or park/schoolyard play equipment, leading to accidental strangulation. Protruding bolts or small gaps between guardrail and slide are examples of snag hazards that can lead to such deaths. Mishaps from waist-area drawstrings were less common.

While data is scarce on people killed when long scarves snag and lead to strangulation and/or dragging trauma, such deaths do sometimes occur. The most infamous example is the death of American dancer Isadora Duncan, who was killed in 1927 in France when her long scarf snagged the axle of the car in which she was riding.

STAY SAFE   Remove hood and neck strings from all children's outerwear, including jackets and sweatshirts. Drawstrings on the waist or bottom of garments should not extend more than 3 inches outside the garment when it is fully expanded.

SOURCE   CPSC, cpsc.gov/Business-Manufacturing/Business-Education/Business-Guidance/Drawstrings-in-Childrens-Upper-Outerwear

ACCIDENTAL STRANGULATION: WINDOW CORDS AND SASHES

ANNUAL US DEATHS   About 12

Every year in the US, about 12 children are killed by accidental strangulation on the cords used to open and close or raise and lower window coverings, such as curtains or horizontal or vertical blinds. Since

1990, more than 200 infants and young children have died from unintentional strangulation involving window cords. They can become wrapped around a child's neck, or the child's head and neck can get trapped in the loop formed by the free end if the cord is made that way.

In 2018, a new voluntary safety standard for stock and custom window coverings cords was approved by the American National Standards Institute in collaboration with the Window Covering Manufacturers Association. The standard recommends that all window coverings be cordless, or use inaccessible cords, or use cords no more than 8 inches long, or use wands instead of cords, and that they have warning labels about the strangulation danger.

STAY SAFE   Keep cords short, secure them higher up than kids can reach, and place cribs, bunk beds, and other children's furniture away from windows. Consider getting cordless window coverings. For more information about childproofing such cords around your house, see "Window Covering Cords Information Center" at cpsc.gov.

SOURCE   CPSC, cpsc.gov/Safety-Education/
Safety-Education-Centers/Window-Covering

ACCIDENTAL SUFFOCATION

ANNUAL US DEATHS   1,064 in 2017

Suffocation occurs when something blocks, fills, or smothers the mouth and nose, preventing breathing. (Intentional suffocation is covered in the **Homicide** and **Suicide** sections.)

Accidental/unintentional suffocation can occur through an occupational mishap (see **On the Job**), but many such mishaps occur while sleeping, whether at home, on vacation, at summer camp, away at school, or in a nursing home. Pillows, sheets, and blankets present particular suffocation dangers, as do plastic sheeting and wrapping, dry cleaning bags, garbage bags, canvas tarpaulins, tents, deflated bounce houses and kiddie pools, or anything else that will not pass air well (or at all) and is large and/or bulky enough to cover a person's nose and mouth.

Extreme youth or advanced age make someone more vulnerable to suffocation, as does any disability or illness, use of medication that causes drowsiness or grogginess, and abuse of alcohol or drugs.

Accidental suffocation is often contributed to by dangers in the surrounding environment, such as an item of furniture or play equipment that is poorly designed or improperly assembled, and/or by a child assuming a posture in an unsafe sleeper or carrier.

Mechanical suffocation is, by far, the leading cause of death for children under age 1, causing 1,064 fatal injuries in 2017. Mechanical suffocation constituted 80 percent of all injury-related mortality cases for infants. Infants who die from mechanical suffocation lose the ability to breathe due to strangulation, or smothering by bed clothes, plastic bags or similar materials. Between 2011 and 2014, unintentional suffocation caused by soft bedding or by infants sleeping on adult mattresses or couch cushions killed 250 infants. Most suffocation deaths of infants and children occur in the home.

STAY SAFE Choose safe bedding for infants and young children; don't let them sleep on adult mattresses or soft bedding. Store safely or dispose of properly any suffocation hazards, such as dry cleaning bags, plastic wrapping and packaging, tarpaulins, and deflated kiddie pools, water slides, and bounce houses. See "Choking and Suffocation Prevention: Children Ages Birth to 19 Years" at health.ny.gov/prevention/injury_prevention/children.

SOURCES A.B. Erck Lambert et al. Sleep-related infant suffocation deaths attributable to soft bedding, overlay, and wedging. *Pediatrics*, May 2019; New York State Department of Health, health.ny.gov/prevention/injury_prevention/children; NSC, nsc.org/community-safety/safety-topics/child-safety/mechanical-suffocation-in-infants

ACCIDENTAL SUFFOCATION: DRY CLEANING BAGS

ANNUAL US DEATHS About 25

In the US, every year, suffocation caused by dry cleaning bags kills about 25 people a year, virtually all of them kids and most of them under the age of 1.

STAY SAFE Heed the warning printed on each dry cleaning bag. Keep them away from children. Do not store clothing inside the bags in your home.

SOURCES CDC, cdc.gov/injury/wisqars/fatal.html; CPSC, cpsc.gov/recall-hazards/suffocation

BATTERIES

A very few people in the US are killed every year by fires and explosions caused by batteries, either spontaneously in the case of certain rechargeable batteries or, in the case of all-electric car battery banks, serious fires ignited by collision.

Additional hazards involving batteries include children swallowing button (or watch) batteries. Button batteries, found in many remote controls and other electronic devices, can appear to young children as a shiny, tasty snack. The battery, if pushed into the nose or swallowed, can get stuck and create a current flow through tissues that generates poisonous sodium hydroxide (lye), which burns a hole in that spot. The result can be serious injury and illness, long-term disability, or even death. Swallowing a button battery kills about 1 child per year in the US, while causing nonfatal injuries in as many as 3,500 people of all ages. Between 1995 and 2010, 14 fatalities involving children ranging in age from 7 months to 3 years were recorded.

Over the 8-year period 2009 through 2016, there were 195 incidents reported in the US of e-cigarette/vaping batteries catching fire or exploding. There were 133 injuries, but no one was killed.

In Houston, Texas, in 2021, 2 men were killed when a self-driving all-electric car crashed and caught fire. The vehicle's battery bank burned stubbornly for 4 hours afterward, repeatedly reigniting after firefighters thought it was extinguished. This happens in electric cars because the damaged batteries continue to hold stored energy until they are finally completely discharged.

Rechargeable lithium batteries that have manufacturing flaws, or are overcharged, or are immersed in conductive salt water, or whose heat vents become clogged can catch fire or explode inside cell phones, laptops, and other devices. Globally, there have been at least 8 deaths in recent years reported from these mishaps. In 1 case in the US, a recycling center was destroyed by a fire caused by an improperly disposed of lithium battery, but no one was killed. E-bike batteries can catch fire while charging.

Large dry-cell batteries can produce sufficient current that a short circuit across their terminals creates sufficient heat to ignite a fire. Always use terminal protector caps to prevent this and dispose of dead dry cells properly. Their inner materials are flammable, even when drained of current.

STAY SAFE   Keep button batteries and devices that use them out of the reach of children. Never try to remove a nonuser-replaceable rechargeable lithium battery from the device in which it is installed; always dispose of the battery and/or the entire device at a special electronics recycling center, never in your household garbage. Fire departments and all-electric vehicle manufacturers are working together for better procedures and equipment to contain and extinguish electric vehicle battery fires. Never tamper with or modify an e-cigarette or vaping pen; pause between puffs to let the unit cool, so that the battery does not overheat. Do not overcharge lithium batteries, and do not block the cooling vents of devices that use them.

SOURCES American Academy of Pediatrics, healthychildren. org/english/safety-prevention/Pages/default.aspx; CDC, cdc. gov/injury/wisqars/fatal.html; FEMA, fema.gov/case-study/ emerging-hazards-battery-energy-storage-system-fires

Asbestos

ANNUAL US DEATHS 12,000 to 15,000

Asbestos, once widely used as fireproofing and heat-insulating material, is a toxic element still found in older residential and other structures, in pipe cladding, and in many devices (especially steam boilers and heating systems). Inhaling asbestos fibers can cause mesothelioma and other fatal lung cancers, as well as asbestosis, a lung disease that causes shortness of breath due to lung tissue damage.

Between 12,000 and 15,000 people a year die in the US from asbestos-related diseases. These diseases take years to develop and more years to kill. The death toll is gradually declining because of nationwide asbestos removal and remediation measures.

Asbestos was widely used in building, plumbing, and other industries until it was banned by the EPA in 1989.

STAY SAFE Asbestos testing and removal require expertly trained personnel using specialized equipment. See "Protect Your Family from Exposures to Asbestos" at epa.gov.

SOURCES CDC, atsdr.cdc.gov/asbestos/health_effects_asbestos. html; CDC, cdc.gov/cancer/mesothelioma; EPA, epa.gov/asbestos; OSHA, osha.gov/asbestos

Carbon Monoxide

ANNUAL US DEATHS 994 in 2019

Carbon monoxide (CO) is a colorless and odorless gas given off as a by-product of combustion from flames and internal combustion engines. Common sources of CO fumes that can cause deaths at home are heating and cooking equipment, especially improperly installed or maintained furnaces, and emergency and portable electrical generators. Each year, about 50,000 people visit emergency rooms due to accidental CO poisoning. Since 2000, the CPSC has recorded about 1,300 deaths from CO poisoning related to portable generators; most fatalities occur during weather-related power outages. In 2021 following extreme cold weather and widespread power failures in Texas, at least 17 people died from CO poisoning related to portable generators.

In 2019 in the US, a total of 994 people died from accidental exposure to gases not involving fires, for a death rate of 0.3 per 100,000 population. At least 430 people a year are killed specifically by carbon monoxide (CO) fumes. In 2017, 131 of those deaths were related to consumer products used in and around the home. Of these, engine-driven tools causes 104 deaths, 95 of which involved generators.

Heating systems were the 2nd leading cause of CO death in 2017: 42 deaths were associated with some type of heating appliance. Gas heating systems caused most of the deaths.

In 2017, males were 73 percent of CO poisoning deaths. Most CO poisoning deaths occur in colder months, with more than half the deaths in November, December, January, and February.

Less often, other gases were deadly: carbon dioxide from evaporating dry ice, for example, and toxic gases used in making and abusing illicit inhalant drugs. Statistics for these deaths are scattered and unreliable.

Because catalytic converters remove virtually all CO from vehicle exhaust emissions, suicides and accidental deaths from this gas source have declined by more than two-thirds over the past few decades.

The symptoms of CO poisoning include burning eyes, scratchy throat, red face, headache, nausea/vomiting, dizziness, lethargy, confusion, shortness of breath, and chest pain. Get to fresh air immediately.

Nonfire gas exposure accidental death risk tends to increase with age. From infancy through age 14, the death rate per million of population for the years 2002 through 2006 was about 0.6. It rose progressively with age, climbing to 5.2 per million—almost 9 times as severe a mortality rate—for people aged 75+.

STAY SAFE   Install battery operated or battery backup CO detectors near every sleeping area in the home. Check them regularly and change the batteries as needed. Have your furnace inspected every year. Never use a portable generator inside the home or garage, even if doors and windows are open. Use generators only outside and more than 20 feet away from doors and windows. In an emergency, evacuate to fresh air immediately and call 911.

SOURCES   CDC NCHS, cdc.gov/nchs/ndi/index.htm; CPSC, cpsc. gov/Safety-Education; NSC, nsc.org/community-safety/safety-topics; NFPA, nfpa.org/News-and-Research

Radon Gas

ANNUAL US DEATHS   21,100 from radon-related cancer

Radon is a radioactive gas produced by the decay of uranium in the Earth's core, soil, and rocks. Radon can build up inside structures such as homes by coming up through the foundation into the cellar. It may then reach living areas on higher floors. Inhaling radon gas is the 2nd leading cause of lung cancer in the US, killing about 21,100 people every year. Cancer can take up to 25 years to develop after exposure to radon; the more radon inhaled over time, the greater the risk of eventually getting lung cancer.

STAY SAFE   One in 15 homes in the US have some radon exposure; the only way to know is to have your home tested. If radon is detected, install radon mitigation measures, usually a fan and ventilation system to vent the gas to the outside air. Install a radon detector.

SOURCES   CDC NCHS, cdc.gov/nchs/ndi/index.htm; NSC, nsc.org/ community-safety

Chemical Poisonings

ANNUAL US DEATHS 65,000

When all deaths due to accidental exposure to a toxic substance are counted, 65,773 people were killed in 2019, for a death rate of 20.04 per 100,000 population. Most were drug-related, either opioid overdoses or other drugs singly or in combination. While many of these deaths occurred at home, the exact number is unknown.

Poisonous substances are common in and around a typical home: laundry pods, cleansers, rodent poison, insecticides, lubricants and solvents, medications, automotive supplies (e.g., antifreeze), drain cleaners, and more. From 2012 to 2018, when the packaging was changed, 8 deaths occurred from ingesting laundry pods.

STAY SAFE Keep all potentially poisonous substances and medications securely away from children. Keep the National Poison Control Center phone number, 1-800-222-1222, on your cell phone and posted in your kitchen. Professional advice about what to do in case of an actual or suspected poisoning is free of charge and is available 24/7/365 by calling or visiting poison.org.

SOURCES AAPCC, aapcc.org/national-poison-data-system; CDC NCHS, cdc.gov/nchs/ndi/index.htm; C.E. Gaw et al. Safety Interventions and Liquid Laundry Detergent Packet Exposures. *Pediatrics*, July 2019.

Dangerous Places

ATTICS

ANNUAL US DEATHS 35 (in attic fires)

Residential attics present householders with a number of potentially deadly hazards. Attic fires kill about 35 people a year in the US and injure another 125; they are notoriously difficult for firefighters to extinguish given their typical poor ventilation and poor access.

Other hazards found in attics include electrocution from frayed or exposed wires and faulty outlets, poisoning from toxic items stored there, tip-overs and fall-ontos from old furniture and/or stacked items, and slips, trips, and falls, especially if the only access is via a ladder.

STAY SAFE Always be careful going up and down the stairs. Make sure there is good lighting and/or bring a flashlight. Don't store flammable or toxic materials in the area.

SOURCE FEMA US Fire Administration, usfa.fema.gov/data/library/research/topics/top_attics.html

BASEMENTS AND CELLARS

> **ANNUAL US DEATHS** About 65 (in basement/cellar fires)

About 65 people a year in the US are killed in fires that start in basements or cellars. Given the number of mechanical and electrical devices present, many associated with the dwelling's utilities, and the amount of poisonous and/or explosive materials that move through pipes or are stored down there, plus all the equipment that generates heat and pressure, and that may electrocute people, the foundation and lowest floor level of any residence can be quite dangerous.

Basements and cellars make good tornado shelters but are not necessarily safe zones during thunderstorms and lightning strikes. Deadly voltages can travel through the dwelling and kill people touching or even standing near utility pipes, wires, and equipment below ground level.

STAY SAFE Always be careful going up and down the stairs. Make sure there is good lighting and/or bring a flashlight. Don't store flammable or toxic materials in the area.

SOURCES FEMA US Fire Administration, usfa.fema.gov/data/library/research/topics/top_attics.html; NFPA, nfpa.org/News-and-Research/Data-research-and-tools/US-Fire-Problem

BATHROOMS

> **ANNUAL US DEATHS** 235,000 people injured in the bathroom; 335 bathroom drownings

Bathrooms kill people in myriad ways. Toilets, shower stalls and bathtubs, hot tubs, sinks, slippery floors, glass doors, medicine cabinet contents and cleaning supplies, sharp objects and electrical devices, even animal attacks have led to human fatalities in the US. The total of injuries that occur in US bathrooms, estimated at about 235,000 for people over age 15, breaks down as about 70 percent from using a bathtub or shower, about 23 percent from using a toilet, about 2 percent from using a sink, and about 5 percent from other causes. More than 80 percent of bathroom-related injuries were caused by slips and falls, mostly while getting in and out of the tub or shower.

An average of 335 people of all ages drown in bathtubs every year in the US. In 2011, 95 children under age 17 drowned in bathtubs.

STAY SAFE Use nonslip bathmats in the tub and shower. Install grab bars. Supervise children closely in the bathroom, especially when bathing.

SOURCES CDC, cdc.gov/injury/wisqars/fatal.html; Nationwide Children's Hospital, nationwidechildrens.org/research/areas-of-research/center-for-injury-research-and-policy/injury-topics/sports-recreation/drowning-prevention

BEDROOMS

ANNUAL US DEATHS About 22,290

Many of the accidental/unintentional deaths that occur at home in the US happen in bedrooms: about 30 percent (10,800) of the total 36,000 yearly in-home fatal slips, trips, and falls, for example. The victim was asleep in almost 2,000 yearly residential building fire deaths. Some deaths are very specific to being in bed: In the US, every year, about 740 people of all ages are killed by falling out of bed. About 960 infants die in bed from suffocation or other airway obstruction. About 590 people a year, on average, are killed in fires started by smoking in bed.

STAY SAFE Don't smoke in bed. Always put your baby to sleep on a flat, firm, horizontal surface, and never co-sleep them with an adult or another child.

SOURCES CDC, cdc.gov/injury/wisqars/fatal.html; NFPA, nfpa.org/News-and-Research/Data-research-and-tools/US-Fire-Problem

GARAGES

ANNUAL US DEATHS 500

Garages and parking lots are dangerous places. About 500 people a year in the US are killed in parking lot and garage motor vehicle accidents, and over 60,000 more are injured. Many of the deaths and accidents take place in and around residential properties.

The motor vehicles parked in a home garage are themselves potentially deadly. Although many of the almost 40,000 people who die in motor vehicle mishaps (including pedestrians and bicyclists) die on the go, some people do die at home from vehicular fall-ontos while working on their car on a jack or lift, or from back-overs in the driveway.

Automatic garage doors can kill people by falling or closing on them if the door or its safety mechanisms malfunction. Children are especially vulnerable. Automatic doors and gates of all kinds, both residential and commercial/industrial, kill several dozen people in the US every year, mostly children, and injure between 20,000 and 30,000 more people each year. Carbon monoxide in exhaust fumes can cause death.

Other causes of death in residential garages involve various items typically stored or used in a home garage, such as toxic chemicals, power tools, and potentially flammable piles of boxes, papers, or oily rags. Dangers include poisoning, electrocution, fires and burns, and tip-overs or fall-ontos. A motor vehicle parked in a garage is a significant fire hazard, due to the combination of fuel, lubricants, and electrical systems it contains. Trips, slips, and falls can happen in garages with lethal consequences, due to wet or icy floors, clutter, or accidents while carrying heavy items into or out of the garage—as when bringing bags of groceries from the car into the house. Icy floors are a particular hazard in wintertime in unheated garages.

STAY SAFE   See the other relevant entries in this section.

SOURCES   CDC, cdc.gov/injury/wisqars/fatal.html; CPSC, cpsc.gov/ Newsroom/News-Releases/1991/CPSC-Safety-Standard-Targets-Garage-Door-Deaths; StaySafe.org, staysafe.org/safety/garage-2

KITCHENS

ANNUAL US DEATHS   953 in 2000

The kitchen can be the deadliest place in the house, with hazards ranging from fires and burns to slips, trips, and falls, to fatal injuries from DIY improvements, repairs, and maintenance, to poisonings by chemicals commonly found in kitchens.

In 2000, mishaps with the various consumer products (housewares and appliances) commonly found in kitchens killed 369 Americans. An average of 550 people a year in the US are killed by cooking fires. About 34 people per year are killed by being scalded by boiling water in home kitchens. Over 100,000 people a year in the US are injured nonfatally in accidents that happen in kitchens. Many of these involve knives.

STAY SAFE   See the relevant safety tips in this section.

SOURCES   CDC, cdc.gov/injury/wisqars/fatal.html; NFPA, nfpa.org/ News-and-Research/Data-research-and-tools/US-Fire-Problem

RESIDENTIAL ELEVATORS

ANNUAL US DEATHS   About 5

From 1997 through 2010, about 5 people per year were killed while they were elevator passengers not at work. A fourth of these were falls down elevator shafts, and another quarter were fatal slips and falls, on the same level, when exiting an elevator. Other types of deadly elevator mishaps included being crushed by a malfunctioning door or trauma suffered when the elevator cable broke and the emergency braking system failed. One of 5 yearly deaths was of a child aged 10 or under. Young children are vulnerable to being crushed by closing elevator doors. A study for 1994 found that 9,800 passengers were hurt badly enough in elevator mishaps to need hospitalization.

Panel of a modern elevator

While a breakdown of these deaths and injuries by nonwork location is not readily available, many of them occur in apartment buildings that have elevators. In addition, some multistory single-family houses are equipped with elevators, which are an aid to the mobility of disabled or elderly residents. Residential elevators in private homes caused an estimated 23 deaths and 4,600 injuries between 1980 and 2019.

STAY SAFE   Proper maintenance of elevators and their safety devices is essential. If you have small children, hold their hand when using any elevator. If the elevator doesn't open at the right level, bounces, makes strange noises, or otherwise just doesn't seem right, do not enter if you are outside and do not exit if you are inside. If inside, use the alarm button and/or your cell phone to summon help. Then be patient. If outside, report the problem immediately to building management if the lobby is staffed or, if necessary, call 911. Install space guards to eliminate the gap between elevator doors and the shaft.

SOURCES CDC, cdc.gov/injury/wisqars/fatal.html; CPSC, cpsc.gov/Newsroom/News-Releases/2019/CPSC-Alert-Protect-Children-from-a-Deadly-Gap-between-Doors-of-Home-Elevators

Defenestration

ANNUAL US DEATHS Estimated 12 children (under age 10)

Defenestration means falling, jumping, or being pushed through a window; the window might or might not be closed at the time. Historically, defenestration was sometimes used as a method of execution. One example is Jezebel from the Old Testament, who was executed for multiple acts of heresy. Another is Miguel de Vasconcelos, a Portuguese nobleman who was executed for treason in 1640.

In national data, these deaths are categorized as falls, but an accurate breakdown of defenestrations is not readily available. As one notorious example, in 1953 an American biological warfare specialist died via defenestration in what some conspiracy theorists allege was a CIA assassination. In the US, as a method of suicide, fewer than 2 percent of people who kill themselves do so by this method. James Forrestal, the first US secretary of defense, died under what some believe are suspicious circumstances, when he "fell"—or was pushed or committed suicide—from a 16th-floor window of the National Naval Medical Center in Bethesda, Maryland, on March 28, 1949.

Intentional self-defenestration as a desperate attempt to try to survive something is sometimes used to escape a bad fire, a sexual assault or kidnapping, an attempted homicide, or to escape from police custody. Perhaps the most infamous example of this in the US was the 9/11 terrorist attacks on the World Trade Center towers in New York City. About 200 people jumped or fell to their deaths; 1 police officer was killed by a falling body.

Accidental defenestration is a significant hazard. Every year, in the US, an average of 12 children aged 10 and under are killed by falling through or out of windows; thousands more are injured. Adults, too, have been killed by accidentally falling through or out of a window, for instance while leaning out to see something, or while washing the window; national statistics are not readily available.

STAY SAFE For child safety around windows, see "Windows are Vital to Survival, but Keep Safety in Mind" at nsc.org. Many jurisdictions require safety guards on windows in apartments with young children; these guards must be removable in an emergency when they adjoin a fire escape. Grownups should also be careful around windows, especially when trying to get a better look at something going on outside, or while washing the outside of the window. Be extremely careful while standing on a stool or ladder right inside a window while doing something to it or something right next to it; this increases the risk of falling through it to your death. A fall from 5 stories or more is almost invariably fatal; a fall from 4 or fewer stories can be fatal if you land on your head or hit something that impales you.

SOURCES   CDC, cdc.gov/injury/wisqars/fatal.html; NSC, injuryfacts. nsc.org; NSC, nsc.org/community-safety/safety-topics/child-safety/window-safety

Driveway Back-Overs

ANNUAL US DEATHS   About 200

Every year in the US, about 200 people are killed by vehicles backing over them due to the driver's blind spot to the rear (blindzone). While some of these deaths occur On the Job, many occur in residential driveways. About half of the deaths are of children under age 5. In over 70 percent of these incidents, the child's parent is behind the wheel. Over 60 percent of backing-over incidents involved a larger-sized vehicle (truck, van, SUV). About 17,000 nonfatal back-over injuries occur in the US every year. About 2,500 of these are of young children.

All vehicles since 2018 are required to have rearview video cameras, but older vehicles may not have them.

STAY SAFE   Always look behind your vehicle before getting in and putting it into reverse; a child, pet, or sharp or bulky object might be present. If your vehicle has a rearview video camera, make sure it is working and the lens is clean and unobstructed; always check the display screen for hazards while backing up.

SOURCE   KidsAndCars.org, kidsandcars.org/how-kids-get-hurt/backovers

Drowning at Home

ANNUAL US DEATHS   Estimated 4,000

About 4,000 people of all ages in the US drown every year; 8,000 others suffer serious injuries, including permanent brain damage, from near-drownings, sometimes from incidents at home in bathtubs or backyard swimming pools. In 2000, for drownings where location was reported, 341 (16.7 percent) occurred in bathtubs, 567 (27.8 percent) occurred in home or public swimming pools (split not available), and 1,135 (55.5

percent) occurred in natural bodies of water (some of which were on homeowner property).

About 300 children ages 14 or younger drowned at home in 2017; drowning is the leading accidental killer of kids ages 1 through 4 and the 2nd most common accidental killer of kids ages 1 through 14. Infants left unattended can drown almost immediately anywhere around the house in anything that can hold an inch of water: sink, toilet bowl, bathtub, shower stall, bathinette, bucket, decorative pond, rain puddle, any type of swimming pool, and even in a flooded basement.

Overall, males are about 4 times as likely to drown as females; people of color are more likely to drown than White people. People of all ages who take prescription drugs that cause drowsiness, or who abuse alcohol and/or drugs, or who have physical or cognitive disabilities are all at increased risk of drowning.

STAY SAFE   See the safety guide "Drowning Facts" at cdc.gov. Never leave infants unattended near any sort of water hazard. Kids should take swimming lessons and learn drownproofing (see drownproofing. com). Don't swim while impaired by drugs or alcohol. Wear a personal flotation device while boating.

SOURCES   CDC, cdc.gov/injury/wisqars/fatal.html; CPSC, cpsc. gov/safety-education/neighborhood-safety-network/toolkits/ drowning-prevention; NSC, injuryfacts.nsc.org/home-and-community/ deaths-in-the-home

Electrocution

ANNUAL US DEATHS   About 180

About 180 people a year in the US are killed by electrocution while at home. About 20 percent, or roughly 36 deaths a year, are caused by worn, damaged, or exposed wiring, either part of the home's electrical infrastructure system (outlets, switches, sockets, etc.) or part of individual devices or appliances and their cords or plugs (toasters, hair dryers, etc.). About 10 percent of home electrocutions are caused by major appliances (refrigerators, dryers, etc.). About 14 percent are caused by the victim or their ladder contacting a high-voltage power line. Power tools cause 9 percent of deaths, while home landscaping, gardening, and farming equipment or activities lead to most of the rest of the electrocution fatalities.

Exposure to electrical hazards also causes about 4,000 serious injuries per year, some at home and some in the workplace.

Besides electrocuting people, electricity causes about 400 additional deaths a year and some 4,000 injuries by starting fires from arcing or hot wires from overloads and/or short circuits. About 75 percent, or approximately 300, of these deaths occur in residences.

Lightning strikes can also electrocute people at home by energizing the conductive metal wires, pipes, and framing of their house or apartment.

STAY SAFE See "Electrical Safety at Home Checklist" at Safety.com and "Electrical Safety in the Home" at nfpa.org. Electrical safety begins with properly grounded circuitry, wiring that is up to modern code, effective lightning protection, and a good system of ground fault circuit interrupters (GFCIs) in your main junction/circuit-breaker box. If in doubt, hire a licensed electrician to do a safety audit of your home. Don't overload electrical outlets or run too many electrical devices on the same circuit. During lightning storms, avoid touching any metal connected to the frame of your house, such as showers and faucets. After storms, be alert for downed power lines and don't go near them. The minimum safe distance from a downed power line is 35 feet.

SOURCES CPSC report, cpsc.gov, Electrocutions Associated with Consumer Products: 2004–2013; Electrical Safety Foundation, esfi. org/home-safety; NFPA, nfpa.org/News-and-Research/Data-research-and-tools/US-Fire-Problem; SafeElectricity.org, safeelectricity.org/public-education/tips

Lead Poisoning

ANNUAL US DEATHS About 412,000 (contributing factor)

Environmental exposure to the poisonous heavy metal lead, often via lead paint in older buildings and in consumer products, is estimated to contribute to about 412,000 deaths in the US every year. More than half of these are lethal cardiovascular events such as strokes and heart attacks caused by lead within the body in even very small doses. These take years to occur.

When children are exposed to lead, even in small doses, they can suffer permanent intellectual impairment. It is estimated that about 1.2 million children in the US suffer from lead poisoning, and only half of them receive effective medical treatment.

New lead paint and leaded gasoline were banned and gradually phased out in the US in the 1970s. Old lead paint remains hazardous unless sealed or removed. Because peeling lead-based paint tastes sweet, young children may eat it. Lead can also build up in a locality's environment due to industrial pollution ("brownfields") or because it is used in water mains. For instance, the drinking water supply in Flint, Michigan, was contaminated by lead. From 2014 to 2019, between 6,000 and 12,000 children there received dangerously high doses.

STAY SAFE   Avoid all contact with lead, which can enter the body by eating, drinking, breathing, and skin contact. Older residences should be checked for lead and remediated if needed. People of all ages who might have been exposed to lead should be tested and treated as appropriate.

SOURCES   CDC https://www.cdc.gov/nceh/lead/prevention/health-effects.htm; R.B. Kaufmann et al. Deaths related to lead poisoning in the United States, 1979–1998. *Environmental Research*, February 2003; B.P. Lanphear et al. Low-level lead exposure and mortality in US adults: a population-based cohort study. *Lancet Public Health*, April 2018.

Running with Scissors

ANNUAL US DEATHS   0 recorded

No record can be found of anyone dying from trauma due to a mishap while running with scissors. Children and adults have been injured this way, in the eyes and even the brain (via the eye socket), but there is no record of fatalities. In 2014 in a sample of injury reports in the US, there were 3 documented cases of injuries from running with scissors. This can be extrapolated to a national injury toll of about 130 people annually.

STAY SAFE   Don't run with scissors or any other sharp instrument.

SOURCE   CDC, cdc.gov/injury/wisqars/fatal.html

Slips, Trips, and Falls

ANNUAL US DEATHS   39,443 in 2019

Slips, trips, and falls at home killed 39,443 people in the US in 2019, for a mortality rate of 12.02 per 100,000 population. Over 3 million people needed emergency room treatment or hospitalization. Many of the victims were elderly; more of the fatalities were male than female. As the US population has aged and the number of people over 65 has increased, the rate of fatal falls per 100,000 senior residents has been going up by about 3 percent per year.

The safest state for fatal falls in the elderly was Alabama, with a

Falling is part of life and, sometimes, a cause of death.

2016 death rate of 24.4 per 100,000. The worst state for seniors dying from falls was Wisconsin, where the rate was 142.7. Nationwide, the 2016 death rate for elderly victims of falls was 61.6 per 100,000 population.

Most of the 529 fatal falls from ladders in 2019 occurred either at home or on a farm, not at a construction site. The exact number killed at home is unknown. In 2019, 1,050 Americans were killed falling out of bed; another 314 were killed falling out of chairs; approximately 65 were killed by falling from other furniture.

Many people who fall at home do so on level surfaces, rather than from ladders or stools or on steps or stairs. In 2019, 793 people died from slipping, tripping, or stumbling over hazards such as poorly secured edges of area/throw rugs or runners, wet or slippery surfaces, clutter left on the floor, pets or pet beds, or even their own feet. Outside around residences, in 2019 there were 171 slip-and-fall fatalities on icy or rainy pavement.

By room, deadly falls were:

Living room	31 percent
Bedroom	30 percent
Kitchen	19 percent
Bathroom	13 percent
Hallways	10 percent

SOURCE   CDC, cdc.gov/falls/data/index.html

FALL PREVENTION FOR OLDER ADULTS

More than 1 in 3 people aged 65 years or older fall each year. Falls are the leading cause of injury-related death among adults aged 65 and older. The age-adjusted fall death rate is 64 deaths per 100,000 older adults. The fear of falling may lead older people to avoid activities such as walking, shopping, or taking part in social activities. This leads to poor fitness and social isolation, factors that can contribute to premature death in the elderly.

To avoid injury or death from a fall, older adults should stay physically active. Changes in vision and hearing can cause a fall; have your vision and hearing tested. Always wear your eyeglasses and hearing aids. If any medications make you sleepy or dizzy, discuss them with your doctor or pharmacist. The dose could be changed or a different drug substituted. Limit the amount of alcohol you drink. Stand up slowly to give your blood pressure time to adjust and prevent wobbliness or dizziness. Use an assistive device such as a cane or walker if you need help feeling steady when you walk.

STAY SAFE Install bannisters, handrails, grab bars, safety gates, and nonslip mats on all stairs and in bathrooms. Improve lighting and install nightlights in poorly illuminated areas. Secure or remove area rugs, runners, power cords, and other floor-level trip hazards. Wear sturdy shoes or slippers with nonslip soles around the house. Wear eyeglasses if needed.

SOURCES CDC, cdc.gov/falls/facts.html; NIH NIA, nia.nih.gov/health/topics/falls-and-falls-prevention

On the Go

The first major entries in this section examine fatalities that involve US transportation modes other than motor vehicle accidents, but do involve both a working crew and paying passengers. These entries are followed by accidents involving cars, trucks, buses, and other motor vehicles.

Most American adults know how to drive a car. But transportation of passengers for the purpose of getting them from point A to point B by the 3 other main modes besides motor vehicles—air, water, and rail —is qualitatively different from automobile travel. These specialized, very heavy types of vehicles require highly trained, expert professional crews, often of much more than just a lone driver, who together operate their vehicle to transport passengers (and their luggage) over significant distances. Their safe operation, and very detailed accident investigations with public final reports, are mandated by tough laws at the state and federal level.

Commercial transportation services are distinct from purely recreational flying or boating, wherein nonexperts operate an aircraft or a watercraft themselves, only occasionally or intermittently, and mostly for fun. Those leisure activities, and their mortality statistics, are covered in this book's **At Play** section. Crewed but purely scenic/tour and travel-for-its-own-sake, by water or by air (cruise ships, duck boats, tourist helicopters, and hot-air balloons), are also covered in **At Play**.

Overall occupational death rates, for the broadest groups of "all aircraft crews" or "all watercraft crews," whether they carry passengers, are cargo-only, and/or only do other work (such as harbor dredging or aerial law enforcement), are also covered separately, in the section **On the Job**.

The first portions of **On the Go** cover air, water, and rail transportation where (almost always) multiperson, expert crews, backed up by professional maintenance, repair, booking/dispatching, and traffic control personnel, operate heavy vehicles, which transport large numbers of passengers or large loads of freight as common carriers, open to everyone who has the fare.

For simplicity, all passenger and freight US railroad operations are covered in **On the Job** and in this section. US cargo ships, which rarely if ever carry passengers, are covered only in **On the Job**. US commercial airlines, which often carry both passengers and cargo, are covered both in **On the Job** and here in **On the Go**.

ANNUAL US DEATHS   See **Fatality Rates** later in this section.

US aviation is categorized as commercial air transport (passengers, all-cargo, and mixed), aerial work operations (such as crop-dusting planes and medevac or news helicopters), and general aviation, which includes private transport (e.g., corporate jets) and recreational flying. This section of **On the Go** covers commercial air transport. Recreational aviation is covered by several entries in the **At Play** section. Aerial work operations and private transport (far more dangerous than commercial air transport) are included in the **On the Job** entries for **Top 10 Deadliest Occupations**, for **Deaths by Broad Cause**, and for **Transportation and Material-Moving Occupations: Air Transportation Workers.** (As with other international activities, including those involving Americans living or traveling abroad, information on international aviation mishap deaths is beyond the scope of this book.)

In the US, the National Transportation Safety Board (NTSB) compiles commercial air transport data in 4 classifications—essentially a 2 × 2 matrix—according to whether the aircraft has 10 or more seats or not, and whether the company offers scheduled flights or not, as follows:
- Large Aircraft, Scheduled Flights (Regular Airlines)
- Large Aircraft, Nonscheduled Flights (Air Charters)
- Small Aircraft, Scheduled Flights (Commuter Airlines)
- Small Aircraft, Nonscheduled Flights (On-Demand Air Taxis, Bush Pilots)

In the US, in 2019, commercial airlines flew their passengers on 813 million individual trips.

Sometimes an individual or small group of people are killed on a flight for reasons other than a plane crashing. Such incidents might include a heart attack, a sudden cabin-pressure blowout from fuselage or window material fatigue, or engine turbine fragmentation (compressor stall), which kills someone in midflight, but does not prevent the pilots from bringing the aircraft to a safe emergency landing.

STAY SAFE   Commercial aircraft safe operations, and safety responsibilities of flight crews, flight attendants, gate agents and ground crews, and of federal security officers, is a very complex subject. Contact the US Federal Aviation Administration (FAA) at faa.gov for more information. All adult air passengers should pay careful attention to safety briefings and brochures. Follow instructions from aircraft crews in case of any security or safety problems. Keep your shoes on during the flight; families should sit together; practice releasing your seat belt; know the brace position; count and remember how many rows you are from the nearest exit. Supervise children closely. Behave safely and appropriately at all times; do not attack flight crew or attempt to leave the aircraft while it is still on the ground. Local law enforcement will be called and the FAA can impose substantial fines.

SOURCES   FAA, faa.gov/data_research/accident_incident; NTSB, ntsb.gov/safety/data/Pages/Data_Stats.aspx

Causes of Fatal Mishaps

ANNUAL US DEATHS   See **Fatality Rates** later in this section.

Based on a study of over 1,000 fatal large commercial aircraft accidents between 1950 and mid-2019, whose causes were determined, the following were the types of mishaps that occurred:

PILOT ERROR

   Improper procedure

   Failed visual flying rules

   Flew into mountain

   Flew too low

   Disoriented

   Descended too soon

   Landing speed too high

   Missed runway

   Out of fuel

   Navigation error

   Used wrong runway

   Midair collision (pilot caused)

MECHANICAL PROBLEM

   Engine failure

   Equipment failure

   Structural Failure

   Design flaw

TERRORISM

   Hijacking

   Shot down

   Bombing

   Pilot suicide

WEATHER
Bad turbulence
Wind shear
Mountain updraft
Bad visibility
Heavy rain
Severe wind
Airfoil icing
Severe thunderstorm
Lightning strike

OTHER
Accident causes in this category included: air traffic control error, ground crew error, overloaded aircraft, improperly loaded cargo, bird strike, fuel contamination, pilot incapacitation, obstruction on runway, midair collision caused by another, fire and/or smoke in cabin or cockpit, and maintenance/repair error.

STAY SAFE   See **In the Air**.

SOURCE   NTSB, ntsb.gov/safety/data/Pages/Data_Stats.aspx

Fatality Rates

ANNUAL US DEATHS   See data in entry.

The long-term fatality rate per million flight hours, for different aircraft-sized classifications of commercial versus general (noncommercial, nonaerial work) flying, were as follows:

Regular and Charter Airlines	0.006
Commuter Airlines, Air Taxis, Bush Pilots	6.10
General Aviation	11.00

Note that in the US the fatality rate for large (usually jet-driven) commercial aircraft is only about 1 one-thousandth of what it is for small (typically propeller-driven land or seaplane) commercial aircraft. The fatality rate for general aviation (corporate and recreational flying) is almost twice what it is for small commercial aircraft.

Over the period 1993 through 2012, when US commercial airlines are separated into those with good safety records versus those with bad safety records, 2 key probability comparisons were:

	ODDS OF 1 FLIGHT YOU'RE ON EXPERIENCING AT LEAST 1 FATALITY	ODDS OF YOURSELF BEING KILLED DURING 1 FLIGHT YOU'RE ON
Good Safety Record	1 in 10 million	1 in 19.8 million
Bad Safety Record	1 in 1.5 million	1 in 2.0 million

STAY SAFE See **In the Air**.

SOURCE planecrashinfo.com

Mishap Survival Rates

For the period 2010 through mid-2019 in the US, the survival rate for passengers aboard any flight carrying at least 19 passengers and suffering a mishap in which at least 2 people died was 27 percent.

The survival rate for passengers aboard a flight carrying at least 19 passengers that was forced to make a water landing ("ditching") in controlled flight was 42 percent. Perhaps the most famous instance of an airliner forced to ditch occurred on January 15, 2009, on New York City's Hudson River. US Airways Flight 1549 (pilots Chesley "Sully" Sullenberger and Jeffrey Skiles) from LaGuardia Airport, bound for Charlotte, North Carolina, hit a flock of birds just after takeoff and lost all engine power. All (100 percent) of the 155 passengers and crew aboard the aircraft survived, although a few were seriously injured.

STAY SAFE See **In the Air**.

SOURCES planecrashinfo.com; the *New York Times*

Airplane crash
in a residential
neighborhood

Large Scheduled Aircraft (Regular Airlines)

ANNUAL US DEATHS 1 in 2019

In 2019 in the US, there was 1 fatality during a total of 19.2 million flight hours on large scheduled airline flights. This is the same death toll for this classification of flights as in 2018.

STAY SAFE See **In the Air**.

SOURCES FAA, faa.gov/data_research/accident_incident; NTSB, ntsb.gov/safety/data/Pages/Data_Stats.aspx

Large Nonscheduled Aircraft (Charter Airlines)

ANNUAL US DEATHS 3 in 2019 (0 in the prior 5 years)

In 2019 in the US, 1 fatal accident killed 3 people on a large nonscheduled (charter) airline flight, during a total of 606,000 flight hours. The previous 5 years had seen no fatalities on this classification of flights.

STAY SAFE See **In the Air**.

SOURCES FAA, faa.gov/data_research/accident_incident; NTSB, ntsb.gov/safety/data/Pages/Data_Stats.aspx

Small Scheduled Aircraft (Commuter Flights)

ANNUAL US DEATHS 2 in 2019

In 2019 in the US, 1 fatal accident killed 2 people on a commuter flight, during a total of 415,000 flight hours.

STAY SAFE See **In the Air**.

SOURCES FAA, faa.gov/data_research/accident_incident; NTSB, ntsb.gov/safety/data/Pages/Data_Stats.aspx

Small Nonscheduled Aircraft (On-Demand Air Taxis)

ANNUAL US DEATHS 32

In 2019 in the US, 12 fatal accidents killed a total of 32 people on air taxi flights, during 3.8 million flight hours.

STAY SAFE See **In the Air**.

SOURCES FAA, faa.gov/data_research/accident_incident; NTSB, ntsb.gov/safety/data/Pages/Data_Stats.aspx

Helicopters

ANNUAL US DEATHS 51 in 2019

As with fixed-wing aircraft, the FAA classifies US helicopter flight operations as either commercial transport, aerial work, or general (private/corporate and recreational). Statistics on fatal helicopter crashes are not generally available broken out on this basis. Aerial work and private/corporate transport helicopter deaths, combined with fixed-wing aerial work and private/corporate transport deaths, are included in the **On the Job** section's entries for **Top 10 Deadliest Occupations** (aircraft cockpit crews) and for **Deaths by Broad Cause** (aircraft). Tourist helicopters, a type of commercial transport operations, are discussed in the entry **Tourist Helicopters** in the **At Play** section.

There are almost 10,000 helicopters in the US. In 2019, 24 fatal helicopter accidents killed a total of 51 people. This amounts to 2.1 deaths per fatal accident. In 2018, 24 fatal crashes killed a total of 55 people, for an average death toll of 2.3 per fatal crash. The 5-year average fatal accident rate was 0.63 per 100,000 flight hours. Using the average death toll between 2018 and 2019 of 2.2 deaths per fatal accident, the death rate was about 1.4 deaths per 100,000 flight hours.

Kobe Bryant, one of the greatest basketball players ever, was killed, along with his young daughter Gianna and 7 others, in a helicopter crash in California in 2020. The helicopter pilot became disoriented after flying into low clouds and mistakenly flew into a hillside.

STAY SAFE See **In the Air**.

SOURCES AINonline, ainonline.com/aviation-news/rotorcraft; FAA, faa.gov/data_research/accident_incident; NTSB, ntsb.gov/safety/data/Pages/Data_Stats.aspxI

Seaplanes and Floatplanes

ANNUAL US DEATHS   Average 4.2

Seaplanes and floatplanes are hybrid transportation vehicles that take off and land on the water but fly through the air. Compared to other aircraft, they involve both similar risks and additional risks to passengers and crew and to bystanders, particularly while the aircraft floats at a dock or a pier, or taxis or lands on water.

Because their runway is the surface of a body of water, seaplanes and floatplanes are exposed to additional risks during inclement weather. The water is generally not marked off and might be occupied by swimmers, watercraft, and/or floating debris, so water operations face greater risk of collision. Another potentially lethal risk is landing on glassy water, where the mirrorlike, smooth surface during a dead calm (no wind or swells) robs pilots of visual clues needed for good depth perception. In a water mishap, evacuees can drown, water rescues take longer to organize, and immersion in cold water can lead to death by hypothermia. Offsetting these risks are that crashes or collisions into or on the water tend to be more survivable than comparable mishaps on land, and crews have more scope to pick spots for emergency landings.

Between 1983 and 1995, a total of 195 fatal seaplane accidents on the water killed a total of 54 people. Another 49 people were badly injured. This gives an average annual US deaths for seaplane operations on the water of 4.2 killed per year.

The leading cause of water-operations mishaps was improper crew technique or procedures (types of pilot error), followed by landing in the water with landing gear extended (usually due to pilot error, but about 20 percent of the time because of mechanical failure), followed by bad weather (including crosswinds or wind shear), and then glassy water on landing.

An example of mechanical failure occurred in 2005 near Miami, when 20 passengers and crew were killed when a seaplane's wing broke off during takeoff due to metal fatigue. In 2020 in New York City, a seaplane pilot misjudged the landing, crashed onto a concrete pier, and killed a passenger. Also in 2020, in Idaho, 8 people were killed when a seaplane collided in the air with another aircraft.

STAY SAFE   Seaplane pilots should consider working to qualify for the FAA's SEAWINGS Pilot Proficiency Award Program; see faasafety.gov.

SOURCES   AvStop.com, avstop.com/seaplane/index.html; FAA, faa.gov/data_research/accident_incident; NTSB, ntsb.gov/safety/data/Pages/Data_Stats.aspx

Ground Crew

Ground crew are an essential part of commercial aviation operations. They are needed to direct aircraft that are taxiing after arrival and prior to takeoff and also to provide fire protection and fight fires, transfer baggage on and off planes, load refreshments and supplies, refuel, clean, maintain/repair, and otherwise service aircraft, and maintain runways, taxiways, jetways, and airstairs. They also deice aircraft and clear snow.

Deaths among ground crew at US airports are very rare. A study for 1983 through 2004 found that about 1 ground crew worker a year was killed in an occupational accident, about two-thirds of which involved aircraft during departure. The types of accidents were primarily collisions between work vehicles and aircraft, contact with moving propellers or landing gear, or trauma from jet blast or fire.

Another study for the period 1985 through 2000 found that about 12 airport ground crew workers were killed when struck by vehicles or aircraft on runways or tarmac, for an average of about 0.75 per year.

STAY SAFE   Follow FAA guidelines for ground safety.

SOURCES   AviationPros.com, aviationpros.com/ground-handling; FAA, faa.gov/airports/airport_safety

Recent Air Disasters

IMPROPERLY LOADED CARGO

ANNUAL US DEATHS   A lethal mishap resulting from improperly loaded cargo on a major airline/cargo carrier flight is a low-probability/high-consequence event.

In 1996, in the Florida Everglades, a passenger airliner from Miami to Atlanta crashed about 10 minutes after takeoff, due to a fire started by hazardous freight in the cargo hold. The NTSB investigation determined that aircraft oxygen generators had been illegally packaged and improperly stowed next to flammable freight by the airline's maintenance subcontractor. A bump during the takeoff or in flight activated an oxygen generator, which created heat, starting a fire, which was then fed by other oxygen generators and other flammable freight. All 110 people aboard died.

One upshot was that the FAA issued more stringent regulations against transporting hazardous cargoes on US aircraft. Another was the requirement that all commercial airliners have smoke detectors and automatic fire-suppression systems in their cargo and checked-baggage holds.

STAY SAFE See **In the Air**.

SOURCES FAA, faa.gov/data_research/accident_incident; NTSB, ntsb.gov/safety/data/Pages/Data_Stats.aspx

MECHANICAL FAILURE

ANNUAL US DEATHS Lethal mechanical failure on a commercial airliner is a low-probability/high-consequence event.

In 1996, TWA Flight 800, a US passenger airliner that had just taken off from JFK for Paris, exploded in the air and crashed into the Atlantic Ocean off Long Island, killing all 230 people aboard. After a long and difficult process for salvage divers to recover enough wreckage to reconstruct what had occurred, the NTSB determined that a mechanical failure due to a design flaw was the primary cause: An electrical spark, created by a short circuit in a wiring bundle that ran through a fuel tank, had ignited jet-fuel vapor in an almost empty fuel tank, causing a catastrophic explosion.

The upshot was twofold: The FAA and the aircraft manufacturer introduced revised design measures to help prevent defective or worn wires from sparking, and to help prevent fuel vapor from igniting via a system that injects inert nitrogen into aircraft fuel tanks as the liquid fuel is used up.

STAY SAFE See **In the Air**.

SOURCE NTSB accident reports, ntsb.gov, Aircraft Accident Report: In-flight Breakup Over the Atlantic Ocean Trans World Airlines Flight 800, July 17, 1996

PILOT ERROR

ANNUAL US DEATHS Lethal pilot error on a commercial airliner is a low-probability/high-consequence event.

In November 2001, an airliner bound for Santo Domingo crashed into a residential neighborhood just after takeoff from JFK. The impact and explosion/fire killed everyone on the plane (251 passengers and 9 crew members), plus another 5 people on the ground. It was the 2nd deadliest plane crash in American history. (The deadliest was in 1979 in Chicago, when an engine fell off an American Airlines plane just 1 second before takeoff, causing a crash that killed all 273 people on board.)

The subsequent NTSB investigation determined the primary mishap was caused by pilot error: The copilot overused the rudder controls when the plane flew through persisting wake turbulence from a 747 that had just taken off. The mechanical overstressing caused the tail's vertical stabilizer to snap off, and the engines also broke off.

STAY SAFE   See **In the Air**.

SOURCE   NTSB, accident reports, ntsb.gov, In-Flight Separation of Vertical Stabilizer American Airlines Flight 587

SEVERE WEATHER

ANNUAL US DEATHS   A lethal weather-related mishap on a commercial airline flight is a low-probability/high-consequence event.

In 1985, a passenger airliner from Fort Lauderdale, Florida, coming in to land at Dallas/Fort Worth, Texas, hit a wind shear microburst (a sudden downdraft near the ground) caused by a thunderstorm and crashed short of the runway. In all, 137 people (8 crew members, 128 passengers, and 1 person on the ground) were killed; 3 crew members and 24 passengers survived with injuries.

An upshot of this disaster was more impetus for the FAA to mandate, by 1993, fully implemented pilot training in immediately recognizing and properly compensating for wind shear microbursts, and sensors at airports and aboard aircraft to sense such dangerous, sudden downdrafts.

STAY SAFE   See **In the Air**.

SOURCE   NTSB, ntsb.gov/safety/data/Pages/Data_Stats.aspx

TERRORISM

ANNUAL US DEATHS   Lethal aircraft-involved terrorism is a low-probability/high-consequence event.

The worst air disaster in US history occurred as part of the Al Qaeda terrorist attacks on September 11, 2001. Four large commercial airliners were hijacked by a total of 19 terrorists. One was crashed into each of the 2 World Trade Center towers in New York City, 1 was crashed into the Pentagon in Arlington, Virginia, and 1 crashed in a field in

World Trade Center attack, September 11, 2001

Somerset County, Pennsylvania. All passengers, crew, and terrorists aboard the 4 planes were killed, for a total of 265 people. A further 2,730 people in and around the 3 buildings, including first responders, were killed. More than 6,000 others were injured.

After the attack, approximately 400,000 people were exposed to toxic contaminants and other factors that raised their risk for physical and mental health conditions. The CDC WTC Health Program began screening, monitoring, and treating these people in 2011. In 2020, 104,223 people were enrolled; 73.4 percent (76,543) were first responders. By 2020, 41,387 enrollees had received treatment for a related illness.

By September 11, 2021, 250 firefighters had died of a WTC-related illness. As of August 31, 2021, the September 11th Victim Compensation Fund had paid 63,559 compensation claims for deceased victims and 3,904 claims for personal injury.

Since this disaster, procedures, technologies, and manpower for the security of air travel in the US and elsewhere, in airports, aboard aircraft, and via no-fly lists, have been very significantly strengthened.

STAY SAFE See **In the Air**. If you see something, say something.

SOURCES FBI 9/11 report, fbi.gov/file-repository/final-9-11-review-commission-report-unclassified.pdf/view; September 11th Victim Compensation Fund, vcf.gov; Uniformed Firefighters Association of New York, ufanyc.org/wtc

Aerial Drones (Business and Professional Unmanned Aerial Vehicles)

ANNUAL US DEATHS Insufficient data

Aerial drones are growing significantly in numbers and in types of uses for business, professional, and commercial purposes in the US and other countries. They are also continually improving in terms of their payload capacity, flight endurance, autonomous capabilities, data downlink baud rates, and—as required by new FAA regulations—their visual and electronic identification and tracking capabilities. Drones have valuable cost-saving and safety advantages in such diverse business applications as surveillance and maintenance inspections for railroad tracks, power lines, and other utilities, in agriculture, in real estate development and marketing, in law enforcement and municipal planning, in photography and filmmaking, in demographic and land-use research and analysis, in delivery services, and beyond.

Industry analysts predict that there will soon be well over 2 million business or hobby/recreational drones owned and operated in the US. It is likely that as more and more drones take to the air, crashes, property damage, pilot and bystander injuries, and even deaths will sometimes occur, just as they occasionally do for larger, manned aircraft.

STAY SAFE Follow all FAA regulations for operator licensing and drone identification, and stay abreast of new regulations. Never let the wider business, professional, and/or commercial purposes behind your use of a drone get the better of placing safety first.

SOURCE FAA, faa.gov/uas

Astronauts and Space Crews

ANNUAL US DEATHS 14 since 1980

Professional astronauts, as opposed to space tourists, are paid employees of either NASA or of a private American corporation, a foreign government space agency, or a non-US corporation. They sometimes are killed in on-the-job mishaps, which might occur during training and equipment testing or during actual space flight.

In the US since 1980, a total of 13 NASA astronauts and 1 Israeli astronaut have died on the job, aboard the *Challenger* and *Columbia* space shuttle disasters. The *Challenger* disaster in 1986 occurred shortly after launch; the *Columbia* disaster occurred in 2003 during reentry. In addition, 1 American privately employed astronaut was killed in 2014, when a suborbital space plane, meant for space tourism, broke up during a test flight. All of these mishaps occurred well within Earth's atmosphere; no American astronauts have yet died in outer space. In the 1960s, a total of 9 NASA astronauts died on the job, either in training-aircraft crashes, in a suborbital near-space X-15 rocket-plane crash, or in the 1967 *Apollo 1* capsule fire on the launchpad during an equipment test.

In 2014, the Virgin Galactic space plane *Enterprise* broke apart during a test flight and killed the copilot. The pilot parachuted to safety. On July 11, 2021, the *Unity 22* aircraft from Virgin Galactic made the first commercial suborbital spaceflight. The 2 pilots and 4 mission specialists, including Sir Richard Branson, landed successfully at SpaceportAmerica in New Mexico after a 90-minute flight.

STAY SAFE   See NASA's website (nasa.gov) for information on astronaut safety and health during training and during space flight.

SOURCE   NASA, nasa.gov/topics/humans-in-space/history

On the Rails

ANNUAL US DEATHS   1,052 in 2019

The US railroad industry consists of 7 major (Class I) railroads, 22 regional systems, and 584 regional or short lines. The industry runs on about 160,000 route-miles of track, employs over 167,000 people, and generates more than $80 billion a year in revenue. In 2019, American freight trains transported almost 1.5 trillion ton-miles of fuels, raw materials, and manufactured goods, all vital to the US and world economies.

America's rails are connected to a continuous, multicontinent system that extends from Canada in the north all the way to the far reaches of South America. In the Americas overall, in 2019, railroads carried passengers on trips covering about 16.65 billion passenger-miles.

In 2018 in the US, 988 people were killed in railroad accidents, including mass transit rail. In 2019, 1,052 were killed, for a year-over-year increase of 6.5 percent. In 2019, an additional 7,867 people were injured nonfatally, showing a decline of 5 percent from 2018.

There are about 210,000 public and private railroad grade crossings in the US. In 2019, about 293 deaths resulted from train-vehicle collisions at grade crossings, compared to 260 in 2018, which was a 13 percent increase.

Trespassing on railroad property resulted in 578 deaths in 2019. Nineteen railroad, supplier, or shipper employees died in 2019 during rail operations. One railroad passenger was killed during train travel in 2019, compared to 6 in 2018. The remainder of fatal mishaps were other, unknown, under investigation, or not specified.

STAY SAFE   A moving train needs a very long distance to stop; a single locomotive can weigh about 100 times more than an automobile. Never trespass on railroad property, which includes all land along tracks, plus rail yards, maintenance facilities, bridges, and tunnels. Never tempt fate and try to outrun a train at a grade crossing, in a vehicle, on foot, or on a bicycle. Always be careful while getting on or off a train, especially if the platform is at any distance below or away from the car vestibule steps; don't run on platforms, as you could trip and fall against or under a moving train. Never go near an electrified line's third rail or overhead trolley wire. If you must walk across tracks, step between the rails, not on them, as they can be very slippery from oil and grease; never go near a track switch, which can move suddenly, without warning. Never climb on a railroad signal mast or interfere with communications antennas and electrical cabinets.

SOURCES   FRA, railroads.dot.gov/railroad-safety/accident-data-reporting-and-investigations; NSC, injuryfacts.nsc.org/home-and-community/safety-topics/railroad-deaths-and-injuries; OLI, oli.org/track-statistics

Accidents by Cause

ANNUAL US DEATHS   See table in entry.

Over the period 2007 through 2013, among the fatalities that were not either suicides or trespassers (each of which was caused by the victim themself), the person or factor deemed responsible for causing the fatality broke down as follows:

Workforce error, equipment failure	8 percent
Passenger or shipper accident or imprudence	39
Public imprudent behaviors	30
Other (falls, assaults, heart attacks, etc.)	23

STAY SAFE   See **On the Rails**.

SOURCES   FTA, transit.dot.gov/regulations-and-guidance/safety/rail-and-bus-safety-data-reports; FRA, railroads.dot.gov/railroad-safety/accident-data-reporting-and-investigations

Accidents by Type

ANNUAL US DEATHS   See table in entry.

Over the period 2007 through 2013, railroad-related deaths broke down by the type of accident or event that produced each fatality as follows:

Grade crossing collisions	6
Collisions with another train or object	13
Derailments	0
Fires on railroad property	1
Other (falls, assaults, heart attacks, etc.)	80

Note that the "Other" category includes all fatalities not caused by moving trains or by fires on railroad property; assaults include attacks by members of the public on railroad workers and on passengers or shippers—muggings, shootings, etc.—as well as violence between co-workers.

STAY SAFE   See **On the Rails.**

SOURCES   FTA, transit.dot.gov/regulations-and-guidance/safety/rail-and-bus-safety-data-reports; FRA, railroads.dot.gov/railroad-safety/accident-data-reporting-and-investigations

SOURCE   FRA, railroads.dot.gov/railroad-safety/accident-data-reporting-and-investigations

Accidents by Victim

ANNUAL US DEATHS   See table in entry.

Over the period 2007 through 2013, the type of victim in railroad accident fatalities broke down as follows:

Railroad workers	2 percent
Passengers	5
Shipping customers	16
General public	77

Shipping customers often handle railcars on their own private industry spurs, after delivery and before pickup by the railroads, especially the loading and unloading of raw materials, fuels, and manufactured goods to and from freight cars. The general public are any persons distinct from railroad workers, passengers, and shippers; of all the general public fatalities, 81 percent were either trespassers or suicides (it can be hard to tell them apart, postmortem), and 19 percent died from other causes, mostly grade crossing collisions.

STAY SAFE See **On the Rails.**

SOURCES FTA, transit.dot.gov/regulations-and-guidance/safety/ rail-and-bus-safety-data-reports; FRA, railroads.dot.gov/ railroad-safety/accident-data-reporting-and-investigations

Employee Deaths by Type of Mishap

ANNUAL US DEATHS   19

In 2019, the 19 rail-related employee deaths on the job occurred in the following types of mishaps:

Derailment	1
Repair facility accident	2
Switching yard: hit or crushed	6
Track work: hit by train	2
Motor vehicle accident	1
Grade crossing collision	1
Tank car cleaning/inspecting	2
Collision on shipper spur	1
Runaway train (brake failure)	3

This gives an annual occupational fatality rate for the approximately 167,000 railroad employees of 11.4 deaths per 100,000 Full-Time Equivalent (FTE) per year.

STAY SAFE Pay careful attention to all safety procedures and bulletins, and use all safety equipment such as hard hats, steel-toed boots, visibility vests, and safety harnesses. Make sure when going on duty that portable radios and lanterns are working and their batteries are fully charged. Always look both ways, and listen, when working on the ground near tracks. Don't operate a train while impaired, ill, or distracted. Take safety training seriously, follow good teamwork, and remember that "Safety First" is not just a slogan—it's a necessity of the railroader's way of life.

SOURCE FRA, railroads.dot.gov/railroad-safety/
accident-data-reporting-and-investigations

Mishap Deaths by Train Type

The 1,052 US railroad-related deaths in 2019 break down by type of
train as follows:

Passenger trains (long-distance, intercity, commuter)	282
Freight trains	597
Mass transit trains	173

STAY SAFE See the entry for **On the Rails**.

Passengers Killed or Injured in Railroad Accidents

ANNUAL US DEATHS Average of 5.7

From 2010 through 2019 in the
US, 57 passengers were killed
in railroad-related accidents, for
an average annual death toll of
5.7 passengers. During this 10-
year period, US rail passengers
completed about 110 million
passenger-miles of train travel
per year. This gives an average
annual fatality rate of 5.2 passenger deaths per 100 million
passenger-miles.

From 2010 through 2019,
15,424 railroad passengers were
injured nonfatally. This is an
average annual nonfatal injury
rate of 1,542 passengers, or on a
passenger-miles basis, it is 1,402
nonfatal injuries per 100 million
passenger-miles.

Insignia of the Alaska Railroad Police

These figures do not include passengers who were the victims
of violent crimes, such as muggings and homicides, perpetrated by
nonemployees, while on railroad property. (Suicides by passengers are
counted in trespasser deaths.)

STAY SAFE   See **On the Rails**.

SOURCES   FRA, railroads.dot.gov/railroad-safety/
accident-data-reporting-and-investigations; FTA, transit.
dot.gov/regulations-and-guidance/safety/rail-and-bus-
safety-data-reports; FRA, railroads.dot.gov/railroad-safety/
accident-data-reporting-and-investigations

Railfans and Railroad Photographers

ANNUAL US DEATHS   No breakout data for this category

Attitudes of railroad corporations and their employees including rail-road police officers vary greatly as to whether railfans (train spotters) and railroad photography enthusiasts are at all welcome around rail-road property or are seen instead as dangerous nuisances. Consult a railroad's website or their public relations department and check rail-fan information sources such as online chat boards to see where you can watch trains in safety and without being harassed or arrested by railroad cops.

Railroad photographers who use drones to obtain aerial views of trains must at all times obey all FAA regulations and must also make sure that their drone does not get so close to the rails, other stationary equipment, or a moving train that a collision could occur, causing prop-erty damage and/or personal injury. In the UK, for instance, on at least 2 occasions, photography drones have hit a popular historical steam preservation train, the world-famous Flying Scotsman.

STAY SAFE   Railfans and railroad photographers should never let their enthusiasm get the better of their common sense about safety around railroad operations. Never enter any railroad private property unless you get permission, which usually requires signing a waiver form absolving the railroad company of any liability if you are injured or killed. Otherwise, you are a trespasser subject to detention by rail-road police. If you watch and/or photograph trains at a public railroad crossing, watch and listen for approaching trains and/or for the cross-ing lights, gates, and bells to activate. Immediately move back to safety until the train has passed, and make sure that another train isn't com-ing from *either* direction before again getting any closer to the rails.

SOURCE   OLI, oli.org/safety-near-trains/walking-safely-near-tracks/
photographer-and-filmmaker-safety

Railroad Police

> **ANNUAL US DEATHS** 2.2

All major railroads have a railroad police force to enforce safety and security on railroad grounds, in their facilities, and on their trains. These law enforcement officers respond to crime scenes, accidents and mishaps, and emergencies to protect the public, the railroad's employees, passengers, and shippers, and to protect railroad property from trespassing, theft, damage and vandalism, sabotage, and terrorism. Railroad police can and will detain perpetrators, turning them over to local law enforcement authorities for criminal prosecution. Rarely, railroad police personnel die while on the job.

The Officer Down Memorial Page (odmp.org) maintains complete records of all US law enforcement officers who have died in the line of duty since 1776. The Railroad Police Line of Duty Deaths page lists 439 railroad police officers killed since American railroads began operations in the 1820s. This gives a very long-term average of 2.2 deaths per year. The causes of death break down as follows:

Murdered	298
Hit by train	95
Other work accidents	42
Heart attacks	4

STAY SAFE Always obey instructions from railroad police officers and never trespass on railroad property. If you see something, say something: If you can quickly find the railroad police telephone number, contact them immediately; otherwise, call 911.

SOURCE Officer Down Memorial Page, odmp.org/search/browse/railroad-police

Amtrak

> **ANNUAL US DEATHS** 58.8 per million train-miles traveled in 2017

Amtrak is a quasi-governmental corporation created in 1971 to take over and continue offering intercity and long-distance passenger rail service. Those types of trains had previously been provided to the public by private, for-profit railroad corporations that kept losing money on their passenger services, or had gone bankrupt.

The overall mishap mortality rate on Amtrak was 41.1 per million train-miles traveled in 2008 and was 58.8 in late 2017. These figures include all deaths, such as Amtrak and freight train crews, train passengers, motor vehicle occupants, trespassers, etc. Fewer than one-tenth of all deaths were of Amtrak passengers.

STAY SAFE See **On the Rails.**

SOURCE FRA, railroads.dot.gov/railroad-safety/accident-data-reporting-and-investigations

Commuter Trains

ANNUAL US DEATHS 10.8 per 100 million passenger-miles traveled from 2000 to 2011

Commuter trains are passenger trains that take people between suburban bedroom communities and their jobs and other activities in a nearby city. In recent times, almost all commuter train services in the US are operated by municipal and/or state authorities, such as New York State's Metro-North (in the New York City area) or the Chicago area's Metra, although the public entity sometimes contracts out the daily running of trains to a private, for-profit corporation.

From 2000 through 2011 in the US, the overall mishap mortality rate on commuter railroads was 10.8 deaths per 100 million passenger-miles traveled.

STAY SAFE See **On the Rails.**

SOURCES FRA, railroads.dot.gov/railroad-safety/accident-data-reporting-and-investigations

Freight Trains

ANNUAL US DEATHS No data available

In 2019 in the US, the overall fatality rate from freight train mishaps was 0.4 per billion ton-miles of cargo transported.

STAY SAFE See **On the Rails.**

SOURCE FRA, railroads.dot.gov/railroad-safety/accident-data-reporting-and-investigations

Heavy Rail Mass Transit

> **ANNUAL US DEATHS** 1.4 per 100 million passenger-miles traveled from 2000 to 2011

Heavy rail mass transit refers to inner-city and/or metropolitan area subway systems and other electrically powered trains that have heavier construction and higher top speeds than light rail. Heavy rail mass transit systems operate only on segregated rights of way, often in tunnels under city streets and/or on elevated trestles above the streets.

From 2000 through 2011 in the US, the overall mishap mortality rate for heavy rail mass transit was 1.4 deaths per 100 million passenger-miles traveled.

STAY SAFE See **On the Rails.**

SOURCE American Public Transportation Association (APTA), apta. com/research-technical-resources/transit-statistics

Railroad Grade Crossing Accidents

> **ANNUAL US DEATHS** Estimated 300

Collisions between motor vehicles and trains usually happen at railroad grade crossings. In such encounters, the motor vehicle and any occupants always lose. The weight of the average freight train can exceed 10,000 tons; the average midsized sedan weighs about 3,350 pounds and the average SUV about 4,800 pounds. Over the years, safety improvements in grade crossing equipment and public education have led to a drop in fatal accidents.

- In 1981, there were 9,461 collisions, resulting in 3,293 injuries and 728 deaths.
- In 1991, there were 5,388 collisions, resulting in 2,094 injuries and 608 deaths.
- In 2001, there were 3,237 collisions, resulting in 1,157 injuries and 421 deaths.
- In 2011, there were 2,061 collisions, resulting in 1,045 injuries and 250 deaths.
- In 2019 (most recent available), there were 2,216 collisions, resulting in 807 injuries and 293 deaths.

GRADE CROSSING ACCIDENTS: WORST STATES

In 2020, California had the most deaths from train-vehicle collisions at railroad grade crossings, with 40 people killed. Florida was next worst, with 20 deaths. Then came Illinois with 15, Texas with 11, and Georgia and Missouri were tied at 9 each.

STAY SAFE   See **On the Rails** and **Railroad Grade Crossing Accidents**.

SOURCE   OLI, oli.org/track-statistics

An NTSB investigator examines a truck involved in a grade crossing collision in California.

TRESPASSER DEATHS: WORST STATES

In 2020, California had the most deaths of pedestrians trespassing on railroad tracks, with 136 people killed. Texas was next worst, with 31 deaths. Then came Florida with 28, Georgia with 26, and Illinois with 24.

OPERATION LIFESAVER, INC.

Operation Lifesaver, Inc. (OLI) is a nonprofit organization, strongly supported by railroads, that engages in public safety education programs for preventing collisions, injuries, and deaths at grade crossings and along tracks. OLI provides videos and promotional materials that increase visibility and awareness about rail safety. These programs include "Stop Track Tragedies" and "Rail Safety Week."

Fully 95 percent of all railroad-related fatalities in the US in recent years were caused either by pedestrians trespassing on tracks or by vehicles trying to beat trains at grade crossings. Since Operation Lifesaver was established in 1972, it has contributed greatly to the significant overall decline in railroad grade crossing deaths, from over 1,200 per year to fewer than 300 per year.

STAY SAFE   Trains can't stop, but you can. (A long, heavy freight train requires about 1 mile to stop.) Trains always have the right of way. At a grade crossing, stop, look both ways, and listen. Stop 15 feet away from flashing red lights, lowered gates, a signaling flagger, or a stop sign. Once you enter the crossing, keep moving. If your vehicle gets stuck on the tracks, immediately abandon your vehicle, rescue any passengers, and flee to a safe distance.

SOURCES   FRA, railroads.dot.gov/railroad-safety/accident-data-reporting-and-investigations; OLI, oli.org/track-statistics

Light Rail

> **ANNUAL US DEATHS**   22.6 per 100 million passenger-miles traveled from 2000 to 2011

Light rail refers to the overall category of (usually electric) streetcars, trolleys, and trams, which usually operate within a single city or a greater metropolitan area. Light rail vehicles sometimes run on a segregated right of way, and sometimes run on local streets amid motor vehicle traffic.

From 2000 through 2011 in the US, the overall mishap mortality rate on light rail lines was 22.6 deaths per 100 million passenger-miles traveled.

STAY SAFE   See **On the Rails.**

SOURCE   American Public Transportation Association (APTA), apta.com/research-technical-resources/transit-statistics

Railroad Museums, Scenic/Tourist/ Cog Lines, Preserved Steam, Dinner/Mystery/Wine Trains

> **ANNUAL US DEATHS**   No data available

No available national database focuses specifically on fatal mishaps in America's many railroad museums, aboard its numerous scenic, tourist, cog (mountain), and historical preservation steam railroads, or involving private (luxury) railcars, or while riding its variously themed dinner, murder mystery, holiday, and wine trains. Such trips are a small minority of all rail travel in the US and they have an excellent safety record.

Those lines that run working steam locomotives must comply with stringent federal steam-boiler safety inspection rules, with periodic safety checks conducted by objective federal experts not affiliated with the railroads they oversee.

Way back in the days when most locomotives in the US were steam locomotives, before about 1950, things were not so benign. Catastrophes involving steam locomotive boilers, due to collisions and derailments, various mechanical problems, excessive wear, and improper maintenance, often produced fatal results. In 1912, in San Antonio, Texas, a steam locomotive being fired up for the day's work after repairs exploded in a roundhouse, killing at least 26 and injuring 40, mostly railroad employees. The blast threw debris up to 7 blocks away and destroyed several houses.

In 1925, in Hackettstown, New Jersey, a steam passenger train derailed on debris at a grade crossing after a violent thunderstorm. The first 2 cars piled up on top of and next to the damaged locomotive, whose boiler spewed superheated live steam that scalded to death 42 passengers and 5 crew and injured 23 others.

STAY SAFE   Obey all safety instructions and signage and follow crew instructions. Never trespass into any areas, such as repair depots, not open to the public. Never venture too close to tracks; never crowd close to or in front of a live locomotive. Always be careful entering and leaving static or operating locomotives, cabooses, railroad cars, and other equipment and displays; avoid hijinks or reckless behavior. Keep a close eye on children

SOURCE   Heritage Rail Alliance, heritagerail.org

Recent Disasters

BRIDGE COLLAPSE

ANNUAL US DEATHS   No breakout data for this category

In 1993, near Mobile, Alabama, an Amtrak passenger train went off a bridge that had just been hit and damaged by an errant tugboat pulling 6 barges of coal and wood. The mishap killed 42 passengers and 5 crew aboard the train; 103 others were injured. This is the worst death toll in the history of Amtrak.

The barge crew were determined by a National Transportation Safety Board (NTSB) investigation to have caused the disaster, due to negligent and incompetent operation of the tug. In postmidnight darkness, in dense fog, the captain was sleeping; the crewman at the helm did not know how to use the boat's radar; the boat had no compass or navigation charts. Disoriented, the helmsman steered into a bayou branch forbidden to barge traffic. When the barge hit the low-lying bridge, it bent the tracks 3 feet out of alignment—but didn't break them, which would have activated a stop signal, halting the train and averting catastrophe. Instead, the train kept going at 70 mph, derailed, crashed into the bayou, and then caught fire. Most victims drowned in the bayou; 2 died of burns.

This disaster was one impetus behind railroads installing more sophisticated track damage sensors, which are better able to activate train stop signals if track becomes misaligned.

STAY SAFE   See **On the Rails.**

SOURCES   FRA, https://railroads.dot.gov/railroad-safety/acci-dent-data-reporting-and-investigations; NTSB, ntsb.gov/safety/data/Pages/Data_Stats.aspx

DERAILED ON INTERSTATE OVERPASS

ANNUAL US DEATHS   No breakout data for this category

In 2017 near DuPont, Washington, an Amtrak passenger train derailed on a curve near a railroad overpass over Interstate 5, killing 3 passengers and injuring another 57 passengers and crew. The lead locomotive and 6 passenger cars plunged down the embankment of the overpass onto I-5, crushing several cars and trucks and spilling hundred of gallons of diesel fuel.

The accident investigation found the train had been going at 78 mph when it derailed, compared to the local track speed limit of 30 mph. The train was making the first regularly scheduled run along a new portion of its route; the engineer was found to not have been adequately familiarized with that part of the railroad beforehand and not adequately trained in the new type of locomotive he was driving. Sound Transit, a regional transportation authority that was responsible for that portion of the tracks used by the Amtrak train, was criticized for not having changed the layout of the curve, which had previously been deemed to be dangerous. Positive Train Control (PTC), subsequently installed on virtually all tracks in the US used by passenger trains, would have prevented this disaster.

STAY SAFE   See **On the Rails.**

SOURCE   FRA, railoads.dot.gov, Federal Railroad Administration Office of Railroad Safety Accident and Analysis Branch Accident Investigation Report HQ-2017-1239

GRADE CROSSING COLLISION

ANNUAL US DEATHS   Average 300

In 2015, along an electrified part of the line, a Metro-North commuter railroad passenger train collided with an SUV at a grade crossing, killing the driver of the vehicle and 5 passengers on the train; 15 others were injured, 7 of them seriously. For an unknown reason, the SUV driver stopped her car on the tracks as the crossing arm descended and hit her vehicle. Witnesses said that, rather than move her car (which was easy to do) or flee to safety on foot, she inspected the vehicle for damage from the crossing arm. The train hit her and her SUV, causing a fire. The wreck also caused the end of a third rail to penetrate the first car of the train, adding to the carnage. Accident investigators blamed the woman for the mishap, but her behavior was never explained. (See the entry **Operation Lifesaver, Inc.**)

STAY SAFE   See **On the Rails**.

SOURCE   OLI, oli.org/track-statistics

HEAD-ON COLLISION

ANNUAL US DEATHS   No breakout data for this category

In 2008 in Chatsworth, California, a Metrolink commuter train collided head-on with a Union Pacific freight train, killing 25 and injuring more than 135. Accident investigators determined that the Metrolink engineer had been distracted by texting while driving the train, and he ran through a stop signal into the path of the freight train.

This disaster was a major factor in convincing Congress to pass a law requiring virtually all US railroads to install Positive Train Control (PTC) technology, which would have stopped the Metrolink train as soon as it ran the red signal, sounded an alarm in the cab, and alerted dispatchers about the engineer's dangerous human error. This deadly crash was also an impetus to many railroads installing video recorders facing *into* the locomotive cab, to document any crew infractions of official railroad safety rules. Most locomotives already had video recorders facing *out of* the cab, to document grade crossing collisions and near misses, as well as trespassing and vandalism incidents.

STAY SAFE   See **On the Rails**.

SOURCE   FRA, railroads.dot.gov/human-factors/elearning-attention/
metrolink-111-collision-chatsworth-sept-2008

IMPAIRED AND DISTRACTED ENGINEERS

ANNUAL US DEATHS   No breakout data for this category

In 1987 in Maryland, an Amtrak passenger train, moving at high speed, hit from behind a string of Conrail locomotives whose engineer had ignored a red signal, went through a switch, and ran onto the main line at slow speed, right in front of the Amtrak train. The resulting rear-end collision killed 14 passengers, the Amtrak engineer, and an Amtrak lounge attendant. The Conrail engineer tested positive for marijuana use. In reaction to this disaster, federal drug and alcohol screening was significantly strengthened for rail and other transportation workers.

In 2015, outside Philadelphia, an Amtrak passenger train derailed while going more than 100 mph on a curve that had a speed limit of 50. The wreck killed 8 passengers and injured over 200 others. The engineer reported being distracted by hearing, over his cab radio, that another train's engineer had just been injured by trackside vandals throwing rocks at his locomotive. This disaster was a major impetus to Amtrak rapidly completing installation of Positive Train Control (see that entry), which would have stopped the train when it first exceeded the speed limit.

STAY SAFE   See **On the Rails**.

SOURCE   FRA, railroads.dot.gov/human-factors/elearning-attention/
amtrak-188-derailment-may-2015

This 2015 Amtrak derailment near Philadelphia killed 8 and injured more than 200, 11 critically.

IN A PASSENGER STATION

ANNUAL US DEATHS   No breakout data for this category

In 2016, in Hoboken, New Jersey, a New Jersey Transit passenger train went too fast entering the Hoboken passenger terminal, where the tracks end. The train crashed through the end-of-rails bumper and hit the station building, causing extensive structural damage and scattering debris. One woman waiting on the platform was killed, and over 100 other people, including the train engineer, were injured. Engineer health problems, leading to inattention, might have been the cause. New Jersey Transit subsequently strengthened medical screenings for train crews and has installed Positive Train Control (see that entry.)

STAY SAFE   See **On the Rails.**

SOURCE   FRA, railroads.dot.gov/human-factors/elearning-attention/port-authority-collision-hoboken-nj-may-2011

RUNAWAY TRAIN

ANNUAL US DEATHS   No breakout data for this category

In 2013, in eastern Québec Province, Canada, a parked and unoccupied Montreal, Maine and Atlantic Railway freight train, consisting of 5 engines and 73 tank cars full of flammable crude petroleum, suffered a sequence of mechanical problems and human error that was truly a perfect storm of rail calamity. Forty-seven people in the town of Lac-Mégantic were killed. This is the worst ever death toll of a railroad disaster since Canadian independence in 1867, and the worst in Canadian history involving a freight train.

Earlier in the evening, the engineer—working the train alone, as a 1-man crew, per the railroad's work agreement—decided to halt the train because of heavy smoke and lubricating oil spray from a mechanical problem in the lead locomotive. He secured the train until morning, parked on the main line tracks, by setting only a few hand brakes. But he left the lead engine running, to provide the air pressure needed to top off the pneumatic brakes' reservoir to keep the whole train's air brakes working. The lead engine was the same one that was having the oil spray problem, but he expected this problem to subside while that locomotive idled for a few hours. He radioed dispatchers to tell them the situation, then took a taxi to a hotel to get some sleep.

But the problem worsened. The lead locomotive caught fire. The Lac-Mégantic fire department responded and extinguished the fire;

the railroad, forbidding the engineer to return to the scene, sent some-one else who was not familiar with air brakes. Exactly who did what next was disputed later, but the engine was shut down. The pneumatic reservoir gradually lost sufficient air pressure for the air brakes to keep holding—and the few hand brakes the engineer had set were not enough to hold the train on the steep downgrade. The train started moving, gained speed, and 7 miles later it derailed at 65 mph, on a curve in the middle of downtown Lac-Mégantic that had a speed limit of only 10 mph. Volatile crude leaked from ruptured tank cars, exploded, and "a tsunami of fire" spread through the central business district's storm sewers and to nearby buildings, including a busy downtown bar. The resulting conflagration killed 47 people.

As a result of the disaster, railroads in Canada and the US made significant safety improvements to work rules, procedures, and routing of dangerous cargoes. The engineer was blamed for not setting enough hand brakes, but inadequate tank car integrity during a serious train crash was also a factor. Regulations were introduced to require the phasing in of a newer, stronger tank car design, with enhanced features for more safely transporting volatile liquids.

STAY SAFE   See **On the Rails.**

SOURCE   Transportation Safety Board of Canada Railway Investigation Report, bst-tsb.gc.ca/eng/enquetes-investigations/rail/2013/R13D0054/R13D0054.html

SPEEDING ON CURVE

ANNUAL US DEATHS   No breakout data for this category

In 2013, near Spuyten Duyvil, New York City (where the Harlem River joins the Hudson River in the Bronx), a Metro-North commuter railroad passenger train derailed on a sharp curve, killing 4 people and injuring at least 61. The train was going at 82 mph where the speed limit was only 30 mph.

The engineer later told accident investigators that he had gone into a daze as his train neared the curve. A subsequent medical examination diagnosed sleep apnea, a serious sleep disorder. Safety improvements following this disaster included more rigorous mandatory health screenings for railroad workers in safety-sensitive jobs, the installation of automatic speed control on that section of Metro-North track, and added impetus for the installation of Positive Train Control (PTC), which would have prevented the mishap by stopping the train when it first exceeded the speed limit before the curve.

STAY SAFE   See **On the Rails**.

SOURCE   FRA, railroads.dot.gov/elibrary/federal-railroad-administration-releases-results-metro-north-safety-review

TRESPASSERS

ANNUAL US DEATHS   No breakout data for this category

In 1999 near Fairfield, Connecticut, a mother and her 4 young children left a YMCA shelter late at night for an unknown reason, carrying school books and toys, and walked together onto an active rail line. While they were on an overpass 2 miles from the shelter, an Amtrak passenger train came and hit them. The woman and 3 of the kids were killed; the other child was very critically injured. The mother's behavior was never explained, but mental illness may have played a role.

STAY SAFE   See **On the Rails**.

SOURCE   *Hartford Courant*

WORKPLACE VIOLENCE

ANNUAL US DEATHS   No breakout data for this category

In May 2021 in San Jose, California, a disgruntled employee of the Santa Clara Valley Transportation Authority (SCVTA) went on a shooting spree at an SCVTA light rail train control center that had adjacent train storage and maintenance facilities, killing 9 co-workers and himself. He reportedly had a history of anger management problems, had talked about harming co-workers at a previous job years before, and had recently expressed dissatisfaction with his job assignments at SCVTA and with their vacation pay policies. In 2016, when he returned from a trip to the Philippines, US customs officers detained him for a secondary inspection and found literature about terrorism in his possession, but apparently San Jose law enforcement was never informed about this by federal officials.

STAY SAFE   See **Active Shooter/Mass Killings**.

SOURCE   nbcnews.com

Positive Train Control (PTC)

Positive Train Control (PTC) is a new, high-tech nationwide system of computer servers, trackside sensors, communications pathways, and onboard train control devices that significantly reduce the possibility of human error as a cause of railroad accidents. PTC was first required by an act of Congress, after a deadly head-on collision between a freight train and a commuter train killed 25 people in California in 2008. The engineer of the commuter train ran a stop signal, apparently while distracted by texting.

PTC is designed to prevent 4 types of potentially fatal railroad mishaps: 2 trains colliding (in either a head-on or rear-end crash), a derailment due to excessive speed, the switching of a train onto the wrong track, and a train intruding dangerously on track work.

PTC is not able to prevent mishaps that result from some forms of mechanical failure, such as derailment due to wheel damage, or break-in-halves due to suboptimal train handling over undulating terrain, but other measures are increasingly providing protection in these areas.

By the end of 2020, virtually all railroads in the US were protected by fully implemented Positive Train Control systems. (Also see **Other Safety Technologies**, below.)

STAY SAFE   Urge Congress and the railroads themselves to continue supporting improvements in safety procedures and technology.

SOURCE   Association of American Railroads (AAR), aar.org/campaigns/ptc

OTHER SAFETY TECHNOLOGIES

Railroads have long used defect detectors, sensors between the rails that detect problems such as dragging equipment or overheated wheel bearings and transmit a warning by radio. Increasingly, train designs incorporate miniaturized sensors linked to computer models, which allow predictive maintenance intervention when a component might soon fail. Track geometry monitoring—via instrumented track vehicles, or specialized cars inserted occasionally in trains, or future sensors built into many trains—is vital to help assure that the rail-wheel interface always works smoothly and safely. Tracks that fall out of gauge and rails that develop cracks or high spots or low spots can cause derailments unless discovered early and fixed quickly. Improved braking systems can activate every car's brakes simultaneously, via electronic control, instead of gradually/sequentially, from the front of the train to the back, via pneumatic control.

Railroads have almost entirely replaced manned cabooses by small digital devices at the end of trains, known as End of Train (EOT) devices or FREDs (Flashing Red End-of-train Devices). These battery powered devices, usually mounted on the trailing coupler of the final car in the train, contain sensors and radio links to the lead locomotive cab. They monitor brake line air pressure and give a warning if a train breaks apart in the middle (a "break-in-half") due to the sometimes violent, conflicting forces of slack and strain that can occur as a long, heavy freight train moves over undulating terrain. The FRED also displays a flashing red light to mark the back of the train to any following train and personnel.

Grade crossing safety is a subject of ongoing research into technologies that might both better prevent drivers from trying to beat a train and also be able to slow or stop any approaching trains if a grade crossing is blocked by a stuck vehicle. Trespasser detectors and active warning systems are also of keen interest to railroads, because of the dangers to pedestrians walking on, across, or near tracks.

Railroads have also been reducing the size of onboard and central dispatcher crews as they replace manned cabooses with FREDs and further automate signal and track switch systems. They have begun researching concepts such as 1-man crews and even unmanned, self-driving trains. These crew reductions are seen by some railroad safety advocates and labor organizations as favoring corporate profits over crew and public safety. Industry executives who seek more such experimentation argue that hypermodern technologies are needed to compete against the trucking industry's cost-cutting moves toward self-driving trucks, while also improving overall railroad safety by further reducing the human factor in operations.

STAY SAFE   Urge Congress and the railroads themselves to continue supporting improvements in safety procedures and technology.

SOURCE   Association of American Railroads (AAR), aar.org/topic/safety

On the Water

Some ships have commercial/professional crew only, some boats have recreational/amateur occupants only, and some vessels have both a professional crew who provide commercial services and passengers. For example, a tugboat/towboat, cargo ship, or fishing factory ship has a professional crew, but (generally speaking) doesn't carry passengers. A small sailboat or motorboat has only amateur, recreational skippers and guests. A cruise ship or ferryboat, in contrast, has both a professional crew and paying passengers.

Mortality statistics for these 2 demographic subpopulations aboard occupied watercraft are usually grouped separately. Crew deaths and passenger deaths tend to occur for somewhat different reasons and in somewhat different ways, even aboard the same vessel, and the 2 statistics are of interest to somewhat different audiences. Those audiences can include, for instance, maritime worker labor unions and US Department of Labor/OSHA safety regulators, versus recreational boating safety patrols by the US Coast Guard. Accordingly, *You Bet Your Life*'s mortality data for persons aboard water-going vessels is subdivided into different sections of this book, as follows: The **At Play** section includes death statistics for the *passengers* only, whether aboard professionally crewed vessels ranging from huge cruise ships to floating casinos, or aboard all-amateur, purely recreational watercraft ranging from jet skis and rowboats to superyachts. The **At Work** section includes death statistics for paid crews on all vessels—whether or not they carry passengers.

This section of **On the Go**, for waterborne vessels that have both professional crews and paying, amateur passengers, is of necessity very brief. In modern times, most passenger-miles of paying water transportation usually begin and end at the same point, primarily for recreation and pleasure. Even local water taxis and passenger ferries are as often used by tourists as they are by people commuting to work. (In contrast, almost all modern US rail and commercial air travel is for passengers to get from point A to a distant, different point B; the return trip is made separately, later on.)

Commercial Divers and Underwater Welders

ANNUAL US DEATHS   6 to 13

Six to 13 commercial divers are killed on the job every year in the US. About 10,000 commercial divers are employed in the American economy, involved in such activities as offshore oil rig and pipeline servicing, underwater salvage, ship inspection, cleaning, and repair, and construction of dams, bridges, and piers, plus scientific research and seafood aquaculture.

Although equipment and procedures for safety have improved significantly in recent years, commercial diving, and especially underwater welding, remain potentially dangerous occupations. Risk of death comes from drowning, hypothermia, and decompression sickness (the bends), plus the hazards of working underwater with hand tools, power tools, and other heavy materials in often constricted spaces with poor-to-zero visibility. Welders also face risks of electrocution or explosion if their equipment is not properly adapted for use underwater.

STAY SAFE Dive site supervisors should make sure that work is never performed while understaffed, or without adequate training, and is not attempted with improper equipment and unsuitable tools. Proper work safety practices must be followed at all times, for dive duration, maximum depth, water temperature, currents and tidal or wave action, and safe decompression schedules. Rescue and emergency resources must be close at hand before beginning a diving work shift.

SOURCE OSHA, osha.gov/pls/imis/AccidentSearch

A nineteenth-century hard hat deep sea diver

Commuter Ferry Crash

ANNUAL US DEATHS See various **On the Job** and **At Play** entries for ships and other watercraft.

In 2003, a New York City Staten Island Ferry vessel with 1,500 passengers aboard, running between Lower Manhattan and Staten Island, crashed at full speed into a concrete maintenance pier on the Staten Island side of New York Harbor, killing 11 and seriously injuring 70. Wind gusts at the time exceeded 40 mph, and the harbor waters were very choppy.

A subsequent investigation determined that the lone pilot had lost consciousness while at the controls, due to taking drugs that can cause drowsiness, and that management had failed to enforce a rule that there must be 2 pilots present in the wheelhouse during docking. The pilot and 4 other members of the crew or ferry service management were charged with offenses such as seaman's manslaughter, making false statements in a medical report, lying to investigators, and obstruction of justice; the pilot and a ferry manager served jail sentences.

STAY SAFE See various **On the Job** and **At Play** entries for ships and other watercraft.

SOURCE   NTSB marine accident report, NTSB.gov, Allision of Staten Island Ferry, Andrew J. Barberi, St. George, Staten Island, New York October 15, 2003

371

ON THE GO

Shipboard Fire and Sinking

ANNUAL US DEATHS   No cumulative data available

One of the worst US maritime disasters in decades occurred in 2019 off the coast of Santa Cruz Island, California. The 75-foot, 3-decked dive boat MV *Conception,* which offered overnight accommodations for passengers and crew, caught fire late at night while 33 passengers and 1 crew member slept below decks. They all died. Five other crew members were able to escape by jumping overboard after radioing the Coast Guard for help. The boat subsequently sank.

Multiple investigations could not determine the exact cause of the fire. The high death toll was blamed on a combination of inadequate crew training, inadequate passenger safety briefings, and the failure to set a night watch.

STAY SAFE   See various **On the Job** and **At Play** entries for ships and other watercraft.

SOURCE   thescubanews.com

On the Road

ANNUAL US DEATHS   36,096 in 2020

Motor vehicle transportation is vital to the functioning of modern US society. Almost 280 million cars, limos, SUVs, vans, motorcycles, buses, and many classes of trucks are on the roads of America, operated by almost 228 million licensed drivers. Some 3.5 million people are employed as truck drivers, transporting raw materials, fuels, and manufactured goods to destinations nationwide. Our economy could not function without motor vehicles. In 2016, US drivers made 186 billion trips, spending 70 billion hours behind the wheel, for a total of 2.62 trillion vehicle-miles traveled (VMT). Most of those trips were completed safely. But not always.

According to the National Highway Traffic Safety Administration (NHTSA), in 2020, 36,096 people were killed in motor vehicle mishaps in the US. Roughly half were drivers, one-sixth were passengers, one-sixth were pedestrians, and one-seventh were motorcyclists.

The number of motor vehicle traffic fatalities rose sharply in 2021. For the first half of the year, 21,450 people died in motor vehicle traffic crashes, an increase of about 16 percent compared to the 17,020 fatalities in the first half of 2020. The percentage increase is the highest half-year rise in the history of data recorded by the NHTSA. The fatality rate for the first half of 2021 increased to 1.34 fatalities per 100 million VMT, up from 1.28 fatalities per 100 million VMT in 2020. An underlying cause of the increase may be a 13 percent increase in VMT in the first half of 2021 over the same period in 2020.

The nonprofit National Safety Council (NSC) takes a slightly different approach to tabulating traffic fatalities and comes up with higher numbers. The NSC counts not only traffic deaths on the road but also deaths in private spaces such as driveways and parking lots; it also counts deaths that occur up to a year after a traffic crash. By that reckoning, car accidents killed 42,060 people in 2020. The NSC estimates the lifetime odds of dying in a motor vehicle crash. Unless otherwise stated, in the discussion of **On the Road** fatalities, NHSTA figures are the primary source.

The causes of fatalities vary, and many can be prevented by more attention to safe driving habits, proper vehicle and highway maintenance, and ever-improving design practices for new cars and trucks and for the roads on which they operate. Human factors contributing to traffic deaths include failure to yield right of way or obey stop signs and traffic lights, failure to wear seat belts, failure of motorcyclists to wear helmets, driving while impaired by drugs and/or alcohol, speeding and reckless driving, rude/aggressive driving and road rage, driving while drowsy, driving while texting/phoning, or driving while distracted by eating/drinking/smoking or by the radio, a passenger, or a pet. Other causes of fatalities include wet or icy roads, poor visibility, engine fires or gas tank explosions, tire blowouts, jammed accelerators, and steering or brake failures.

The resulting mishaps include head-on collisions, rear-enders, T-boners, rollovers, wrong-way driving, jackknifings, railroad grade crossing accidents, plunging off embankments or into water or driving through floods, hitting buildings or obstructions, hitting pedestrians or bicyclists, and hitting animals.

US traffic deaths have seen a generally downward trend over time. More than 51,000 people died on the roads in 1980, even though the total VMT was just under half that of 2019. Better automotive and road safety designs, smarter crashworthiness engineering (such as airbags and ABS), stronger laws and wider public education about wearing seat belts, complying with speed limits, not texting or phoning while driving, and not driving under the influence have all contributed to the improved statistics. The fatality rate went up, however, from 1.06 per 100 million VMT in the first half of 2019 to 1.25 in the first half of 2020, even though the total miles driven declined by 16 percent (meaning there were fewer cars on the road to hit each other). This increase is mostly attributed to the COVID-19 pandemic: less driving, but more unsafe driving.

STAY SAFE   Always follow safe driving habits. Always wear your seat belt. Keep your vehicle properly maintained.

SOURCES   AAA, https://aaafoundation.org/research; NSC, nsc.org/nsc-membership/injury-facts; NHTSA, crashstats.nhtsa.dot.gov

Driver Risks

AGGRESSIVE DRIVING (ROAD RAGE)

ANNUAL US DEATHS   Average of 26,000

Aggressive driving, commonly called road rage, is a general term that includes many bad behaviors: making speed racing challenges to other drivers, tailgating, cutting people off, blocking them from changing lanes, weaving in and out of traffic impatiently, excess use of horn and flashing headlights, rude gestures, trying to run someone off the road, intentionally ramming or bumping another vehicle, and/or stopping, getting out, and engaging in verbal and physical altercations with other drivers. Reportedly, 37 percent of serious road rage incidents involve a firearm.

Aggressive driving contributed to two-thirds of traffic fatalities in the US. This amounts to about 800,000 deaths over the past 30 years, or an average of over 26,000 fatalities per year. Most incidents are triggered by males under age 39, especially male teenagers.

About 30 people per year on average are intentionally murdered in acts of road rage in the US. In May 2021, a 6-year-old boy was shot and killed during a road rage incident in Los Angeles.

Insurance industry studies show that the small number of drivers who are chronically aggressive behind the wheel cause the majority of all serious traffic accidents in the US. Road rage is often listed as an exemption in vehicle insurance policies, leading to zero coverage for road rage offenses.

STAY SAFE   Always drive courteously. Do not respond aggressively to another driver who is being rude. Stay in your car with doors locked and windows closed, try to defuse the situation, and if necessary drive to a police station or other well-traveled public area that has witnesses and sources of aid.

SOURCES   AAA Foundation for Traffic Safety technical report, aaafoundation.org/category/driver-behavior-performance; III, iii.org/fact-statistic/facts-statistics-aggressive-driving

CRASHES INTO COMMERCIAL BUILDINGS AND OTHER NOT-IN-TRAFFIC MISHAPS

ANNUAL US DEATHS   Estimated 500

A study of US data from 2014 through 2021 found that almost 500 people per year were killed when a motor vehicle crashed into a commercial building. Restaurants and storefronts accounted for almost two-thirds of these deaths. Drunk driving, distracted driving, speeding, or mistaking the brake for the accelerator (pedal error) were the leading causes. Pedal error is a particular culprit in cars hitting commercial establishments, because many drivers park their vehicle facing in, very near the building.

Combined with another study for the US in 2008 through 2011 of all not-in-traffic fatal vehicle mishaps, the data shows that about 1,100 additional people per year were killed by other accidents on private property. About 600 deaths were pedestrians or bicyclists; the other 500 were vehicle occupants. Most of these accidents occurred on private roads or in parking lots, parking garages, or driveways. They include vehicles that crashed into residential buildings or fell on someone. An average of 93 people per year are killed in vehicle roll-aways (driver absent, parking brake not on).

STAY SAFE   Never drive while impaired, distracted, or drowsy, and always wear a seat belt. Don't speed. When leaving a parking spot, lot, or garage, avoid pedal error. Always use the parking brake on hills.

SOURCES   NHTSA, nhtsa.gov/crash-data-systems/fatality-analysis-reporting-system; Storefront Safety Council, storefrontsafety.org/statistics.html

DISTRACTED DRIVING

ANNUAL US DEATHS   About 2,800 in 2018

Taking your eyes off the road for any reason, for only 5 seconds at 55 mph, amounts to driving blind for 120 yards—the length of a football field, including both end zones. This is exactly as dangerous as it sounds, yet it happens all too often on US roads. There are far too many ways, documented in too many fatal accident reports, for a driver to distract themselves, or be distracted, while their vehicle is moving down the road. They fall into 3 main categories:

- Visual: taking your eyes off the road
- Manual: taking your hands off the wheel
- Cognitive: taking your mind off driving

They include texting, making phone calls, listening to the radio or music, talking, arguing, or even fighting with a fellow passenger, a child, or a pet; eating, drinking, or smoking; being upset emotionally; personal grooming (fixing hair, putting on makeup, painting nails, changing clothes); reading a printed item; watching TV or a video; having sex.

In the US in 2018, over 2,800 people were killed—about 8 people every day—and an estimated 400,000 were injured in crashes involving a distracted driver. About 1 in 5 of the people who died in crashes involving a distracted driver were walking, riding a bike, or otherwise outside a vehicle. This death toll is almost certainly an underestimate, because many states don't track many types of distracted driving in their accident reports.

In the US in 2018, 25 percent of the distracted drivers involved in fatal crashes were young adults aged 20 to 29. Drivers aged 15 to 19 were more likely to be distracted than drivers aged 20 and older among drivers in fatal crashes. Among these drivers, 8 percent of drivers aged 15 to 19 were distracted at the time of the crash. Nine percent of all teens who died in motor vehicle crashes were killed in crashes that involved distracted driving. For teen drivers, having 1 passenger in the car doubles the risk of a lethal accident. If 2 or more passengers are in the car with a teen driver, the risk of death *quintuples*.

Studies and surveys consistently show that the effects of driver distraction on risk are staggering. Making a phone call raises the odds of crashing by 12 times. Reading something, crying, or being angry raises the odds by 10 times. Having a child in the car with you can be up to 12 times more distracting than talking on a phone, which is itself a major distraction. Texting while driving, banned in almost every state, raises the risk of a crash by a frightening *23 times*.

Driving while ill is also a type of distraction. Driving with a bad cold or flu reduces driver reaction times and situational awareness as dangerously as if the person were violating the legal limit for alcohol.

STAY SAFE Always wear a seat belt. Hands on the wheel, eyes on the road.

SOURCES CDC, cdc.gov/transportationsafety/ Distracted_Driving/index.html; NHTSA, nhtsa.gov/risky-driving/distracted-driving; Safer America, safer-america.com/ distracted-driving-fatalities-in-the-u-s-a-interactive-map

FATIGUE/DROWSINESS

ANNUAL US DEATHS 697 in 2019

A CDC study in 2016 found that one-third of US drivers get less than the 7 hours of nightly sleep experts recommend. The NHTSA estimates that in 2017, 91,000 police-reported crashes involved drowsy drivers, leading to an estimated 50,000 people injured and nearly 800 deaths. In 2019, fatigued or drowsy drivers caused 697 highway deaths. These estimates are conservative, and up to 6,000 fatal crashes each year may be caused by drowsy drivers.

Most fatal drowsy-driver mishaps occur in the late afternoon or after midnight. About 4 percent of all drivers report that they have nodded off behind the wheel at least once in the recent weeks before being surveyed. Men are 3 times as likely as women to say this happened to them.

Going too long without sleep can impair your ability to drive in the same way as drinking too much alcohol can. Being awake for at least 18 hours is the same as having a blood alcohol concentration (BAC) of 0.05 percent. Being awake for at least 24 hours is equal to having a BAC of 0.10 percent, higher than the legal limit (0.08 percent BAC) in all states.

STAY SAFE Get enough sleep. Don't drive if drowsy. If you feel sleepy while behind the wheel, pull over and take a nap for 10 to 20 minutes or change drivers.

SOURCES CDC, www.cdc.gov/sleep/about_sleep/drowsy_driving. html; NHTSA, nhtsa.gov/crash-data-systems/fatality-analysis-report-ing-system; Safer America, safer-america.com/car-accident-statistics

FIRES

ANNUAL US DEATHS 560 in 2018

In 2018 in the US, a total of 212,500 vehicle fires killed 560 people (not counting firefighters responding to the scene); this is a fatality rate per fire of about 0.26 percent. The total number of vehicle fires has declined 60 percent since 1980, largely due to safer motor vehicle and highway design and construction.

An automobile suffers an engine fire on the Masschusetts Turnpike.

Four out of 5 deaths were male. Collisions were the leading cause of fatal vehicle fires. Most fires originated in the engine area, running gear, or wheel area. Fuel tank or fuel line fires caused 12 percent of all car fire deaths. In more than two-thirds of car fire deaths, what started burning first was the fuel or another flammable liquid.

STAY SAFE Safe driving habits are the best way to avoid vehicle fires. At gas stations, always follow posted safety procedures while refueling. Be careful at all times in and around motor vehicles if you smoke, are welding or grinding, or are otherwise using an open flame or other ignition source.

SOURCES NFPA, nfpa.org/News-and-Research/Data-research-and-tools/US-Fire-Problem/Vehicle-fires; NHTSA, nhtsa.gov/crash-data-systems/fatality-analysis-reporting-system

POLICE PURSUITS

ANNUAL US DEATHS Average of 350

Between 1994 and 2002 in the US, during police pursuits of fleeing criminals, a total of 3,965 motor vehicles were involved in 2,654 fatal collisions, and a total of 3,146 people were killed. Of this death toll, 1,088 were not in the vehicle that was fleeing; some were law enforcement officers, and some were innocent people caught up in the chase. These pursuits were often at night, at very high speeds, and/or on local roads. This is equivalent to an average death toll of 350 people per year. On average, 1.2 people were killed in each fatal crash.

An analysis by *USA Today* found that between 1979 and 2013, more than 5,000 bystanders and passengers were killed in police car chases; tens of thousands more were injured. Bystanders and passengers in the chased cars were nearly half of all the fatalities. Most bystanders were in their own cars and were killed when they were hit by a fleeing driver. Since 2016, at least 30 people have been killed in precision immobilization technique, or PIT, maneuvers used by police to stop a fleeing vehicle.

STAY SAFE If you hear or see a police pursuit in progress, pull your vehicle well over, and slow down or stop. If you are on foot in the street, get out of the road immediately.

SOURCES NHTSA, nhtsa.gov/crash-data-systems/fatality-analysis-reporting-system; *USA Today*, usatoday.com/story/news/2015/07/30/police-pursuits-fatal-injuries/30187827/

ROADWORK ZONES

ANNUAL US DEATHS 842 in 2019

Crashes in roadwork zones in the US killed 842 people in 2019. Of these, 135 were road construction workers and 707 were passing motorists, bicyclists, or pedestrians. This was an increase of 11 percent over such fatalities in 2018. Workers acting as flaggers were most at risk, with a death rate of 40.9 per year per 100,000 Full-Time Equivalent (FTE) employed as flaggers.

Roadwork zones are construction sites that are also major intrusions into active thoroughfares. Heavy construction vehicles, such as cement mixers or loaded flatbeds, need to enter and exit the flow of traffic in order to access the work site. Bulldozers, excavators, graders, and pavers are also massive but very slow-moving, and they too present deadly collision hazards to passing motorists and work crews.

Overall, occupational fatalities of roadwork crews over the period 2011 through 2017 broke down as 76 percent due to motor vehicle accidents of some kind. In 60 percent of these the worker was hit by a vehicle that was driving past the work zone. In about 25 percent of them, the worker was run over by a backing construction vehicle.

For all fatal work zone crashes in 2019, 24 percent were rear-enders, and speeding was a factor in 31 percent. Construction vehicles, either hit by a car or hitting a worker, were involved in 33 percent.

STAY SAFE Obey all posted work zone speed restrictions. Exercise extreme caution when driving through the work zone. Courteously obey all instructions from flaggers about slowing or stopping. Allow work vehicles and workers on foot to move safely around the site and near or into the active roadway as needed.

SOURCE NHTSA, nhtsa.gov/crash-data-systems/
fatality-analysis-reporting-system

SPEED AND SPEEDING

ANNUAL US DEATHS 9,478 in 2019

Speeding was a significant contributing factor in 9,478 road deaths, or 26 percent of the 36,096 total road deaths, in 2019 in the US.

The number of traffic fatalities rose as the speed limit did, even for deadly accidents that did not involve going above the speed limit. The faster any vehicle moves—whether or not it's obeying the local speed limit at the time—the more kinetic energy it has. The more likely it is to skid on curves or in poor road conditions, the longer it takes to stop, the more difficult it is to control in general—and the more destructive are any impacts in a mishap.

In 2019, for deaths in which going above the speed limit was a factor, the breakdown by applicable speed limit at the crash site was:

35 mph or less	2,306	24 percent
40 to 50 mph	2,613	28
55 mph or more	4,526	48

In 2019, for deaths where speeding over the limit was not a factor, the breakdown still showed more fatal accidents occurring when/where higher speed limits were allowed:

35 mph or less	5,117	19 percent
40 to 50 mph	7,268	27
55 mph or more	14,060	53

STAY SAFE Don't speed. Slow down if traffic or road and visibility conditions demand it. Drive attentively and courteously. Always wear a seat belt. Never drive while impaired.

SOURCE IIHS, iihs.org/topics/fatality-statistics

STOPPED ON THE SHOULDER

ANNUAL US DEATHS Estimated 4,000

A study of US motor vehicle fatalities from the 1980s found that 11.1 percent were caused by a collision between a moving vehicle and a vehicle that had stopped on the shoulder of an interstate, parkway, or highway. The leading reason for the stop was a mechanical problem, such as engine failure or to change a flat tire. Sometimes a police vehicle was also stopped, either to pull over a vehicle for a violation or to assist a disabled vehicle. Such collisions are particularly dangerous because occupants often exit the vehicle and stand or crouch next to it. Alcohol abuse and drowsy driving were the main contributing factors. This type of mishap was found to concentrate in the hours from midnight to 6 a.m.

Based on the number of US road fatalities in recent years, this 11.1 percent statistic would imply that about 4,000 people a year die in such collisions.

STAY SAFE   Try never to stop on the shoulder of a high-speed thoroughfare. If you must, stay inside the vehicle, phone for help, and wait for a highway patrol, tow truck, or Good Samaritan to render assistance. If possible, drive slowly, with your flashers on, to the nearest exit, despite having a flat tire or mechanical problem. Pull over only when it is safe to do so.

SOURCE   The SURVIVE Group, survivegroup.org

USE OF SEAT BELTS

ANNUAL US DEATHS   10,521 in 2019

When fatal crash reports indicate whether the victims were wearing seat belts, analyses conclusively prove that wearing them saves lives: The use of lap/shoulder belts cuts the risk of death in half for people in the front seats.

In 2019, for those driving, 8,306 fatalities occurred while wearing seat belts, and 7,077 while not. For passenger victims, 2,215 people were killed while wearing seat belts, and 2,245 while not. Surveys show Americans wear seat belts about 90 percent of the time. So, the 10 percent of people who didn't wear seat belts accounted for fully 46 percent of all traffic accident deaths.

The combination of driving drunk and not wearing a seat belt can be particularly deadly. In 2015, 60 percent of teen drivers killed on the road while they were drunk were also not wearing seat belts.

STAY SAFE   Always wear a seat belt. Slow down if traffic or road and visibility conditions demand it. Drive attentively and courteously. Never drive while impaired.

Lethal Risk by Type of Road User

ANNUAL US DEATHS   36,096 in 2019

In 2019, motor vehicle accident deaths broke down by type of user of the
roads as follows:

Automobile occupants	12,420	35.7 percent
Pickup and SUV occupants	9,621	27.7
Large truck occupants	679	2.0
Motorcyclists	5,014	14.4
Bicyclists	843	2.4
Pedestrians	6,205	17.8

STAY SAFE   Drive attentively and courteously. Always wear a seat
belt. Never drive while impaired.

SOURCE   IIHS, iihs.org/topics/fatality-statistics

Automobiles

DRIVER OR PASSENGER

ANNUAL US DEATHS   See data in entry.

In 2016, 74 percent of all passenger vehicle occupants killed on the
roads were the drivers. The other 26 percent of victims were passengers. This seeming imbalance results in part because many crashes involved cars carrying just the driver.

STAY SAFE   Drive attentively and courteously. Always wear a seat
belt. Never drive while impaired, drowsy, or distracted.

SOURCE   IIHS, iihs.org/topics/fatality-statistics

BY AGE OF VICTIM

> **ANNUAL US DEATHS** See data in entry.

In 2019, motor vehicle crash death rates per 100,000 people, for the different age groups, were:

Under 13	1.6
13 to 15	3.0
16 to 19	11.7
20 to 24	16.9
25 to 29	15.5
30 to 34	13.2
35 to 39	12.7
40 to 44	11.6
45 to 49	11.8
50 to 54	12.5
55 to 59	13.4
60 to 64	11.8
65 to 69	11.6
70 to 74	11.6
75 to 79	14.1
80 to 84	17.1
Over 84	11.0

These age-specific traffic death rates (per 100,000 of population in that age group) are at best only a proxy for age-specific death rates per 100 million vehicle miles traveled (VMT). The latter, more informative statistic is not available—because miles traveled each year are not broken out by age group.

Drivers under 25 tend to be more reckless and aggressive. Older drivers tend to suffer decline of driving faculties. Kids have lower death rates, in part because they make fewer trips, usually sit in the (safer) back seat, and/or gain additional crash survivability by riding in child car seats. Occupants over age 75 are more likely to die of injuries of a given severity suffered in a crash.

STAY SAFE Drive attentively and courteously. Always wear a seat belt. Never drive while impaired, drowsy, or distracted.

SOURCE IIHS-HLDI, iihs.org/topics/fatality-statistics

BY AGE OF DRIVER

ANNUAL US DEATHS   See data in entry.

Complete national data on the age of drivers involved in fatal motor vehicle accidents in the US is not available. This is because, unlike those people who are killed, for which there is always a death certificate that indicates age, in a significant portion of all deadly crashes 1 or more drivers involved survive, and their age is not reported to vital statistics records.

The age groups of drivers disproportionately likely to cause fatal motor vehicle accidents are those under age 20 and those over age 85. In 2013, teenage drivers caused about 11 percent of all accidents while comprising only 7 percent of the US population. Teenagers are usually inexperienced drivers, and data shows that at all ages someone is most likely to have a crash during their first few months behind the wheel. Teenagers are also particularly prone to all the bad habits that lead to dangerous driving: speeding and driving recklessly/aggressively, not wearing seat belts, driving while distracted, and driving while under the influence of drugs and/or alcohol.

The elderly are more likely to cause road accidents in part because of failing vision and hearing, generally worsened cognition and reaction time, and the possible influence of prescription drugs. In 2019, 7,214 people 65 and older were killed in traffic crashes, accounting for 20 percent of fatalities although this age group is only 16 percent of the total population. Older drivers made up 20 percent of all licensed drivers in 2019 and were 15 percent of all drivers involved in fatal crashes.

Automobile accidents are nothing new. This one is from 1918.

STAY SAFE   Drive attentively and courteously. Always wear a seat belt. Never drive while impaired, drowsy, or distracted. Teenagers and the elderly should take particular care.

SOURCES   IIHS-HLDI, iihs.org/topics/fatality-statistics; NHTSA, nhtsa.gov/data/national-center-statistics-and-analysis

BY SEX OF DRIVER

> **ANNUAL US DEATHS**   See data in entry.

While there are almost exactly as many male as female licensed drivers in the US, male drivers cause significantly more fatal accidents than females. Males drive more vehicle miles traveled (VMT), they on average drive more carelessly and/or aggressively, and they tend to cause more severe accidents when accidents do occur.

In 2018, 25,841 male drivers were involved in fatal traffic accidents of all kinds, compared to 10,676 female drivers. This splits as 71 percent male drivers, 29 percent female drivers.

STAY SAFE   Drive attentively and courteously. Always wear a seat belt. Never drive while impaired, drowsy, or distracted.

SOURCE   IIHS-HLDI, iihs.org/topics/fatality-statistics

BY SEX OF VICTIM

> **ANNUAL US DEATHS**   See data in entry.

The total number of victims killed in a year on the roads, tallied by their sex, is not the same statistic as the number of drivers involved in fatal accidents that year, tallied by sex. In 2018, the split of motor vehicle crash deaths by sex of the victim was 71 percent male and 29 percent female, as it also was in 2019.

The number of drivers killed is yet another, different statistic. Males comprised 71 percent of all automobile drivers killed (another coincidence), 48 percent of automobile passenger deaths, 97 percent of large-truck driver deaths, 71 percent of large-truck passenger deaths, 69 percent of pedestrian deaths, 86 percent of bicyclist deaths, and 91 percent of motorcyclist deaths.

STAY SAFE   Drive attentively and courteously. Always wear a seat belt. Never drive while impaired, drowsy, or distracted.

SOURCE   IIHS-HLDI, iihs.org/topics/fatality-statistics

Because the relative population numbers differ greatly among races/ethnicities in the US, the best statistic for analyzing motor vehicle accidents is fatality rate per 100,000 of population within that race or ethnicity. The most recent such study used data from 2006.

The fatality rates per 100,000 were:

Hispanic	12.27
White	12.50
Black	12.31
Native American	31.17
Asian	4.00
Pacific Islander	13.90

Many factors combined to cause such disparate outcomes: different habits as to drunk or drugged driving, speeding, wearing seat belts or motorcycle helmets, and jaywalking, as well as other demographics that are known to affect road fatality rates, such as level of income and level of education.

The above death rates are for all types of road users combined. When pedestrians killed by motor vehicles are split out separately, the racial disparities are proportionately much greater, even though the death rates per 100,000 of overall population, for this single type of accident, are (of course) themselves always much lower than those rates for all road accidents combined. These on-foot death rates for 2018, available only for the more populous US ethnicities, were:

White	1.8
Hispanic	3.6
Black	2.9

One explanation for this disparity is the wider variation between ethnic groups in whether pedestrians had been drinking, relative to the proportions of drunk driving between different ethnic groups.

STAY SAFE   Drive attentively and courteously. Always wear a seat belt. Never drive while impaired, drowsy, or distracted.

SOURCES   CDC, cdc.gov/injury/wisqars/fatal.html; NHTSA, nhtsa.gov/crash-data-systems/fatality-analysis-reporting-system

WORST AND BEST STATES

> **ANNUAL US DEATHS**   See data in entry.

Death rates per 100,000 of population allow meaningful comparison of road fatality risks between states with different population sizes. However, this data does not take account of the amount of driving done per person in each state, which also affects death tolls on the roads, but vehicle miles traveled (VMT) are not broken out by state.

For 2019, the worst states and their road deaths per 100,000 citizens were:

Mississippi	22.3
South Carolina	20.6
Alabama	19.5
Wyoming	19.2
New Mexico	18.7

The best states were:

New York	4.8
Massachusetts	5.2
Rhode Island	5.6
New Jersey	6.3
Minnesota	6.8

STAY SAFE   Drive attentively and courteously. Always wear a seat belt. Never drive while impaired, drowsy, or distracted.

SOURCE   IIHS-HLDI, iihs.org/topics/fatality-statistics

WORST CITIES

> **ANNUAL US DEATHS**   See data in entry.

The 5 deadliest cities in the US for road mishap fatalities, and their 2018 death rates per 100,000 of residents, were:

Baton Rouge, LA	23.1
St. Louis, MO	18.1
Savannah, GA	15.7
Detroit, MI	15.3
Dallas, TX	14.5

STAY SAFE Drive attentively and courteously. Always wear a seat belt. Never drive while impaired, drowsy, or distracted.

SOURCE IIHS-HLDI, iihs.org/topics/fatality-statistics

BY HOLIDAY

ANNUAL US DEATHS See data in entry.

Based on data for 2019, the number of people killed on each of the 4 main long weekend or week-long holiday periods were 594 for the Fourth of July period; 448 for Labor Day; 454 for Thanksgiving weekend, and 799 for Christmas through New Year's Day. These numbers will vary each year, in part because every calendar year can have more or fewer days in each of its big holiday periods, depending on which day of the week the actual holiday falls on. Holidays are especially dangerous times to be on the roads, partly because there are more vehicles out there, and partly because holidays see peaks in drunk driving.

In 2019, these 4 main holiday reporting periods covered a total of 21 days during which a total of 2,295 people were killed, for an average daily holiday death toll of 109.3. For comparison, the death total across all of 2019 was 36,096. This means the average daily nonholiday death toll was 98.3. So, on average, being on the road during a holiday period was 11.2 percent more deadly than on the other days of the year.

STAY SAFE Be careful driving on holidays. Always wear a seat belt. Never drive while impaired, drowsy, or distracted.

SOURCE Autoinsurance.org, autoinsurance.org/deadliest-holidays-to-drive

BY DAY OF WEEK

ANNUAL US DEATHS See data in entry.

Data for 2019 shows that the deadliest day of the week on US roads is Sunday, which accounted for 17 percent of all traffic fatalities (even though each of the 7 days comprises only 14.3 percent of a week). Next most deadly were Friday and Saturday, each with 16 percent of the total weekly deaths. The other days, Monday through Thursday, each accounted for only 13 percent of the total.

More children were killed in motor vehicle crashes on Saturdays in 2019 (20 percent) than on any other day of the week.

STAY SAFE   Be particularly careful driving on weekends, and never drive while impaired, drowsy, or distracted.

SOURCE   IIHS-HLDI, iihs.org/topics/fatality-statistics

BY TIME OF DAY

ANNUAL US DEATHS   See data in entry.

Based on data for 2019, traffic fatalities peak in the stretch of the day that includes evening rush hour, dinnertime, and after-dinner drinking. The percent of each day's total road deaths is spread around the clock as follows:

Midnight to 3 a.m.	11 percent
3 a.m. to 6 a.m.	8
6 a.m. to 9 a.m.	10
9 a.m. to noon	9
Noon to 3 p.m.	13
3 p.m. to 6 p.m.	16
6 p.m. to 9 p.m.	17
9 p.m. to midnight	15

This data is averaged across the whole calendar year. There is a seasonal pattern to the single most dangerous driving period: during and just after dusk and sunset, coming between 8 p.m. and midnight in spring and summer, and between 4 p.m. and 8 p.m. from October through March.

STAY SAFE   Be particularly careful driving in the evening and at night. Always wear a seat belt. Never drive while impaired, drowsy, or distracted.

SOURCE   IIHS-HLDI, iihs.org/topics/fatality-statistics

BY MONTH OF YEAR

ANNUAL US DEATHS   See data in entry.

Based on data over the years 2007 through 2017, September is the worst month of the year for getting killed in a traffic accident, with a rate of death of 1.15 per 100 million VMT.

The safest month, measured by death rate, is March, at 0.97 fatalities per 100 million VMT.

Looking instead at total number of people killed over this 11-year data period, August is the worst, at 32,678, and February is the best, at 23,764.

Why the different results for deaths per VMT versus total death counts? August sees the most driving (the most total VMT) and the most crowded roads, so it has the most vehicles hitting each other and pedestrians, plus August has 31 days. February, in the depths of winter, sees the least driving, plus it has only 28 or 29 days.

The rates of death per 100 million VMT for each calendar month are:

January	1.02
February	0.99
March	0.97
April	1.00
May	1.06
June	1.07
July	1.10
August	1.11
September	1.15
October	1.12
November	1.12
December	1.08

STAY SAFE   The safest month in which to drive is March. Always wear a seat belt. Never drive while impaired, drowsy, or distracted.

SOURCE   AAA Safety Foundation, aaafoundation. org/2020-traffic-safety-culture-index

ALTERNATIVE FUEL VEHICLES

ANNUAL US DEATHS   No data available

Alternative fuel (hybrid, electric, hydrogen, propane, natural gas) motor vehicles are likely to appear on US roads in larger numbers in the future. Some of the risks to their occupants and other people differ from conventional vehicles. Because some lack internal combustion engines under their hood and/or have heavy battery or fuel-cell banks in their trunk, their weight distribution and frame structure can vary from conventional vehicles. Some are quieter in operation, posing added risk because they can be harder to hear coming. By storing and harnessing energy differently, some could, in a worst case, kill people in different ways.

Electric cars are quickly emerging as an alternative to gasoline-powered automobiles.

For all-electric and hybrid gas/electric systems, the potential for battery fires and electric shocks is a concern. So far, however, the aggregate risk of fatalities from these systems appears no more severe than for conventional gasoline or diesel vehicles. The overall risk of death from crashworthiness appears to be similar to conventional designs. Safety systems are being developed that compensate for the quieter engine sound when running on electric motors, including sensors and AI that detect pedestrians and either brake the vehicle, emit a warning noise, or both.

Hydrogen, carried as a quickly vaporizing, odorless, highly flammable, and deeply cold liquid, presents special concerns for fire, asphyxiation, and frostbite. Hydrogen leak alarms, fire/explosion suppression systems, and passenger compartment ventilation, all activated post-impact, are being developed. Some of the aggregate public health risks from hydrogen fuel are expected to be more than offset because hydrogen is very clean burning. Its use may reduce deaths from lung diseases that are triggered by air pollution.

Similar considerations apply to liquid natural gas (LNG) power. Over 200,000 LNG vehicles are on the road in the US. Only 1 fuel-related death, attributed to human error, has been reported since these vehicles were introduced in the 1970s. Propane-powered vehicles, such as forklifts and golf carts, also have an excellent safety record.

STAY SAFE   Be aware of the special fueling and driving/handling characteristics of your alternative energy vehicle. Always practice safe driving habits, and never drive while impaired.

SOURCES   NFPA, nfpa.org/News-and-Research/Data-research-and-tools/US-Fire-Problem/Vehicle-fires; NHTSA, nhtsa.gov/crash-data-systems/fatality-analysis-reporting-system

BRIDGES

ANNUAL US DEATHS   Average of 3.2 (bridge collapses only)

Fatal traffic accidents specific to driving on a bridge, such as a vehicle plunging from a bridge, are not broken out separately in available data. Typically, those deaths are included in statistics on overall road mishap deaths, either within broader categories such as driving off an embankment or plummeting into water, or simply "other."

Over the 40-year period from 1967 to 2007, 6 traffic bridge collapses in the US killed a total of 128 motorists, for an average annual death toll of 3.2 people killed.

STAY SAFE   Be alert on bridges for structural failures, obey speed limits, and try to avoid driving into a collapse disaster scene.

SOURCE   III, iii.org/fact-statistic/facts-statistics-highway-safety

CLASS OF ROAD

ANNUAL US DEATHS   See data in entry.

In 2019, road accident deaths broke down by class of road as follows:

Interstates and freeways	6,133	17.1 percent
Other major roads	18,806	52.4
Minor roads	10,928	30.5

Another breakdown is by urban versus rural locations:

Urban	19,595	54.5 percent
Rural	16,340	45.5

STAY SAFE   Drive attentively and courteously. Always wear a seat belt. Never drive while impaired, drowsy, or distracted.

SOURCE   IIHS-HLDI, iihs.org/topics/fatality-statistics

HIGHWAYS OF DEATH

ANNUAL US DEATHS   2,049 in 2016

The Top 5 most dangerous interstate highways in the US, based on fatalities per mile from 2010 to 2015, are:

1. Interstate 4 between Tampa and Daytona Beach: 132 miles, 1.41 fatalities per mile
2. Interstate 45 connecting Dallas, Houston, and the Gulf of Mexico: 285 miles, 1.24 fatalities per mile
3. Interstate 17 between Phoenix and Flagstaff: 146 miles, 1.03 fatalities per mile
4. Interstate 30 from Fort Worth to North Little Rock: 367 miles, 1.03 fatalities per mile
5. Interstate 95 from Miami to Houlton, Maine: 1,926 miles, 0.89 fatalities per mile

STAY SAFE   Drive attentively and courteously. Always wear a seat belt. Never drive while impaired, drowsy, or distracted.

SOURCE   NHTSA, nhtsa.gov/crash-data-systems/ fatality-analysis-reporting-system

HIT-AND-RUN

ANNUAL US DEATHS   2,049 in 2016

In 2016 in the US, 2,049 people were killed by hit-and-run drivers, for 5.5 percent of all motor vehicle fatalities that year. This percentage has shown a steady upward trend from 2009, when it was 3.8 percent.

STAY SAFE   Always look both ways. Cross at the green, not in between. Avoid walking or biking on road shoulders, especially at night.

SOURCES   AAA Foundation for Traffic Safety 2018 research brief, A. Benson et al. *Hit-and-Run Crashes: Prevalence, Contributing Factors and Countermeasures*; NHTSA, nhtsa.gov/crash-data-systems/ fatality-analysis-reporting-system

INTERSECTION CRASHES

ANNUAL US DEATHS   10,011 in 2018

Every year about 25 percent of all traffic fatalities are attributed to intersections, including signalized intersections, roundabouts, crossover intersections, unsignalized intersections (the most common type in the US), and other designs. In 2018, out of 36,835 total traffic fatalities, 10,011 involved an intersection. Of those, 6,737 involved unsignalized intersections (stop sign, yield sign, uncontrolled). Pedestrian fatalities at unsignalized intersections totaled 979; 220 bicyclists were killed.

STAY SAFE   Obey all traffic signals, signs, and right-of-way rules while driving.

SOURCE   FHWA, highways.dot.gov/research/research-programs/safety/intersection-safety

NUMBER OF INVOLVED VEHICLES PER MISHAP

ANNUAL US DEATHS   See data in entry.

In 2019, out of 36,096 fatal road mishaps, 19,257, or 53.3 percent, were single-vehicle accidents. The other 16,839, or 46.7 percent, were multi-vehicle accidents.

Pileups involving huge numbers of vehicles get loads of media attention but are very rare. Since 1990, only about 15 fatal pileups involving more than 100 vehicles have occurred. The number of people killed per incident ranged from 1 or 2 (the most common result) up to a record of 17 in a very deadly massive pileup that occurred in Coalinga, California, in 1991, as the result of a huge dust storm on Thanksgiving. The record for the number of vehicles involved in a pileup is 216, set in Los Angeles in 2002 during thick fog—no one was killed.

STAY SAFE   Drive attentively and courteously. Always wear a seat belt. Never drive while impaired, drowsy, or distracted.

SOURCE   IIHS-HLDI, iihs.org/topics/fatality-statistics

NUMBER OF PEOPLE KILLED PER ACCIDENT

ANNUAL US DEATHS   See data in entry.

In 2019, 36,096 people were killed in 33,244 different fatal accident events. This is an average of 1.09 people killed per lethal road mishap. In almost all cases, if anyone is killed, it is only 1 person.

STAY SAFE   Drive attentively and courteously. Always wear a seat belt. Never drive while impaired, drowsy, or distracted.

SOURCE   IIHS-HLDI, iihs.org/topics/fatality-statistics

RACING: AMATEUR STREET

ANNUAL US DEATHS Average of 100

A study for 1998 through 2001 found that during that 4-year period, 399 people were killed in the US in street racing crashes, for an average of 100 per year. Three-quarters of the deaths were of occupants in 1 or more of the vehicles being raced. The remaining quarter were occupants of other vehicles, race spectators, or other pedestrians and bicyclists.

Street racing fatalities were heavily concentrated on urban roads, and usually occurred in the hours around midnight. Postcrash accident investigation reports show that they were almost 6 times as likely to have happened at a speed of at least 65 mph, as were other fatal motor vehicle accidents. The drivers were more likely to be teenagers with prior driving violations and previous traffic accidents.

STAY SAFE Don't do it.

SOURCES NHTSA, nhtsa.gov/crash-data-systems/fatality-analysis-reporting-system; S. Knight et al. The fast and the fatal: street racing fatal crashes in the United States. *Injury Prevention*, February 2004.

RACING: DEMOLITION DERBIES

ANNUAL US DEATHS Near 0

Anecdotally, afficionados of this sport say that very few people have ever been killed at a demolition derby in the US in the last 50 years. There was 1 fatal accident in 2019 in Montana, in which a spectator was killed when one of the competing vehicles went out of control.

Demolition derbies became popular in the US after World War II. There are about 5,000 of them a year, mostly at county fairs and folk festivals. It isn't surprising that they can be dangerous, since the whole idea is for a group of motor vehicle drivers to intentionally crash into each other until only 1, the winner, remains operable. The contestants ram each other as violently and as damagingly as possible. All glass is removed in advance from the windows as a safety precaution, and the derbies are held on flooded dirt fields where the muddy conditions limit maximum speeds. Chronic neck and back pain from repeated whiplash is a common complaint among dedicated demolition derby drivers.

STAY SAFE Safer to watch than to participate

RACING: PROFESSIONAL

ANNUAL US DEATHS   Average of 2.5

Professional car and motorcycle racing is a very dangerous sport. Since the dawn of motor vehicles around 1900, hundreds of drivers, pit crews, and spectators have been killed in races around the world. Since 2000, about 50 American racing pros have been killed, either in practice or during races. While deaths have occurred among almost every class of racing vehicle, they were concentrated in 13 fatalities among stock car drivers, 10 among sprint drivers, and 9 among dragsters. Nondriver deaths have been greatly reduced in recent years due to improved grandstand and pit area designs, more effective safety barriers, and better racing rules. NASCAR made major upgrades to safety after Dale Earnhardt was killed in 2001. Globally, fewer than a dozen American professional motorcycle racers have been killed during races since 2000.

STAY SAFE   Follow safety instructions for spectators at car races.

SOURCES   NASCAR, nascar.com; Motor Sports Memorial, motorsportmemorial.org

RED-LIGHT RUNNING

ANNUAL US DEATHS   846 in 2019; 2,884 in 2017

In 2019, 846 people were killed in crashes that involved red-light running. Of those killed, 46 percent were passengers or people in other vehicles, just over 5 percent were pedestrians or bicyclists, and just over 35 percent were the drivers who ran the red light. In 2017, 2,884 crash deaths that occurred at signalized intersections were caused by a driver running a red light.

Safety studies show that red-light cameras can reduce fatal red-light running in large cities by 21 percent and the rate of all fatal crashes at signalized intersections by 14 percent.

STAY SAFE   Encourage local government to install red-light cameras. Use good judgment when approaching signalized intersections to avoid entering when the light turns yellow. Drive defensively. Pedestrians and bicyclists should wait a few seconds and look both ways to make sure all vehicles have come to a complete stop before moving through an intersection.

SOURCES AAA Foundation for Traffic Safety, aaafoundation.org/2020-traffic-safety-culture-index; FHWA, safety.fhwa.dot.gov/intersection; IIHS-HLDI, iihs.org/topics/fatality-statistics

SELF-DRIVING/AUTONOMOUS VEHICLES

ANNUAL US DEATHS Insufficient data

Autonomous motor vehicles, which use artificial intelligence computer programming fed by optical and radar sensors to drive themselves when the human operator chooses not to, are currently being evaluated on US roads. In early 2021, there were an estimated total of over 1,400 self-driving cars and trucks being built, tested, and sold by several US and foreign companies. The most advanced systems, called Level 5, are able to operate completely without any human intervention but are still in development.

As of 2021, the National Highway Safety Administration (NHTSA) has investigated 29 cases of Tesla crashes, including at least 12 deaths from 10 crashes involving vehicles operated in automated driving mode. The crashes have been linked to factors such as the AI system not properly interpreting the broad white side of a tractor trailer and then driving into it, or the human safety driver neglecting their job during a test run, or a self-driving car speeding, skidding on a curve, hitting a tree, and bursting into flames while no one was sitting at the wheel.

Tesla estimates that the fatality rate of its self-driving vehicles has been 0.3 per 100 million VMT, which is about one-third the rate for human drivers. Self-driving cars are predicted by their makers to be much safer overall than cars driven by human operators: The AI system can't get drunk or use drugs; won't become drowsy or fatigued; and can't be distracted by texting/phoning or other sources.

Cybersecurity experts worry that malevolent hackers might be able to interfere with the AI systems and cause crashes on purpose, whether or not a person is behind the wheel at the time.

STAY SAFE If you travel in a self-driving vehicle, make sure a qualified human sits alertly behind the wheel. Always wear a seat belt. Never drive while impaired, drowsy, or distracted.

SOURCES NHTSA, nhtsa.gov/crash-data-systems/fatality-analysis-reporting-system; Tesla, tesla.com/VehicleSafetyReport

TOWING TRAILERS

ANNUAL US DEATHS   Average of 440

Motor vehicle accidents involving passenger vehicles towing a trailer kill an average of 440 people a year in the US. Trailer types include towed vehicles, boat trailers, hayrides, parade floats, moving-van trailers ("u-hauls"), flatbeds with various loads, and other utility trailers.

Jackknifing is a serious added risk, which passenger vehicle drivers might not know how to avoid, since a car/trailer combination handles in unpredictable ways. Lost or shifting cargo on the trailer is another unfamiliar danger.

STAY SAFE   Know the trailer's weight limits and never overload it. Make sure any cargo is very securely fastened. Make sure the trailer brake and signal lights are in working order. Be extra cautious if the trailer does not have its own braking system. Do not make sudden stops or sharp turns or lane changes while towing a trailer. Always wear a seat belt, never speed, and never drive while impaired.

SOURCE   NHTSA, nhtsa.gov/crash-data-systems/ fatality-analysis-reporting-system

This large 1938 trailer was designed to be pulled by an International tow car.

TYPES OF MISHAP

ANNUAL US DEATHS See data in entry.

Motor vehicle accidents can happen in different ways. Some are more frequent and/or more potentially fatal than others. Data for 2019 in the US showed this breakdown of various types of collisions and noncollision fatalities:

Pedestrian	7,700
Another vehicle	16,700
Head-ons	5,000
Rear-enders	2,900
Angle collisions (T-bones, etc.)	7,500
Sideswipes	1,300
Railroad train	117
Bicycle	1,100
Hit horse or horse-drawn wagon	100
Hit a deer or moose	440
Fixed obstruction or other object	10,400
Noncollisions	3,100

These numbers add up to about 56,000, which is substantially more than 2019's total of 36,096 US motor vehicle deaths. The difference is because some fatal mishaps involved more than 1 type of incident before the vehicles involved came to rest, such as a car hitting a pedestrian, then swerving into another car, and then rolling over.

In 2016 in the US, 7,488 people died in vehicle rollovers. The 2019 data does not break this type of incident out separately. Rollovers can happen in different ways even in a single-vehicle mishap, such as by an overly sharp turn while speeding, or by overcompensating for a skid on a wet or icy road.

STAY SAFE Drive attentively and courteously. Always wear a seat belt. Never drive while impaired, drowsy, or distracted.

SOURCES IIHS-HLDI, iihs.org/topics/fatality-statistics; NSC, injuryfacts.nsc.org/home-and-community/safety-topics/ deaths-by-transportation-mode

TYPES OF FATAL INJURY

ANNUAL US DEATHS   See data in entry.

Fatal injuries in road accidents usually result when the human body hits something, or something hits the body. The object(s) can be part of the victim's own vehicle and/or another vehicle; or a fixed obstruction such as a tree or pole, building, abutment, or road sign; a pedestrian, bicyclist, animal, or debris in the road; or the road surface itself (the pavement). The wounds are usually either impact injuries (blunt force trauma) or penetrating (sharp force trauma). They can be inflicted when the person is ejected, when they are crushed, or when someone wearing a seatbelt suffers high g-forces in a very sudden stop, a hard rear-ending, or a violent rollover. In some cases, the person might be drowned, burned to death, or even electrocuted. Multiple injuries are not uncommon in lethal highway mishaps.

Seat belts reduce the risk of death by 45 percent. People not wearing seat belts are 30 times as likely to be ejected from the vehicle in a collision.

Hard data on road deaths broken down by type of injury is scarce, in part because police accident reports and coroner death certificates are usually prepared separately and not matched later. Many traffic deaths are simply recorded as due to massive trauma, and an autopsy is never performed. The most common types of serious injury in motor vehicle accidents are:

- Soft tissue injuries, especially whiplash
- Cuts, scrapes, and puncture wounds that can cause severe bleeding and/or organ damage
- Head injuries, especially concussions, fractured skulls, and brain bleeds
- Internal injuries, such as broken ribs, collapsed lungs, internal bleeding, organ damage
- Limb injuries, such as violent flail injuries or traumatic amputations
- Spine injuries, including broken necks or severed spines

Some of these injuries can cause immediate death, while in other cases the victim succumbs later.

STAY SAFE   If you are hurt in a car crash, get medical help immediately. Head injuries can be especially insidious: Brain bleeds not found by a CT scan and treated right away can take hours to produce overt symptoms and can lead to death.

SOURCES   IIHS-HLDI, iihs.org/topics/fatality-statistics; NHTSA, nhtsa.gov/crash-data-systems/fatality-analysis-reporting-system

WEATHER CONDITIONS

ANNUAL US DEATHS   5,376

Weather conditions are a contributing factor in about 22 percent of the 6 million motor vehicle accidents, from minor to fatal, that occur in the US every year. Weather-related factors contributing to traffic fatalities include impaired visibility from rain, snow, sleet, hail, smoke, mist, or fog; longer braking distances or skidding out of control on wet or icy roads; rollovers from overcompensating for skids; blow-overs caused by high winds; and driving into flooded areas. It can take up to 10 times as long to stop a motor vehicle on wet or icy pavement.

Over the 10-year period ending with 2016, an annual average of 5,376 people were killed in weather-related road mishaps in the US. This averages to 16 percent of all road mishap fatalities during that period. The breakdown by type of bad weather was as follows:

Wet pavement	76 percent
Rain	46
Snow/sleet	13
Icy pavement	10
Snowy/slushy pavement	10
Fog	9

These add up to more than 100 percent because more than 1 type of bad weather was sometimes involved.

STAY SAFE   Be extra careful, if you must drive in bad weather. Listen to and respect travel advisories. If visibility is impaired or if road conditions become dangerous due to thick fog, black ice, or local flooding, pull over, stop, and do not get out until conditions improve to the point it is safe to continue. Turn around, don't drown. If your vehicle goes into a skid, steer into the skid if possible, and pump the brakes if you do not have ABS (antilock braking system) until you recover or come to a stop; do not overcontrol your vehicle.

SOURCES   AAA Foundation for Traffic Safety technical report, aaa.com, B.C. Tefft, *Motor Vehicle Crashes, Injuries, and Deaths in Relation to Weather Conditions, United States*, 2010–2014; FHWA, ops.fhwa.dot.gov/weather/q1_roadimpact.htm

WRONG-WAY CRASHES

ANNUAL US DEATHS   2,008 in 2014

Fatal wrong-way crashes typically happen on divided highways or access ramps and involve high-speed head-on or sideswipe crashes. In 2014, 2,008 people were killed in wrong-way driving crashes. Six in 10 of these crashes involved an alcohol-impaired driver.

In 1988, a wrong-way crash in Carrollton County, Kentucky, involving an intoxicated pickup truck driver and a former school bus carrying a church youth group, killed 27 people. The crash led to improved standards for school buses, including an increased number of emergency exits.

In 2009, a wrong-way crash on the Taconic State Parkway near Mount Pleasant, New York, killed 8 people, including the wrong-way driver, her daughter, her 3 nieces, and 3 passengers in the oncoming vehicle.

STAY SAFE   Don't drive drunk.

SOURCES   AAA Foundation for Traffic, aaa.com/2021/03/heading-the-wrong-way-with-wrong-way-driving; NTSB, ntsb.gov/safety/data/Pages/Data_Stats.aspx

ATVs

ANNUAL US DEATHS   259 in 2019

In 2019, 259 people in the US were killed while riding ATVs on public roads. Three-quarters of these mishaps were single-vehicle only. Virtually all the dead were not wearing a helmet when they were killed.

ATVs are more likely than any other vehicle type to kill their occupant(s) in a rollover. Drunk driving is also especially common among ATV fatalities, with a blood alcohol concentration (BAC) of 0.08 percent or more in 46 percent of deadly mishaps and a BAC of 0.15 percent or more in 34 percent of all deaths.

A different study from 2016 found 591 ATV fatalities in the US that year, including those that occurred off public roads. About 15 percent of the dead were children under age 16.

STAY SAFE   Wear a helmet, don't drive drunk, and avoid rollovers. Passengers should never ride on ATVs designed for a driver alone. Young people need adequate adult supervision.

SOURCES   CPSC, cpsc.gov/Safety-Education/Safety-Education-Centers/ATV-Safety-Information-Center/Death-Associated-With-ATVs-by-State-; IIHS-HLDI, iihs.org/topics/fatality-statistics/detail/motorcycles-and-atvs

Bicycles

ANNUAL US DEATHS   712 (bicycle versus motor vehicle) and 377 (other bicycle mishaps) in 2019

In 2019, 712 bicyclists were killed in the US in accidents involving motor vehicles.

Another 377 people were killed in other bicycle mishaps, such as hitting another bicycle, hitting an obstruction, or going off a cliff or embankment.

This is an increase of 6 percent over 2018 and an increase of 37 percent since 2010, when only 793 total deaths were reported. The National Safety Council estimates the lifetime risk of a bicyclist being killed in an accident at 1 in 3,825.

STAY SAFE   Always wear a helmet. Make sure brakes and tires are in good condition. Equip the bike with visibility gear, including a helmet light, reflectors, blinking taillights and headlight, and tires with reflective sidewalls. Wear high-visibility apparel.

SOURCES   IIHS-HLDI, iihs.org/topics/fatality-statistics/detail/bicyclists; NHTSA, nhtsa.gov/road-safety/bicycle-safety

E-BIKES

ANNUAL US DEATHS   Average of 3.3

Electrically motorized bicycles (e-bikes) killed 10 people in the US from 2017 to 2019. This is a death rate averaging 3.3 people per year. The risks of e-bikes are similar to those of e-scooters and hoverboards (see those entries below). In addition to rider deaths, e-bike batteries have caused deadly fires, including 3 deaths in 2021 just in New York City.

STAY SAFE   Always wear a helmet, equip the e-bike with visibility devices, and obey all traffic safety laws. Follow the manufacturer's instructions for charging and storage of e-bikes.

SOURCES   CPSC, cpsc.gov, report, Micromobility Products-Related Deaths, Injuries, and Hazard Patterns: 2017–2019; NCCSIR,

Buses, Taxis, Limos

ANNUAL US DEATHS   About 50 (bus accidents); about 16 in crashes
and 30 in robberies and carjackings

Bus travel is an extremely safe way to get around, second only to airline travel for its low rate of death per trip or miles traveled.

About 50 people a year, both drivers and passengers, are killed in the US in crashes of buses of all kinds. Most of the drivers involved did not show any risky driving behaviors, such as speeding or driving drunk. The number of deaths per 100,000 bus crashes (45) is only 17.9 percent of the rate of death per 100,000 car crashes (251).

Approximately 16 taxi and limousine drivers are killed in crashes in the US every year. About twice that number are murdered in robberies or carjackings.

On September 23, 2005, a bus evacuating nursing home residents in the path of Hurricane Rita caught fire near Wilmer, Texas. Twenty-three out of 44 passengers died.

In March 2011, a speeding bus leaving a casino in Connecticut just before dawn swerved and collided with a metal pole that sliced off the top of the vehicle. Seventeen people were killed.

On October 6, 2018, a stretch limousine crashed in Schoharie, New York. In all, 20 died: the driver, all 17 passengers, and 2 pedestrians in a nearby parking lot.

This Metro DC bus T-boned a car on a Washington street.

STAY SAFE Ride only in buses that are properly maintained and operated by trained drivers. Be wary of budget bus lines.

SOURCES Federal Motor Carrier Safety Administration (FMCSA), fmcsa.dot.gov/safety/data-and-statistics/large-truck-and-bus-crash-facts; NHSTA, nhtsa.gov/fatality-analysis-reporting-system-fars/ trucks-fatal-accidents-tifa-and-buses-fatal-accidents-bifa

SCHOOL BUSES

ANNUAL US DEATHS Average of 28.1

School buses are the safest vehicles on the road. A child is much safer taking a bus to and from school than traveling by car. Although 4 to 6 school-age children die each year on school transportation vehicles, that's less than 1 percent of all traffic fatalities nationwide.

From 2007 through 2016, 281 school-age children were killed in motor vehicle accidents involving transportation to or from school or school-related activities. This is an average of 28.1 young people killed per year. Some died inside when the school bus crashed, while others were killed while walking on the street, riding a bike, or riding in another vehicle. In addition, 1,001 nonstudents were killed in accidents involving school buses during that same 10-year period.

In total, over the 10-year period, 1,282 deaths were related to school buses. Of the 118 people killed within school transportation vehicles, 50 were drivers and 68 were passengers. Of the 216 student pedestrians and bicyclists killed, 163 were hit by a school vehicle and 52 were hit by another vehicle. Among the nonstudents killed in accidents involving school vehicles, 902 were occupants of other vehicles; 46 were pedestrians or bicyclists.

Among school-age pedestrian fatalities, 33 percent were ages 5 through 7, 35 percent were ages 8 through 13, and 32 percent were age 14 or above.

STAY SAFE Never pass a stopped school bus showing a stop sign or flashing its lights. Children should wait for the bus at least 6 feet from the curb. Children should never walk behind a school bus and should cross the street in front of the bus by at least 10 feet. If a child drops something near the school bus, they should tell the driver right away. The child shouldn't try to pick it up, because the driver might not be able to see them.

SOURCE SA, crashstats.nhtsa.dot.gov, School-Transportation-Related Crashes

Construction and Earthmoving Vehicles

> **ANNUAL US DEATHS** Estimated 566

Construction and earthmoving vehicles, massively heavy and slow-moving, need to travel along public roads to reach their work sites. Then they need to move around the construction site, sometimes intruding again onto public sidewalks and/or roadways. This can lead to deadly accidents, involving them and their drivers, with other construction equipment, or with passing vehicles, bicyclists, or pedestrians.

Building sites are dangerous for many reasons. Occupational death statistics for the construction professions don't always list transportation (vehicle) accidents as a separate category. Besides the mishaps any vehicle is prone to, workers around construction sites might get killed by falling from a vehicle, being struck by a machine's moving arm or blade, or getting hit by something falling from a vehicle. Caught-betweens—getting crushed between a work vehicle and some obstruction—are another type of accidental death that is not always broken out separately.

Construction crane mishaps killed an average of 42 people per year in the US during the period 2011 through 2017. Of these, 13 percent, or about 5 or 6 people a year, died in transportation crashes.

From 1992 through 2007, 829 construction workers were killed in dump truck accidents in the US. This is an average of about 51 deaths per year.

Bulldozer accidents killed about 9 people per year. Three-fourths of these happened when the dozer rolled over while being loaded or unloaded from its carrying trailer.

Excavators killed about 57 people per year.

Loaders killed about 50 people per year.

Cement mixers killed about 357 people per year.

STAY SAFE Exercise extreme caution in and around heavy equipment at construction sites. Always wear a hard hat, safety boots, work gloves, and a high-visibility reflector vest. Pedestrians should stay under a sheltered walkway or cross the street. Drivers should obey flaggers and work zone speed rules.

SOURCES BLS, bls.gov/opub/mlr/2013/article/an-analysis-of-fatal-occupational-injuries-at-road-construction-sites-2003-2010.htm; CDC NIOSH, cdc.gov/niosh/docs/2011-119/default.html; OSHA, osha.gov/motor-vehicle-safety/construction

E-Scooters

Electric scooters killed 27 people in the US between 2017 and 2019, for an average of 9 deaths per year. Most fatal accidents occurred when the person riding the e-scooter hit or was hit by a motor vehicle, a pedestrian, or an obstruction. Automobile collision was the leading culprit. In addition, however, scooter brake failures were sometimes reported.

Jimi Heselden, who acquired the Segway company in 2009 from founder Dean Kamen, accidentally rode his personal transporter over a cliff and was killed in 2010. The Segway company ceased manufacturing personal transporter scooters in 2020.

STAY SAFE   Always wear a helmet and obey all traffic safety laws.

SOURCES   CPSC, cpsc.gov/Newsroom/News-Releases/2021/Injuries-Using-E-Scooters-E-Bikes-and-Hoverboards-Jump-70-During-the-Past-Four-Years; the *New York Times*, nytimes.com/2020/06/24/business/segway-pt-discontinued.html

Farm, Ranch, and Agricultural Vehicles

Motor vehicle mishaps account for 1 out of every 4 farmworkers killed by all causes on the job. Modern farm, ranch, and other agricultural business activities are heavily dependent on tractors, combine harvesters, utility terrain vehicles (UTVs), and other specialized motor vehicles. Their operation is especially hazardous because they sometimes lack road-legal safety features such as seat belts and rollover prevention systems. Their operators often work alone, unsupervised, far from good emergency medical care.

Recent data indicates that an overall annual US deaths from farm equipment crashes of 5.2 people per 100,000 full-time equivalent farmworkers. There were 2,038,000 FTE workers employed in US agricultural production in 2018, implying a death toll that year of about 106.

Tractor rollovers are the leading cause of fatal farmworker injuries. They killed an average of 96 people per year from 1992 to 2007. Agricultural UTV accidents, also primarily rollovers, killed about 15 people per year. The overall annual US deaths declined slightly from 2007 to 2018 due to safer vehicle design and farming practices.

An informative study of farm vehicle deaths, broken down by vehicle or machinery type, goes back to the early 1980s: Tractors accounted for

69 percent of all agricultural machinery fatalities, augers and elevators 3 percent, combines 2 percent, hay balers 2 percent, brush hogs and mowers 2 percent, loaders/skidders 2 percent, corn pickers 1 percent, and other or unknown 20 percent.

STAY SAFE Agricultural businesses should upgrade to tractors and other equipment that have modern safety features, especially seat belts and rollover prevention systems.

SOURCES BLS, bls.gov/opub/ted/2020/a-look-at-workplace-safety-in-agriculture.htm; CDC NIOSH, cdc.gov/niosh/oep/agctrhom. html; CPSC Farm Injury Resource Center, farminjuryresource.com/crashworthiness-rollovers

First Responder Vehicles

ANNUAL US DEATHS Estimated 92

Unfortunately for themselves and the public they serve while riding their ambulances, fire trucks, or police cars to the scene of an emergency, America's 2,000,000 first responders face the same risks of motor vehicle mishaps as do other drivers and passengers. Their flashing lights and blaring sirens don't grant full immunity from being involved in deadly crashes—especially because, on a call-out, they need to get there *fast*.

In 2017, 41 law enforcement officers were killed in line-of-duty police vehicle crashes. That same year, 18 firefighters died in traffic accidents involving their fire trucks. On average, 12 percent of on-duty firefighter fatalities occur each year while responding to or returning from incidents, with the majority of fatalities resulting from vehicle crashes. Vehicle collision is the 2nd leading cause of firefighter fatalities.

An average of 33 people were killed per year in the US in crashes involving ambulances. (The breakdown among ambulance crews, their patients/passengers, and other people is not available.) This adds up to about 4.6 deaths per 100,000 first responders.

STAY SAFE Constant safety training, and due caution and good teamwork on the job, can mitigate the inherent risks of these dangerous yet very rewarding, essential professions.

SOURCES EMS.gov report, A National Perspective on Ambulance Crashes and Safety; FEMA, usfa.fema.gov/operations/ops_vehicle.html; NHTSA, nhtsa.gov/crash-data-systems/fatality-analysis-reporting-system

Golf Carts

ANNUAL US DEATHS   Estimated 10 to 20 in 2020

Golf cart accidents injure an estimated 15,000 people in the US every year. People are killed, but hard data on the fatality rate is not available. Anecdotal news reports indicate that the number killed in 2020 was probably between 10 and 20. Types of fatal golf cart mishaps are similar to some for regular vehicles: hitting a pedestrian, an obstruction, or another cart or a car, causing a rollover or occupant ejection.

About 70 percent of all accidental golf cart injuries serious enough to need medical attention occur on golf courses or other sports fields, while 15 percent occur on public roads (mainly in Sun Belt retirement communities); 15 percent occur elsewhere. About 30 percent of all such injuries are to children.

Golf carts are potentially dangerous because they can attain considerable speeds but lack some of the safety equipment found in most other 4-wheeled vehicles, such as seat belts and side doors. In addition, their drivers might (wrongly) think that the activity they're engaged in is just a relaxing game of golf, not operating a potentially lethal motor vehicle. This can lead to carelessness.

STAY SAFE   Always exercise care when operating or riding in a golf cart. Never drive drunk, do not race them, do not overload them, do not make very sharp turns, and at all times stay on paved golf cart paths or roads.

SOURCES   J.E. Castaldo et al. Analysis of death and disability due to golf cart crashes in The Villages, Florida: 2011–2019. *Traffic Injury Prevention*, July 2020; CDC, cdc.gov/injury/wisqars/fatal.html; Texas Department of Insurance, tdi.texas.gov, Golf Cart Safety Fact Sheet

Hoverboards

ANNUAL US DEATHS   Average of 1.3

Hoverboard accidents killed 4 people between 2017 and 2019. This is an annual death toll of 1.3 people. Dangers are similar to those for e-scooters (see above). In addition, hoverboards sometimes suffer battery fires. In 1 Pennsylvania case in 2017, a hoverboard battery burst into flames, igniting a house and killing 2 children.

STAY SAFE   Always wear a helmet and obey all traffic safety laws. Follow manufacturer instructions carefully to avoid battery fires.

Landscaping (Groundskeeping) Vehicles

ANNUAL US DEATHS   Estimated 126

Landscaping (groundskeeping) is a dangerous line of work, comparable in risk of being killed on the job to being a farmer or a miner. About 200 people per year die in work-related accidents in the landscaping occupation. This is equivalent to annual US deaths of 25 per 100,000 FTE workers. A significant fraction of these death are vehicle-related.

About 20 groundskeepers per year are killed in highway traffic accidents while driving a motor vehicle as part of their job. An additional 16 per year are killed in off-road motor vehicle mishaps, mostly involving tractors or rider-mowers.

Many people who are not professional groundskeepers also take care of their lawns, gardens, trees, and properties on a DIY basis. They are exposed to the same risks of death as the professionals, such as being hit by a falling tree, falling from a tree, or being electrocuted by contact with a power line. About 90 people a year are killed in rider-mower accidents, many of them rollovers.

STAY SAFE   Follow all safety instructions from equipment manufacturers. Be cautious operating equipment on wet or unstable ground, on slopes or near embankment edges, and while making sharp turns, to avoid rollovers. Exercise extreme caution when cutting trees or branches, climbing in trees, working near overhead electrical lines, or using power tools and extension cords. Wear a hard hat, work boots, and a high-visibility safety vest. In hot weather, stay hydrated and avoid overexertion.

SOURCES   CDC NIOSH, cdc.gov/niosh/docs/2008-144/default.html; OSHA, osha.gov/riding-mowers

Lift Vehicles: Forklifts and Aerial Lifts (Cherry Pickers)

ANNUAL US DEATHS   Estimated 111

Forklifts, aerial lifts (cherry pickers), and other business/industrial cargo or personnel lift vehicles present many special safety challenges. They can expose lifted workers to electrocution or other hazards, as well as to being thrown or dropped from a height. Some, especially those designed for lighter lifts and more local transport, are powered by compressed propane or electric batteries (see **Alternative Fuel Vehicles**). Many lifts are operated both indoors, say in a warehouse or on a factory floor, as well as outdoors, such as in a busy trucking terminal or materials handling/storage yard, or at a construction site or next to a utility pole.

Aerial lift vehicles also frequently operate next to heavily trafficked public roads. All types of lifts sometimes shuttle through building exits and entrances, raising and carrying, then lowering and unloading their payloads, over and over again. All these work areas tend to be congested, with people and with other vehicles moving around. Aerial lift vehicles can have limited sight lines, due to stacks of goods and machinery needed for the work site's daily operations.

Forklifts and aerial lifts are also subject to the special risk of their lift mechanism failing catastrophically, as well as the danger inherent in having a high and shifting center of gravity, which can lead to vehicle tip-overs. Opportunities abound for lift vehicles, or their passengers, to hit someone, or touch or ram something dangerous such as a power line or live steam pipe, or for workers to fall from a height, sometimes fatally.

A study of data from 1992 through 1999 found that an average of 26 workers died each year from using cherry pickers and other personnel aerial lifts. About one-third of deaths were electrocutions, one-third were worker falls, and the remainder were primarily platform tip-overs, boom collapses, or vehicle run-overs or back-overs. Another study found that an average of 85 people per year are killed in cargo forklift accidents. The most frequent mishap, causing 24 percent of deaths, was a tip-over.

STAY SAFE Work-site safety training and supervision are essential, as are use of hard hats, harnesses, and other personal protective equipment. Care must always be taken while refueling/recharging. Never exceed your lift's weight limit, don't drive it with the platform raised, and exercise extreme caution near energized power lines and other overhead obstructions.

Forklifts often operate in confined factory and warehouse floor spaces. This one backed into a hole.

SOURCES CDC NIOSH, cdc.gov/niosh/injury; The Center for Construction Research and Training, cpwr.com/research/data-center/data-reports; McCue Safety Products, mccue.com/blog/forklift-accident-statistics; OSHA, osha.gov/powered-industrial-trucks

Mining and Logging Vehicles

ANNUAL US DEATHS   Estimated 354

In 2019, there were 24 mining fatalities from all causes. In 2020, there were 29 miner work-related deaths. About half of these fatalities were caused by a mine vehicle accident. With 330,000 miners employed, this was an annual vehicle-mishap death rate of just under 5 per 100,000 workers.

For the logging/forestry industry, with 50,000 employees nationwide, a total of 1,700 loggers were killed on the job between 2000 and 2018. About 20 percent were killed in vehicle mishaps. This is an annual vehicle mishap death rate of about 36 per 100,000 workers.

Logging and mining are dangerous occupations. Sometimes accidents with gigantic dragline excavators, immense dump trucks and bulldozers, or timber harvesters, forwarders, and stake flatbeds—highly specialized motor vehicles—can kill. (Across the entire US workforce, transportation incidents were by far the single largest category of work-related fatalities in recent years.)

The mining industry, working with the US Department of Labor's Mine Safety and Health Administration (MSHA), has in recent years worked with mine operators and workers to greatly improve safety. This has significantly reduced on-the-job deaths from all causes, especially from vehicle collisions and powered haulage accidents.

STAY SAFE   Always wear a seat belt.

SOURCES   BLS, bls.gov/iag/tgs/iag10.htm#fatalities_injuries_and_illnesses; CDC NIOSH, cdc.gov/niosh/mining/topics; Mine Safety and Health Administration, msha.gov/data-reports/fatality-reports/search

Motorcycles

ANNUAL US DEATHS   5,014 in 2019

Riding a motorcycle carries more risk than driving in an enclosed vehicle. Motorcycles are much less crashworthy and give their riders far less physical protection. They are harder for other drivers to see, and with only 2 or at most 3 wheels, they have less stability than vehicles with 4 or more wheels. In addition, operating a motorcycle calls for specialized skills, especially when roads are wet or icy.

In 2019, 5,014 people in the US died in motorcycle mishaps. One-third of these did not have a valid driver's license. This is twice the rate for fatal mishaps of passenger cars being operated without a license. One-third of motorcycle deaths occurred in single-vehicle mishaps, while two-thirds involved at least 1 other vehicle. Bike passenger deaths accounted for less than 6 percent of the overall motorcycle death toll; 90 percent of any passengers killed were female.

In 2018, the rate of death per 100 million vehicle miles traveled (VMT) for motorcyclists, a good measure of comparative risk, was 24.83, *almost 27 times* that for automobiles. Measured across motorcycle ownership instead of motorcycle-miles traveled, in 2018 the fatality rate was 57.52 per 100,000 registered motorcycles owned. This is about 2.5 times the fatality rate for other motor vehicles owned. The apparent discrepancy in relative risks—2.5 times versus 27 times—is because the average motorcycle is driven many fewer miles than the average car.

Expressed differently, motorcyclists rack up a mere 0.6 percent of total annual national VMT put in on the road, yet they account for fully 14 percent of all traffic deaths. One motorcycle accident is 4 times as likely to injure or kill its driver and any passenger than is a car or SUV accident. The National Safety Council estimates the lifetime odds of a motorcyclist dying in a crash at 1 in 899.

STAY SAFE   Take a motorcycle safety class. Always wear a good helmet, even if not required by law in your jurisdiction, and make sure any passenger you carry wears their helmet, too. Obey all traffic laws, and never drive drunk, drugged, or distracted.

SOURCES   IIHS-HLDI, iihs.org/topics/fatality-statistics/detail/motorcycles-and-atvs; III, iii.org/fact-statistic/facts-statistics-motorcycle-crashes; NHTSA, nhtsa.gov/road-safety/motorcycles; NSC, injuryfacts.nsc.org/motor-vehicle/road-users/motorcycles/

ALCOHOL/DRUG USE

In 2019, a driver whose blood alcohol concentration (BAC) was above the legal limit of 0.08 percent was involved in 29 percent of all fatal motorcycle crashes. In fully 18 percent of all fatal motorcycle crashes, a driver BAC of at least 0.15 percent (*extremely* intoxicated) was involved.

Reliable data on drug involvement in fatal motorcycle crashes is hard to come by, for the same reasons as discussed under **Automobiles**. As autopsies show time and again, many drunk drivers are simultaneously also under the influence of 1 or more recreational or prescription drugs.

STAY SAFE   Take a motorcycle safety class. Always wear a good helmet, even if not required by law in your jurisdiction, and make sure any passenger you carry wears their helmet, too. Obey all traffic laws, and never drive drunk, drugged, or distracted.

SOURCES   IIHS-HLDI, iihs.org/topics/fatality-statistics/detail/motorcycles-and-atvs; III, iii.org/fact-statistic/facts-statistics-motorcycle-crashes

BY AGE

In 2019, US motorcycle fatalities broken down by age of victim were 27 percent for ages under 29, 19 percent for ages 30 to 39, 16 percent for ages 40 to 49, and 38 percent for over age 50.

STAY SAFE   Take a motorcycle safety class. Always wear a good helmet, even if not required by law in your jurisdiction, and make sure any passenger you carry wears their helmet, too. Obey all traffic laws, and never drive drunk, drugged, or distracted.

SOURCES   IIHS-HLDI, iihs.org/topics/fatality-statistics/detail/motorcycles-and-atvs; III, iii.org/fact-statistic/facts-statistics-motorcycle-crashes

BY SEX

In 2019, 91 percent of all US motorcycle accident fatalities were male; 9 percent were female. For the males, 99 percent were the driver and only 1 percent were a passenger. For the females, 40 percent of those killed were the driver and 60 percent were a passenger. Put differently, 90 percent of the passengers killed were female, while 96 percent of the drivers killed were male.

STAY SAFE   Take a motorcycle safety class. Always wear a good helmet, even if not required by law in your jurisdiction, and make sure any passenger you carry wears their helmet, too. Obey all traffic laws, and never drive drunk, drugged, or distracted.

SOURCES   IIHS-HLDI, iihs.org/topics/fatality-statistics/detail/motorcycles-and-atvs; III, iii.org/fact-statistic/facts-statistics-motorcycle-crashes

MONTH OF YEAR

In 2019, US motorcycle accident fatalities broke down by month as follows:

January	3 percent
February	4
March	7
April	8
May	12
June	13
July	13
August	13
September	12
October	8
November	5
December	3

The death toll was concentrated in the middle months of the calendar year, which have the warmest weather, when more motorcycle trips are made and more miles are traveled.

STAY SAFE Take a motorcycle safety class. Always wear a good helmet, even if not required by law in your jurisdiction, and make sure any passenger you carry wears their helmet, too. Obey all traffic laws, and never drive drunk, drugged, or distracted.

SOURCES IIHS-HLDI, iihs.org/topics/fatality-statistics/ detail/motorcycles-and-atvs; III, iii.org/fact-statistic/ facts-statistics-motorcycle-crashes

A biker down with an injury to the shoulder and neck

TIME OF DAY

In 2019, US motorcycle accident deaths broke down by time of day as follows:

Midnight to 3 a.m.	6 percent
3 a.m. to 6 a.m.	3
6 a.m. to 9 a.m.	8
9 a.m. to noon	8
Noon to 3 p.m.	19
3 p.m. to 6 p.m.	24
6 p.m. to 9 p.m.	17
9 p.m. to midnight	13

Compared to the time of day of motor vehicle deaths overall, motorcycle deaths were more likely to occur in the afternoon and less likely to occur at night.

STAY SAFE   Take a motorcycle safety class. Always wear a good helmet, even if not required by law in your jurisdiction, and make sure any passenger you carry wears their helmet. Obey all traffic laws, and never drive drunk or drugged.

SOURCES   IIHS-HLDI, iihs.org/topics/fatality-statistics/ detail/motorcycles-and-atvs; III, iii.org/fact-statistic/ facts-statistics-motorcycle-crashes

WEEKDAY VERSUS WEEKEND

In 2018, 51.9 percent of US motorcycle deaths occurred on a weekday, and 48.1 percent occurred on weekends. Thus weekends, which comprise 28.6 percent of the days of the week (2 out of 7), account for 70 percent more than their fair share of a week's total fatalities.

STAY SAFE   Take a motorcycle safety class. Always wear a good helmet, even if not required by law in your jurisdiction, and make sure any passenger you carry wears their helmet, too. Obey all traffic laws, and never drive drunk, drugged, or distracted.

SOURCES   IIHS-HLDI, iihs.org/topics/fatality-statistics/ detail/motorcycles-and-atvs; III, iii.org/fact-statistic/ facts-statistics-motorcycle-crashes

CLASS OF ROAD AND URBAN VERSUS RURAL

In 2019, US motorcycle accident deaths broke down by class of road being traveled as follows:

Interstates and freeways	12 percent
Other major roads	54
Minor roads	34

Broken down differently, the split between urban and rural locations for fatal crashes was 62 percent urban and 38 percent rural.

STAY SAFE Take a motorcycle safety class. Always wear a good helmet, even if not required by law in your jurisdiction, and make sure any passenger you carry wears their helmet, too. Obey all traffic laws, and never drive drunk, drugged, or distracted.

SOURCES IIHS-HLDI, iihs.org/topics/fatality-statistics/ detail/motorcycles-and-atvs; III, iii.org/fact-statistic/ facts-statistics-motorcycle-crashes

HELMET AND PROTECTIVE CLOTHING USE

A 2017 survey found that about two-thirds of motorcyclists used a helmet that complied with the Department of Transportation (DOT) safety standards. When going along high-speed roads such as highways or freeways, or riding in heavy traffic, helmet use increased to 80 or 90 percent of all motorcyclists.

Wearing a helmet reduces the death rate for motorcyclists involved in accidents by about 40 percent. This statistic understates the importance of wearing a helmet, because helmet use also significantly reduces the severity of nonfatal accident injuries, especially permanent brain and neck/spine damage. An Australian study showed that wearing motorcycle pants, jacket, boots, and/or gloves all greatly reduced the need for hospital admissions after crashes.

STAY SAFE Take a motorcycle safety class. Always wear a good helmet, even if not required by law in your jurisdiction, and make sure any passenger you carry wears their helmet, too. Obey all traffic laws, and never drive drunk, drugged, or distracted.

SOURCES IIHS-HLDI, iihs.org/topics/fatality-statistics/ detail/motorcycles-and-atvs; III, iii.org/fact-statistic/ facts-statistics-motorcycle-crashes

TYPE OF MOTORCYCLE

In 2019, a breakdown of motorcycle accident fatalities by type of motorcycle was as follows:

Cruiser or standard	31.3 percent
Touring	21.1
Sport touring	0.8
Sport	11.7
Supersport	21.1
Off-road	1.3
Other or not reported	12.7

Within this data, drivers over age 30 accounted for 82 percent of deaths for cruiser or standard bikes, and 93 percent for touring bike deaths, but only 48 percent for supersport deaths.

STAY SAFE   Take a motorcycle safety class. Always wear a good helmet, even if not required by law in your jurisdiction, and make sure any passenger you carry wears their helmet, too. Obey all traffic laws, and never drive drunk, drugged, or distracted.

SOURCES   IIHS-HLDI, iihs.org/topics/fatality-statistics/detail/motorcycles-and-atvs; III, iii.org/fact-statistic/facts-statistics-motorcycle-crashes

Engine Size

In 2019, a breakdown of engine size in fatal motorcycle accidents was as follows:

Under 1,000 cc	46 percent
1,001 to 1,400 cc	14
Over 1,400 cc	34

Within this data, 92 percent of touring bikes were in the most powerful class, while almost all supersports were in the least powerful class.

STAY SAFE   Take a motorcycle safety class. Always wear a good helmet, even if not required by law in your jurisdiction, and make sure any passenger you carry wears their helmet, too. Obey all traffic laws, and never drive drunk, drugged, or distracted.

SOURCES   IIHS-HLDI, iihs.org/topics/fatality-statistics/
detail/motorcycles-and-atvs; III, iii.org/fact-statistic/
facts-statistics-motorcycle-crashes

DIRT BIKES

ANNUAL US DEATHS   Not broken out from motorcycle deaths

A dirt bike is a lightweight motorcycle with a low-power engine, intended for driving on dirt roads and rough terrain and equipped with special knobby tires. They can be used for trail riding and/or to compete in motocross and supercross events. Splashing through mud ("mudding") is favored by many dirt bike riders.

Annual national death statistics are included in the broader category of all motorcycles with engine displacement under 1,000 cc. A study for 2003 showed that 245 people age 18 or younger were killed that year in dirt bike accidents. Many of those killed were not wearing helmets. Since then, many states have passed helmet laws and laws setting a minimum age for operating a dirt bike and requiring adult supervision.

Dirt bikes are not as lethal in accidents as are ATVs, in part because the latter are 4-wheel vehicles that weigh much more than dirt bikes and can go faster. One study found that people who were injured in a mishap had a roughly 50 percent greater chance of dying if they had been riding an ATV compared to a dirt bike.

STAY SAFE   Obey laws about minimum operator age and adult supervision. Always wear a helmet and protective clothing. Take a dirt bike safety course. If choosing between an ATV and a dirt bike for off-road activities, be aware that dirt bikes are about two-thirds less lethal in a bad accident.

SOURCE   IIHS-HDLI, iihs.org/topics/fatality-statistics/detail/
motorcycles-and-atvs#motorcycle-type-and-engine-size

MOPEDS

ANNUAL US DEATHS   Not broken out from motorcycle deaths

Mopeds are a type of motorcycle with a low-power engine (under 250 cc engine displacement for a 2-person moped, and under 50 cc for a single-seater), but various models can hit top speeds of 30 to 60 mph—and kill people in mishaps. Moped deaths in the US are included in motorcycle deaths in the under-1000 cc engine category, which also includes many more powerful true motorcycle makes and models.

A study in Finland found an annual death rate of 7.5 per 100,000 people due to moped crashes, implying moderate risk in the US, too, although moped ownership and use are low here.

STAY SAFE Mopeds are not toys. Always wear a helmet, drive safely, never operate a moped while drunk or impaired, and respect the speeds they can attain. Be especially cautious off roads, on unpaved, muddy, rough, or rugged terrain.

SOURCE N. Airaksinen et al. Comparison of injury severity between moped and motorcycle crashes: a Finnish two-year prospective hospital-based study. *Scandinavian Journal of Surgery*, March 2016.

Pedestrians

ANNUAL US DEATHS 6,205 in 2019

In 2018, 6,227 pedestrians were killed in traffic in public thoroughfare areas in the US, or about 1 death every 85 minutes. Approximately 3,000 people on foot were killed in parking lots, in driveways, or on private roads. About 29 percent of pedestrians killed in car accidents in 2016 were hit because the vehicle's driver did not properly yield the right of way at a crosswalk, stop sign, or traffic light.

Per trip, pedestrians are 1.5 times more likely than passenger vehicle occupants to be killed in a car crash. The lifetime odds of dying in a pedestrian incident (car crash or something else) are 1 in 543.

Almost half (47 percent) of crashes that resulted in a pedestrian death involved alcohol for the driver and/or the pedestrian. One in every 3 fatal pedestrian crashes involved a pedestrian with a blood alcohol concentration of at least 0.08, and 17 percent involved a driver with a blood alcohol concentration of at least 0.08. Some fatal pedestrian crashes involved both.

In 2017, pedestrians aged 65 and older accounted for 20 percent of all pedestrian deaths. One in every 5 children under the age of 15 killed in traffic crashes were pedestrians.

American roads are designed for traffic speed, not pedestrians, leading to many intersections where high-speed roads cross. A pedestrian hit by a car going 23 mph has a 10 percent chance of dying; at 32 mph, the chance of dying is 25 percent; at 50 mph, the chance of dying is 75 percent.

STAY SAFE   Look both ways before crossing the street. Cross streets at designated crosswalks or intersections whenever possible. Walk on a sidewalk or path instead of the road. If you must walk on the shoulder, walk facing traffic. Don't walk while wearing earbuds or looking at a phone. Wear high-visibility clothing. At night, carry a flashlight and wear reflective clothing. Work in your community to implement lower speed limits, speed bumps, and other traffic-calming measures in pedestrian areas.

SOURCES   AAA Foundation for Traffic Safety, B.C. Tefft et al. *Examining the Increase in Pedestrian Fatalities in the United States*, 2009–2018 (Research Brief); FHWA, safety.fhwa.dot. gov/intersection; NHTSA, nhtsa.gov/crash-data-systems/fatality-analysis-reporting-system

CHILDREN

ANNUAL US DEATHS   4,074 in 2016

Children are especially vulnerable to being killed in a motor vehicle mishap. They are hard for drivers to see. Because they're small, they're more vulnerable to deadly injury. They tend to play carelessly and run into traffic.

The rate of motor vehicle crash deaths per million children younger than 13 has decreased 80 percent overall since 1975. Safer vehicles and seat belt and child restraint laws are very effective, but crashes still cause 1 of every 4 unintentional injury deaths among children. Most crash deaths occur among children traveling as passenger vehicle occupants. Restraining children in rear seats instead of front seats reduces fatal injury risk by about three-quarters for children up to age 3, and almost half for children ages 4 to 8.

In 2019, 73 percent of child motor vehicle crash deaths were passenger vehicle occupants, 16 percent were pedestrians, and 4 percent were bicyclists. Child pedestrian and bicyclist deaths have declined by 92 and 93 percent, respectively, since 1975. In 2018, 52 children died in overheated vehicles due to heat stroke. About 83 children under the age of 5 are killed per year when a motor vehicle inadvertently backs over them. Over 60 percent of those deaths are caused by an SUV or truck.

In 2016, a total of 4,074 children and teenagers died in motor vehicle mishaps in the US.

STAY SAFE   Always watch out for children before starting your car and while driving. Protect young children in properly installed child car seats. Older children should be properly restrained in the back seat. Never leave a child alone in a parked car, especially on a hot day.

SOURCES   CDC, cdc.gov/transportationsafety/child_passenger_safety/cps-factsheet; IIHS-HDLI, iihs.org/topics/fatality-statistics/detail/children

RVs and Camper Vans

> **ANNUAL US DEATHS**   Average of 26

RV road accidents kill an average of 26 people in the US every year. This equates to an average fatality rate of 0.44 per 100 VMT, which is only one-third the death rate across all motor vehicles.

RVs do present special risks. These vehicles have unusual driving and handling characteristics, yet do not require special training or permits to operate them. They are very large and heavy, are less maneuverable, have long braking distances, and have blind spots. Their high sides and high centers of gravity make them prone to wind blow-overs and other rollovers. They might be further imbalanced by carrying a heavy load on the roof and/or by towing a trailer or car. They can also be difficult to evacuate in case of an onboard fire.

STAY SAFE   Choose an RV that meets Federal Motor Vehicle Safety Standard 208. Take an RV driving and safety course. Be aware of the height and weight of your vehicle. Be wary of and cautious about the unusual handling characteristics. Carry a fire extinguisher in the kitchen area, and be especially careful while cooking, smoking, or using any open flame.

SOURCES   FEMA, usfa.fema.gov/data/statistics/reports/snapshot-rv-fires.html; NHTSA, nhtsa.gov/crash-data-systems/fatality-analysis-reporting-system

Sanitation and Snow Removal Vehicles

> **ANNUAL US DEATHS**   107 in 2017

Jurisdictions ranging from towns to cities to counties to states, and many private companies, operate different types of specialized vehicles whose crews work to clean up after us and after snowstorms.

In 2017 in the US, 107 garbage truck and recycling vehicle crew members were killed on the job in motor vehicle mishaps. This was equivalent to a death rate of 33 per 100,000 workers, making it a very dangerous occupation. Hard data on deaths in snowplow and street-sweeper accidents is scarce, but anecdotal information shows only a handful of workers, and of other motorists or pedestrians, are killed in accidents with these vehicles every year.

STAY SAFE Motorists should exercise courtesy, care, and patience whenever driving behind or passing any sanitation vehicle, street cleaning vehicle or snowplow especially in bad weather. They are extremely heavy and have blind spots, dangerous working parts, and long braking distances. Do not pass a snowplow on a highway.

SOURCES BLS, bls.gov/ooh/building-and-grounds-cleaning/grounds-maintenance-workers.htm; CDC NIOSH, cdc.gov/niosh/docs/97-110/default.html; NHTSA, nhtsa.gov/crash-data-systems/fatality-analysis-reporting-system

Snowmobiles

ANNUAL US DEATHS Estimated 200

Based on states reporting snowmobile data, every year in the US, about 200 people are killed in snowmobile accidents. The leading causes are speeding, alcohol impairment, inexperience, and poor judgment.

STAY SAFE When driving a snowmobile, stick to stable and even terrain, exercise care appropriate to your level of experience, dress appropriately for very cold/wet weather conditions, and bring emergency survival supplies. Plan your route in advance, let someone else know where you are going, and carry communications gear to call for help in wilderness areas. Do not speed, and never drive drunk. Take a snowmobile safety course.

SOURCE NHTSA, nhtsa.gov/crash-data-systems/fatality-analysis-reporting-system

It is difficult to enforce speed limits on snowmobiles.

Steel, Oil, and Gas Industry Vehicles

ANNUAL US DEATHS Estimated 60

Specialized transportation vehicles are used to extract the raw materials from the ground and to service associated stationary machinery, then move the very hot or volatile products to and within plant and refinery sites. These vehicles also build up and service furnace coke stockpiles and slag waste dumps, endless pipelines, and huge liquid and gas storage tanks. Workers can get killed in motor vehicle accidents at the extraction sites or around the mills and cracking towers and while providing support services or out on the highways.

About 100,000 people are employed at blast furnaces and steel mills in the US. Estimates put the annual occupational death toll from all causes at about 50. Some of these deaths are due to motor vehicle accidents. Blast furnaces and steel mills use specialized heavy vehicles to transport intensely hot molten metal in heat-resistant tubs. They use reinforced flatbeds to move big cast ingots still so hot that they glow brightly even in broad daylight. A breakdown of iron and steel industry fatalities by exact cause is not available. The overall annual US deaths is about 50 per 100,000 workers.

About 450,000 people are employed in the oil and gas extraction and support industry in the US. In 2014, on-site cranes, forklifts, and winch trucks were involved in 12 fatal accidents; roadway vehicle incidents killed another 18 oil and gas workers. This death toll of 30 equates to an annual motor vehicle occupational accident death rate of 6.7 per 100,000 workers.

STAY SAFE Always use safety equipment, including seat belts, and obey all company safety procedures. Always get enough sleep, and never drive drunk or impaired.

SOURCES CDC NIOSH, cdc.gov/niosh/about/strategicplan/trauoil. html; OSHA, osha.gov/oil-and-gas-extraction

Trucks

ANNUAL US DEATHS 4,119 in 2019

In 2019 in the US, 4,119 people were killed in accidents involving large trucks, 11 percent of all deaths on the road that year. Seventy-four percent of all large-truck accident deaths involved tractor trailers of varying lengths and total weights, some of them double-articulated (hauling two trailers). These big rigs and their cargos vary, from refrigerated trailers of frozen foodstuffs, to tankers of hazardous fuels or milk, to boxy trailers filled with all sorts of manufactured goods, to car transporters carrying 12+ autos or SUVs, to hoppers of grain or sand, to flatbeds loaded with steel I-beams.

In 2019, American truckers and trucks carried 12 billion tons of cargo, fully 72.5 percent of all freight in the US. At $800 billion per year of freight billing revenue, trucking employs 3.5 million drivers. It can be a dangerous profession when proper safety protocols are neglected.

Large trucks can weigh 20 times or more than the average car. This gives them significantly longer braking distances, with huge kinetic energy and impact force in a mishap. Big rigs have blind spots, make very wide turns, and handle very differently from cars, SUVs, and light trucks. Their high road clearances, big side areas, and articulated structures also make them vulnerable to accidents that most other vehicles can avoid: jackknifings, blow-overs, drive-unders, lost, shifting, or leaking cargo, multi-axle brake failure, and catastrophic tire blow-outs or even tire fires. The portion of all fatal large-truck crashes that involves rollovers is 47 percent, which is twice the percentage of rollovers among fatal crashes involving all motor vehicles combined. Four percent of large trucks involved in fatal mishaps (1 in every 25) were carrying hazardous materials, such as combustible liquids, explosives, inflammable solids, acids, or radioactive waste.

Of the 2019 fatalities, 16 percent were truck occupants, 67 percent were in other vehicles, and 15 percent were motorcyclists, bicyclists, or pedestrians. This data implies an annual occupational death rate of 19 truckers per 100,000 employed. There has been an upward trend recently, with the overall death toll from large-truck crashes rising 31 percent from 2009, while the number of truck occupants killed in crashes rose 51 percent.

SOURCES IIHS-HDLI, iihs.org/topics/fatality-statistics/detail/large-trucks; NHTSA, nhtsa.gov/crash-data-systems/fatality-analysis-reporting-system; OSHA, osha.gov/trucking-industry

CLASS OF ROAD, TIME OF DAY, DAY OF WEEK, DRIVER GENDER

Because large trucks are commercial vehicles, the patterns of their fatal mishaps differ from those of passenger vehicles. More big-rig crashes happen during business hours on weekdays.

By class of road, 32 percent of fatal crashes involving large trucks occurred on interstates and freeways, 52 percent on other major roads, and 15 percent on minor roads.

By time of day, 31 percent of mishap deaths involving large trucks occurred between 6 a.m. and noon, compared to only 18 percent for other vehicle deaths, while only 18 percent of large-truck crashes occurred between 6 p.m. and midnight, compared to 34 percent for other vehicle deaths.

The most dangerous day of the week for fatal large-truck crashes was Thursday. Only 25 percent of all such mishaps occurred on weekends, contrasted with 34 percent of other types of fatal motor vehicle accidents happening on weekends.

Women make up 6 percent of all US large-truck drivers, but women large-truck drivers are involved in only 3 percent of all fatal crashes.

STAY SAFE   Respect the huge weight disparity, blind spots, longer braking distances, and wider turning radii of large trucks. Truckers should always wear seat belts, not speed, never drive under the influence, and always get enough rest between shifts. Trucking companies should maintain and never overload their vehicles.

On the Job

> **ANNUAL US DEATHS**   5,333 in 2019

In 2019 in the US, 5,333 people died on the job, in a national workforce totaling about 152,400,000 people. This was an annual US deaths due to occupational fatalities of 3.5 per 100,000 full-time equivalent (FTE) employees. About 2.8 million more people were injured on the job seriously enough to need some medical attention and/or to lose time from work.

The US Bureau of Labor Statistics (BLS) gathers data that unbundles the overall occupational death rate into subcomponents, regarding number of FTE workers and number of deaths, in 2 ways: Standard Occupational Classification (SOC) codes, and North American Industry Classification System (NAICS) codes. In published reports, jobs or industries are grouped together when they are deemed comparable as to the types of tasks performed, the types of environment(s) in which they are performed, and the types of equipment and materials involved; such similarities mean their risks of fatal occupational injuries are also comparable. These groupings of like with like help produce more statistically credible information, smoothing out the noise that can come from overly small sample cells. The groupings also compress the myriad line items that reports to the public would otherwise contain. (The 2 coding systems have some overlap, and the same job or industry might be listed in both but be treated differently.)

The SOC looks at what specific type of work task a person might do, such as be a manager or a motor vehicle operator, while the NAICS look at what specific type of industry or business their employer is in, such as textile production or publishing. The various **On the Job** entries below look at death risk along 1 or the other of these 2 dimensions of the US economy. Both are needed to understand the relative risks of death when comparing different sorts of jobs, at a desk, in a factory, or in the field, in different industries, such as resource extraction, finished goods, knowledge workers, and personal services.

People who die while at their place of work, but from natural causes not specifically related to their job activities, are excluded from the BLS tally of occupational fatalities. For instance, someone who dies of a heart attack or a stroke in their office is not considered an occupational fatality—even if that heart attack or stroke was brought on, in part, by the stress of their job.

Note also that the various **On the Job** entries give the death rates just for workers in that industry. Death rates that also include visitors/customers/clients who get killed while patronizing these industries are included in the entries for **On the Go** or **At Play**. For instance, the **On the Job** death statistics for railroads includes only railroad employees; the **On the Go** death statistics for railroads includes employees, passengers, authorized visitors, and trespassers. **On the Go** for cement mixers includes everyone killed in a cement mixer crash, whether on a public road or at a construction site; **On the Job** for construction includes only the drivers and site workers.

STAY SAFE　See the entries below for the different related job subgroups.

SOURCE　BLS, bls.gov/data/#injuries

The Role of OSHA

The US Occupational Safety and Health Administration (OSHA), along with the many state-level counterparts, provides information, training, inspections, and enforcement of measures to reduce the incidence and severity of workplace injuries and occupational illnesses and deaths. OSHA has an annual budget of almost $600 million. Together, OSHA and the states employ almost 20,000 work site inspectors.

OSHA continually promulgates and updates standards of all sorts for on-the-job safety and health. According to the over 33,000+ inspection reports that compliance officers, tasked to monitor over 8 million work sites, filed in fiscal year 2019, the 10 most frequently cited violations of these standards were as follows:

Fall protection, in construction

Hazard communication standards (signage, etc.), general industry

Respiratory protection, general industry

Scaffolding requirements, construction

Ladders, construction

Hazardous energy (electricity, live steam, compressed air, intense heat), general industry

Heavy motor vehicles, general industry

Fall protection, training requirements

Eye and face protection

Machinery and machine guarding, general requirements

OSHA's efforts, though seemingly an expense for American taxpayers and for businesses and industries across the country, have in fact yielded very significant benefits in lives not sacrificed and in income (to worker families) and work hours (to employers) not lost. Over the past 50 years, despite significant growth in the US population and expansion of gross domestic product (GDP), on-the-job deaths have been cut by 60 percent, and nonlethal occupational injuries and illnesses have been cut by 75 percent.

SOURCE OSHA, osha.gov/data

Top 10 Deadliest Occupations

ANNUAL US DEATHS See data in entry.

The 10 occupations in the US with the worst on-the-job mortality, as measured by 2019 deaths per 100,000 FTE workers, were:

Fishing and hunting	145.0 per 100,000
Logging	68.9
Aircraft cockpit crews	61.8
Roofers	54.0
Construction trade helpers	40.0
Refuse and recyclables collectors	35.2
Truck drivers and traveling salespeople	26.8
Structural iron and steelworkers	26.3
Farmers, ranchers, agricultural	23.2
Grounds maintenance	19.8

STAY SAFE See individual job entries.

SOURCE BLS, bls.gov/iif/oshcfoi1.htm

Deaths by Employed-by-Another versus Self-Employed

ANNUAL US DEATHS See data in entry.

In 2019 in the US, overall deaths on the job broke down as 4,240 "employed-by-another" (wage and salaried workers) versus 1,093 "self-employed."

How many people were really self-employed, among all Americans who had jobs at all, is vexed by data classification issues. Estimates of de facto self-employment range from 10 percent to 30 percent of the overall national workforce. One reason for this uncertainty is that the US Bureau of Labor Statistics counts as "employed-by-another" anyone who works for an incorporated business that they themselves own—even though, in reality, since they own their own business and are often the only employee, they would generally be considered to be self-employed.

STAY SAFE   See individual job entries.

SOURCE   BLS, bls.gov/iif/oshcfoi1.htm

Deaths by Age

ANNUAL US DEATHS   See data in entry.

In 2019 in the US, overall deaths on the job broke down by age range as follows:

Under 16	17
16 to 17	17
18 to 19	50
20 to 24	325
25 to 34	866
35 to 44	967
45 to 54	1,082
55 to 64	1,212
65 or over	793

While overall (population mortality) death rates are well known to vary by age, it is not possible to give reliable age-specific occupational death rates, because the denominator of this fraction, which needs an accurate breakdown of FTE worker counts by age, is not readily available.

STAY SAFE   See individual job entries.

SOURCE   BLS, bls.gov/iif/oshcfoi1.htm

Deaths by Race/Ethnicity

ANNUAL US DEATHS   See data in entry.

In 2019 in the US, overall deaths on the job, estimated FTE workers, and implied annual race-specific death rate broke down by race or ethnicity as follows:

	DEATHS	WORKERS	EST. DEATH RATE PER 100,000 FTE
White	3,297	97.7 million	3.4
Black	634	21.2	3.0
Latino	1,088	27.7	3.9
Native American	30	1.6	1.9
Asian	181	9.8	1.8
Pacific Islander	14	1.6	0.9
Multiracial	22	3.2	0.7
Race not reported	67	—	—

The 2019 US workforce of 162,800,000 FTE people was about 60 percent White, 17 percent Latino, 13 percent Black, 6 percent Asian, 1 percent Native American, 1 percent Pacific Islander, and 2 percent multiracial.

The race-specific occupational death rates are merely illustrative estimates. Relative comparisons are less meaningful due to distortions from different demographic and occupational distributions within each of the different race/ethnicity groups. In addition, small death counts reduce statistical credibility for the derived annual US deaths for smaller ethnic/racial groups.

STAY SAFE   See individual job entries.

SOURCE   BLS, bls.gov/iif/oshcfoi1.htm

Deaths by Sex

ANNUAL US DEATHS   4,896 (males) and 437 (females) in 2019

In 2019 in the US, overall deaths on the job broke down as 4,896 males and 437 females. The total workforce broke down as 53.2 percent males and 46.8 percent females. The total workforce of 152,400,000 FTE workers thus split as 81,077,000 males and 71,323,000 females.

This gives the sex-specific occupational death rate per 100,000 as

The demands of World War II brought many women onto the factory floor.

6.0 for males and 0.61 for females. This disparity, as with other sex-specific death rate comparisons, is generally because men tend to get more physically dangerous jobs to begin with, and men as a group take more risks, act more aggressively, and do more things while impaired by drugs and/or alcohol.

STAY SAFE   See individual job entries.

SOURCE   BLS, bls.gov/iif/oshcfoi1.htm

Deaths by Broad Cause

ANNUAL US DEATHS   See data in entry.

There are many ways that someone might get killed on the job. The US Bureau of Labor Statistics groups all reported occupation-related and workplace fatalities according to 6 broad categories. These categories, and the total number of American worker deaths in each, are as follows:

Transportation incidents	2,122
Falls, slips, trips	880
Workplace violence	841
Contact with objects or equipment	732
Exposure to toxins or extreme environments	642
Fires and explosions	99

Each of the above 6 broad categories, other than fires and explosions (which is self-explanatory), is further broken down into several sub-categories, as appropriate. See the next few **On the Job** entries for deaths split out into the various related subcategories.

Death rates are not given for each of these separate subcauses of occupational death (let alone for the even smaller sub-subcauses below), because detailed death-count breakdowns (the denominators of the fractions) themselves suffer from the mathematical problem of increasing statistical noise due to smaller and smaller sample sizes, plus the practical problem of coding errors and ambiguities when finely subdividing aggregate reported data.

STAY SAFE   See individual job entries.

SOURCE   BLS, bls.gov/iif/oshcfoi1.htm

Contact with Objects or Equipment

ANNUAL US DEATHS   See data in entry.

In 2019 in the US, the rate of death on the job from contact with objects or equipment (struck, caught, compressed, crushed) was 0.48 per 100,000 FTE workers. The deaths in this broad cause of death broke down approximately as follows:

Struck by falling object or equipment	241
Struck by nontransport vehicle	205
Caught in running equipment or machinery	93
Crushed in collapsing structure, equipment, or material	83
Caught in or compressed by equipment or object	27
Struck by discharged or flying object	26
Other/unspecified	57

STAY SAFE   See individual job entries.

SOURCE   BLS, bls.gov/iif/oshcfoi1.htm

WOOD CHIPPERS

ANNUAL US DEATHS   Average of 3

Wood chippers are a good example of the many ways a worker can be killed from contact with objects and equipment. Between 1982 and 2016, 113 workers were killed in mobile hand-fed wood chipper–related incidents. While performing chipper-related tasks, the victims were struck by in 57 cases, caught in in 41 cases, hit by in 7 cases, electrocuted by in 4 cases, fell in in 2 cases, and had heat stroke in 2 cases.

STAY SAFE Follow all safety protocols while operating a wood chipper. Don't get caught in the feed wheel control device.

SOURCE J. Zhu and K. Gelberg. Occupational fatal injuries associated with mobile hand-fed wood chippers. *American Journal of Industrial Medicine*, October 2018.

Industrial smoke and fumes make for a toxic workplace.

Exposure to Toxins or Extreme Environments

ANNUAL US DEATHS See data in entry.

In 2019 in the US, the rate of death on the job from exposures to toxins or extreme environments was 0.42 per 100,000 FTE workers. The deaths in this broad cause of death broke down approximately as follows:

Toxic exposures: inhaled	320
Electrocution	166
Toxic exposures: other	59
Extreme heat or cold	53

STAY SAFE See individual job entries.

SOURCE BLS, bls.gov/iif/oshcfoi1.htm

Falls, Slips, and Trips

ANNUAL US DEATHS See data in entry.

In 2019 in the US, the rate of death on the job from falls, slips, and trips was 0.58 per 100,000 FTE workers. The deaths in this broad cause of death broke down approximately as follows:

Fall to lower level	479
Fall on same level	146
Fall through surface or opening	95
Fall due to structure/equipment collapse	37

STAY SAFE   See individual job entries.

SOURCE   BLS, bls.gov/iif/oshcfoi1.htm

Transportation Incidents

ANNUAL US DEATHS   1.4 per 100,000 FTE workers

In 2019 in the US, the rate of death on the job from transportation incidents was 1.4 per 100,000 FTE workers. The deaths in this broad cause of death broke down approximately as follows:

MOTOR VEHICLE MISHAP ON ROADWAY	
Hit another vehicle	729
Hit object	325
Noncollision (jackknife, rollover)	212
MOTOR VEHICLE MISHAP OFF ROADWAY	236
PEDESTRIAN HIT BY VEHICLE	262
Aircraft	152
Boat, ship, ferry	63
Train	47

STAY SAFE   See individual job entries.

SOURCE   bls.gov/iif/oshcfoi1.htm

Workplace Violence

ANNUAL US DEATHS   841 in 2019

Workplace violence is now the 3rd leading cause of workplace death, after traffic accidents and falls. In 2019, 841 people died in workplace violence incidents. Of these, 454 worker deaths were homicides, while 307 workers committed suicide at work.

For women workers, workplace homicide was the 2nd leading cause of job death after traffic accidents, accounting for 20 percent of their work-related deaths. Domestic violence in the workplace means that women were more than 5 times more likely to be killed by a relative or domestic partner at work than men.

The workplace homicides were mostly in retail trade (84 deaths), accommodations and food services (57 deaths), public administration (51 deaths), transportation and warehousing (45 deaths), and healthcare and social services (32 deaths).

White workers experienced 43 percent of workplace homicides; Hispanic workers experienced 16 percent of homicides. Black workers experienced 28 percent of workplace homicides, while representing only 13 percent of total employment. Homicides were 20 percent of workplace deaths, or 127 out of 634 deaths, for Black workers but only 6 percent of deaths, or 197 out of 3,297, for White workers and 7 percent of deaths, or 74 out of 1,088, for Hispanic workers.

STAY SAFE   Report concerns about fellow workers to supervisors.

SOURCE   AFL-CIO Safety and Health Office, aflcio.org/issues/workplace-health-and-safety, *Death on the Job: The National and State-by-State Profile of Worker Safety and Health in the United States*, 30th edition, May 2021.

Architects and Engineers

ANNUAL US DEATHS   43 in 2019

About 3.2 million Americans were employed in 2020 as architects, engineers or engineering technologists and technicians, or drafters, surveyors, and mappers across the US economy. In 2019, 43 of them were killed while at work. This is an annual US death rate of about 1.3 per 100,000.

STAY SAFE   Pay attention to all work safety procedures such as fire drills and active-shooter lockdown briefings. When appropriate, use all personal protective equipment. Exercise due caution when in and around work vehicles, both on and off the road, and when near heavy machinery, equipment, toxic substances, and stacks of material that might collapse. Be careful not to trip or fall on stairs or ladders, or by railings, or on roofs and parapets. Bosses should ensure that all OSHA and state safety standards are met.

SOURCE   BLS, bls.gov/iif/oshcfoi1.htm

Arts, Entertainment, Hobbies, and Recreation Industries

ANNUAL US DEATHS   90 in 2019

In 2020 in the US, 1.7 million people were employed in the arts, entertainment, and recreation industries, providing the public with all aspects of everything from live performances and events in theaters and concert venues, to museums and exhibits, to all sorts of recreational activities, amusement parks, and hobby or leisure time interests. In 2019, 90 of them were killed on the job. This is an annual US deaths of about 5.3 per 100,000.

STAY SAFE Pay attention to all work safety procedures such as fire drills and active-shooter lockdown briefings. Use all personal protective equipment. Exercise due caution when in and around motor vehicles, trains, ships, and aircraft, both on and off the road (or on the rails, on the water, or in the air), and whenever near other heavy machinery, equipment, sources of high heat or deep cold, toxic substances, and stacks of material that might collapse. Be careful not to trip or fall on lift platforms, stairs, ladders, fire escapes, and railings, or on roofs and parapets. Make sure to wear a respirator or work in a well-ventilated area when around any smoke, fumes, industrial cleaning products, insecticides, or other toxic inhalants. Bosses should ensure that all OSHA and state safety standards are met.

SOURCE BLS, bls.gov/iif/oshcfoi1.htm

Gun Ranges and Sports Shooting

ANNUAL US DEATHS 118 range deaths from 2004 to 2015

About 2,100 professionally managed gun ranges are in the US, used by individuals for practice shooting and by competitive teams for action-shooting sports events. While suicides and fatal accidental injuries are not unknown among visitor-users to these ranges, it is reportedly very rare for a gun range employee or a professional action-shooting athlete to be killed on the job at the range by firearms crimes or mishaps.

In 2014 at a popular gun range in Arizona, a 9-year-old girl accidentally shot and killed the instructor who was teaching her how to shoot an Uzi.

A study of firearm suicides in 16 states identified 118 deaths at public ranges from 2004 to 2015, for a rate of 0.12 per million population. Extrapolating from that number for the country as a whole, there would have been roughly 35 shooting range suicides per year.

STAY SAFE Always treat and handle firearms as if they are loaded. Keep their muzzle pointed in a safe direction at all times, and never touch the trigger until you are ready to open fire. Be especially careful with semiautomatic or automatic firearms, as they can be ready to fire again immediately after you have fired previous rounds. Make sure that very young shooters receive thorough safety training and adult supervision; do not let beginners operate any weapon whose recoil they are not strong and skilled enough to safely control.

SOURCES   C. Barber et al. Suicides at shooting ranges. *Crisis*, January 2021; Wikipedia

Sporting Goods, Hobby, Book, and Music Stores

ANNUAL US DEATHS   6 in 2019

In 2020 in the US, 464,900 people were employed in retail outlets that sold products and equipment and gave advice to the public for sports and athletics gear and apparel, musical instruments and recorded music, books and magazines, or hobbies ranging from crafts and needlepoint to model railroading. In 2019, 6 of them were killed on the job. This is an annual US death rate of about 1.29 per 100,000.

STAY SAFE   Pay attention to all work safety procedures such as fire drills and active-shooter lockdown briefings. Use all personal protective equipment. Exercise due caution around stacks of material that might collapse. Bosses should ensure that all OSHA and state safety standards are met.

SOURCE   BLS, bls.gov/iif/oshcfoi1.htm

Stunt Doubles

ANNUAL US DEATHS   Estimated 1

The US film industry employs stunt doubles to take the place of actors during the production of dangerous scenes. The stunts might involve driving a motorcycle, flying an aircraft, jumping from a height, falling into deep/rough water, or being involved in a (simulated) fire or explosion. Such employment tends to be episodic, since different film projects call for different stunt skills, and stunt performers sometimes need to resemble specific cast members.

Somewhere between 1,000 and 10,000 stunt doubles work actively enough to get credit for at least 1 movie or TV show a year for several years. About 10 of them have been killed in the past decade-plus while performing a stunt on set. This is equivalent to an annual occupational death rate of very roughly 10 to 100 per 100,000 FTE.

STAY SAFE Pay attention to all work safety procedures and follow directions of stunt coordinators. Bosses should ensure that all OSHA, state, and industry safety standards are met.

SOURCE Stephen Follows Film Data and Education, stephenfollows.com/stunt-performers

Construction Industry

ANNUAL US DEATHS 1,007 in 2018

During 2018 in the US, 1,007 people were killed across the construction industry's various trades, crafts, and businesses. Of these, 338 workers died in falls, 250 in transportation incidents, 173 from exposure to toxic substances or harmful environments, 169 from deadly contact with objects or equipment, 16 in fires or explosions, and 61 were the victims of murder or animal attacks.

In 2019, 7,492,200 people were employed in construction. This is equivalent to an annual US deaths of about 13.5 per 100,000.

STAY SAFE Teamwork and attention to detail are keys to Safety First on the job. Always dress for the weather, and use personal protective equipment such as a hard hat, work gloves, steel-toed boots, high-visibility safety vest, and safety harness or personal flotation device whenever required. Specialized workers such as welders, electricians, steelworkers, crane operators, and cement mixer drivers should always follow the safety procedures needed to avoid burns, electrical shocks, falls, crush injuries, or vehicle crashes.

SOURCE BLS, bls.gov/iif/oshcfoi1.htm

Buildings

ANNUAL US DEATHS 185 in 2018

In 2018, 185 US workers were killed while constructing residential, commercial, institutional, and commercial buildings. About 65 percent of these deaths were on residential construction projects. The leading cause of death (89) was a fall. Transportation mishaps killed 39, exposure to toxic substances or dangerous environments killed 31, and harmful contact with objects or equipment killed 26.

STAY SAFE See **Construction Industry**.

SOURCE BLS, bls.gov/iif/oshcfoi1.htm

Caisson and Tunnel Workers: Decompression Sickness

> **ANNUAL US DEATHS** Rare with modern technology to mitigate decompression hazards

Commercial divers are not the only workers subject to getting decompression sickness, also known as the bends. Caisson and tunnel workers ("sandhogs") who enter and exit compressed air environments—needed to keep out groundwater or poisonous gas that might otherwise flood the work site—are also exposed to the risk of decompression sickness. Putting the human body under water or air pressure causes nitrogen gas to dissolve more into the blood, and relieving that pressure too quickly causes the nitrogen to form bubbles in the bloodstream. Decompression sickness is extremely painful, can lead to permanent injuries, and can be fatal. Treating a victim involves placing them back under pressure, either in a hyperbaric chamber (for divers) or by immediate return to the high-pressure work site (if dry), followed by treatment of symptoms and a more gradual, safe return to normal atmospheric pressure.

Hard data is scarce, but occupational deaths due to decompression sickness are very rare in the US in recent years. This is due to improvements in both worker safety procedures, and in use of construction methods and machinery that avoid the need for humans to spend much time in environments subject to high atmospheric pressure.

STAY SAFE Understand and observe the rules for preventing the bends.

SOURCE CDC NIOSH, cdc.gov/niosh/topics/decompression

Confined Spaces

> **ANNUAL US DEATHS** Estimated 147

Between 2011 and 2018, 1,030 workers in the US were killed due to occupational mishaps in confined spaces. The most deaths occurred among construction workers (especially plumbers and pipelayers) and agricultural workers. Such small and poorly ventilated areas, difficult to enter and exit and not designed for continuous occupancy, include railroad cars, tanker trucks, storage tanks and hoppers, grain silos and bins, wells, cisterns, septic tanks and manure pits, sumps, sewers, manholes, dumpsters, crawl spaces, trenches and excavations, ditches, foundations, drains, tunnels, mines, and caves.

It is helpful to spell out the nature of these risks. This is especially important as do-it-yourselfers can be exposed to some of these same mortal dangers in confined spaces.

STAY SAFE   Be aware of, and take care to avoid, all of the following hazards:

- Falling, especially if the space is wet, icy, and/or poorly lit
- Breathing toxic fumes, such as carbon monoxide, hydrogen sulfide, methane, sewer gas, or solvents, which can accumulate in cramped, enclosed spaces
- Asphyxiation due to lack of oxygen, even if the oxygen has been displaced by a gas that is not itself poisonous. One example is carbon dioxide, which is heavier than air and is given off by dry ice, which is sometimes very wrongly used to freeze pipes to be able to work on them without their liquid contents flooding
- Being engulfed, crushed, and/or suffocated by stored material, such as grain
- Fiery explosions caused by a buildup of grain dust or combustible fumes
- Being caught in machinery, especially powered conveyors or grain augers
- Drowning when working in confined spaces around running or stationary liquids, or when rainwater or groundwater might accumulate

SOURCES   BLS, bls.gov/iif/oshcfoi1.htm; OSHA, osha.gov/confined-spaces

Heavy and Civil Engineering Projects

ANNUAL US DEATHS   178 in 2018

In 2018 in the US, 178 people died while working on heavy construction and civil engineering projects. The breakout was 34 workers killed in water and sewer line projects, 15 in oil and gas pipelines, and 35 in power and communications lines and towers construction. Another 73 died working on roads and bridges, and about 21 were killed during other heavy construction projects.

Of the total deaths, 87 were due to transportation crashes, 35 due to lethal contact with objects or equipment, 31 due to toxins or extreme environments, and 16 due to falls from a height.

STAY SAFE   See **Construction Industry**.

SOURCE   BLS, bls.gov/iif/oshcfoi1.htm

Concrete pumps and concrete mixers are deployed to many construction sites.

Poured Concrete Foundation and Structure Contractors

ANNUAL US DEATHS 28 in 2018

In the US in 2018, 28 construction industry workers were killed during concrete pouring operations for foundations and structures. The leading specific cause of death was falls, which killed 9 people.

STAY SAFE See **Construction Industry**.

SOURCE BLS, bls.gov/iif/oshcfoi1.htm

Structural Steel and Precast Concrete Contractors

ANNUAL US DEATHS 18 in 2018

During 2018 in the US, 18 construction workers were killed building the steel frames of buildings and structures, or while working with precast (preformed, prestressed) concrete components. The leading cause of death was harmful contact with objects or equipment, which killed 8 workers.

STAY SAFE See **Construction Industry**.

SOURCE BLS, bls.gov/iif/oshcfoi1.htm

Other Exterior and Structural Contractors

ANNUAL US DEATHS Some 170 in 2018

During 2018 in the US, 20 people were killed while working on non-steel structural framing, 28 were killed working on exterior masonry, 106 were killed working on roofing, 7 were killed working on siding, and 7 were killed during other structural and exterior work. Across these activities, 107 people died in falls, 30 died from lethal contact with objects and equipment, 23 died from exposure to toxic substances or extreme environments, and 10 died from transportation crashes.

STAY SAFE See **Construction Industry**.

SOURCE BLS, bls.gov/iif/oshcfoi1.htm

Electricians and Electrical Workers

ANNUAL US DEATHS 79 in 2016; 166 in 2019

In 2016 in the US, electricians had an annual occupational death rate of 10.0 per 100,000 workers, for 79 fatal injuries out of about 790,000 FTE electricians employed. Their most common fatal accident was falls/slips/trips, not being electrocuted.

In 2016 in the US, electrical power-line installers and repairers had an annual occupational death rate of 14.6 per 100,000, for 21 fatalities out of about 144,000 FTE power-line workers employed. The most common cause of death was overexertion and heat stroke, not being electrocuted.

In 2019 there were 166 electrical fatalities. Contact with or exposure to electric current was the cause of death in only 3 percent of cases.

STAY SAFE Always work with safety buddies, wear all protective equipment and harnesses, and make sure work areas are properly grounded and/or insulated. Be extremely careful around anything that is or might be energized.

SOURCES BLS, bls.gov/iif/oshcfoi1.htm; Electrical Safety Foundation International (ESFI), esfi.org/workplace-safety/workplace-injury-fatality-statistics

Explosives Workers, Ordnance Handling Experts, and Blasters

> **ANNUAL US DEATHS** About 100

In 2017 in the US, 5,850 people were employed working with explosives in demolitions, construction, mining and quarrying, and other clearing of earth and rock or structures and debris. While national worker death data does not split out this category from other extraction and construction workers killed in fires and explosions, a report by the state of Nevada indicated that about 1 in 40 worker deaths from all causes in all industries there were due to explosives workers being killed on the job. This translates into a national occupational death rate of very roughly 100 explosives workers killed per year.

STAY SAFE Always be extremely careful when handling explosive materials and while blasting. Use the proper amount of explosive given the job to be done, allowing for the space in which the concussive force and associated solid debris will spread. Always keep a safe distance, in a protected area, before setting off detonations. Follow all safety measures regarding stray electrical currents and radio transmissions that might trigger a premature blast.

SOURCE BLS, bls.gov/iif/oshcfoi1.htm

Demolition experts used high explosives to trigger a controlled explosion to bring down this dam.

Plumbers and Heating, Ventilation, and Air-Conditioning (HVAC) Contractors

ANNUAL US DEATHS 73 in 2018

In 2018 in the US, 73 plumbers and HVAC technicians were killed working on construction projects. The leading cause of death was falls, which killed 24. Then came transportation crashes, which killed 22.

STAY SAFE See **Construction Industry**.

SOURCE BLS, bls.gov/iif/oshcfoi1.htm

Site Preparation, Building Interior Finishing, and Other Specialty Contractors

ANNUAL US DEATHS 178 in 2018

During 2018 in the US, deaths among building interior finishing trades broke down as 23 among drywall and insulation workers, 30 among plastering and painting workers, 8 among flooring workers, 3 among tile layers, 26 among finishing carpenters, and 7 among other finishing trades.

Deaths among site preparation contractors were 81, and among other specialty trades were 42. The leading causes of these deaths were lethal contact with objects and equipment, which killed 44, and transportation crashes, which killed 41.

STAY SAFE See **Construction Industry**.

SOURCE BLS, bls.gov/iif/oshcfoi1.htm

Community, Religious, and Social Services Professions

ANNUAL US DEATHS 31 in 2019

About 2.7 million Americans were employed in 2020 in counseling, therapy, social worker, probation officer, or clergy and religious jobs across the US economy. In 2019, 31 of them were killed while at work. This is an annual US deaths of about 1.1 per 100,000.

STAY SAFE   Be aware of the potential for client violence. Get trained in risk assessment, safety planning, verbal de-escalation techniques, and nonviolent self defense.

SOURCE   BLS, bls.gov/iif/oshcfoi1.htm

Clergy

> **ANNUAL US DEATHS**   10 to 25 percent lower than that of general population

A meta-study (study of studies) of data for the years 1959 through 2000 in the US found that Judeo-Christian clergy (priests, ministers, rabbis, nuns, and monks) had mortality rates between just over 10 percent to more than 25 percent *below* those for the average population.

This significant mortality advantage can be explained as the effect of some beneficial aspects of leading a religious life: peace of mind and a well-ordered existence, abstaining from various high-risk activities and vices, a strong sense of community and of service to others and to God, and ready access to healthcare and safety programs, with higher denominational authority oversight.

STAY SAFE   Becoming a member of the clergy is good for longevity.

SOURCE   K.J. Flannelly et al. A review of mortality research on clergy and other religious professionals. *Journal of Religion and Health*, March 2002.

Education

> **ANNUAL US DEATHS**   24 in 2019

About 8.9 million Americans were employed in 2020 as teachers or teaching assistants in all grades from preschool through university, or as tutors, instructors, or special educators, or as librarians, curators, archivists, or museum technicians across the US economy. In 2019, 24 of them were killed while at work. This is an annual US deaths of about 0.27 per 100,000.

STAY SAFE   Pay attention to all safety procedures such as fire drills and active-shooter lockdown briefings. Be aware of the potential for student violence.

SOURCES   BLS, bls.gov/iif/oshcfoi1.htm; National Center for Education Statistics, nces.ed.gov/programs/coe/indicator/a01#

Farming, Ranching, and Agriculture

ANNUAL US DEATHS   457 in 2018

In 2018, in the US, 457 people were killed while doing jobs producing crops and/or livestock. Of these, about 25 died of exposure to dangerous toxins or extreme environments, while 274 were killed in transportation incidents, 167 by fatal contact with objects or equipment, 62 by electrocution, 35 by falls, and 14 by fires or explosions. (These numbers include some double-counting of deaths from multiple causes.) Of those killed, 250 were in crop production, 161 were in animal ranching and aquaculture, and the rest, 46, were in support services.

In 2019, about 956,500 workers were employed across these occupations. This is equivalent to a death rate of 48 per 100,000 FTE employees.

STAY SAFE   Teamwork and attention to detail are keys to Safety First on the job. Always dress for the weather. Use personal protective equipment such as a hard hat, respirator, work gloves, steel-toed boots, high-visibility safety vest, and safety harness whenever required.

SOURCE   BLS, bls.gov/iif/oshcfoi1.htm

Grain Bin Entrapments and Explosions and Other Confined Spaces in Agriculture

ANNUAL US DEATHS   50 to 60

Once grain crops are harvested on farms, they need to be gathered into a network of more centralized storage facilities, pending distribution by road or rail. This storage is accomplished in grain bins, usually round, tall structures made of metal and/or concrete. These bins present special safety hazards to workers who need to enter them: entrapments leading to crush injuries or asphyxiation.

Another serious risk is a violent grain silo explosion, if dust mixes with air and then is exposed to a spark or flame. There were 12 grain dust explosions in the US in 2018, at grain elevators, feed mills, or ethanol plants; only 1 person was killed, due to greatly improved explosion safety provisions in recent years.

The number of grain bin workers killed on the job ranges from 20 to 25 per year. Entrapments are more likely to occur when weather has been cold and wet, causing the grain to tend to stick together in the bin, form hidden voids, and then collapse under the weight of a worker standing on the of the stored grain.

Another 20 to 25 people are killed every year in other confined spaces found on farms and ranches, such as manure pits. In August 2021 in Ohio, 3 brothers were killed in a manure pit when they were overcome by fumes while fixing a faulty pump. Manure pits can cause multiple fatalities when 1 person after another attempts to rescue the original victim.

About 956,500 people worked in agricultural production in 2019, making the annual confined-spaces death rate, overall, about 4.7 per 100,000.

STAY SAFE Always wear a safety harness and lifeline and have a safety observer-buddy on site. Turn off all equipment and machinery before entering the grain bin or other confined space. Never smoke or carry an open flame in areas prone to grain dust buildup.

SOURCES BLS, bls.gov/iif/oshcfoi1.htm; Cleveland.com, cleveland. com/nation/2021/08/ohio-brothers-die-after-being-overcome-by-fumes-in-manure-pit.html; D.M. Hallam et al. Manure pit injuries: Rare, deadly, and preventable. *Journal of Emergencies, Trauma, and Shock*, 2012; Purdue University, https://www.grainnet.com/ article/229517/purdue-university-report-grain-bin-entrapments-fall-7.9-in-2020#:~:text=Purdue%20University%20Report%3A%20 Grain%20Bin%20Entrapments%20Fall%207.9%25%20in%20 2020,-Date%20Posted%3A%20Mar&text=A%20minimum%20of %2035%20grain,and%2015%20were%20non%2Dfatal.

First Responders

> **ANNUAL US DEATHS** 231 in 2019; significantly higher during the COVID-19 pandemic

About 9.6 million Americans were employed in 2020 as protective services workers, a vital area of the US economy involving frontline first responders (firefighters, police and corrections officers, and bailiffs) and their supervisors (corporals and sergeants), plus fish and game or animal control wardens, private detectives, security guards, crossing guards, and transportation screeners and monitors.

In 2019, 231 of them were killed while at work. This is an annual US deaths of about 2.4 per 100,000.

STAY SAFE Pay attention to all work safety procedures such as fire drills and active-shooter lockdown briefings. When appropriate, use all personal protective equipment, especially bulletproof vests in the protective and security services, and follow any applicable protocols regarding sanitization, sterility, and quarantine for any pathogens or hazmats. Be extremely careful if your job calls for carrying and sometimes brandishing or otherwise using a firearm. Bosses should ensure that all OSHA and state safety standards are met. Get vaccinated and observe all safety protocols during public health emergencies.

SOURCES   BLS, bls.gov/iif/oshcfoi1.htm; CDC NIOSH, cdc.gov/ niosh/programs/epr/risks.html; Y.H. Chen et al. Excess mortality associated with the COVID-19 pandemic among Californians 18–65 years of age, by occupational sector and occupation: March through November 2020. *PLoS* One, June 2021.

Firefighters

ANNUAL US DEATHS   62 while on duty in 2019; significantly higher during the COVID-19 pandemic

In 2020 in the US, 395,000 people were professionals employed in firefighting, fire prevention, and arson investigation. In 2019, 25 of them were killed on the job. This is an annual US deaths of about 6.1 per 100,000 for professional fire protection personnel.

As of 2018, there were about 750,000 volunteer firefighters in the US. In 2019, 34 of them were killed on duty. This is an annual US deaths of about 4.5 per 100,000 for volunteer firefighters.

In addition, in 2019, 3 wildland firefighters were killed fighting forest, brush, and grass fires.

The total of all firefighters killed on the job in 2019 comes to 62. The causes of the fatalities were:

Stress or overexertion	36
Vehicle crashes	6
Caught or trapped	5
Struck by object	5
Structure collapse	2
Exposure	2
Fall	2
Assault	1
Other	3

Firefighters at Ground Zero following the terrorist attacks of September 11, 2001

In 2019, 37 firefighters died from activities related to an emergency incident. An additional 33 died from heart attacks. Eighteen died from activities at a fire scene and 12 died from activities at a nonfire scene. Responding to or returning from emergency incidents killed 8 firefighters. Five died in training activities.

In 2010, 90 firefighters died on duty. The trend has been downward ever since, for an overall reduction of 17.7 percent by 2019.

STAY SAFE   Follow any applicable protocols regarding sanitization, sterility, and quarantine for exposure to pathogens or hazmats. Bosses should ensure that all OSHA and state safety standards are met. Get vaccinated and observe all safety protocols during public health emergencies.

SOURCES   BLS, bls.gov/iif/oshcfoi1.htm; Y.H. Chen et al. Excess mortality associated with the COVID-19 pandemic among Californians 18–65 years of age, by occupational sector and occupation: March through November 2020. *PLoS One*, June 2021; FEMA, apps. usfa.fema.gov/firefighter-fatalities

Law Enforcement

ANNUAL US DEATHS   More than 345 in 2021

In 2020 in the US, of the 1.44 million people employed in law enforcement and corrections, 265 died in the line of duty. Of those, 42 were shot, 40 died in traffic-related fatalities, and 183 died of other causes. As of November 2021, at least 345 officers had died in the line of duty: 52 were shot, 52 died in traffic-related fatalities, and 241 died of other causes, including COVID-19 contracted as essential workers.

STAY SAFE   Be careful out there. Observe all department safety protocols. Wear a bullet proof vest. Drive carefully and pay attention to roadside safety. Wear personal protective equipment where appropriate. Get vaccinated and observe all safety protocols during public health emergencies.

SOURCES   BLS, bls.gov/iif/oshcfoi1.htm; Y.H. Chen et al. Excess mortality associated with the COVID-19 pandemic among Californians 18–65 years of age, by occupational sector and occupation: March through November 2020. *PLoS One*, June 2021; FBI, fbi.gov/services/cjis/ucr/leoka; National Law Enforcement Officers Memorial Fund, nleomf.org/memorial

Fishing, Hunting, and Trapping

ANNUAL US DEATHS   33 in 2018

In 2018 in the US, 33 people were killed while working in fishing, hunting, and trapping. Of these, 6 died of exposure to dangerous toxins or extreme environments, while 23 were killed in transportation incidents.

In 2019, about 36,700 workers were employed in this occupation. This is equivalent to a death rate of 90 per 100,000 FTE employees.

STAY SAFE   Always dress for the weather. Use personal protective equipment such as goggles, work gloves, steel-toed boots, high-visibility safety vest, and safety harness or personal flotation device whenever required. Exercise appropriate caution with firearms, traps, and poisons.

SOURCE   BLS, bls.gov/iif/oshcfoi1.htm

Forestry and Logging

ANNUAL US DEATHS   84 in 2018

In 2018 in the US, 84 people were killed while working in forestry and logging. Of these, 49 had fatal accidents involving objects or equipment and 28 others died in transportation incidents.

In 2019, about 70,100 workers were employed in this occupation. This is equivalent to a death rate of 120 per 100,000 FTE employees.

STAY SAFE   Teamwork and attention to detail are keys to Safety First on the job. Always dress for the weather. Use personal protective equipment such as a hard hat, work gloves, steel-toed boots, high-visibility safety vest, and safety harness whenever required.

SOURCE   BLS, bls.gov/iif/oshcfoi1.htm

Government, Federal

ANNUAL US DEATHS 111 in 2019

In 2020 in the US, 2.83 million people were employed by the federal government in some capacity, including civilians who worked in the Department of Defense. In 2019, 111 of them were killed on the job. This is an annual US death rate of about 3.96 per 100,000.

STAY SAFE Pay attention to all work safety procedures. Bosses should ensure that all OSHA safety standards are met.

SOURCE BLS, bls.gov/iif/oshcfoi1.htm

Government, State and Local

ANNUAL US DEATHS 315 in 2019

In 2020 in the US, 19.8 million people were employed by state or local governments all around the country, in a very wide variety of jobs. In 2019, 315 of them were killed on the job. This is an annual US death rate of about 1.59 per 100,000.

STAY SAFE Pay attention to all work safety procedures. Bosses should ensure that all OSHA, state, and local safety standards are met.

SOURCE BLS, bls.gov/iif/oshcfoi1.htm

Healthcare Industry

ANNUAL US DEATHS 56 in 2019; 3,549 in 2021, mostly due to COVID-19 infections

About 9.6 million Americans were employed in 2020 as doctors, dentists, pharmacists, optometrists, dieticians, surgeons, various types of therapists, veterinarians, nurses and LPNs, paramedics and EMTs, sonographers and radiologists, medical technicians and technologists, acupuncturists, chiropractors, and speech pathologists across the US economy. In 2019, 56 of them were killed while at work. This is an annual US deaths of about 0.58 per 100,000.

During 2020, the first full year of the pandemic, 3,607 US healthcare workers died of COVID-19. Nurses were 32 percent of all deaths; patient support aides were 20 percent; physicians were 17 percent; medical first responders were 7 percent. Worldwide, the WHO estimates some 115,000 healthcare workers died from January 2020 to May 2021.

STAY SAFE   Pay attention to all work safety procedures such as fire drills and active-shooter lockdown briefings. When appropriate, use all personal protective equipment. Observe all hygiene and anti-infection protocols. Get vaccinated.

SOURCES   BLS, bls.gov/iif/oshcfoi1.htm; the *Guardian*, theguardian.com/us-news/series/lost-on-the-frontline; WHO, who.int/news/item/20-10-2021-health-and-care-worker-deaths-during-covid-19

Healthcare Support

> **ANNUAL US DEATHS**   56 in 2019; 3.549 in 2021, mostly due to COVID-19 infections

About 4.8 million Americans were employed in 2020 in healthcare support, a broad area of the US economy involving home health and personal care aides, nursing assistants and patient care aides, occupational and physical therapy aides, plus massage therapists, dental and medical assistants, pharmacy and veterinary assistants, phlebotomists, and other healthcare support jobs.

In 2019, 56 of them were killed while at work. This is an annual US deaths of about 0.79 per 100,000.

As of February 12, 2022, 999,628 nursing home staff members had contracted COVID-19; 2,303 died.

STAY SAFE   When appropriate, use all personal protective equipment. Observe all hygiene and anti-infection protocols. Get vaccinated.

SOURCES   BLS, bls.gov/iif/oshcfoi1.htm; the *Guardian*, theguardian.com/us-news/series/lost-on-the-frontline

Hospitality and Related Services

> **ANNUAL US DEATHS**   188 in 2019

In 2020 in the US, 11.4 million people were employed in providing people with places to stay, such as hotels, motels, and hostels, and/or preparing food to eat at establishments such as restaurants, dining rooms, snack bars, takeout counters, food trucks and carts, and fast-

A bartender in New York City, c. 1947, apparently with little to fear

food outlets. In 2019, 188 of them were killed on the job. This is an annual US death rate of about 1.04 per 100,000.

STAY SAFE   Pay attention to all work safety procedures. Bosses should ensure that all OSHA and state safety standards are met.

SOURCE   BLS, bls.gov/iif/oshcfoi1.htm

Bartenders

ANNUAL US DEATHS   31 in 2018

In 2019 in the US, 654,700 people were employed as bartenders in bars, clubs, restaurants, hotels, lounges, beer gardens, and other drinking establishments. In 2018, 31 of them were killed on the job, 77 percent (24 bartenders) due to workplace violence. This is an annual occupational death rate of about 11.8 per 100,000.

STAY SAFE   Be aware of the risks of workplace violence such as drunken brawls or robberies.

SOURCE   BLS, bls.gov/iif/oshcfoi1.htm

Food Preparation and Serving

ANNUAL US DEATHS   99 in 2019; higher during the COVID-19 pandemic

About 6.6 million Americans were employed in 2020 in food preparation and food serving across the US economy. They were chefs and cooks or other food preparers, wait staff, hosts, dishwashers, and fast food and counter workers, and they worked in dining rooms, cafeterias, restaurants, coffee shops, bars and lounges, or were the frontline supervisors of these workers. In 2019, 99 of them were killed while at work. This is an annual US deaths of about 1.5 per 100,000.

A study in California found that line cooks in restaurants had a mortality risk ratio of 1:6 during the first 10 months of the COVID-19 pandemic in 2020.

STAY SAFE   Pay attention to all work safety procedures. When appropriate, use personal protective equipment. Get vaccinated and observe all safety protocols during public health emergencies.

SOURCES   BLS, bls.gov/iif/oshcfoi1.htm; Y.H. Chen et al. Excess mortality associated with the COVID-19 pandemic among Californians 18–65 years of age, by occupational sector and occupation: March through November 2020. *PLoS One*, June 2021

Hotel, Motel, and Hostel Guests

ANNUAL US DEATHS   No reliable statistics; death by carbon monoxide poisoning may claim some 70 lives per year in these accommodations.

Managers of hotels, motels, and hostels are understandably reluctant to report deaths, both from natural causes and suicide, among their guests; reliable data is therefore scarce. Anecdotally, casino hotels have a higher than average suicide rate. More people kill themselves every year in Las Vegas than in any other place in America. For that reason, most Las Vegas hotel rooms do not have balconies, and the windows open only an inch or 2.

Carbon monoxide poisoning can kill guests in hotels, motels, and resorts. From January 1, 2005, to December 31, 2018, 905 guests were poisoned by CO in 115 identified incidents, including 22 fatalities. Children represented 16 percent of those poisoned and 27 percent of fatalities. Most poisonings were caused by heaters fueled by natural gas and could probably have been prevented by an in-room carbon monoxide alarm.

STAY SAFE Look over a hotel before staying there; if it appears unsafe due to shabby maintenance or high crime, stay somewhere else. Always review the fire safety information on the back of your room door and locate your path of evacuation in case of a fire or other emergency. For help with severe depression or any thoughts of self-harm, contact the National Suicide Prevention Lifeline at 1-800-273-8255 (TALK) or suicidepreventionlifeline.org.

SOURCES N.B. Hampson et al. Carbon monoxide poisonings in hotels and motels: The problem silently continues. *Preventive Medicine Reports*, August 2019; National Suicide Prevention Lifeline, suicidepreventionlifeline.org

Legal Professionals

ANNUAL US DEATHS   11 in 2019

About 1.2 million Americans were employed in 2020 in legal, paralegal and legal support, judicial, and title examiner jobs across the US economy. In 2019, 11 of them were killed while at work. This is an annual US deaths of about 0.92 per 100,000.

STAY SAFE Be aware of the potential for client and workplace violence.

SOURCE BLS, bls.gov/iif/oshcfoi1.htm

Management and Supporting Occupations

ANNUAL US DEATHS   380 in 2019

About 18.6 million Americans were employed in 2020 in management jobs across the US economy. In 2019, 380 of them were killed while at work. This is an annual US deaths of about 2.0 per 100,000.

STAY SAFE Pay attention to all work safety procedures such as fire drills and active-shooter lockdown briefings. Be aware of the potential for client and workplace violence.

SOURCE bls.gov/iif/oshcfoi1.htm

All Desk Jobs

ANNUAL US DEATHS   See data in entry.

People who sit at least 6 hours a day while at work were found to have odds of dying within 15 years that were 40 percent higher than people who sit no more than 3 hours a day. Compared to people who sit and watch television for 1 hour or less each day, people who sit and watch television for more than 7 hours a day are at 1.61 times greater risk of all-cause mortality, 1.85 times greater risk of cardiovascular mortality, and 1.22 times greater risk of cancer mortality. Moderately vigorous daily exercise did not completely offset the effects of prolonged sitting.

STAY SAFE   Stand up and move around every hour if you sit a lot. Get regular physical activity for at least 30 minutes each day. Consider purchasing a sit-stand desk.

SOURCE   C.E. Matthews et al. Amount of time spent in sedentary behaviors and cause-specific mortality in US adults. *American Journal of Clinical Nutrition*, February 2012.

Management and Business and Financial Operations

ANNUAL US DEATHS   409 in 2019

About 27.1 million Americans were employed in 2020 across the US economy in a very broad range of jobs that were management in nature or were focused on the broad-level business and financial operations of their employer. They were found in virtually every industry, every business, every office in the country. Most of the time, most of them "flew a desk" or "pushed a pencil." Some of them did at times leave their desks, to visit clients and customers, or to go to factory floors, or to check on separate locations of branches, stores, and outlets, or to enter the field to inspect a mine, a construction site, a transportation hub, or a warehouse.

All these activities exposed them to various risks of death on the job. While they might not have worked with heavy machinery or spent as much time driving or riding in heavy vehicles, they often drove in traffic as part of their work. They were exposed to the risk of dying in an occupation-related transportation crash. They used stairs, elevators, and escalators and went through automatic doors, which could cause lethal falls or other deadly injuries. They could get caught in, struck by, or crushed under machinery, or have a heavy object fall on them, or be electrocuted or asphyxiated. They were exposed to workplace interpersonal violence, even sometimes being murdered by a co-worker or an outsider. All of them could be exposed to toxic substances or to extreme environments such as very hot or cold places (weather, or equipment like furnaces and freezers). Fires and explosions could happen, and kill people, literally anywhere.

In 2019, 409 of them were killed while at work. This is an annual US deaths of about 1.5 per 100,000 FTE.

STAY SAFE   Pay attention to all work safety procedures such as fire drills and active-shooter lockdown briefings. Be aware of the potential for client and workplace violence. When appropriate, use personal protective equipment. Be careful not to trip or fall on stairs or ladders. Supervisors should ensure that all OSHA and state safety standards are met.

SOURCE   BLS, bls.gov/iif/oshcfoi1.htm

Office and Administrative Support Occupations

ANNUAL US DEATHS   92 in 2019

About 15.6 million Americans were employed in 2020 across the US economy in a very broad range of office and administrative support jobs or as first-line supervisors. These workers could be found everywhere from banks to courts to casinos to hotels, from libraries to warehouses to intermodal terminals to rail yards, and from utilities to legal and medical offices and beyond. They held all sorts of office, office support, and clerical positions, including receptionists and switchboard operators, filing and billing and accounting clerks, payroll and financial clerks, tellers and customer service reps, desk clerks and reservations clerks, couriers and cargo agents, dispatchers, meter readers, secretaries of every variety, mail clerks and mailroom workers, proofreaders and desktop publishers.

The nature and locations of such work often exposed them to the risks of getting killed on the job: working with the public (who might murder them), being near toxic chemicals, driving or riding in work vehicles, and working close to heavy equipment.

Like other employed people, they also often used stairs, elevators, and escalators and went through automatic doors, which could cause lethal falls or other deadly injuries. They could have a heavy object fall on them or be electrocuted or asphyxiated. They might be exposed to extremely hot or cold environments. Fires and explosions could happen and kill them, literally anywhere.

In 2019, 92 of them were killed on the job. This is an annual US deaths of about 0.59 per 100,000 FTE.

STAY SAFE   Pay attention to all work safety procedures such as fire drills and active-shooter lockdown briefings. Be aware of the potential for client and workplace violence. Pay attention to all work safety procedures such as fire drills and active-shooter lockdown briefings. Be aware of the potential for client and workplace violence. When appropriate, use personal protective equipment. Be careful not to trip or fall on stairs or ladders. Supervisors and bosses should ensure that all OSHA and state safety standards are met.

SOURCE   BLS, bls.gov/iif/oshcfoi1.htm

Minerals Extraction Industries

ANNUAL US DEATHS   128 in 2018

Minerals extraction industries employed a total of 684,600 people in the US in 2019. This broad category includes oil and gas drilling, oil and gas extraction, coal mining, metal ore mining, and quarrying. A total of 128 people were killed on the job in these occupations in 2018. This implies a death rate of about 19 per 100,000.

STAY SAFE   Teamwork and attention to detail are keys to Safety First on the job. Always dress for the weather, and use personal protective equipment such as a hard hat, work gloves, steel-toed boots, high-visibility safety vest, and safety harness or personal flotation device whenever required.

SOURCE   US Department of Labor, Mine Safety and Health Administration (MSHA), msha.gov/data-reports/fatality-reports/search

Pennsylvania coal miners, 1942

Coal Mining

ANNUAL US DEATHS   11 US deaths in 2018

In 2018 in the US, 11 people were killed in the coal mining industry. Of them, 4 were reported killed in surface mining and 6 in underground mining.

STAY SAFE See **Minerals Extraction Industries**.

SOURCE US Department of Labor, Mine Safety and Health Administration (MSHA), msha.gov/data-reports/fatality-reports/search

Oil and Gas Drilling

ANNUAL US DEATHS 23 in 2018

In 2018 in the US, 23 people were killed in oil and gas drilling, of whom 10 died in transportation.

STAY SAFE See **Minerals Extraction Industries**.

SOURCE US Department of Labor, Mine Safety and Health Administration (MSHA), msha.gov/data-reports/fatality-reports/search

Oil and Gas Extraction

ANNUAL US DEATHS 81 in 2018

In 2018 in the US, 81 people were killed on the job in these occupations. Of the total, 38 died in transportation incidents, 18 by fatal contact with objects or equipment, 6 each in falls and in toxic exposures, and 13 in fires and explosions.

STAY SAFE See **Minerals Extraction Industries**.

SOURCE BLS, bls.gov/iif/oshcfoi1.htm

Metal Ore Mining

ANNUAL US DEATHS 4 in 2018

In 2018 in the US, 4 people were killed in mining for metal ores such as gold, silver, copper, nickel, lead, and zinc.

STAY SAFE See **Minerals Extraction Industries**.

SOURCE US Department of Labor, Mine Safety and Health Administration (MSHA), msha.gov/data-reports/fatality-reports/search

Quarrying

ANNUAL US DEATHS   17 in 2018

In 2018 in the US, 17 people were killed in the quarrying of nonmetals such as limestone and marble, sand, gravel, clay, and potash.

STAY SAFE   See **Minerals Extraction Industries**.

SOURCE   BLS, bls.gov/iif/oshcfoi1.htm

Sales and Related Occupations

ANNUAL US DEATHS   240 in 2019

About 14.2 million Americans were employed in 2020 across the US economy in a broad range of jobs related to sales. They included people who sold retail goods, wholesale and manufacturing goods, spare parts, advertising, insurance, securities and financial services, travel packages, and real estate, or who were cashiers, counter clerks, rental clerks, models, demonstrators, promoters, telemarketers, news and street vendors, door-to-door salespeople, or who were their first-line supervisors.

The nature and location of such work often exposed them to the risks of getting killed on the job, such as working with the public (who might murder them) or driving or riding in work vehicles (which might crash). Like other employed people, they often used stairs, elevators, and escalators and went through automatic doors, which could cause lethal falls or other deadly injuries. They could have a heavy object fall on them, be electrocuted or asphyxiated, or be exposed to extreme heat or cold while out of doors. Fires and explosions could happen and kill them, literally anywhere.

In 2019, 240 of them were killed on the job. This is an annual US death rate of about 1.7 per 100,000 FTE.

STAY SAFE   Pay attention to all work safety procedures such as fire drills and active-shooter lockdown briefings. When appropriate, use personal protective equipment. Be careful not to trip or fall on stairs, ladders, fire escapes, or railings. Supervisors should ensure that all OSHA and state safety standards are met.

SOURCE   BLS, bls.gov/iif/oshcfoi1.htm

Sciences, Applied

ANNUAL US DEATHS   15 in 2019

About 1.6 million Americans were employed in 2020 as research scientists, conservationists, economists, psychologists and sociologists, health and safety specialists, or scientific technicians across the US economy. In 2019, 15 of them were killed while at work. This is an annual US deaths of about 0.94 per 100,000.

STAY SAFE   Pay attention to all work safety procedures such as fire drills and active-shooter lockdown briefings. When appropriate, use personal protective equipment, especially bulletproof vests in the protective and security services, and follow any applicable protocols regarding sanitization, sterility, and quarantine for any pathogens or hazmats. Exercise due caution when in and around work vehicles, both on and off the road, and when near heavy machinery, equipment, sources of high heat or deep cold, toxic substances, and stacks of material that might collapse. Be careful not to trip or fall on lift platforms, stairs, ladders, fire escapes, railings, or on roofs and parapets. Wear a respirator or work in a well-ventilated area when around smoke, fumes, industrial cleaning products, insecticides, or other toxic inhalants. And of course, be extremely careful if your job calls for carrying and sometimes brandishing or otherwise using a firearm. Bosses should ensure that all OSHA and state safety standards are met. Get vaccinated and observe all safety protocols during public health emergencies.

SOURCE   BLS, bls.gov/iif/oshcfoi1.htm

Service Occupations

ANNUAL US DEATHS   762 in 2019; significantly higher during the COVID-19 pandemic

About 22.9 million Americans were employed in 2020 across the US economy in a broad range of jobs that were service and support occupations, in industries ranging from healthcare to law enforcement and protection, to food prep and serving, to cleaning and maintenance or pest control, to personal care (from dog walkers to hairdressers to personal trainers to undertakers). The nature and location of such work often exposed them to the risks of getting killed on the job: working with the public (who might murder them), using toxic chemicals (that might poison them), driving or riding in work vehicles (which might crash), and working with heavy equipment (that might squash or mangle them).

Like other employed people, they often used stairs, elevators, and escalators and went through automatic doors, which could cause lethal falls or other deadly injuries. They could have a heavy object fall on them or be electrocuted or asphyxiated. They might get attacked by an animal or be exposed to extremely hot or cold environments (bad weather, or equipment like ovens and deep fryers, or walk-in food freezers). Fires and explosions could happen and kill them, literally anywhere—especially if they were a firefighter.

In 2019, 762 of them were killed on the job. This is an annual US death rate of about 3.3 per 100,000 FTE.

STAY SAFE Pay attention to all work safety procedures such as fire drills and active-shooter lockdown briefings. When appropriate, use personal protective equipment, especially bulletproof vests in the protective and security services, and follow any applicable protocols regarding sanitization, sterility, and quarantine for any pathogens or hazmats. Exercise due caution when in and around work vehicles, both on and off the road, and when near heavy machinery, equipment, sources of high heat or deep cold, toxic substances, and stacks of material that might collapse. Be careful not to trip or fall on lift platforms, stairs, ladders, fire escapes, railings, and on roofs and parapets. Wear a respirator or work in a well-ventilated area when around smoke, fumes, industrial cleaning products, insecticides, or other toxic inhalants. And of course, be extremely careful if your job calls for carrying and sometimes brandishing or otherwise using a firearm. Bosses should ensure that all OSHA and state safety standards are met. Get vaccinated and observe all safety protocols during public health emergencies.

SOURCES BLS, bls.gov/iif/oshcfoi1.htm; Y.H. Chen et al. Excess mortality associated with the COVID-19 pandemic among Californians 18–65 years of age, by occupational sector and occupation: March through November 2020. *PLoS One*, June 2021

Cleaning and Maintenance

ANNUAL US DEATHS 333 in 2019; higher during the COVID-19 pandemic

About 5.1 million Americans were employed in 2020 across the US economy in jobs related to performing, or directly supervising, building and grounds cleaning and maintenance activities. They were janitors, housekeepers, or maids, landscapers, lawn mowers, or tree trimmers, exterminators, or other building cleaning and grounds maintenance staff. In 2019, 333 of them were killed while at work. This is an annual US deaths of about 6.5 per 100,000.

STAY SAFE Pay attention to all work safety procedures. When appropriate, use all personal protective equipment and follow any applicable protocols regarding sanitization, sterility, and quarantine for any pathogens or hazmats. Exercise due caution when in and around work vehicles, both on and off the road, and when near heavy machinery, equipment, sources of high heat or deep cold, toxic substances, and stacks of material that might collapse. Be careful not to trip or fall on lift platforms, stairs, ladders, fire escapes, railings, and on roofs and parapets. Wear appropriate protective gear when around smoke, fumes, industrial cleaning products, insecticides, pesticides, and other toxic inhalants. Bosses should ensure that all OSHA and state safety standards are met. Get vaccinated and observe all safety protocols during public health emergencies.

SOURCES BLS, bls.gov/iif/oshcfoi1.htm; Y.H. Chen et al. Excess mortality associated with the COVID-19 pandemic among Californians 18–65 years of age, by occupational sector and occupation: March through November 2020. *PLoS One*, June 2021

BUILDING CLEANING AND PEST CONTROL

ANNUAL US DEATHS 63 in 2019

In 2020 in the US, 3.64 million people were employed in building cleaning and pest control. In 2019, 63 of them were killed on the job. This is an annual US deaths of about 1.7 per 100,000.

STAY SAFE Pay attention to all work safety procedures such as fire drills and active-shooter lockdown briefings. Wear appropriate protective gear around all chemicals and other toxic inhalants. Use personal protective equipment and follow any applicable protocols regarding sanitization, sterility, and quarantine for any pathogens or hazmats. Exercise due caution when in and around work vehicles, both on and off the road, and toxic substances. Be careful not to trip or fall on lift platforms, stairs, ladders, fire escapes, railings, and on roofs and parapets.

SOURCE BLS, bls.gov/iif/oshcfoi1.htm

GROUNDS MAINTENANCE

ANNUAL US DEATHS 97 in 2019

In 2020 in the US, 1.44 million people were employed in grounds maintenance. In 2019, 97 of them were killed on the job. This is an annual US death rate of about 6.7 per 100,000.

STAY SAFE Pay attention to all work safety procedures. When appropriate, use personal protective equipment and follow any applicable protocols regarding sanitization, sterility, and quarantine for any pathogens or hazmats. Exercise due caution when in and around work vehicles, both on and off the road. Wear appropriate protective gear when around smoke, fumes, fertilizers, pesticides, insecticides, herbicides, and other toxic inhalants.

SOURCE BLS, bls.gov/iif/oshcfoi1.htm

Information Industry

ANNUAL US DEATHS 31 in 2019

In 2020 in the US, 2.86 million people were employed in information industries, either providing the infrastructure to process data and distribute information via the internet, broadcast radio and TV, or other telecommunications, or via movies and video and audio recordings, or generating the actual news, entertainment, and cultural content that was transmitted. In 2019, 31 of them were killed on the job. This is an annual US death rate of about 1.1 per 100,000.

STAY SAFE Pay attention to all work safety procedures such as fire drills and active-shooter lockdown briefings.

SOURCE BLS, bls.gov/iif/oshcfoi1.htm

Computers and Math

ANNUAL US DEATHS 15 in 2019

About 5.6 million Americans were employed in 2020 as computer software, web, and information analysts, programmers, or designers, or as actuaries, mathematicians, or statisticians across the US economy. In 2019, 15 of them were killed while at work. This is an annual US deaths of about 0.27 per 100,000.

STAY SAFE   Pay attention to all work safety procedures such as fire drills and active-shooter lockdown briefings.

SOURCE   BLS, bls.gov/iif/oshcfoi1.htm

Insurance and Finance

ANNUAL US DEATHS   21 in 2019

In 2020 in the US, 6.55 million people were employed in insurance, banking, and employee benefits sales, administration, risk management, funds collection, investing and lending or disbursing, and various supporting activities. In 2019, 21 of them were killed on the job. This is an annual US death rate of about 0.32 per 100,000.

STAY SAFE   Pay attention to all work safety procedures such as fire drills and active-shooter lockdown briefings.

SOURCE   BLS, bls.gov/iif/oshcfoi1.htm

Installation, Maintenance, and Repair Occupations

ANNUAL US DEATHS   438 in 2019

About 4.60 million Americans were employed in 2020 across the US economy in a broad range of jobs involving the installation, maintenance, and repair of all sorts of equipment, machinery, and other gear needed by businesses and homes nationwide. They were either workers, or their first-line supervisors, working on computers, automated tellers, office machines, vending machines; communications and radio, avionics and other electrical or electronic devices, engines and motors, or other parts and equipment used in aircraft, motor vehicles, industries, or utilities; HVAC technicians and wind turbine technicians, machinery mechanics, appliance repairers, locksmiths, commercial divers, riggers, mobile home installers, and power and communications line installers/maintainers.

Such work exposed them to the risks of getting killed on the job: workplace interpersonal violence, using toxic fuels and lubricants, driving or riding in work vehicles, falling from a high place, and working with heavy equipment.

Like other employed people, they used stairs, elevators, ladders, or lift platforms and went through automatic doors and gates, which

Exotic repairs and maintenance: Coyote's Flying Saucer Retrievals and Repair Service, Ocotillo, California

could cause lethal falls or other deadly injuries. They could have a piece of heavy machinery land on them or be electrocuted or asphyxiated. They might be exposed to extremely hot or cold environments. Fires and explosions could happen and kill them, literally anywhere.

In 2019, 438 of them were killed on the job. This is an annual US death rate of about 9.5 per 100,000 FTE.

STAY SAFE   Pay attention to all work safety procedures such as fire drills and active-shooter lockdown briefings. When appropriate, use personal protective equipment and follow any applicable protocols regarding sanitization, sterility, quarantine, and hazmats. Be cautious in and around work vehicles and when near heavy machinery, equipment, sources of high heat or deep cold, toxic substances, and stacks of material that might collapse. Be careful not to trip or fall on lift platforms, towers, stairs, ladders, railings, and on roofs and parapets. Make sure to wear appropriate protective gear when around smoke, fumes, industrial cleaning products, pesticides, coatings (paints), or other toxic inhalants. Bosses should ensure that all OSHA, industry, and state safety standards are met.

SOURCE   BLS, bls.gov/iif/oshcfoi1.htm

Vehicle and Mobile Equipment

ANNUAL US DEATHS 155 in 2019

About 1.74 million Americans were employed in 2020 across the US economy in a broad range of jobs involving the maintenance and repair of transportation vehicles and road-mobile heavy equipment. They were workers or first-line supervisors, working on the different mechanical parts (including bodies and glass, pumps and valves), engines and motors, and various electric/electronic or hydraulic systems in cars, trucks, buses, aircraft, cranes, cherry pickers, and the like.

Such work exposed them to the risks of getting killed on the job: workplace interpersonal violence, using toxic fuels and lubricants, driving or riding in work vehicles, falling from a high place, and working with heavy equipment.

Like other employed people, they used stairs, elevators, ladders, or lift platforms and went through automatic doors and gates, which could cause lethal falls or other deadly injuries. They could have a piece of heavy machinery land on them or be electrocuted or asphyxiated. They might be exposed to extremely hot or cold environments. Fires and explosions could happen and kill them, literally anywhere.

In 2019, 155 of them were killed on the job. This is an annual US death rate of about 8.9 per 100,000 FTE.

STAY SAFE   Pay attention to all work safety procedures such as fire drills and active-shooter lockdown briefings. When appropriate, use personal protective equipment and follow any applicable protocols regarding sanitization, sterility, quarantine, and hazmats. Be cautious in and around work vehicles, and when near heavy machinery, equipment, sources of high heat or deep cold, toxic substances, and stacks of material that might collapse. Be careful not to trip or fall on lift platforms, towers, stairs, ladders, railings, or on roofs and parapets. Make sure to wear appropriate protective gear when around smoke, fumes, industrial cleaning products, coatings (paints), lubricants, pesticides, or other toxic inhalants. Bosses should ensure that all OSHA and state safety standards are met.

SOURCE   BLS, bls.gov/iif/oshcfoi1.htm

Personal Care and Services

ANNUAL US DEATHS   61 in 2019

About 3.40 million Americans were employed in 2020 around the US economy as personal care workers. Their activities ranged from working with animals to working in funeral homes, casinos, theaters, hotels and travel agencies, or barbershops and beauticians, to being childcare workers or personal trainers or residential advisors or other recreation and fitness workers.

In 2019, 61 of them were killed while at work. This is an annual US death rate of about 1.8 per 100,000.

STAY SAFE   Pay attention to all work safety procedures. Get vaccinated and wear personal protective equipment as appropriate.

SOURCE   BLS, bls.gov/iif/oshcfoi1.htm

Private Household Staff

ANNUAL US DEATHS   1 in 2019

In 2020 in the US, about 200,000 private households employed at least 1 person in and around the premises to do work related to running the household, as cooks, maids, nannies, gardeners, handymen, drivers, and others. In 2019, only 1 of them was killed on the job. Hard data is scarce, but assuming on average 1.5 FTE workers per household, this is an annual US death rate of about 0.33 per 100,000.

STAY SAFE   Pay attention to all work safety procedures. Use personal protective equipment as needed. Wear appropriate protective gear around cleaning products, insecticides, pesticides, herbicides, and other toxic inhalants.

SOURCE   BLS, bls.gov/iif/oshcfoi1.htm

Production Occupations

ANNUAL US DEATHS   245 in 2019

About 7.59 million Americans were employed in 2020 across the US economy in a broad range of jobs as either workers or first-line supervisors involved in manufacturing, fabricating, molding and shaping, rigging and assembling, finishing and bonding, sewing and pressing, cutting and punching, machining, polishing and buffing, welding and soldering, pouring and casting, mixing and blending, baking and roasting, printing, laundering, painting and packaging, inspecting and testing, or otherwise making or processing a myriad of manufactured goods ranging from metals, plastics, furniture and clothing to food, books, tires, jewelry, medical and dental appliances, and much more.

The nature and location of such work often exposed them to the risks of getting killed on the job: working with the public, using toxic chemicals, driving or riding in work vehicles, and working with heavy equipment.

Like other employed people, they used stairs, elevators, and escalators and went through automatic doors, which could cause lethal falls or other deadly injuries. They could have a heavy object fall on them or be electrocuted or asphyxiated. They might be exposed to extremely hot or cold environments. Fires and explosions could happen and kill them, literally anywhere.

In 2019, 245 of them were killed on the job. This is an annual US death rate of about 3.2 per 100,000 FTE.

STAY SAFE Pay attention to all work safety procedures such as fire drills and active-shooter lockdown briefings. When appropriate, use all personal protective equipment, and follow any applicable protocols regarding sanitization, sterility, cleanliness, and hazmats. Exercise due caution when in and around work vehicles, both on and off the road, and when near heavy machinery, equipment, sources of high heat or deep cold, toxic substances, and stacks of material that might collapse. Be careful not to trip or fall on lift platforms, stairs, ladders, fire escapes, railings, or on roofs and parapets. Make sure to wear a respirator or work in a well-ventilated area when around any smoke, fumes, industrial cleaning products, insecticides, particulates, adhesives and paints, or other toxic inhalants. Bosses should ensure that all OSHA and state safety standards are met.

SOURCE BLS, bls.gov/iif/oshcfoi1.htm

Chemical Manufacturing

ANNUAL US DEATHS   16 in 2019

In 2020 in the US, 848,800 people were employed as chemists, technicians, machine operators and tenders, fillers and packagers, and so on, by different types of chemical plants, turning raw organic and inorganic chemicals into useful outputs. The products manufactured ranged from pharmaceuticals to pesticides and fertilizers, to artificial and synthetic fibers and filaments, to various cleaning products, paints and coatings, resins, and more.

In 2019, 16 of them were killed on the job. This is an annual US death rate of about 1.89 per 100,000.

STAY SAFE   Pay attention to all work safety procedures. Use all personal protective equipment. Exercise due caution near heavy machinery, equipment, sources of high heat or deep cold, toxic substances, and stacks of material that might collapse. Wear appropriate protective gear. Bosses should ensure that all OSHA, industry, and state safety standards are met.

SOURCE   BLS, bls.gov/iif/oshcfoi1.htm

Paper Product Manufacturing (Paper Mills)

ANNUAL US DEATHS   9 in 2019

In 2020 in the US, 356,800 people were employed in manufacturing various paper products, turning wood into pulp, processing the pulp into sheets, and then cutting, shaping, and laminating the sheets into finished paper products. In 2019, 9 of them were killed on the job. This is an annual US death rate of about 2.52 per 100,000.

STAY SAFE   Pay attention to all work safety procedures. Use all personal protective equipment. Exercise due caution near heavy machinery, equipment, sources of high heat or deep cold, toxic substances, and stacks of material that might collapse. Wear appropriated protective gear. Bosses should ensure that all OSHA, industry, and state safety standards are met.

SOURCES   BLS, bls.gov/iif/oshcfoi1.htm; OSHA, osha.gov/pulp-paper

Shipyard and Dock Workers

ANNUAL US DEATHS   Estimated 7.5

Employees in US shipyards and on America's docks and piers work in a very specialized and potentially dangerous environment. They build, repair, and load and unload ships, boats, barges, and ferries. More than any average land-based construction site, or a rail/truck intermodal terminal, these waterborne transportation vehicles are gigantic, very tall, massively heavy, and working on them exposes people to the added risks of rough seas and drowning. Yet 99 percent of the volume of overseas trade enters and leaves the US by ship.

About 105,500 people were employed in US private-sector shipyards in 2015, plus another roughly 60,000 workers in federal government–owned shipyards (operated by the US Navy), spread across 26 states bordering coastal and inland waters. Their annual occupational death rate averaged 4.0 per 100,000 over the period 2011 through 2017. Between 2011 and 2017, there were at least 45 fatal accidents (4.0 per 100,000) among shipyard workers.

In 2017 in the US, about 98,000 people were employed in marine terminals and port operations. (Before the introduction of modern-era shipping containers and heavy mobile lift-and-shift cranes, they were called longshoremen and stevedores.) Their annual occupational death rate averaged 15.9 per 100,000 over the period 2011 through 2017—fully 5 times that of the overall US workforce.

STAY SAFE   See entries above for **Construction Industry, Installation, Maintenance, and Repair Occupations**, and **Transportation and Material-Moving Occupations**.

SOURCES   BLS, bls.gov/iif/oshcfoi1.htm; CDC NIOSH, cdc.gov/niosh/programs/cmshs/shipyards.html; OSHA, osha.gov/ship-building-repair

Textile Mills

ANNUAL US DEATHS   1 in 2019

In 2020 in the US, 95,100 people were employed by textile mills, turning natural or artificial fibers into yarn or fabric that is then cut and sewn to become apparel, sheets and towels, and other useful items of cloth. In 2019, only 1 of them was killed on the job. This is an annual US death rate of about 1.05 per 100,000.

STAY SAFE   Pay attention to all work safety procedures. Use personal protective equipment as needed. Exercise due caution near heavy machinery, equipment, sources of high heat or deep cold, toxic substances, and stacks of material that might collapse. Wear appropriate protective gear. Bosses should ensure that all OSHA, industry, and state safety standards are met.

SOURCES   BLS, bls.gov/iif/oshcfoi1.htm; OSHA, osha.gov/textiles

This welder is making boilers in a World War II–era defense plant.

Welding

ANNUAL US DEATHS   About 60

Occupational welding, cutting, and brazing activities in the US have an annual US deaths of approximately 60 workers each year, or about 10 per 100,000 workers. Causes of death include electrocution from arc welding and fires and explosions from gas welding. America's roughly 550,000 welders perform one of the most dangerous trades in the entire construction industry.

STAY SAFE   Know and observe safe welding practices.

SOURCES   BLS, bls.gov/iif/oshcfoi1.htm; OSHA, osha.gov/welding-cutting-brazing

Wood Product Manufacturing (Sawmills)

ANNUAL US DEATHS   24 in 2019

In 2020 in the US, 403,300 people were employed in manufacturing various wood products, starting with logs and then sawing, planning, and otherwise shaping and finishing them into lumber, flooring, wooden crates and boxes, roof trusses, prefabricated structures, and more. In 2019, 24 of them were killed on the job. This is an annual US deaths of about 5.96 per 100,000.

STAY SAFE   Pay attention to all work safety procedures. Use personal protective equipment as needed. Exercise due caution near heavy machinery, equipment, sources of high heat or deep cold, toxic substances, and stacks of material that might collapse. Wear appropriate protective gear. Bosses should ensure that all OSHA, industry, and state safety standards are met.

SOURCES BLS, bls.gov/iif/oshcfoi1.htm; OSHA, osha.gov/sawmills

Real Estate and Property Rental and Leasing

> **ANNUAL US DEATHS** 87 in 2019

In 2020 in the US, 2.24 million people were employed in helping others to appraise, buy, sell, or rent real estate and other tangible assets, and/or providing follow-on property management services. In 2019, 87 of them were killed on the job. This is an annual US deaths of about 3.9 per 100,000.

STAY SAFE Pay attention to all work safety procedures.

SOURCE BLS, bls.gov/iif/oshcfoi1.htm

Retail Trade

> **ANNUAL US DEATHS** 291 in 2019

In 2020 in the US, 15.6 million people were employed in retail trade, selling (and often installing and/or repairing) finished merchandise, in small individual quantities, to the general public, either in stores or by other marketing methods such as mail order catalogs, door-to-door salesmen, or via the internet. In 2019, 291 of them were killed on the job. This is an annual US death rate of about 1.9 per 100,000.

STAY SAFE Pay attention to all work safety procedures such as fire drills and active-shooter lockdown briefings. Be aware of the potential for customer and workplace violence.

SOURCE BLS, bls.gov/iif/oshcfoi1.htm

Food and Beverage Stores

> **ANNUAL US DEATHS** 54 in 2019

In 2020 in the US, 3.15 million people were employed in retail trade at food and beverage stores, selling these products to the general public.

In 2019, 54 of them were killed on the job. This is an annual US death rate of about 1.7 per 100,000.

STAY SAFE　Pay attention to all work safety procedures such as fire drills and active-shooter lockdown briefings. Be aware of the potential for customer and workplace violence. Use all personal protective equipment. Exercise due caution near stacks of material that might collapse.

SOURCE　BLS, bls.gov/iif/oshcfoi1.htm

Elevators and Escalators

ANNUAL US DEATHS　About 20 per year for elevators, 2.4 for escalators

For elevators not in residences and for escalators in general, death statistics are broken into 3 categories: workers involved in constructing or maintaining buildings and their elevators and escalators, passengers riding them while at work, and passengers riding them while not at work.

Many nonworkplace elevators are residential (e.g., in apartment buildings or private houses). Many nonworkplace escalators are non-residential (e.g., in stores and malls, sports arenas, or mass transit stations).

Based on BLS data for 1992 through 2009, the average annual US deaths for workers involved in constructing/maintaining building elevators was 15; for escalators, it was 0.4. Based on CPSC data for 1997 through mid-2010, the average annual US deaths for passengers at work was 5 for elevators and 0.2 for escalators. For passengers not at work, it was 5 for elevators; for escalators, it was 2. Falls down elevator shafts can be much deadlier than trip-and-fall mishaps on escalators. Elevator doors and elevator cars moving between floors present extra risks of deadly accidents. Elevator shafts in use can be especially dangerous workplaces due to the moving elevator car and also the moving counterweight.

STAY SAFE　For recommendations about elevator and escalator safety for riders, see National Elevator Industry (neii.org).

SOURCE　BLS, bls.gov/iif/oshcfoi1.htm

Gas Stations

ANNUAL US DEATHS   51 in 2019

In 2020 in the US, 937,500 people were employed in gas stations as cashiers, clerks, and attendants. In 2019, 51 of them were killed on the job. This is an annual US deaths of about 5.44 per 100,000.

STAY SAFE   Pay attention to all work safety procedures, including fire and response procedures to attempted crimes. Be aware of the potential for customer violence.

SOURCE   BLS, bls.gov/iif/oshcfoi1.htm

LATE-NIGHT VIOLENCE AT CONVENIENCE AND LIQUOR STORES AND GAS STATIONS

ANNUAL US DEATHS   78

A study from 2007 indicated that nearly half of all retail trade workers killed on the job died as a result of violent crime in late-night establishments. The breakdown of such homicides that year was 39 convenience store workers, 32 gas station workers, and 7 liquor store workers.

STAY SAFE   Pay attention to all work safety procedures, including fire and response procedures to attempted crimes. Be aware of the potential for customer violence.

SOURCE   A. Altizo and D. York. Robbery of Convenience Stores. Guide No. 49 (2007), ASU Center for Problem-Oriented Policing, popcenter.asu.edu

Motor Vehicle and Parts Dealers

ANNUAL US DEATHS   58 in 2019

In 2020 in the US, 1.95 million people were employed at showrooms and lots in selling cars, SUVs, trucks, motorcycles, and RVs to the general public. Their jobs ranged from sales to financing and registration support to maintenance and repair. In 2019, 58 of them were killed on the job. This is an annual US death rate of about 2.97 per 100,000.

STAY SAFE   Pay attention to all work safety procedures. Use personal protective equipment. Exercise due caution when in and around motor vehicles. Be aware of the potential for customer and workplace violence.

SOURCE BLS, bls.gov/iif/oshcfoi1.htm

Shopping Carts

ANNUAL US DEATHS Comprehensive data unavailable, but it is relevant that some 20 children riding in shopping carts are hit and killed by automobiles.

Comprehensive data is not readily available on how many children are killed by shopping cart accidents in the US every year. However, about 20 children are killed every year when riding in shopping carts that are hit by cars. About 24,000 children are injured each year by shopping cart accidents, badly enough to need emergency room treatment; about 4,000 of these spend at least 1 night in the hospital. The most common mishaps are for the child to fall out of the cart or for the cart to tip over onto the child. The most common location of the injury is the child's head and face.

In at least 1 instance, an infant was killed when the baby carrier he was in was placed on a shopping cart but not secured by the safety straps. The infant fell out of the cart when it hit a speed bump in the parking lot. In another mishap, a young child not strapped in stood up, lost his balance, fell over backward, hit his head, and was killed.

STAY SAFE For child safety around and in shopping carts, see "Shopping Cart Safety" at nationwidechildrens.org. At the store, always select carts that are in good repair, roll well and evenly, and have fully intact and working safety harness straps and buckles. Never let your child stand in the cart or ride in or under the cargo area. Never put 2 kids in a cart together or let a child push a cart. Stay with your child at all times. Exercise all due caution around moving traffic in the parking lot. If your infant is in a portable car seat or baby carrier, use the cart's safety straps to secure the carrier to the cart, and make sure the baby is secured in the seat or carrier.

SOURCE Nationwide Children's Hospital, nationwidechildrens.org

Vending Machines

ANNUAL US DEATHS 2.2

Tipping or rocking a vending machine to try to get something out of it, either for free or because the machine ate your money and didn't disgorge your merchandise, is risky. In the US, vending machines kill an average of 2.2 people every year by tip-overs that crush the victim.

Vending machines are ubiquitous throughout the world.

Vending machines are very heavy, the glass in front can shatter and inflict lethal lacerations, and they carry line current that can electrocute you in a mishap.

STAY SAFE Never shake, tip, rock, or kick a vending machine, or climb on top of one. Never break into one or reach into one with your arm. If the machine takes your money and doesn't dispense your purchase, call the machine complaint number.

SOURCES CDC, cdc.gov/injury/wisqars/fatal.html; CPSC, cpsc. gov/Newsroom/News-Releases/1996/CPSC-Soda-Vending-Machine-Industry-Labeling-Campaign-Warns-Of-Deaths-And-Injuries

Professional Sports

ANNUAL US DEATHS   Average of less than 1

In the US in 2016, professional sports workers (athletes, coaches, umpires, and referees) had an overall death rate of 11.7 per 100,000. Overall, elite athletes live significantly longer than the general population. Male athletes have a 31 percent lower risk of all-cause mortality; female athletes have a 49 percent lower risk of all-cause mortality.

Active players in the 4 major sports leagues (NBA, NFL, NHL, MLB) have an overall low death rate. As of 2016, of the 53,400 male athletes who historically played in the 4 leagues, only 205 (0.38 percent) died while active. Many of the deaths were accidental (mostly car crashes).

STAY SAFE   Always wear appropriate safety gear. Avoid head injuries.

SOURCES   S. Lemez et al. Early death in active professional athletes: Trends and causes. *Scandinavian Journal of Medicine & Science in Sports*, May 2016; A. Runacres et al. Health consequences of an elite sporting career: long-term detriment or long-term gain? A meta-analysis of 165,000 former athletes. *Sports Medicine*, February 2021.

BASE Jumping

ANNUAL US DEATHS   383 worldwide deaths since 1981

BASE jumping is considered the world's most dangerous sport. Athletes jump from a fixed object with a parachute pack they then open to try to land safely or jump wearing a wingsuit and fly to land. BASE stands for the things they can jump from: Buildings, Antenna masts, Spans (bridges), or Earth (cliffs). Given the relatively low altitude from which they start their jump, the risk of death or injury is 43 times that of parachuting from an airplane. A total of 383 people worldwide have been killed BASE jumping since 1981. A study in Norway found 1 death per every 2,317 jumps. Presumably, each active BASE-jumping competitor makes several jumps per year.

STAY SAFE   BASE jump at your own risk. Make sure all equipment is in good working order.

SOURCE   Base Fatality List, bfl.baseaddict.com/list; Basejumper.com; K. Soreide et al. How dangerous is BASE jumping? An analysis of adverse events in 20,850 jumps from the Kjerag Massif, Norway. *Journal of Trauma*, May 2007.

Professional Boxing

ANNUAL US DEATHS   Approximately 9 to 10

Between 1890 and 2011, an estimated 1,604 boxers worldwide died as a direct result of injuries sustained in the ring, for a historic average of 13 deaths a year. Between 1950 and 2007 worldwide, 339 boxers died; 64 percent of deaths were associated with knockout and 15 percent with technical knockout. Around the world today, on average 13 boxers a year die as a direct result of injuries in the ring. Almost all boxing-related fatalities result from traumatic brain injury.

Boxing deaths get a lot of publicity, but boxing has a fatality rate of 0.13 deaths per 1,000 participants per year. This compares favorably with other high-risk sports such as scuba diving, hang gliding, and college football.

SOURCES L.C. Baird et al. Mortality resulting from head injury in professional boxing: case report. *Neurosurgery*, 2010; Death Under the Spotlight: The Manuel Velazquez Collection, ejmas.com/jcs/velazquez/jcsart_svinth_0711.html; R.L. Heilbronner et al. Neuropsychological consequences of boxing and recommendations to improve safety: a National Academy of Neuropsychology education paper. *Archives of Clinical Neuropsychology*, 2009.

Major League Baseball

ANNUAL US DEATHS   431 from 1986 to 2020

Deaths on the field among active players in Major League Baseball are very rare. However, retired players are more likely to die prematurely. From 1986 to 2020 among 2,708 former players, 431 died, for an annual US deaths of about 0.46 percent. Looked at historically, however, MLB players have a lower mortality rate. A study that looked at deaths among all MLB players from 1979 to 2013 (10,451 players in all) found that the mean age at death was 77.1 years. Compared to all US males, MLB players had significantly lower mortality rates from all causes (24 percent) and for most underlying causes of death excluding neurodegenerative disease. Compared to 20-year-old US males, MLB players can expect almost 5 additional years of life.

On opening day 1996, umpire John McSherry collapsed behind the plate and died. His death was only the second to occur in the course of a major league game. The first was the Cleveland Indians' Ray Chapman, who was struck and killed by a pitch during a game in 1920.

STAY SAFE   Live a healthy lifestyle.

SOURCES   V.T. Nguyen et al. All-cause and cause-specific mortality among Major League Baseball players. *JAMA Internal Medicine*, September 2019; J.M. Saint Onge et al. Major League Baseball players' life expectancies. *Social Science Quarterly*, July 2008.

National Basketball Association

ANNUAL US DEATHS   Elevated risk of sudden cardiac death but, overall, no special risk

Pro basketball players in the US have an above-average rate of dying young from sudden cardiac death, usually related to hypertrophic cardiomyopathy. The taller the player, the greater the risk; the risk is greatest for the tallest Black players. Overall, however, among the cohort of NBA players from its inception in 1946 to April 2015 (481 active and 3,504 former players), the median age at death was 81.3 years.

STAY SAFE   Have routine physicals, including an EKG.

SOURCE   J.A. Martínez et al. Mortality of NBA players: risk factors and comparison with the general US population. *Applied Sciences*, February 2019.

National Football League

ANNUAL US DEATHS   517 between 1986 and 2020 (retired players who died prematurely)

Deaths on the field among active players in the NFL are very rare. However, retired players are more likely to die prematurely. From 1986 to 2020 among 3,419 retired players, 517 died, for an annual US deaths of about 0.44 percent. Compared to professional baseball players, professional football players have an increased risk of death. A recent study looked 3,419 NFL players and 2,708 MLB players over the 30-year period from 1979 to 2013. In that time, 517 NFL players died, and 431 MLB players died. The NFL players had a 26 percent higher mortality rate. They also had 3 times the likelihood of dying from a neurodegenerative condition and 2.5 times the risk of dying from heart disease. (See also **Chronic Traumatic Encephalopathy**.)

The only NFL player to die on the field during a game was Chuck Hughes, a wide receiver for the Detroit Lions, who died of a heart attack on October 24, 1971, in a game against the Chicago Bears. The Bears won.

STAY SAFE   Wear a top-quality helmet and get regular medical checkups.

SOURCES   CDC NIOSH; V.T. Nguyen et al. Mortality Among Professional American-Style Football Players and Professional American Baseball Players. *JAMA Network Open*, May 2019; B. Walia et al. Age at League Entry and Early All-Cause Mortality among National Football League Players. *International Journal of Environmental Research and Public Health*, December 2021.

National Hockey League

> **ANNUAL US DEATHS**   2 on-ice deaths (US and Canadian) over 12 years

Ice hockey averaged 2 on-ice deaths over 12 years among about 1,000 NHL players in the US and Canada, for an annual US death rate of 16.7 per 100,000.

STAY SAFE   Wear top-quality protective gear and get regular medical checkups.

SOURCE   C. Mueller. Ranking Sports from Least to Most Dangerous: Includes NFL, NBA, NHL and Soccer. *Bleacher Report*, October 3, 2011.

Soccer

> **ANNUAL US DEATHS**   70 worldwide over 12 years

Professional soccer has suffered 70 deaths on the field worldwide over 12 years. A survey by FIFA, soccer's worldwide governing body, indicated there were about 113,000 professional soccer players globally. This suggests an annual US death rate of about 5.16 per 100,000. These statistics do not include professional women soccer players.

STAY SAFE   Wear top-quality protective gear and get regular medical checkups.

SOURCE   CDC, cdc.gov/nchs/fastats/accidental-injury.htm

Transportation and Material-Moving Occupations

> **ANNUAL US DEATHS**   1,481 in 2019

About 10.6 million Americans were employed in 2020 across the US economy, as either workers or frontline supervisors, either directly or in support roles, in a broad range of jobs that involved transporting, moving, conveying, carrying, shifting or shipping people and/or goods. They were drivers or operators of every sort of transportation mode or vehicle ranging from airplane flight crews and air traffic controllers to locomotive engineers to taxi and limo and ambulance drivers and truck or bus drivers, to crane and dredge operators, refuse and recyclable collectors, to refueling, cleaning, and parking workers, to stockers, order fillers, machine feeders and off-bearers, passenger attendants, and other workers involved in the transportation or moving of material.

The nature and location of such work often exposed them to the risks of getting killed on the job: working with the public, using toxic chemicals, driving or riding in work vehicles that might crash or sink, carrying loads and working with equipment that might squash or maim them.

Like other employed people, they used stairs, elevators, and escalators and went through automatic doors, which could cause lethal falls or other deadly injuries. They could have a heavy object fall on them or be electrocuted or asphyxiated. They might be exposed to extremely hot or cold environments. Fires and explosions could happen and kill them, literally anywhere.

In 2019, 1,481 of them were killed on the job. This is an annual US death rate of about 14.0 per 100,000 FTE.

STAY SAFE Pay attention to all work safety procedures. Use all personal protective equipment. Exercise due caution when in and around motor vehicles, trains, ships, and aircraft, both on and off the road (or on the rails, on the water, or in the air), and whenever near other heavy machinery, equipment, sources of high heat or deep cold, toxic substances, and stacks of material that might collapse. Be careful not to trip or fall on lift platforms, stairs, ladders, fire escapes, railings, or on roofs and parapets. Make sure to wear a respirator or appropriate protective gear when around any smoke, dust, fumes, industrial cleaning products, insecticides, or other toxic inhalants. Bosses should ensure that all OSHA, industry, and state safety standards are met.

SOURCE bls.gov/iif/oshcfoi1.htm

Air Transportation Workers (Commercial and Aerial-Work Aviation)

ANNUAL US DEATHS 85 in 2019

In 2020 in the US, 274,000 people were employed in commercial aviation (both air transportation and ground support), including common carrier, commercial, corporate, air taxi, or charter fixed-wing large and small airplanes and helicopters, providing services for passengers, freight and packages, medevac, crop dusting, wildfire fighting, tourism, newscasting, and land-use and rail, road, and utility-line inspection/maintenance flights. In 2019, 85 of them were killed on the job. This is an annual US death rate of about 31.0 per 100,000.

Aviation mortality statistics are covered in more detail in the section **On the Go: In the Air**.

STAY SAFE Pay attention to all work safety procedures. Use personal protective equipment. Exercise due caution when in and around aircraft and whenever near heavy machinery, equipment, sources of high heat or deep cold, toxic substances, and stacks of material that might collapse. Be careful not to trip or fall on lift platforms, stairs, ladders, fire escapes, railings, or on roofs and parapets. Bosses should ensure that all OSHA, FAA, and state safety standards are met.

USAF cargo loaders at work

SOURCES BLS, bls.gov/iif/oshcfoi1.htm; FAA, faa.gov/data_research/accident_incident

Material-Moving Workers (Ground)

ANNUAL US DEATHS 238 in 2019

In 2020 in the US, 4.41 million people were employed in moving, carrying, shipping, or shifting goods and materials on the ground: inside and/or outside fixed structures and enclosures (factories, warehouses, private homes) and/or into/out of road, sea, or air vehicles while stationary. In 2019, 238 of them were killed on the job. This is an annual US death rate of about 5.4 per 100,000.

STAY SAFE Pay attention to all work safety procedures. Use personal protective equipment. Exercise due caution when in and around motor vehicles and other heavy machinery, equipment, sources of high heat or deep cold, toxic substances, and stacks of material that might collapse. Be careful not to trip or fall on lift platforms, stairs, ladders, fire escapes, railings, or on roofs and parapets. Bosses should ensure that all OSHA and state safety standards are met.

SOURCE BLS, bls.gov/iif/oshcfoi1.htm

Motor Vehicle Operators

ANNUAL US DEATHS   1,091 in 2019

In 2020 in the US, 5.75 million people were employed in operating (driving and/or crewing) different kinds of motor vehicle transportation and direct support occupations. In 2019, 1,091 of them were killed on the job. This is an annual US death rate of about 19.0 per 100,000.

Motor vehicle operator mortality statistics are covered in more detail in **On the Go: On the Road.**

STAY SAFE   Pay attention to all work safety procedures. Use personal protective equipment. Drive carefully; never drive drunk, impaired, or drowsy. Bosses should ensure that all OSHA, FHA, and state safety standards are met.

SOURCE   BLS, bls.gov/iif/oshcfoi1.htm

Railroad Workers

ANNUAL US DEATHS   11 in 2019

In 2020 in the US, 107,000 people were employed in freight, passenger, and mass transit rail operations and direct support. In 2019, 11 of them were killed on the job. This is an annual US death rate of about 10.3 per 100,000.

Full railroad mortality statistics are covered in more detail in the section **On the Go: On the Rails.**

STAY SAFE   Pay attention to all work safety procedures. Use personal protective equipment. Exercise due caution when in and around motor vehicles and trains and whenever near other heavy machinery, equipment, sources of high heat or deep cold, toxic substances, and stacks of material that might collapse. Be careful not to trip or fall on lift platforms, stairs, ladders, fire escapes, railings, or on roofs and parapets. Wear a respirator or appropriate protective gear around any smoke, fumes, dust, industrial cleaning products, insecticides, or other toxic inhalants. Bosses should ensure that all OSHA, FRA, and state safety standards are met.

SOURCES   BLS, bls.gov/iif/oshcfoi1.htm; FRA, railroads.dot.gov/railroad-safety/accident-data-reporting-and-investigations

Commercial Ships and Boats

> **ANNUAL US DEATHS** 13 in 2019

In 2020 in the US, 83,000 people were employed in crewing and servicing various types of passenger, cargo, and vehicle ships, boats, and ferries. In 2019, 13 of them were killed on the job, up slightly from the annual average of 12.4 for the years 2011 through 2017. This is an annual US death rate, in 2019, of about 15.7 per 100,000 FTE. (Passenger ship and recreational boating mortality statistics are covered in the **At Play** section.) In 2017 in the US, there were 13 deaths specifically among tugboat or towboat crews. Six were related to operations, while 7 were due to natural causes.

Waterborne and amphibious vessels serve a myriad of commercial purposes. Some carry passengers on day trips or longer voyages, for recreation and/or for transportation. Others carry only cargo. Their basic structure is optimized for 1 particular type of freight, whether it be solid bulk or break bulk like grain, ore, coal, or gravel; or bulk liquids, such as on a gasoline or oil tanker or a liquefied natural gas carrier; or brand-new motor vehicles, such as on a roll-on-roll-off (RO-RO) carrier; or 10,000 or more loaded shipping containers on a huge and mostly hollow container ship.

These vessels traverse 4 basic types of bodies of water, depending on what routes they generally transit, and for what purposes. These are navigable rivers (and canals), such as the Mississippi or Hudson or the Houston Shipping Channel; large lakes, such as the Great Lakes; coastal routes, such as between New York Harbor and the Port of Miami; and oceangoing routes, such as Boston to Liverpool. The size, engine power, and overall design of each vessel tends to be customized and optimized for the type of route it is intended to ply, depending on whether the water is freshwater or saltwater, how deep or shallow it is, how narrow or winding the channel might get, how high the waves and strong the winds during storms.

A number of different mishaps can occur aboard a ship, some of which might kill individuals, some of which might cause the loss of all hands—or somewhere in between. Ships sometimes sink, in severe storms or due to springing bad leaks or colliding with other ships or big icebergs, they catch fire, explode, run aground, hit piers or buoys or canal locks, and more. Individual crew members are killed by natural causes (e.g., heart attacks or pandemics), falling or being washed overboard and drowning, being electrocuted, being caught in shipboard fires, being asphyxiated by toxic fumes in enclosed spaces, being caught in or crushed by equipment such as winches and cranes, or due to trauma from collisions, slips and falls, suicides, workplace violence, and more.

In 2015 the American cargo ship *El Faro* sank with all 33 members of its crew after sailing into a Category 3 hurricane. This was the worst US maritime disaster since 1985.

STAY SAFE Pay attention to all work safety procedures including fire, lifeboat, and man-overboard drills. Use personal protective equipment and a flotation device. Exercise due caution when near heavy machinery, equipment, sources of high heat or deep cold, toxic substances, and stacks of material that might collapse. Bosses should ensure that all OSHA, US Coast Guard, Department of Transportation Maritime Administration, and state safety standards are met.

SOURCES BLS, bls.gov/iif/oshcfoi1.htm; US Department of Transportation Maritime Administration, maritime.dot.gov/data-reports; NTSB accident report, ntsb.gov, Sinking of US Cargo Vessel SS El Faro Atlantic Ocean, Northeast of Acklins and Crooked Island, Bahamas, October 1, 2015; USCG, dco.uscg.mil/Our-Organization/ Assistant-Commandant-for-Prevention-Policy-CG-5P

US Military Service

ANNUAL US DEATHS 1,485 in 2010

The United States military is primarily part of the US Department of Defense (DoD). DoD's forces consist of officers and enlisted personnel, each serving as either active duty, reserve, or National Guard. Each serves in a specific branch: the US Army, the US Air Force, the US Coast Guard (after 9/11, part of the Department of Homeland Security), the US Marine Corps, the US Navy, or the new US Space Force (organized as a separate branch under the USAF).

Serving military men and women can have any one of many different job specialties. Some of these extend beyond the unique frontline war-fighting capabilities not encountered in civilian employment, such as fighter pilot or tank driver. Active duty, reserve, and National Guard members often work in DoD's extensive tail of logistics and support services, healthcare, construction, training, and administration, or they might be a commander or assigned to a command staff. Any one of these volunteers might from time to time be deployed into combat operations somewhere in the world, leading to some of them being killed in action (KIA).

Soldiers, sailors, airmen, and marines could also be killed occupationally (while wearing the uniform) due to other causes of death. The most recent published data that tabulates all US military deaths on an *annual* basis, rather than cumulatively over all the years of the different named operations of America's recent "Forever Wars," is for the calendar year 2010: 456 were killed in combat, 424 in accidents, 39 were murdered, 238 had fatal illnesses, and 289 committed suicide. None that year were counted as killed by terrorists while *not* combat-deployed; 39 deaths were listed as pending or unknown causes. The total of these deaths was 1,485 in 2010.

A breakdown of total active duty officers plus enlisted personnel as of March 2020 was:

US Army	475,000
US Navy	335,000
US Air Force	329,000
US Marine Corps	185,000
US Coast Guard	41,000
US Space Force	4,800
Total	1,369,800

This gives a very rough estimate of an annual military service occupational death rate of about 108 per 100,000 FTE serving personnel.

STAY SAFE "Follow your training" is as true as ever. For help coping with PTSD, call the 24/7/365 free, confidential Veteran Crisis Line at 1-800-293-8255. VA coverage is not needed.

SOURCES BLS, bls.gov/iif/oshcfoi1.htm; Congressional Research Service (CRS) report, crsreports.congress.gov, Trends in Active-Duty Military Deaths Since 2006; Department of Defense, dod.gov, Casualty Status report

Utilities

ANNUAL US DEATHS 22 in 2019

In 2020 in the US, 549,000 people were employed in all aspects of producing and then delivering, or removing and then disposing of, utilities ranging from electricity (generated by coal, natural gas, or nuclear power), to home/office cooking and heating gas, water and steam supplies, to sewage removal and treatment. Their jobs included everyone from billing clerks to line pole and manhole maintenance crews. In 2019, 22 of them were killed on the job. This is an annual US death rate of about 4.0 per 100,000.

STAY SAFE Pay attention to all work safety procedures. Use personal protective equipment. Exercise due caution when in and around motor vehicles and other heavy machinery and equipment. Be careful not to trip or fall on lift platforms, stairs, ladders, fire escapes, railings, or on roofs and parapets. Bosses should ensure that all OSHA and state safety standards are met.

SOURCES BLS, bls.gov/iif/oshcfoi1.htm; OSHA, osha.gov/power-generation

Wholesale Trade

ANNUAL US DEATHS 178 in 2019

In 2020 in the US, 5.9 million people were employed in wholesale trade, selling raw materials or finished goods to other businesses for further processing and/or eventual resale. In 2019, 178 of them were killed on the job. This is an annual US death rate of about 3.0 per 100,000.

STAY SAFE Pay attention to all work safety procedures. Use personal protective equipment. Exercise due caution when in and around motor vehicles and whenever near stacks of material that might collapse. Bosses should ensure that all OSHA and state safety standards are met.

SOURCE BLS, bls.gov/iif/oshcfoi1.htm

Beverage and Tobacco Products

ANNUAL US DEATHS 8 in 2019

In 2020 in the US, about 88,000 people were employed in making, bottling, and shipping soft drinks, wine and beer, and distilled spirits or cigarettes, cigars, chewing tobacco, and pipe tobacco. In 2019, 8 of them were killed on the job. This is an annual US death rate of about 9.09 per 100,000.

STAY SAFE Pay attention to all work safety procedures. Use personal protective equipment. Exercise due caution when in and around motor vehicles and when near heavy machinery, equipment, sources of high heat or deep cold, toxic substances, and stacks of material that might collapse. Be careful not to trip or fall on lift platforms, stairs, ladders, fire escapes, railings, or on roofs and parapets. Bosses should ensure that all OSHA and state safety standards are met.

SOURCE bls.gov/iif/oshcfoi1.htm

> **ANNUAL US DEATHS**   See individual entries.

Every American, one way or another, participates in play, recreational, leisure, and hobby activities. Gatherings of people, strenuous physical activities, and anything needing supplies and equipment carries some degree of risk of injury and even death. But it is important to have fun, relax, let off steam, pursue your hobby interests and passions, and keep physically fit, all in ways that enrich your life and enhance your life expectancy without resulting in tragedy.

Note that some of the activities treated in this section involve working with or watching or otherwise relying on professionals. The lethal risks for all such professional players are detailed in **On the Job,** whereas **At Play** focuses on the visitors, spectators, customers, patrons, and amateur participants, who engage in the activities during their leisure time. **At Play** also includes at-home hobbies, but household chores and home maintenance lethal risks are included in the **At Home** section.

Note also that being **At Play** (and also being **At Home**) doesn't involve the same sort of carefully clocked, measured, and regulated activities as being **On the Job** or **On the Go**. On the job, people typically work set hours, sometimes even punching a time clock; each employer needs to keep careful records for payroll and other purposes. While people are traveling, their vehicle, whether owned personally or operated by a commercial common carrier, has an odometer, and companies may have maintenance crews who keep careful logs. Also, government regulatory bodies such as OSHA and the NTSB issue safety guidelines, inspect for compliance, and investigate any fatalities.

Government reporting requirements allow reliable annual US death rates to be calculated, based on good total estimates for total units in the rate calculation's denominator, like 100 million Vehicle Miles Traveled (VMT), or 100,000 Full-Time Equivalent (FTE) workers employed. While virtually everyone in the US relaxes, unwinds, and enjoys themselves some of the time, very few of us—at least when out of school—keep tallies of cumulative time elapsed, or even just number of visits, during our various indoor and outdoor leisure exposures. For organized high school and college sports, records of how many students join which team for a school year are detailed enough, in some but not all cases, to calculate statistically meaningful, national annual US death rates per 100,000 yearly participants. For some sports, such as golf, accurate records of participants are kept, so death rates per outing can be determined.

To evaluate and compare different recreational lethal risks, the best statistics to look at vary, depending on what data is readily available. Sometimes it is the total number of people killed in a year by (during) each different activity, sport or game, hobby, or pastime. Sometimes it is an annual US death rate per 100,000.

At Play Aloft

Aerial Drones (Unmanned Aerial Vehicles)

ANNUAL US DEATHS   None yet reported in the US

Flying aerial drones—also known as unmanned aerial vehicles (UAVs)—for fun is now a very popular hobby, but drones are not mere toys. They have many commercial uses, such as photography and film-making, industrial inspection, agriculture, and real estate. They also have important military applications in conflict zones. Armed drones have made weapon strikes to kill enemy terrorists—tragically, they have sometimes also killed civilians by mistake or as collateral damage.

Although no fatal accidents involving a commercial or recreational drone have been reported in the US (yet), there have been hundreds of crashes, and they have on occasion caused serious injury. In 2014, a drone carrying mistletoe as a Christmas promotion hit a news photographer in the face and cut off the tip of her nose. Drones have hit athletes and spectators at sporting events on several occasions, causing wounds that needed stitches. In Cape Town, South Africa, in 2016, a drone flown by amateurs crashed through a 5th-floor window and hit a man in the head. In Quebec in 2017, a small plane with 8 people aboard hit a drone as the plane came in for a landing; by a miracle, no one was hurt, and the plane landed safely with minor damage. At least twice in the UK, recreational train-spotting drones have crashed into trains, again with no injuries.

STAY SAFE   Learn how to operate your drone safely, and never go beyond your level of skill and experience or the drone's capabilities. Be extremely cautious of spinning blades and any other moving parts. Never fly a drone at a person on the ground for any reason. Operating a drone around airplanes, helicopters, and airports is dangerous and illegal. Unauthorized operators may be subject to stiff fines and criminal charges, including possible jail time. The FAA is enacting stricter regulations for both commercial and recreational drone licensing, identification, and operation. Report illegal drone operations to local law enforcement.

SOURCES FAA, faa.gov/uas; III, iii.org/fact-statistic/
facts-statistics-aviation-and-drones

BASE Jumping (Amateur)

ANNUAL US DEATHS From 1995 through 2005, 40 per 100,000 jumps

The acronym BASE stands for some of the high points on land from which the participant jumps, wearing a parachute that they open on the way down or wearing a wingsuit: Building, Antenna tower, Span (bridge), or Earth (cliff edge). Because none of these starting points is even close to the altitude, thousands of feet higher, from which skydivers jump out of airplanes, BASE jumping, as own by the death statistics, is truly the world's most dangerous sport.

BASE jumpers use only 1 canopy, with no reserve chute; if the main chute fails, there is no time to deploy a reserve. The fatality rate in a study of over 20,000 BASE jumps from 1995 through 2005 was about 40 per 100,000 jumps, which is equivalent to about 1.7 per 100 participant-years. About 400 jumps per 100,000 resulted in significant injury. Cause of death was usually massive trauma from hitting an object on the way down, and/or the ground at the end of the jump. Reason for death was usually the late opening or nonopening of the canopy, or being thrown badly off course by a high crosswind.

The National Park Service has banned BASE jumping in US national parks. The BASE Fatality List records 417 deaths worldwide for BASE jumping from April 1981 through September 2021.

STAY SAFE Follow all safety precautions and make a will before BASE jumping.

SOURCES Base Fatality List, bfl. baseaddict.com/list; Basejumper.com; K. Soreide et al. How dangerous is BASE jumping? An analysis of adverse events in 20,850 jumps from the Kjerag Massif, Norway. *Journal of Trauma*, May 2007.

BASE jumping from an antenna tower: a dangerous way to play

Wingsuit Flying

ANNUAL US DEATHS   14 in 2013; see entry for detailed data

Wingsuit flying involves falling through the air wearing a suit equipped with inflatable wing areas that stretch between the participants arms and legs. These wings generate lift, which, with a bit of luck plus considerable skill plus nerves of steel, enable gliding flight similar to that of a flying squirrel. Beginners typically start by BASE jumping, and later graduate to wingsuit flying from an airplane or helicopter. A reserve parachute might or might not be used as backup.

The rate of death for wingsuit flying is about 1 in every 500 jumps, or 200 per 100,000, which is about 400 times the fatality rate from USPA-affiliated conventional skydiving. In 2013 alone, 14 people died from wingsuit flying. One reason that wingsuits are so dangerous is their aerodynamics: They are prone to various added instabilities and failures that cannot occur with parachute canopies.

STAY SAFE   Follow all safety measures and write your will before attempting wingsuit flying.

SOURCE   SkyAboveUs, skyaboveus.com/extreme-sports/
Extreme-Sports-Wingsuit-Flying

Blimps and Hot-Air Balloons

ANNUAL US DEATHS   Very low; see data in entry.

A blimp is a sausage-shaped aircraft that gets its structural rigidity from the pressure of the lighter-than-air gas (usually helium) that inflates it; this contrasts with dirigibles, which get their shape from extensive metal frameworks. Blimps got their name in the 1920s, when the US Navy held a fly-off between 2 competing designs, the A-limp and the B-limp, and the B-limp won.

The worst blimp crashes by far occurred in the 1930s, when 2 US Navy blimps, the *Akron* and the *Macon*, crashed into the sea in bad weather, between them killing 75 crewmen. In 1960, the US Navy blimp *Reliance*, used for radar surveillance, crashed into the Atlantic Ocean off New York and 18 crewmen were killed. Since then, there have been very few blimp crashes, in part because very few blimps have been flown. Goodyear blimps have crashed 4 times since 1960, due to bad weather or mechanical problems; in the most recent crash, in 2011, the blimp caught fire and the pilot was killed.

A hot-air balloon is a lighter-than-air aircraft consisting of a large bag filled with hot air to make it rise in the air, typically carrying a wicker basket, or gondola. The source of heat for the hot air is an open flame from burning propane. Despite the open flame and their lack of steering, hot-air balloons are very safe. All pilots must be licensed by the FAA. Balloon rides are a popular recreational and tourism activity. Dozens of hot-air balloon festivals take place around the country every year.

During the 12-year period from 2000 through 2011, the NTSB reported 169 hot-air balloon crashes. Of these crashes, 78 (46 percent) occurred during hot-air balloon tours involving 519 occupants, of which 94 (18 percent) suffered minor injuries, 91 (18 percent) sustained serious injuries, and 5 (1 percent) died.

In 2016, in Texas, a hot-air balloon caught fire and crashed, killing all 16 aboard; this was described as the deadliest hot-air balloon disaster in the US in decades. Between 1964 and 2014, there were 70 other fatalities in hot-air balloon mishaps. The causes included high winds, bad weather, and hitting high-voltage wires. Five of those fatalities occurred between 2000 and 2011. In 2021, all 5 passengers, including the pilot, were killed in Albuquerque, New Mexico, when their hot-air balloon drifted into power lines, which set the gondola on fire, and the balloon then crashed into an intersection.

STAY SAFE Fly only in a hot-air balloon that is licensed and inspected and with a licensed pilot. Pay attention to all safety instructions before and during flight. If you see power lines or obstacles that are closer

This **DirecTV** blimp is typical of present-day commercial blimps and dirigibles, which use low-flammability helium for lift rather than lighter but highly explosive hydrogen.

than you feel is comfortable, speak up and let the pilot know. Aviation safety is a vital and complex subject beyond the scope of this book. See the US Federal Aviation Administration (FAA) website at faa.gov.

SOURCES S.B. Ballard et al. Hot-air balloon tours: crash epidemiology in the United States, 2000–2011. *Aviation, Space, and Environmental Medicine*, November 2013; CNN, cnn.com/2016/07/30/us/texas-hot-air-balloon-crash/index.html; the *New York Times*, nytimes.com/2021/06/26/us/hot-air-balloon-crash-albuquerque.html; NTSB, ntsb.gov/safety/data/Pages/Data_Stats.aspx

Cable Cars (Funiculars, Inclines, and Aerial Trams)

ANNUAL US DEATHS Estimated 5 since 2000

In the US since 2000, about 5 people have been killed in cable car (funicular, incline, aerial tram) accidents. Only 1 person was killed in each incident.

About 20 US cities have a funicular system of some kind. Some cable cars, such as the famous ones in San Francisco, move through city streets in traffic, while others, such as those in Pittsburgh, move up steep escarpments on rails, while still others, such as the one that fell in May 2021 in Italy, killing 14 occupants, move through the air suspended from cables held up by towers.

In 2016, the operator of a San Francisco cable car died of injuries he suffered when his cable car was hit by an alleged drunk driver; in 2017, a pedestrian was struck and killed by a San Francisco cable car; the previous San Francisco cable car pedestrian fatality had been in 2008; about 1 rider or pedestrian is injured nonfatally per month. In May 2021, someone was killed when he fell through a faulty railing near the top of a Pittsburgh incline, and there have been suicides by jumping from incline cars. New York City's scenic Roosevelt Island aerial tram, popular with tourists, has never had a fatality since it opened in 1976.

STAY SAFE Do not hang on to the outside of a cable car. Obey all safety warnings of operating crew and safety signage. Be careful hopping on and off cable cars.

SOURCES Roosevelt Island Operating Corporation, rioc.ny.gov/170/The-New-Tram; SFGate, sfgate.com/bayarea/article/Cable-car-accidents-a-fiery-explosion-on-SF-s-10824642.php; Wikipedia, Aerial Lift

Dirigibles (Zeppelins)

> **ANNUAL US DEATHS**   0

Dirigibles—also called zeppelins, in honor of the German engineer who pioneered them—are sausage-shaped aircraft with rigid metal frameworks (in contrast to nonrigid blimps) to give them shape and structural stability. Lift comes from large bags, or bladders, filled with lighter-than-air gas. Large gondolas, with cockpit, passenger and crew accommodations, and engines for forward propulsion, are attached under the main body of the dirigible.

Dirigibles were attractive in the early years of twentieth-century aeronautics because they offered comfort (even luxury), very long flight ranges, even being able to cover transoceanic distances at speeds much faster than passenger liners. However, spectacular disasters (including the *Hindenburg* explosion of May 6, 1937) and the advent of less expensive, lighter-weight, easily stored and much safer blimps, led to a general abandonment of dirigibles decades ago.

Germany's zeppelins, used as war aircraft in World War I and as passenger aircraft afterward, were filled with highly flammable hydrogen gas because Germany did not have its own supply of helium, which was embargoed by the US. The explosion and fire on the *Hindenburg* as it docked in New Jersey at the end of the first-ever scheduled trans-Atlantic passenger zeppelin flight—3 days from Frankfurt to New York City—was due to a combination of hydrogen gas and the flammable coating used on the aircraft's skin. Thirty-five people aboard and 1 on the ground died; most of the other 62 people on the flight were seriously burned. The origin of the fire was never determined.

STAY SAFE   Don't use hydrogen to achieve lighter-than-air flight.

SOURCES   FAA, faa.gov/data_research/accident_incident; Zeppelin History, zeppelinhistory.com

Hang Gliding and Paragliding

> **ANNUAL US DEATHS**   3 (out of 3,000 to 6,000 participants) but recently higher

Data from the US Hang Gliding and Paragliding Association (USHPA) indicates that from 2000 through 2016, the death count from these sports averaged about 3 fatalities per year, out of an active community of about 3,000 to 6,000 participants. This is an annual US deaths of between 50 and 100 per 100,000 hang gliders/paragliders. There were 10 deaths in 2015 and 8 in 2016, suggesting that in more recent years, participants might be taking greater risks relative to their experience, the capabilities of their equipment, and wind/weather conditions. In 2021, a paraglider and a single-engine small plane collided midair in Texas. Both the pilot and the paraglider were killed.

The main difference between hang gliding and paragliding is that a hang glider has a metal frame, which gives the airfoil some shape, whereas a paraglider has no frame and is shaped by the airflow, like a parachute canopy. For hang gliders, most accidents result from excessive speed, whereas for paragliders, most accidents result from wind or thermals not adequate enough to provide lift.

STAY SAFE Conventional skydiving is significantly safer for anyone who wants to personally experience falling and flying through the sky.

SOURCES CNN, cnn.com/2021/12/22/us/texas-small-plane-paraglider-collision-deaths/index.html; US Hang Gliding and Paragliding Association, ushpa.org/page/fatalities

Home-Built Aircraft

ANNUAL US DEATHS Under 50

Over 33,000 amateur home-built aircraft in the US are licensed by the FAA. While some are built entirely from scratch, following plans and making the parts from raw materials, most are assembled from commercially available kits. Home-builts must be registered with and inspected by the FAA before an airworthiness certificate is issued. The pilot must be FAA-licensed. The pilot must then flight-test the home-built in a specified nonpopulated area, and only if this testing period is successful may the craft carry any passengers. The same annual FAA safety inspections are subsequently required as for other small aircraft.

FAA and NTSB data indicate that home-built aircraft have almost exactly the same accident rate as professionally manufactured aircraft. Another indication of the similarity of risks between home-builts and commercial production aircraft is that the insurance rates are very similar for both. However, 18 percent of all US aircraft accidents result in fatalities, while about 25 percent of accidents in home-builts result in fatalities. Pilot miscontrol is the most common cause of a fatal home-built aircraft crash. (See **Private and Recreational Aviation**.)

STAY SAFE See the US Federal Aviation Administration (FAA) website at faa.gov.

SOURCES   Experimental Aircraft Association, eaa.org/
eaa/about-eaa/eaa-media-room/eaa-news-releases/2021-10-
28-experimental-category-fatal-accident-total-drops-again-
finishes-under-not-to-exceed-total; FAA, faa.gov/data_research/
accident_incident; NTSB, ntsb.gov/safety/data/Pages/Data_Stats.aspx

Kite Flying and Kite Fighting

ANNUAL US DEATHS   No cumulative data available

Kite flying can sometimes turn deadly. This is especially true for the
sport of kite fighting. In kite fighting, the kite's string is coated with
glass shards, and/or the kite is equipped with razor blades, designed
to cut the string of another kite, which thereby "loses" the aerial kite
dogfight.

Kite fighting is popular in India and Pakistan. In India in 2016, 2
boys and 1 man were killed when their throats were cut by glass-coated
fighting-kite strings. In Pakistan in 2003, 5 children were killed in a
4-month period the same way, leading to a crackdown by national police. In 2007, also in Pakistan, at least 11 people were killed at a large
annual kite flying/fighting festival, some by sharp strings and some by
electrocution or by falling off roofs.

Data on any kite flying or kite fighting deaths in the US is not
available.

STAY SAFE   Avoid kite fighting, as it is extremely dangerous to you
and to bystanders. Never use a metal kite string; never fly a kite near
electrical wires or power lines; if your kite hits a live wire, do not touch
the string or the kite!

SOURCES   American Kitefliers Association,kite.
org/about-kites/kite-safety; CBS News, cbsnews.com/
news/11-dead-at-pakistani-kite-festival

Model Planes

ANNUAL US DEATHS   No cumulative data available, but see events
in entry.

Flying model airplanes, powered by propellers or even jet turbines, and
either tethered or (more usually) remote-controlled, can be dangerous.
On rare occasions, model aircraft have hit people hard enough to kill
them.

In 1979 in New York City, 2 people in the stands at Shea Stadium were hit by a remote-controlled model airplane being flown from the field during a halftime show. One died and the other received a concussion. In 2013, also in New York City, a 19-year-old was killed when a powerful model helicopter he was flying went out of control and cut off the top of his head.

STAY SAFE As with full-sized manned aircraft, model airplanes should be operated safely or not at all. Do not fly in bad weather or too close to bystanders. For essential precautions and valuable guidelines, see the Safety Program document from the Academy of Model Aeronautics (modelaircraft.org).

SOURCES Academy of Model Aeronautics, modelaircraft.org; the *New York Times*, nytimes.com/2013/09/07/nyregion/flying-model-aircraft-comes-under-scrutiny-after-fatal-accident-in-brooklyn-park.html

Model Rocketry

ANNUAL US DEATHS 4 from 1992 through 2016

During the years 1992 through 2016 in the US, 4 people are reported to have been killed as a result of model rocketry. All of them were electrocuted while trying to retrieve spent rockets that had become entangled in power lines. The National Association of Rocketry (NAR) counts over 80,000 members. Those interested in getting into model rocketry should read the Civil Air Patrol's guide "Model Rocketry." Consider joining a club in your area.

STAY SAFE Model rocketry safety is a vital and complex subject and is beyond the scope of this book. See the National Association of Rocketry's 45-page downloadable PDF slide show, "Safety in Sport Rocketry," for a detailed set of important safety guidelines for launching and recovering model rockets.

SOURCES Civil Air Patrol. gocivilairpatrol.com/programs/aerospace-education/programs/model-rocketry; National Association of Rocketry, nar.org

Private and Recreational Aviation

ANNUAL US DEATHS 233 in 2019

Private or general aviation includes all nonmilitary aviation that is not flying for hire (such as a scheduled airliner or an air taxi) or flying that performs work (such as crop dusting or fighting forest fires). The types of aircraft involved in private aviation are generally small, and can range from amateur/home-built aircraft (FAA-designated "experimentals"), to ultralights, helicopters, and single-engine and multi-engine fixed-wing propeller planes and jets, built and sold by major aircraft companies, all of them used for personal transportation and for the sheer enjoyment of flight. However, very wealthy people have been known to buy very large jetliners, such as 747s, for their exclusive personal use. In the US, privately owned/flown hot-air balloons and blimps are also included in the FAA's overall category of general aviation.

In 2019 in the US, 233 people were killed in general aviation crashes. There are about 220,000 aircraft active in American general aviation, so this is an annual US deaths of about 105.9 per 100,000 active aircraft. Losses of aircraft control, especially stalls, are the leading cause of fatal mishaps.

STAY SAFE Private/general aviation safety is a vital and complex subject and is beyond the scope of this book. See the US Federal Aviation Administration (FAA) website at faa.gov.

SOURCES Experimental Aircraft Association, eaa.org/eaa/about-eaa/eaa-media-room/eaa-news-releases/2021-10-28-experimental-category-fatal-accident-total-drops-again-finishes-under-not-to-exceed-total; FAA, faa.gov/data_research/accident_incident

Skydiving

> **ANNUAL US DEATHS**   15 in 2019 (in 3.3 million jumps)

During 2019 at the US Parachute Association's 220 affiliated drop zones around America, skydivers made 3.3 million jumps. Of these, 15 led to the parachutist's death, for an annual US deaths of 0.45 per 100,000 jumps. In tandem jumps, where a beginner skydiver is strapped to an expert who controls the parachute, the mortality rate improves to 0.2 per 100,000 jumps.

In 2019, nonfatal injuries occurred from skydiving at the rate of 1 for every 1,310 jumps.

STAY SAFE Safe skydiving begins with professional ground-school training. Parachutes should be packed and inspected by professionals. The aircraft from which you jump should be well maintained by ground crews and crewed by professional pilots and jumpmasters. If you are a recreational skydiver, only jump onto an established drop zone, and

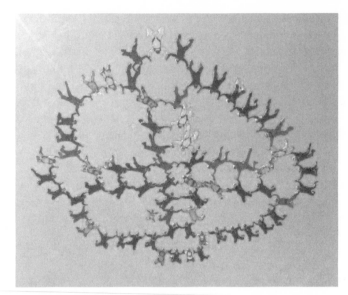

Skydiving is always more fun with others.

only when current and forecast wind and weather conditions are safe. Beginners should consider tandem jumping. Aviation safety is a vital and complex subject beyond the scope of this book. See the US Federal Aviation Administration (FAA) website at faa.gov.

SOURCES   FAA, faa.gov/data_research/accident_incident; United States Parachute Association (USPA), uspa.org/Discover/FAQs/Safety

Soaring (Glider Planes)

ANNUAL US DEATHS   5 to 10 (out of some 15,000 active pilots)

Soaring is done within the enclosed cockpit of a rigid-airframe, fixed-wing glider aircraft. Usually, the glider is first brought up to altitude and speed either using its own auxiliary engine (a motor glider) or by being tethered behind a powered towplane that pulls it down the runway and into the sky. (Gliders are sometimes also launched from cliff edges. In some countries a very long winch cable pulls the glider down the runway fast enough for it to get airborne.) The glider pilot then turns off the engine or releases the tow tether and gains further lift by riding thermal updrafts.

In the US, there are about 5 to 10 soaring pilot deaths per year, out of about 15,000 active soaring pilots. This is an annual US deaths of about 33 to 67 per 100,000 active soaring pilots. In 2006, 7 soaring pilots were seriously injured in nonlethal crashes.

STAY SAFE Take soaring lessons at a professional flight school. Motor gliders can be safer than unpowered gliders, because the engine can be restarted if wind and lift die or the aircraft gets too close to the ground. Aviation safety is a vital and complex subject beyond the scope of this book. See the US Federal Aviation Administration (FAA) website at faa.gov.

SOURCES FAA, faa.gov/data_research/accident_incident; Soaring Safety Foundation, soaringsafety.org/accidentprev/ntsbreports.html

Ultralight Aircraft

ANNUAL US DEATHS About 60

Data from 1987 (most recent available) indicated that there were about 20,000 ultralight aircraft in the US and about 60 fatalities per year due to mishaps. This is an annual US deaths of 300 per 100,000 ultralights.

In the US, ultralights are not as rigorously licensed and regulated by the FAA as are other aircraft, which has allowed pilots with less training and experience to fly them.

STAY SAFE Private/general aviation safety is a vital and complex subject and is beyond the scope of this book. See the US Federal Aviation Administration (FAA) website at faa.gov, or contact your nearest FAA-certified flight school.

SOURCES FAA, faa.gov/data_research/accident_incident; NTSB, ntsb.gov/safety/data/Pages/Data_Stats.aspx

Athletics, Exercise, and Sports

ANNUAL US DEATHS See individual entries.

This section of **At Play** covers school, college, and amateur sports. Professional sports, and those professionals responsible for overseeing student and adult amateur leisure activities (coaches, trainers, and others), are covered in the **On the Job** section.

Over 1,000,000 Americans of all ages are treated every year in emergency departments across the country for injuries related to physical training (PT), working out, getting exercise, and/or playing sports. Each year, whether through the wrong sort of overenthusiasm or via freak accidents, a few hundred of them die. The overall gains to US society of this risky business are large: enhanced health and well-being, longer life expectancies, and invaluable economic activity (jobs). The gains surely outweigh any losses: the occasional medical bills, lost time to recover from simple sprains or bad concussions, and even (rarely, in comparison) tragic deaths. Striving for more safety yields lasting benefits to everyone, while those efforts also make sports, athletics, and fitness much more fun and relaxing for all.

Human beings, young and old alike, need to exercise for fitness, health, and happiness. Current guidelines call for 150 minutes of physical activity a week. Not many Americans meet even this minimal standard (just over 20 minutes a day), and 8.3 percent of deaths in the US each year are attributed to inadequate levels of physical activity. While some of us hate exercise (it feels like work), some of us love it. One way or another, many of us manage to exercise frequently, in all sorts of ways: indoors and out, with or without special equipment, individually and/or on teams, in practice, in friendly games, and in keen competition. Some of us benefit from the help and guidance of parents, teachers, coaches, instructors, athletics directors, personal trainers, managers, group classes, umpires, referees, and league officials.

A number of national and local organizations are devoted to improving the safety and health of exercise, sports, and athletics through safer equipment, as consumer products, and through the whole process, as an experience for participants. These bodies, either government- or community-sponsored, conduct research and publish information about injury and fatality statistics.

At the federal level, relevant entities include the National Safety Council (NSC), the Centers for Disease Control and Prevention (CDC), and the Consumer Product Safety Commission (CPSC). They also include various parts of state and local governments, academic bodies such as the University of North Carolina's National Center for Catastrophic Sports Injury Research (NCCSIR), some public school districts, plus scouting and other youth organizations, YMCA's, religious organizations, parent activist groups, professional bodies such as the National Athletic Trainer's Association (NATA), trade associations and chambers of commerce, and even plaintiffs' law firms.

In some cases, because of their more organized, institutional sponsorship, laws about reporting juvenile hospital visits, and everyone's natural instinct to protect our children, the best data on mortality is available for school- and college-age youths: in after-school junior and peewee leagues, at summer camps, among recreational leagues, or through primary and secondary schools, colleges, and universities.

Mortality statistics for postschool, adult nonprofessionals, say in scratch/pickup sports or in weekend and after-work leagues, tend to be incomplete or nonexistent. Using whatever data is readily available, each sport or activity gets its own entry in this section.

Baseball

(**ANNUAL US DEATHS**   7 from 2007 to 2015)

A study of US youth athlete deaths for 2007 through 2015 found that 7 fatalities occurred while playing baseball in this 9-year period. Sudden cardiac death was the most common reason for these fatalities. Another study for high school and college baseball between 1982 and 2019 found the annual US death rate as follows: high school players 0.22 per 100,000 yearly participants and college players 0.64 per 100,000 yearly participants.

STAY SAFE   Parents, coaches, and adult supervisors should put player safety before intense competitiveness. Players should avoid overaggressive play and unnecessary roughness. Wear a properly fitting batting helmet. Be sure to stay well hydrated; avoid overexertion in hot weather, which can lead to heatstroke.

SOURCE   Stanford Children's Health, stanfordchildrens.org/en/topic/default?id=sports-injury-statistics-90-P02787

Basketball

(**ANNUAL US DEATHS**   16 from 2007 to 2015)

A study of US youth athlete deaths for 2007 through 2015 found that most sudden deaths while playing a sport occurred in basketball, with 16 deaths in this 9-year period. Sudden cardiac death was the most common cause of these fatalities. Reasons for this surprising result include the continual very fast-paced nature of the game, often played outdoors in hot weather, and the close, crowded, aggressive maneuvering of opposing players who wear no protective equipment. Among nonfatal injuries, concussions are a particular concern because the cumulative effect of several concussions can lead to depression and sharply increases the risk of neurodegenerative disease and suicide. In youth basketball, the rate of concussion for girls is 3.93 times the rate for boys.

Another study of high school and college basketball between 1982 and 2019 found annual US death rates as follows: high school boys 0.70 per 100,000 yearly participants, high school girls 0.16, college boys 7.12, and college girls 0.98.

STAY SAFE   Parents, coaches, and adult supervisors should put player safety before intense competitiveness. Players should avoid overaggressive play and unnecessary roughness. Be sure to stay properly hydrated; avoid overexertion in hot weather, which can lead to heatstroke.

SOURCES NCCSIR, nccsir.unc.edu/reports; Stanford Children's Health, stanfordchildrens.org/en/topic/default?id=sports-injury-statistics-90-P02787

Bounce Houses and Other Inflatable Amusement Structures

ANNUAL US DEATHS 12 from 2003 to 2013

In the US from 2003 to 2013, 12 people were reported killed by bounce houses ("moon bounces") or other inflatable amusement structures, including slides and climbing walls. Another 113,000 people needed emergency room treatment for injuries in that 11-year period. This is a death rate averaging 1.1 people per year. Deaths resulted from head and spine injuries (bounce house), falling (slide), or drowning in the case of inflatables wrongly used as water floats. In 2021 in Australia, 6 children were killed when a bounce house was lifted into the air by wind and then fell 32 feet.

STAY SAFE Do not erect the bounce structure on a windy day. Never use it as a water float, and always make sure it is firmly anchored on land. Care should be taken when jumping or climbing to avoid head or spine injuries. Adult supervision is necessary at all times.

SOURCES CPSC report, cpsc.gov, Estimated Number of Injuries and Reported Deaths Associated with Inflatable Amusements, 2003–2013; *People*, people.com/human-interest/sixth-child-dies-after-bouncy-castle-tragedy

A large bounce house

Bowling

(**ANNUAL US DEATHS**   4 to 8)

Data for 2005 and 2006, updated in 2019, indicate that about 4 to 8 people per year are killed while bowling in the US, as a direct result of the game and its equipment. Most of these deaths took place in people's homes or were due to the recent rise of extreme bowling or both. ("Xtreme bowling" is played in poor lighting, with the ball thrown from very high-risk postures, such as the bowler lying on their belly.)

Only 1 death a year on average is reported in traditional commercial bowling alleys. In Wisconsin in 2005, 2 employees were crushed while fixing malfunctioning machinery. One person was killed in a fight at a bowling alley in New Jersey in 2013, when he was intentionally hit in the head by a bowling ball.

STAY SAFE   Traditional family bowling is much safer than Xtreme bowling or Jet Bowling. Bowling balls should never be used as weapons.

SOURCE   OSHA, osha.gov/pls/imis/AccidentSearch

Boxing, Amateur

(**ANNUAL US DEATHS**   Average of 3)

Amateur boxing kills an average of about 3 people per year, mainly because of head injuries or complications of overexertion. This is about one-third of the fatality rate among professional boxers, a sport that sees an average of 9 or 10 participants killed each year.

STAY SAFE   Always use all sanctioned safety equipment. Use common sense during training and practice. Box competitively only at organized, refereed bouts, and throw in the towel before suffering a serious injury or complications of overexertion. It is advisable to have a medical exam before beginning to box, and periodically thereafter.

SOURCES   Association of Ringside Physicians, ringsidearp.org/ consensus-statements; R.L. Heilbronner et al. Neuropsychological Consequences of Boxing and Recommendations to Improve Safety: A National Academy of Neuropsychology Education Paper. *Archives of Clinical Neuropsychology*, February 2009; *WBN: World Boxing News*, worldboxingnews.net/2019/10/22/seven-sports-dangerous-boxing

Cheerleading

ANNUAL US DEATHS 7 from 1982 through 2011

A study of high school athletics for 1982 through 2011 found that cheerleading had only 7 deaths in this 19-year period. However, cheerleading also causes some serious, even catastrophic, injuries every year, especially head and neck injuries due to falls.

Another study for high school and college cheerleading between 1982 and 2019 found annual US death rates as follows: high school girls 0.38 per 100,000 yearly participants and college girls 0.00 per 100,000 yearly participants.

Among nonfatal injuries, concussions are a particular concern because the cumulative effect of several concussions can lead to depression, neurodegenerative disease, and a sharply increased risk of suicide.

STAY SAFE Parents, coaches, and adult supervisors should put cheerleader safety before intense competitiveness. Do not use faulty or defective equipment. Be sure to stay properly hydrated; avoid overexertion in hot weather, which can lead to heatstroke.

SOURCE NCCSIR, nccsir.unc.edu/reports

Cross-Country

ANNUAL US DEATHS 0.24 to 0.47 per 100,000

A study of high school and college cross-country between 1982 and 2019 found annual US death rates as follows: high school boys 0.34 per 100,000 yearly participants, high school girls 0.22, college boys 0.47, and college girls 0.24.

STAY SAFE Parents, coaches, and adult supervisors should put runner safety before intense competitiveness. Be sure to stay properly hydrated; avoid overexertion in hot weather, which can lead to heatstroke.

SOURCE NCCSIR, nccsir.unc.edu/reports

Equestrian Sports

ANNUAL US DEATHS   3.86 per 100,000

A study of college equestrian sports between 1982 and 2019 found an annual US death rate for college girls of 3.86 per 100,000 yearly participants.

STAY SAFE   Parents, coaches, and adult supervisors should put rider safety before intense competitiveness. Wear a properly fitted riding helmet.

SOURCE   NCCSIR, nccsir.unc.edu/reports

Exercise Equipment, Gyms, and Fitness Clubs

ANNUAL US DEATHS   6.3

A study for 1990 through 2007 showed that an average of 6.3 people per year were killed in accidents while using free weights or weight machines. Another study for 2003 through 2012 showed that an average of 3 people per year were killed in accidents involving treadmills.

To put these numbers into perspective, tens of millions of Americans use gyms and fitness clubs. While about 3.7 million people a year are injured seriously enough while exercising or playing sports to need an emergency room visit, accidental death from exercise equipment is uncommon.

Sudden cardiac death (SCD) during hard exertion is a different type of lethal risk. While the overall data on this hazard does not break out exercise equipment or gym and fitness club usage as separate categories, a study found an annual rate of death from SCD varying from 0.75 per 100,000 among young adult males and 0.13 among young adult females to 6 per 100,000 among middle-aged males.

Offsetting the risks of exercising is the significant risk of premature death for people who do not get enough exercise. In 2019, before the effects of COVID-19, almost 3 million people in the US died for 1 reason or another. Of these, a study showed that about 8.3 percent, or about 250,000 Americans, died prematurely from sedentary lifestyles that year.

STAY SAFE   Avoid distractions such as using your cell phone while exercising. Use all safety devices, such as the pull-out clips that will stop a treadmill if you move back too far. Do not exceed your capabilities, especially with free weights, and always have a buddy spot you. Stay hydrated and do not overheat. Do not do strenuous exercise while injured, ill, or exhausted. It is a good idea to have a medical evaluation first, if you are new to working out. Get some lessons from a personal trainer or the fitness club's staff, especially for exercise machines with which you are not familiar.

SOURCES D. Aune et al. Physical activity and the risk of sudden cardiac death: a systematic review and meta-analysis of prospective studies. *BMC Cardiovascular Disorders*, July 2020; CPSC, cpsc.gov/Research—Statistics/NEISS-Injury-Data; S. Carlson et al. Percentage of deaths associated with inadequate physical activity in the United States. *Preventing Chronic Disease*, March 2018.

Field Hockey

ANNUAL US DEATHS 0.10 per 100,000

A study of high school field hockey between 1982 and 2019 found an annual US death rate among high school girls of 0.10 per 100,000 yearly participants.

STAY SAFE Parents, coaches, and adult supervisors should put player safety before intense competitiveness. Players should avoid overaggressive play and unnecessary roughness. Do not use faulty or defective equipment. Wear all protective equipment and make sure it fits properly. Be sure to stay properly hydrated; avoid overexertion in hot weather, which can lead to heatstroke.

SOURCE NCCSIR, nccsir.unc.edu/reports

Football

ANNUAL US DEATHS 12 (but see entry)

A study of US youth athlete deaths for 2007 through 2015 found that 7 fatalities occurred while playing football in this 9-year period. In 2020, 8 fatalities were recorded. Sudden cardiac death was the most common reason for these fatalities. Heatstroke kills about 2 high school football players per year. Another study found that about 12 high school and college football players die each year from injuries during practice or competition. Besides cardiac arrest and heatstroke, catastrophic brain injury was another leading cause.

A study of high school and college football between 1982 and 2019 found annual US death rates as follows: high school boys 1.05 per

100,000 yearly participants and college boys 3.95 per 100,000 yearly participants.

STAY SAFE Parents, coaches, and adult supervisors should put player safety before intense competitiveness. Players should avoid overaggressive play and unnecessary roughness. Do not use faulty or defective equipment. Wear all appropriate protective gear and be sure it fits properly. Be sure to stay properly hydrated; avoid overexertion in hot weather, which can lead to heatstroke. If a congenital heart or circulatory problem, or asthma, diabetes, or other health problem might be present, seek medical advice before giving permission to play football.

SOURCES NCCSIR, nccsir.unc.edu/ reports; STAT News, statnews.com/2021/11/25/ next-boy-up-kids-continue-to-die-on-high-school-football-fields

Golf

ANNUAL US DEATHS Errant balls killed about 3 persons worldwide in recent years; other risks exist.

The game of golf itself is not inherently dangerous, but its equipment and its environment can lead to lethal injuries. These occur predominantly from accidents involving golf carts, players being struck by lightning, or attacks from animals such as alligators.

Errant golf balls and flying heads or broken shafts from faulty clubs injure an estimated 40,000 golfers per year, professional and amateur combined, including at least 3 deaths worldwide in the past few years. Isolated deaths have also occurred on golf courses around the globe, from pesticide toxicity, a falling tree branch, stroke, being hit by a teed-off ball, being killed in a brawl over game delay, or drowning.

STAY SAFE Call "fore" before hitting and take care when you hear someone call "fore." Make sure your clubs are in good repair, and never smash or break one in frustration. Try to not unduly delay following golfers while searching for balls; be courteous and polite at all times on the golf course.

SOURCES CDC, cdc.gov/injury/wisqars/fatal.html; CPSC, cpsc.gov/ Research—Statistics/NEISS-Injury-Data; GolfSupport, golfsupport. com/blog/sports-related-injuries-golf-more-dangerous-than-rugby

Gymnastic floor exercises

Gymnastics

ANNUAL US DEATHS   0.84 to 1.81 per 100,000

A study of high school and college gymnastics between 1982 and 2019 found annual US death rates as follows: high school boys 0.84 per 100,000 yearly participants and college girls 1.81 per 100,000 yearly participants.

Among nonfatal injuries, concussions are a particular concern because the cumulative effect of several concussions can lead to depression, neurodegenerative disease, and a sharply increased risk of suicide.

STAY SAFE   Parents, coaches, and adult supervisors should put gymnast safety before intense competitiveness. Always have a spotter. Do not use faulty or defective equipment. Be sure to stay properly hydrated. Competitive gymnasts should be monitored for eating disorders.

SOURCE   NCCSIR, nccsir.unc.edu/reports

Handball

ANNUAL US DEATHS   No data available

Statistics for deaths that occur from playing handball are not readily available, but broad studies in the US and some other countries have found that among cases of sudden cardiac death in school-age youths, handball ranked very low compared to such other sports as football, basketball, soccer, or baseball.

STAY SAFE   Any form of intense exertion can trigger sudden cardiac death in a young person. A medical evaluation is wise before a parent or guardian gives permission to play strenuous sports.

SOURCE   NCCSIR, nccsir.unc.edu/reports

Horseshoes and Lawn Darts

ANNUAL US DEATHS   Rare fatal casualties

In North America each year the game of horseshoes causes several hundred head injuries, such as concussions, from errant throws. In 2017 in Canada, 2 people were killed this way. The game of lawn darts was banned in the US and Canada after a 7-year-old boy was killed in California by a thrown lawn dart in 1987.

STAY SAFE   Be very careful when you throw a horseshoe, as it can become a lethal projectile. Never throw it at a person or a pet. Bystanders should stay well clear while contestants make their pitches.

SOURCES   CPSC, cpsc.gov/Newsroom/News-Releases/1997/ Following-Recent-Injury-CPSC-Reissues-Warning-Lawn-Darts-Are-Banned-and-Should-Be-Destroyed; The Daily Bonnet, dailybonnet. com/horseshoes-related-injuries-soar-2017

Ice Hockey

ANNUAL US DEATHS   0.83 (high school) and 0.69 (college) per 100,000

A study of high school and college ice hockey between 1982 and 2019 found annual US death rates as follows: high school boys 0.83 per 100,000 yearly participants and college boys 0.69 per 100,000 yearly participants.

STAY SAFE   Parents, coaches, and adult supervisors should put player safety before intense competitiveness. Players should avoid overaggressive play, unnecessary roughness, and brawls. Do not use faulty or defective equipment. Wear properly fitted helmets, mouth guards, face masks, and protective padding, even during practice. Be sure to stay properly hydrated.

SOURCE   NCCSIR, nccsir.unc.edu/reports

Ice Skating

ANNUAL US DEATHS About 12 outdoors (2018 to 2019); rink deaths are very rare.

In the winter of 2018 to 2019 in the US, about 12 people were killed while walking or skating on lake or river ice, as a result of falling through thin ice and dying of drowning, hypothermia, and/or blunt force trauma. Deaths at ice skating rinks are very rare but do occasionally occur in the US, due to lethal head injuries from falls or heart attacks from overexertion. About 50,000 American ice skaters are injured seriously enough each year to need medical attention.

STAY SAFE Be very wary at all times of falling through the ice. Watch out for open or flowing water or rotten ice, which are warnings that nearby ice is thin. If the ice is not so thick that it appears black even in broad daylight, stay away! Beware of unseasonably warm weather, which can lead to unexpected thin ice. Obey all posted signs and buoys warning of dangerous ice. Avoid high speed or attempting maneuvers that exceed your level of ice-skating skill and experience.

SOURCES III, iii.org/fact-statistic/facts-statistics-sports-injuries; Lake Ice, lakeice.squarespace.com/skating-hazards

Inline Skating, Roller Skating, and Rollerblading

ANNUAL US DEATHS 20 to 30

In the US in recent years, about 20 or 30 people per year, mostly youngsters, died from accidents while using roller skates or inline skates (rollerblades). The fatal injuries often resulted from uncontrolled falls (including down stairs) or collisions (including with moving cars). About 38,155 kids under age 14 need medical attention each year for roller-skating injuries.

STAY SAFE Always wear a snug-fitting helmet with chin strap, skate only on smooth, paved, level surfaces, and never skate near moving traffic or at night. Young children should be closely supervised by a responsible adult.

SOURCE Safe Kids USA, safekids.org/research-report/ready-ride-keeping-kids-safe-wheels

Jai Alai

> **ANNUAL US DEATHS**   No data available

Jai alai is called the world's fastest sport because the ball can move at almost 200 mph. The US has few jai alai frontons (courts), but the sport is popular in Spain and Latin American countries. In the US, professional jai alai is often used for legal pari-mutuel betting, and amateur jai alai lessons are also offered. Several professional players have been killed during play, although hard data is scarce.

STAY SAFE   Amateur athletes can find much safer ways to get exercise and have fun than playing jai alai.

SOURCE   Wikipedia, Jai Alai

Lacrosse

> **ANNUAL US DEATHS**   0.07 to 2.18 per 100,000, depending on player group

A study of high school and college lacrosse between 1982 and 2019 found annual US death rates as follows: high school boys 0.64 per 100,000 yearly participants, high school girls 0.07, and college boys 2.18.

Among nonfatal injuries, concussions are a particular concern because the cumulative effect of several concussions can lead to depression, an increased risk of neurodegenerative disease, and a sharply increased risk of suicide.

STAY SAFE   Parents, coaches, and adult supervisors should put player safety before intense competitiveness. Players should avoid overaggressive play and unnecessary roughness. Do not use faulty or defective equipment. Wear helmets, face masks, and protective padding, make sure they fit properly, and always use them, even during practice. Be sure to stay properly hydrated; avoid overexertion in hot weather, which can lead to heatstroke.

SOURCE   NCCSIR, nccsir.unc.edu/reports

Martial Arts

ANNUAL US DEATHS Few catastrophic injuries reported—but may be underreported

Martial arts (unarmed combat) sports such as karate, tai chi, and tae-kwondo are enjoyed by millions of children and adults in the US. While injury rates of 41 to 133 per 1,000 participant-years have been reported in studies, very few catastrophic injuries have been reported among nonprofessionals. Some studies indicate that injuries are more likely among those over 18 than among juveniles and more likely among the more experienced fighters. The rate of injury tends to accelerate among players who trained or competed for more than a few hours a week. It is possible that the rate of concussions, a serious problem with athletes of all ages, is underreported for amateur martial arts.

STAY SAFE Choose a martial arts trainer with credible professional experience and a good safety record. Be especially cautious regarding the possibility of blows to the head, which can cause concussions whose bad effects over time are cumulative and may be irreversible.

SOURCE American Academy of Pediatrics, healthychildren.org/ English/healthy-living/sports/Pages/Martial-Arts.aspx

Miniature Golf (Putt-Putt), Driving Ranges, and Batting Cages

ANNUAL US DEATHS 0 in any given year

Deaths during these types of sports swinging practice are rare. In 2020 in Florida, 2 young children were killed while playing miniature golf when a truck went out of control and veered onto the course.

Deaths at golf driving ranges are likely to be very rare but are not readily available broken out from deaths while playing rounds at golf courses. (See **Golf**.) One leading cause of golfing deaths, golf cart accidents, is presumably unlikely at driving ranges, as are deaths from overexertion, animal attacks, or drowning. Rare events such as being hit and killed by a golf ball or errant club head, or being struck by lightning, are still present.

In 1994, a 12-year-old boy in Brooklyn was killed when a ball from a pitching machine hit him hard in the chest, causing his heart to stop, a condition known as commotio cordis. The very few other recorded deaths have been from brain injuries. In 1997, a 17-year-old boy was killed in Los Angeles when a batted ball richoted in the batting cage and hit him in the head.

STAY SAFE   In batting cages, obey all safety instructions and signage. Wear a batting helmet and appropriate footwear. Observe limits on anyone else in the cage while hitting, adjust the pitching machine to the batter's height and skill level, and turn the machine off when not ready to swing.

SOURCES   The *Baltimore Sun*, baltimoresun.com/news/bs-xpm-1994-07-17-1994198146-story.html; *Los Angeles Times*, latimes.com/archives/la-xpm-1997-06-25-me-6842-story.htm

Mixed Martial Arts (MMA)

ANNUAL US DEATHS   About 15 worldwide in recent years

About 15 professional mixed martial arts (MMA) fighters have been reported killed in recent years in various bouts in the US and other countries. As of early 2019, none had died in bouts of the largest promotion company, Ultimate Fighting Championship. A 2006 study indicated that the overall risk of injury in MMA was comparable to professional boxing, which suffers several deaths worldwide every year.

STAY SAFE   The best way to avoid the lethal risks of MMA is to do something else instead.

SOURCES   G.H. Bledsoe et al. Incidence of injury in professional mixed martial arts competitions. *Journal of Sports Science and Medicine*, July 2006; R.P. Lystad et al. The Epidemiology of Injuries in Mixed Martial Arts: A Systematic Review and Meta-analysis. *Orthopedic Journal of Sports Medicine*, January 2014; Ultimate Fighting Championship, ufc.com/about/sport

Paintball

ANNUAL US DEATHS   Only 2 reported—ever

Hard data on deaths from paintball injuries is scarce. The only 2 known deaths were caused by blunt force trauma inflicted when improperly sealed carbon dioxide propellant canisters took off as projectiles. In 2018 in Atlanta, a carload of pranksters fired paintballs at a gas station. A teen hit by the pellets fired back with a handgun, killing a 3-year-old in the car.

USAF airmen use recreational paintball guns in a training exercise.

STAY SAFE   Always use all safety equipment. Make sure all paintball guns are in good repair and that all propellant cannisters are properly resealed each time they are refilled. Don't fire paintball guns off the range.

SOURCE   Sadler & Company, sadlersports.com/blog/preventing-paintball-injuries/

Playground Equipment

ANNUAL US DEATHS   15 to 20

About 15 to 20 children each year in the US are killed while playing on playground equipment, either at home, at school, or in public parks and play areas. The primary causes of death are lethal head injuries due to falls, and accidental strangulations. Monkey bars and other climbing equipment are the most frequent playground items involved in serious accidents. In 1 US study, over 500,000 playground injuries needed medical attention in 2002 alone. Male children suffer almost two-thirds of all serious injuries from play equipment. Two-thirds of deaths occur on home playground equipment.

STAY SAFE   Never let children play without responsible adult supervision, and make sure they only use equipment that is age-appropriate for them, especially regarding maximum height off the ground (under 4 feet for kids under 5, under 8 feet for kids aged 5 through 12). All elevated play areas need proper guardrails, and slides should have a bar across the top so that children need to sit down before starting to slide. Any hooks, bolts, or protrusions on equipment should be enclosed or otherwise protected, to avoid the risk of catching clothing; children should be forbidden from playing on equipment if their clothing has drawstrings or hoods than can snag on something and choke them. Children should not be allowed to stand or perform dangerous stunts on swing sets. The ground around all play equipment should be thoroughly cushioned to prevent serious trauma from slips and falls; either thick rubber matting or at least a 12-inch layer of loose fill such as mulch or sand is advisable. Make sure that all equipment is properly maintained; if weak, faulty, or rusted, forbid any use until repaired.

SOURCES CDC, cdc.gov/traumaticbraininjury/prevention.html; CPSC, cpsc.gov/content/Injuries-and-Investigated-Deaths-Associated-with-Playground-Equipment-2009-to-2014

Racquet Sports

ANNUAL US DEATHS No data available

Deaths among players for all combined racquet sports, such as squash, racquetball, badminton, and tennis, are not readily available. Overall injury rates have been studied and were found to be relatively low among all major sports, exercise, and recreational activities. Thus, by inference, players of racquet sports are comparatively safe from fatal consequences.

Using US data for 3.73 million emergency room visits due to sports and recreational injuries for the years 2013 through 2019, racquet sports ranked a low 20th out of the 26 different sports and activities analyzed, causing only 25,844 injuries, or 0.69 percent of the total. In comparison, several activities, including football and bicycling, each contributed more than 10 percent each to the total.

STAY SAFE A UK study of the benefits of exercise on longevity looking at 80,000 adults showed that racquet sports are an especially effective route toward better health. People who played racquet sports had a 47 percent lower risk of all-cause mortality compared with those who played no racquet sports and a 56 percent lower risk of death from heart disease. A Danish study that followed 8,577 participants for up to 25 years found that tennis players live an average 9.7 years longer than people who don't exercise.

SOURCES National Safety Council, NSC.org; P. Oja et al. Associations of specific types of sports and exercise with all-cause and cardiovascular-disease mortality: a cohort study of 80 306 British adults. *British Journal of Sports Medicine*, 2017; P. Schnohr et al. Various leisure-time physical activities associated with widely divergent life expectancies: The Copenhagen City Heart Study. *Mayo Clinic Proceedings*, 2018.

Roller Derby

ANNUAL US DEATHS Very rare

Deaths in modern times during or because of roller derby competition, whether professional or amateur, are very rare. In 2012, a woman in Iowa died as a result of an injury. Historically, in the 1880s 2 players died shortly after finishing a 6-day roller derby endurance marathon.

STAY SAFE Use all personal protective equipment. Do not compete while injured or exhausted. Avoids fights and brawls.

SOURCE Wikipedia, History of Roller Derby

Rowing (Crew)

ANNUAL US DEATHS 0.04 to 0.08 per 100,000, depending on participant group

A study of high school and college rowing between 1982 and 2019 found annual US death rates as follows: high school boys 0.08 per 100,000 yearly participants, high school girls 0.04, college boys 0.35, and college girls 0.33.

STAY SAFE Parents, coaches, and adult supervisors should put rower safety before intense competitiveness. Do not use faulty or defective equipment. Be sure to stay properly hydrated; avoid overexertion in hot weather, which can lead to heatstroke.

SOURCES NCCSIR, nccsir.unc.edu/reports; USRowing, usrowing.org/sports/2016/6/2/900_132107062339971607.aspx

Rugby

ANNUAL US DEATHS Estimated 1.4 worldwide; not widely played in the US

Anecdotally, at least 14 rugby players have been killed in the sport around the world in the decade ending in 2018. Rugby isn't widely played in the US at the school, amateur, or professional level. Worldwide, both school-age children and adult amateurs and pros have been killed during rugby games. Causes of death included heart attacks from overexertion, fatal brain bleeds from head trauma, and broken necks.

STAY SAFE   Safer contact sports (including American football) can be played by all age groups, at all competitive levels, in the US. A 2011 study by Auckland University of Technology (New Zealand), using 30 years of data, found that American football had 75 percent fewer catastrophic injuries than rugby.

SOURCE   Brain Injury Law Center, brain-injury-law-center.com/blog/head-injuries-rugby-vs-football

Running, Jogging, and Footraces

ANNUAL US DEATHS   Estimated 6

About 2 million people a year in the US participate in long-distance running races (marathons and half-marathons). Death is very rare. A study looking at the incidence and outcome of cardiac arrest among these runners in the 10 years between 2000 and 2010 found that of 10.9 million runners, 59 (51 of them men) died, for an incidence rate of 0.54 per 100,000 participants, or approximately 6 per year. In a study of sudden cardiac death among marathon runners over a 30-year period, 4 deaths were documented among 215,413 runners, for a prevalence of only 0.002 percent, far lower that the general risk for premature death.

Reliable aggregate data on deaths during and shortly after jogging or running, specifically due to the physiological stresses of exercise, are difficult to separate from deaths due to external or random events, such as a jogger being hit by a car. The other side of this same statistical conundrum is that deaths due to some accidents, such as a pedestrian being hit by a car, are difficult to identify reliably as happening while the victim was exercising. Police reports and death certificates often omit such details.

The preliminary release of 1 study, which indicated that too much running impaired life expectancy, did *not* pass subsequent peer review, due to a flawed methodology. In addition, an unknown number of the occasional deaths due to cardiovascular events (heart attacks and strokes), which occurred during or soon after exercising, would inevitably have happened anyway, whatever the victim was doing at the time.

STAY SAFE   Use common sense about the duration and the speed of your jog or run. Have a medical evaluation before you start any new strenuous exercise program. Be very careful when in or near traffic; don't "jay jog" or "jay run." Wear high-visibility clothing, dress for the weather, stay hydrated, and be especially wary of heatstroke during hot weather.

SOURCES   J.H. Kim et al. Cardiac arrest during long-distance running races. *New England Journal of Medicine*, 2012; B.J. Maron et al. Risk for sudden cardiac death associated with marathon running. *Journal of the American College of Cardiology*, 1996.

Scooters, Kick

ANNUAL US DEATHS   1 reported in 2013

In 2013 in the US, unpowered kick scooters (foot scooters) caused 52,500 injuries and 1 death.

STAY SAFE   Always wear a snug-fitting helmet with chin strap, ride only on smooth, paved surfaces, and never ride near moving traffic or at night. Young children should be closely supervised by a responsible adult.

SOURCE   CDC, cdc.gov/injury/wisqars/fatal.html

Skateboards

ANNUAL US DEATHS   40

In the US, an average of 40 people are killed every year in unpowered skateboard accidents. The main causes of death were being hit by a car or going downhill too fast and losing control. About 125,000 skateboard users were injured in 2018.

STAY SAFE   Always wear a snug-fitting helmet with chin strap, along with knee and elbow pads, wrist guards, and appropriate footwear. Avoid skateboarding in or near traffic or on rough ground; a skateboarding park or paved yard area is much safer. Learn how to fall correctly. Keep your speed and any stunts you attempt within your level of skill and experience. Never hang on to a bicycle, car, or other vehicle.

SOURCES   Safety First Skateboarding, skateboardsafety.org/injury-statistics; Stanford Children's Health, stanfordchildrens.org/en/topic/default?id=sports-injury-statistics-90-P02787

Ski Team

> **ANNUAL US DEATHS**   0.33 to 9.06 per 100,000, depending on participant group

A study of high school and college skiing between 1982 and 2019 found annual US deaths as follows: high school girls 0.33 per 100,000 yearly participants, college boys 9.06, and college girls 5.48.

STAY SAFE   Parents, coaches, and adult supervisors should put skier safety before intense competitiveness. Do not use faulty or defective equipment and make sure it fits properly.

SOURCE   NCCSIR, nccsir.unc.edu/reports

Soccer

> **ANNUAL US DEATHS**   6 from 2007 to 2015

A study of US youth athlete deaths for 2007 through 2015 found that 6 fatalities occurred while playing soccer in this 9-year period. Sudden cardiac death was the most common reason for these fatalities.

A study of high school and college soccer between 1982 and 2019 found annual US death rates as follows: high school boys 0.32 per 100,000 yearly participants, high school girls 0.14, college boys 0.87, and college girls 0.49.

Among nonfatal injuries, concussions are a particular concern because the cumulative effect of several concussions can lead to depression, neurodegenerative disease, and a sharply increased risk of suicide. In youth soccer, the rate of concussion for girls is 1.68 times the rate for boys.

STAY SAFE   Parents, coaches, and adult supervisors should put player safety before intense competitiveness and be aware of the risk of concussion. Players should avoid overaggressive play and unnecessary roughness. Do not use faulty or defective equipment. Be sure to stay properly hydrated; avoid overexertion in hot weather, which can lead to heatstroke.

SOURCE   NCCSIR, nccsir.unc.edu/reports

Softball

ANNUAL US DEATHS   0.01 per 100,000

A study of high school and college softball between 1982 and 2019 found an annual death for high school girls of 0.01 per 100,000 yearly participants.

STAY SAFE   Parents, coaches, and adult supervisors should put player safety before intense competitiveness. Players should avoid overaggressive play and unnecessary roughness. Do not use faulty or defective equipment. Wear a batting helmet. Be sure to stay properly hydrated; avoid overexertion in hot weather, which can lead to heatstroke.

SOURCES   NCCSIR, nccsir.unc.edu/reports; Stanford Children's Health, stanfordchildrens.org/en/topic/default?id=sports-injury-statistics-90-P02787

Stickball

ANNUAL US DEATHS   No data available

Two different team sports are called stickball. One is an ancient game still played among Native American tribes of the northern US and Canada. It is a forerunner of lacrosse (see above entry). In earlier times, stickball had very flexible rules and was typically played between teams of hundreds of players, with opposing goals set hundreds of yards or even several miles apart. In modern times, a more formalized version is played, using safety equipment such as helmets and body pads. Hard data on fatal injuries is not readily available.

The other game called stickball is most commonly played on urban streets or in playgrounds and is roughly analogous to softball or baseball. Sawn-off broom handles or other sticks are used as improvised bats, and the ball is usually a soft pink rubber Pennsy Pinky, Spalding (spaldeen), or the equivalent. Bases can be manhole covers, fire hydrants, or whatever else is available and agreed to by the participants. Instead of having a catcher, the ball is bounced off a wall behind the batter. Hard data is not available, but fatal injuries might occur from being hit by a motor vehicle, or a hard collision with another player, or an obstruction such as a parked car, telephone pole, or the stoop of a building.

STAY SAFE   Players are much safer playing in parks or playgrounds instead of on the street, where hazards include moving traffic and dangerous fixed obstructions.

SOURCES   Wikipedia, Stickball

Swimming Pools

ANNUAL US DEATHS   3,536

Between 2005 and 2015, an average of 3,536 people per year fatally drowned in swimming pools—residential, private club, and public/municipal—in the US. Three-quarters of these occurred at residential pools; one-sixth of all pool drownings occurred in aboveground pools. Two-thirds of all fatalities involved children under 3 years of age. In a few cases, drowning was caused by entrapment against a pool filter suction drain.

Between 2016 and 2018, there were 1,178 reported drowning incidents, leading to 1,191 fatalities, associated with pool or spa submersions involving children under 15 years of age.

Besides deaths, from 2011 through 2013 an average of 4,900 people needed emergency care for swimming pool injuries; more than half of them needed hospitalization or transfer to a long-term/rehab care facility. Some of these were near-fatal drownings that caused permanent brain damage, leading to major treatment costs and serious loss of quality of life.

STAY SAFE   Young children should receive certified professional swimming instruction before being allowed in any pools; adults should also get drownproofing training. Flotation devices are advisable for anyone who is not a strong swimmer. Four-sided fencing is needed to prevent unsupervised swimming. Adult supervision at all times is highly advisable. Never go in or near swimming pools while impaired from alcohol or drugs. Do not swim or play near underwater pool filter suction drains. Always know the depth of the water before you jump in; never jump headfirst at the shallow end, as this can cause severe brain and spine injuries on impact with the pool bottom. When diving from a board, first make sure that no person or object is too close to where you will land.

SOURCE   CDC, cdc.gov/drowning/data/index.html

Swim Team

> **ANNUAL US DEATHS**   0.14 to 2.61 per 1,000, depending on participant group

A study of high school and college swimming between 1982 and 2019 found annual US deaths as follows: high school boys 0.22 per 100,000 yearly participants, high school girls 0.14, college boys 2.61, and college girls 0.81.

STAY SAFE   Parents, coaches, and adult supervisors should put swimmer safety before intense competitiveness. Do not swim alone. Be sure to stay properly hydrated.

SOURCES   NCCSIR, nccsir.unc.edu/reports

Tennis

> **ANNUAL US DEATHS**   0.04 to 0.35 per 100,000, depending on participant group

A study of high school and college tennis between 1982 and 2019 found annual US death rates as follows: high school boys 0.08 per 100,000 yearly participants, high school girls 0.04, college boys 0.35, and college girls 0.33.

STAY SAFE   Parents, coaches, and adult supervisors should put player safety before intense competitiveness. Be sure to stay properly hydrated; avoid overexertion in hot weather, which can lead to heatstroke.

SOURCE   NCCSIR, nccsir.unc.edu/reports

Interstate Tramways
Tennis player, 1935

Track and Field

> **ANNUAL US DEATHS** 0.05 to 0.53 per 100,000, depending on partici-
> pant group

A study of high school and college track and field between 1982 and 2019 found annual US death rates as follows: high school boys 0.30 per 100,000 yearly participants, high school girls 0.05, and college boys 0.53.

STAY SAFE Parents, coaches, and adult supervisors should put athlete safety before intense competitiveness. Be sure to stay properly hydrated; avoid overexertion in hot weather, which can lead to heatstroke.

SOURCE NCCSIR, nccsir.unc.edu/reports

Trampolines

> **ANNUAL US DEATHS** 6 since 2015

In the US in 2018, trampolines at home and at trampoline parks sent about 110,000 people to emergency rooms. At least 6 deaths were reported since 2015. Injuries varied from minor sprains to broken limb bones to serious brain and spine damage. About three-fourths of the injuries occurred when 2 people were jumping together. Children are the most vulnerable to trampoline injury or death, as their small bodies are not able to withstand the high-impact stresses that a trampoline can cause, especially if they hit the ground, the frame or springs, or another person.

The American Academy of Pediatrics recommends parents never purchase trampolines for homes and never allow their children to jump on trampolines at someone else's home.

STAY SAFE Children under 6 should not use trampolines. Allow only 1 person at a time onto the trampoline. Avoid trampolines in poor repair or that lack safety nets to confine the occupant to the central, padded bouncing area. Avoid attempting maneuvers that are beyond your skill level and may result in deadly head or neck trauma.

SOURCES American Academy of Orthopedic Surgeons,
aaos.org, Position Statement Trampolines and Trampoline Safety;
Stanford Children's Health, stanfordchildrens.org/en/topic/
default?id=sports-injury-statistics-90-P02787

Tree Houses and Playhouses

ANNUAL US DEATHS   About 3,000 injuries but very few fatalities

Studies of injury and death data from the 1990s indicate that tree house and playhouse accidents send almost 3,000 people to emergency rooms every year and kill a small number, primarily by falls from a height. Other potentially fatal injuries can result from collapse of the tree or supporting structure of the tree house or from being struck by lightning.

STAY SAFE   Tree houses should only be built in sturdy trees or on sturdy, well-maintained supporting platforms. Railings should always be provided to prevent accidental falls. Occupants should be instructed to never jump from the tree house or playhouse. Do not occupy a tree house or playhouse when lightning strikes in the area are possible.

SOURCES   CDC, cdc.gov/injury/wisqars/fatal.html; Nationwide Childrens' Hospital, nationwidechildrens.org/research/areas-of-research/center-for-injury-research-and-policy/injury-topics/home-safety/tree-house-safety

Volleyball

ANNUAL US DEATHS   0.01 (high school females) to 0.42 (college females) per 100,000

A study of high school and college volleyball between 1982 and 2019 found annual US death rates as follows: high school girls 0.01 per 100,000 yearly participants and college girls 0.42 per 100,000 yearly participants.

STAY SAFE   Parents, coaches, and adult supervisors should put player safety before intense competitiveness. Be sure to stay properly hydrated; avoid overexertion in hot weather, which can lead to heatstroke.

SOURCE   NCCSIR, nccsir.unc.edu/reports

Walking for Exercise (Power Walking)

ANNUAL US DEATHS   No data available

As with jogging and running (see above entry), the risk of death from exercise walking, or power walking, is difficult to quantify separately from those deaths that occur during any form of walking, whether from external factors (such as being hit by a car) or internal conditions (such as heart attack or stroke). Studies comparing walking speed with subsequent life expectancy indicate that someone who can only walk slowly has a higher risk of subsequent death from cardiovascular disease. A 2018 study of nearly 475,000 participants over 7 years found that brisk walkers had greater longevity than slow walkers, regardless of BMI. The life expectancy of brisk-walking women ranged from 87 to 88 years, while the life expectancy of slow-walking women was 72 to 85. Among men, the brisk walkers had a life expectancy of 85 to 87 years, while the life expectancy of the slow-walking men was 65 to 81. Brisk walking is defined as 100 or more steps per minute, or about 4 miles an hour.

STAY SAFE Use common sense about walking safely, whether casually or for exercise. Have a medical evaluation before you start any new strenuous exercise program. Be very careful when in or near traffic; don't jaywalk. Dress for the weather, stay hydrated, and be especially wary of heatstroke during hot weather. Wear high-visibility clothing at night. If you can only walk slowly, seek the advice of a medical professional about possible cardiovascular or musculoskeletal problems. Walking more often, and/or faster, tends to further improve health and increase longevity.

SOURCES NIH NIA, nia.nih.gov/health/exercise-physical-activity; F. Zaccardi et al. Comparative relevance of physical fitness and adiposity on life expectancy: A UK Biobank observational study. *Mayo Clinic Proceedings*, 2019.

Water Polo

ANNUAL US DEATHS 0.26 to 5.46 per 100,000, depending on participant group

A study of high school and college water polo between 1982 and 2019 found annual US death rates as follows: high school boys 0.57 per 100,000 yearly participants and high school girls 0.26. For college boys the rate was 5.46 per 100,000 yearly participants.

STAY SAFE Parents, coaches, and adult supervisors should put player safety before intense competitiveness. Players should avoid overaggressive play and unnecessary roughness. Wear protective gear and do not use faulty or defective equipment. Be sure to stay properly hydrated.

SOURCE NCCSIR, nccsir.unc.edu/reports

Wrestling

> **ANNUAL US DEATHS**   0.37 (high school males) and 2.37 (college males) per 100,000

A study of high school and college wrestling between 1982 and 2019 found annual US death rates as follows: high school boys 0.37 per 100,000 yearly participants, and college boys 2.37 per 100,000 yearly participants.

STAY SAFE   Parents, coaches, and adult supervisors should put wrestler safety before intense competitiveness. Wear all protective gear and make sure it fits properly. Be sure to stay properly hydrated; avoid overexertion in hot weather, which can lead to heatstroke. Coaches should be aware of the potential for eating disorders among wrestlers.

SOURCE   NCCSIR, nccsir.unc.edu/reports

Attractions and Tourism

Amusement Parks, Carnivals, and Fairs

> **ANNUAL US DEATHS**   4 or 5 related to amusement park rides

Between 1987 and 2004, an average of about 4 or 5 deaths per year were related to amusement park rides in the US, counting both workers and visitors (but excluding **Water Parks**—see separate entry). Some rides were permanently installed at large theme parks, some were placed at shopping malls, while other rides were mobile, such as at traveling fairs and carnivals. Causes of death included workers falling or being crushed, visitors ignoring ride safety instructions, park negligence, and other causes such as heart attacks, suicides, police use of force, or lightning strikes.

The rides causing or associated with deaths included roller coasters, whirligigs, Ferris wheels, and trains.

Although a comprehensive national database of deaths related to rides at amusement parks, carnivals, and fairs does not yet exist, the death rate in more recent years appears to have been reduced to below the level found in the 1987 to 2004 studies. Isolated deaths do still occur.

In 2017, among permanent, fixed-site theme parks in the US and Canada, about 100 people, many of them children, were injured badly enough on rides to need immediate hospitalization for at least 24 hours;

another 1,000 or so injuries required less extensive medical aid. That same year, people took over 1.7 billion rides, showing that the odds of being seriously injured are well under 1 in 1 million.

Another study found that between 1990 and 2010, an average of about 4,400 children per year needed emergency room visits because of injuries by amusement park rides. The most injurious types of rides were found to be carousels, smaller roller coasters, and bumper cars.

Perhaps the worst amusement park disaster of modern times occurred in 1984, when a fire killed 8 teenagers in the Haunted Castle attraction at Six Flags Adventure Park in New Jersey. An investigation found that the fire began by someone using a lighter in the dark as a flashlight, accidentally setting flammable props ablaze. In the aftermath, tougher safety laws and inspections, better fire protection systems, and other improved safety measures have been introduced in most states.

STAY SAFE Obey all safety instructions and safety signs. Do not go on a ride if you do not satisfy the ride's height and weight requirements, or if you have a health issue that the ride could exacerbate. Keep your head, arms, and legs inside the seat area at all times, and never take off your seat belt or stand up while the ride is in motion. To avoid neck injuries, always look straight ahead and keep your head up. If you have any questions, ask the ride operator; before you get on the line, watch the ride so you know what to expect. Take breaks between strenuous rides such as extreme roller coasters. Never go on a ride that appears to be poorly maintained, irresponsibly operated, or is otherwise unsafe.

SOURCES R.J. Braksiek and D.J. Roberts. Amusement park injuries and deaths. *Annals of Emergency Medicine*, 2002; International Association of Amusement Parks and Attractions, iaapa.org/safety-security/amusement-ride-safety; R.T. Loder and J.R. Feinberg. Emergency department visits secondary to amusement ride injuries in children. *Journal of Pediatric Orthopedics*, 2008; Ostroff Injury Law, ostrofflaw.com/blog/how-safe-is-that-ride

Circuses

ANNUAL US DEATHS About 1 US

Since 1990 only about 3 circus workers have been killed during performances. Two were aerialists who fell and received fatal injuries when equipment failed, and 1 was stomped on and thrown by an elephant.

Data on recent deaths of visitors attending circuses is not readily available, but deaths appear to be very rare and presumably are caused in ways similar to those at other large audience gatherings: trips and falls, heart attacks, drunken hijinks gone tragically wrong.

On July 6, 1944, in Hartford, Connecticut, a fire broke out during a performance under the big top of the Ringling Bros. and Barnum & Bailey Circus. Between 6,000 and 8,000 people were in the audience; 167 were killed and more than 700 were injured, making this the worst circus disaster ever.

STAY SAFE See **Spectator Accidents at Sports Stadiums and Arenas**. Do not harass circus animals or trespass into performance rings or backstage areas.

SOURCES Circus Arts Conservatory, circusarts.org; Stewart O'Nan, *The Circus Fire: A True Story of an American Tragedy* (Anchor, 2001).

Cruise Ships

ANNUAL US DEATHS Average of 17

A study of global data on fatalities aboard ocean and river cruise ships (which in general provide overnight accommodation to passengers), for the 20-year period 2000 through 2019, found that a total of about 340 passengers who were US residents died while aboard cruise ships, or in port during onshore excursions, somewhere in the world. This is an average of 17 deaths per year. Main causes of death were falls (overboard or onto a lower deck), natural causes such as heart attacks, or suicides.

Since many of these cruises begin and/or end in 1 country but include transits through international waters and port visits in other countries, it is impractical to separate the death statistics into US and non-US territory and territorial waters.

STAY SAFE Take seriously the mandatory lifeboat drills and all other safety instructions from crew members. Do not for any reason lean too far over railings or become so intoxicated that you attempt dangerous stunts. Use hand sanitizer or wash your hands often to avoid norovirus and other contagious illnesses.

SOURCES Cruise Ship Deaths, cruiseshipdeaths.com; T.W. Heggie and T. Burton-Heggie, Death at sea: passenger and crew mortality on cruise ships. *International Journal of Travel Medicine and Global Health*, 2020.

Duck Boats

ANNUAL US DEATHS   About 1.8

Since 1999 in the US and Canada, at least 41 people have been killed in duck boat mishaps. Duck boats are amphibious "DUKW" vehicles (a GMC model designation), originally built for combat in World War II, and are popular with tourists. Passengers can board them at a convenient land-based pickup point, then be taken nonstop for a sightseeing boat ride, and then be disembarked back on land. However, a powered vehicle able to travel on the street, on wheels, and on the water, using a propeller or water jet, exposes people to the risks of both. In 2015, duck boats hit and killed 7 pedestrians and motorists in 3 different traffic accidents in Boston, Philadelphia, and Seattle; duck boats have serious blind spots, because the driver sits well back from the bow/front. A total of 35 passengers have died in duck boat sinkings from collision with a barge, or catastrophic flooding due to mechanical failure, or because of high winds and rough waves. In 1999, 13 people were killed when a duck boat sank in Arkansas. In 2018, 17 were killed in a duck boat sinking in Missouri.

STAY SAFE   Do not ride on a duck boat that is not well maintained or that is not crewed by disciplined professionals. Don a personal flotation device upon boarding and keep your head and arms inside the boat at all times. If the boat appears decrepit or is overcrowded, or if the weather is poor, find a safer way to spend your leisure time and money.

SOURCE   AP News, apnews.com, Duck boats linked to more than 40 deaths since 1999.

A duck boat in the Branson, Missouri, tourist mecca

Scenic and Harbor Cruises, Swan Boats, Party and Casino Boats, Dinner Cruises

> **ANNUAL US DEATHS** Breakout data unavailable, but see boating entries in **On the Job** and **At Play** sections

Scenic, tourist, and party/dinner boats are enjoyable ways to spend some time on the water. Generally, these excursions take up part of a single day and don't offer overnight accommodation to passengers. Hard data for deadly mishaps on these specialized excursions is included in the broader annual US statistics for commercial (**On the Job**) and recreational (**At Play**) boating fatalities.

STAY SAFE Follow crew safety instructions and signage.

SOURCES BLS, bls.gov/iif/oshcfoi1.htm; Cruise Ship Deaths, cruiseshipdeaths.com

Scenic and Sightseeing Transportation

> **ANNUAL US DEATHS** 8 in 2019

In 2020 in the US, 25,200 people were employed in providing scenic and sightseeing tours and travel to the public, usually as same-day round-trip packages aboard buses, steam trains, boats, and helicopters, with an emphasis on entertainment and leisure rather than on fast and efficient transportation between 2 different points. In 2019, 8 of them were killed on the job. This is an annual US deaths of about 31.7 per 100,000.

STAY SAFE Pay attention to all work safety procedures. Use all personal protective equipment. Exercise due caution when in and around motor vehicles, trains, ships, and aircraft, both on and off the road (or on the rails, on the water, or in the air). Bosses should ensure that all OSHA, industry, and state safety standards are met.

SOURCE BLS, bls.gov/iif/oshcfoi1.htm

Space Tourism

> **ANNUAL US DEATHS** Emerging activity

As of May 2021, only a handful of tourists worldwide have ever gone into space, and none of them have died during their space excursions. (The number of astronauts and cosmonauts who have died during space missions runs into double digits; see **Astronauts and Space Crews**.)

A Virgin Galactic spacecraft, designed for passenger and space tourism service

Trips into space for tourists were provided by the Russian Space Agency between 2001 and 2010. Such trips will soon start to be offered again by Russia. Commercial space tourism began in 2021 in the US with suborbital and orbital flights from Virgin Galactic, Blue Origin, and SpaceX.

The cost of a trip is typically tens of millions of dollars, and a journey into space requires a significant amount of prior physical screening and conditioning, equipment training, and safety instruction. The experience is said to be out of this world!

STAY SAFE Proceed at your own risk and expense. Space tourists are exposed to the same potentially mortal dangers as working astronauts: explosive rocket launching, hard vacuum, severe cold, intense radiation in orbit, onboard fires, hull breaches from space junk or meteors, and fiery reentry to landing or splashdown.

SOURCES Blue Origin, blueorigin.com/new-shepard/fly; SpaceX, spacex.com/human-spaceflight; Virgin Galactic, virgingalactic.com

Space Colonization

ANNUAL US DEATHS Future activity

As NASA notes on their website about space colonization, this idea has recently moved from pure science fiction into the realm of possibility. Prolonged occupation of the International Space Station (ISS) is seen by proponents as a demonstrator and precursor for longer-term and even permanent habitation further away from Earth. The first destinations for humans to colonize other heavenly bodies would be the moon, then Mars.

The potentially mortal dangers of human space colonization are considerable: the hard vacuum, extreme cold, and intense radiation in space and on other worlds; the need to either transport from Earth, or find within a very harsh alien environment, every single item needed for survival; the constant risk of a lethal equipment failure or hull breach; not to mention the extreme emotional/psychological stresses and strains of very long confinement with very few fellow humans in a very small space.

It would certainly be the adventure of a lifetime, though that lifetime might be cut short in numerous ways!

STAY SAFE   Boldly go—entirely at your own risk.

SOURCE   NASA, nasa.gov/topics/moon-to-mars

Submarines: Tourist and Personal

ANNUAL US DEATHS   1 murder reported (Denmark, 2017)

Apparently, the only death aboard a privately owned submarine in modern times was a murder in Denmark in 2017. The murder was not related to the operational safety of the vessel.

Personal submarines (properly termed "submersibles") allow a small number of people, usually 2 or 3, to dive down to considerable undersea depths for exploration and extreme ecotourism, while staying warm and dry and breathing normal air at 1 atmosphere. Their maximum depth rating ("crush depth" or "collapse depth") can vary, but often falls between 500 and 3,300 feet, depending on the vessel. Some submersibles are designed for deep submergence and can go to depths of 7,500 feet or more.

Tourist submarines are similar to personal submarines, but some have seating capacity and life support capabilities for up to 8 passengers plus the pilot.

Both types of vessel are typically designed with large, clear viewing bubbles over their passenger compartment, allowing an unobstructed, panoramic view of the surrounding ocean. They have auxiliary pods attached for the necessary machinery and equipment and for battery powered maneuvering motors. Entry and exit are accomplished on the surface, either through a top hatch or via a clamshell design, with handholds and steps for safety. Some models are small enough to fit in the tender storage compartment of luxury yachts. Submerged endurance, both of the environmental support system and of the power and propulsion (at a cruising speed of about 4 knots), can be as long as 8 to 10 hours, with much longer reserve/backup oxygen supplies for emergencies. Many models are equipped with floodlights and with sonar, and some are equipped with manipulator arms for sampling and salvaging.

STAY SAFE   Proceed at your own risk. Personal and tourist submarines/submersibles are exposed to the same types of potentially mortal dangers as other submarines: becoming stranded below the surface and running out of air, suffering an electrical fire at depth, springing a leak and flooding and drowning, colliding with a ship or whale or the seafloor, or going out of control and falling through crush/collapse depth and imploding.

SOURCE   Triton Submarines, tritonsubs.com/whytriton

Tourist Helicopters

ANNUAL US DEATHS   24 in 2019; 19 in 2020; no breakout data available for tourist helicopters

In 2019 in the US, there were 24 fatal helicopter crashes overall, killing a total of 51 people. In 2020, 19 fatal crashes killed 35 people. An exact breakdown between tourist helicopter flights (including backcountry ski-lift choppers) and others is not readily available.

Tourist helicopter crashes tend to occur in areas where they spend the most cumulative hours in the air, such as Hawaii, the Grand Canyon, and New York City. Pilot error, inadequate maintenance, and bad weather cause most helicopter mishaps.

STAY SAFE   Make sure that any tourist or ski-lift helicopter operator you ride with provides an experienced pilot and a properly maintained aircraft. If the operation seems disorganized, undisciplined, or ramshackle, go elsewhere. Check the weather forecast beforehand, and do not fly (even if a pilot is willing to) if rain, snow, thunderstorms, high winds, or poor visibility are predicted.

SOURCES   FAA, faa.gov/data_research/accident_incident; United States Helicopter Safety Team (USHST), ushst.org/reports

Water Parks, Aquariums, and Sea Life Parks

ANNUAL US DEATHS   No cumulative data available, but see events detailed in entry

Millions of people every year safely enjoy their visits to public water parks, aquariums, and sea life exhibition/encounter parks. Deaths by visitors at water parks in the US are very rare but not entirely unknown. Most are from natural causes, such as heart attacks. On occasion, the rides themselves can cause traumatic injury. Public waterslides cause over 4,200 serious injuries per year, such as concussions, broken bones, and lacerations, when a patron collides with another patron or the slide structure, or falls out of the slide, or is otherwise hurt. Deadly microbe infections can be spread via a ride's water supply if the water is not properly filtered and sanitized. Several drowning deaths occurred in past years at water park tsunami and whirlpool rides, but those rides have been discontinued. In 2016, a raft going over the 169-foot waterslide at Schlitterbahn Water Park in Kansas City, Kansas, went airborne and hit a metal pole supporting a safety net, decapitating a 10-year-old boy. The slide was later demolished and the park was closed.

Waterworld at Action Park in Vernon, New Jersey, opened in 1978 as one of the first modern water parks. It featured a number of popular but badly designed rides. In combination with inept management, this led to a poor safety record that included at least 6 deaths. Personal injury lawsuits led to the closing of the park in 1996.

In July 2021 at Adventureland Park in Altoona, Iowa, an 11-year-old boy was killed and 3 others injured, 1 critically, on the Raging River ride when the raft flipped over. In 2016 at the same park, an employee working on the same ride was killed when he fell onto the ride's conveyor belt, causing a traumatic brain injury.

Deaths at public aquariums and sea life parks are also extremely rare. The orca named Tilikum killed 3 people in separate incidents over several years at SeaWorld Orlando before being retired; 2 victims were site employees and 1 was a trespasser.

STAY SAFE Obey all safety instructions and signage. Respect the sea creatures and never harass them.

SOURCES *Des Moines Register,* desmoinesregister.com/story/ news/2021/07/15/adventureland-park-accident-raging-river-ride-failed-earlier-state-report-says-michael-jaramillo/7983886002; Methodshop, methodshop.com/most-dangerous-action-park-rides; Wikipedia, Amusement Park Accidents

Recreational Boating

ANNUAL US DEATHS   767 in 2020

In 2020, 767 people in the US were killed in recreational boating accidents, a 25 percent increase over the 613 killed in 2019. Another 3,191 were injured. Where cause of death was known, 75 percent of fatal boating accident victims drowned while not wearing a personal flotation device. In 247 accidents, at least 1 person was struck by a propeller, for 39 deaths. A full 80 percent of all boaters who drowned were in vessels under 21 feet long.

The overall annual US death rate for recreational boating in 2019 was 5.2 per 100,000 registered recreational water vessels. There were 11,879,000 such vessels in the US in 2019.

The overall annual US death rate in 2020 for recreational boating was 6.5 per 100,000 registered recreational water vessels. This rate represents a 25 percent increase from the 2019 fatality rate of 5.2 deaths per 100,000 registered recreational vessels. There were 11,838,188 such vessels in the US in 2020.

Compared to 2019, the number of accidents increased 26.3 percent, the number of deaths increased 25.1 percent, and the number of injuries increased 24.7 percent.

Major contributing factors in boating fatalities are boating while intoxicated, excessive speed, not looking out properly, operator inexperience, operator inattention, machinery and equipment failure, bad weather, and violating the rules of the road. Hard data is scarce on reasons for the drowning fatalities, but they can occur due to the victim falling overboard, or while swimming and diving off the boat, or due to the boat sinking, colliding, or running aground.

STAY SAFE Always comply with state and federal (US Coast Guard) safety regulations and navigation rules. Make sure the boat's hull is sound and watertight, and all machinery and equipment are in good working order. If the vessel is mechanically powered, be very careful of the propeller(s) or water jet(s). Handle all fuel safely: avoid smoking, prevent dangerous fume buildups, and carry fire extinguishers. Never go boating while intoxicated or impaired. Use nautical charts and other reference materials to familiarize yourself with the waters in which you plan to go boating, and take charts with you. Do not speed or create a dangerous wake. Maintain a proper lookout for other boats, swimmers, and obstructions at all times. Be careful to avoid running aground in shallow areas, and to avoid hitting an obstruction, whether submerged, floating, or above the water such as a dock or a bridge abutment. Avoid rocks, dams and weirs, and rapids and waterfalls. Carry effective communications gear, radar, and depth-finding equipment whenever appropriate. Check a detailed weather forecast, and only go out when conditions are safe given your experience and your boat's capabilities.

SOURCES National Association of State Boating Law Administrators (NASBLA), nasbla.org/home; USCG, uscgboating.org/statistics/accident_statistics.php

Fatal Accidents by Day of Week

ANNUAL US DEATHS   See table in entry.

Most fatal boating accidents occur on the weekend.

Sunday	121
Monday	52
Tuesday	55
Wednesday	69
Thursday	73
Friday	85
Saturday	158

SOURCE   USCG, uscgboating.org/statistics/accident_statistics.php

Fatal Accidents by Time of Day

ANNUAL US DEATHS   See table in entry.

The majority of fatal boating accidents occur after breakfast time and before dinnertime:

12:00 a.m. to 2:30 a.m.	20
2:31 a.m. to 4:30 a.m.	11
4:31 a.m. to 6:30 a.m.	12
6:31 a.m. to 8:30 a.m.	18
8:31 a.m. to 10:30 a.m.	41
10:31 a.m. to 12:30 p.m.	63
12:31 p.m. to 2:30 p.m.	64
2:31 p.m. to 4:30 p.m.	96
4:31 p.m. to 6:30 p.m.	106
6:31 p.m. to 8:30 p.m.	96
8:31 p.m. to 10:31 p.m.	48
10:31 p.m. to 11:59 p.m.	13
Time not reported	25

SOURCE   USCG, uscgboating.org/statistics/accident_statistics.php

Fatal Accidents by Month

ANNUAL US DEATHS   See table in entry.

The number of US deaths in 2019 from boating accidents, broken down by month, is as follows:

Month	
January	16
February	17
March	33
April	37
May	68
June	119
July	119
August	70
September	58
October	30
November	32
December	14

The majority of deaths occurred in the summer, when weather and water temperatures are warmest throughout the country, and when the most people go boating.

SOURCE   USCG, uscgboating.org/statistics/accident_statistics.php

A US coastguardsman rescues a 10-year-old boater.

Fatal Accidents by Personal Flotation Device Worn or Not

> **ANNUAL US DEATHS** See table in entry.

Among boating accident fatalities in 2019 in the US, for which it was reported whether or not the victim was wearing a personal flotation device (life jacket or other device), the breakdown for the top 5 causes of death was as follows:

	WORN	NOT WORN
Drowning	57	362
Trauma	35	49
Cardiac arrest	5	12
Carbon monoxide	0	3
Hypothermia	2	2

Wearing a personal flotation device (PFD) significantly improves the outcome of any serious boating accident and greatly reduces the risk of drowning.

Hard data is scarce on reasons for the drowning fatalities, but these mishaps can occur due to the victim falling overboard, or while swimming and diving off the boat, or due to the boat sinking, colliding, or running aground.

STAY SAFE To meet USCG requirements, a recreational vessel must have a USCG-approved life jacket for each person aboard. The life jacket must be the appropriate size for the intended user, must be appropriate for the intended activity, and must be in good and serviceable condition.

SOURCE USCG, uscgboating.org/statistics/accident_statistics.php

Fatal Accidents by Main Cause

> **ANNUAL US DEATHS** See tables below.

The primary contributing factor of recreational boating deaths in the US in 2019 breaks down as follows:

Operation of vessel	263
Loading	57
Equipment failure	28
Environment	96
Miscellaneous	169

DEATHS FROM MACHINERY OR EQUIPMENT FAILURE

Equipment failure	4
Hull failure	6
Machinery failure	18
Engine failure	14
Steering system failure	4
Auxiliary equipment failure	2
Other	2

DEATHS FROM OPERATION OF VESSEL

Alcohol use	113
Drug use	2
Excessive speed	22
Failure to vent fumes	0
Improper lookout	26
Inadequate navigation lights	1
Navigation rules violation	21
Operator inattention	36
Operator inexperience	39
Restricted vision	0
Sharp turn	3
Starting engine in gear	0

Alcohol or drugs were major factors in more than 1 in 6 of the 613 recreational boating deaths in the US in 2019. Virtually all deaths caused by dangerous vessel operation were preventable.

DEATHS FROM LOADING OF VESSEL

Improper anchoring	4
Improper loading	24
Overloading	17
People on gunwale, bow, or transom	12

DEATHS FROM ENVIRONMENT

Congested waters	0
Dam/lock	5
Force of wave/wake	12
Hazardous waters	48
Missing navigation aid	0
Weather	31

DEATHS FROM MISCELLANEOUS OTHER CAUSES

Carbon monoxide poisoning	2
Ignition of fuel or vapor	0
Sudden medical condition	16
Other	39
Unknown/not reported	112

STAY SAFE   Always comply with state and federal (US Coast Guard) safety regulations and navigation rules. Make sure the boat's hull is sound and watertight, and all machinery and equipment are in good working order. If the vessel is mechanically powered, be very careful of the propeller(s) or water jet(s), and handle all fuel safely, avoid smoking, prevent dangerous fume buildups, and carry fire extinguishers. Never go boating while intoxicated or impaired. Use nautical charts and other reference materials to familiarize yourself with the waters in which you plan to go boating, and take charts with you. Do not speed or create a dangerous wake. Maintain a proper lookout for other boats, swimmers, and obstructions at all times. Be careful to avoid running aground in shallow areas, and to avoid hitting an obstruction, whether submerged, floating, or above the water such as a dock or a bridge abutment. Avoid rocks, dams and weirs, and rapids and waterfalls. Carry effective communications gear, radar, and depth-finding equipment whenever appropriate. Check a detailed weather forecast, and only go out when conditions are safe given your experience and your boat's capabilities.

SOURCE   USCG, uscgboating.org/statistics/accident_statistics.php

Fatal Accidents by Vessel Activity

ANNUAL US DEATHS See table below.

Deaths by vessel activity in 2019:

Boating/relaxation	323
Fishing	198
Fueling	0
Hunting	8
Racing	3
Repairs	7
Starting engine	0
Swimming/snorkeling	34
Towed watersports	15
Towing	1
Whitewater	19
Other	5

STAY SAFE Always comply with state and federal (US Coast Guard) safety regulations and navigation rules. Make sure the boat's hull is sound and watertight, and all machinery and equipment are in good working order. If the vessel is mechanically powered, be very careful of the propeller(s) or water jet(s). Handle all fuel safely, avoid smoking, prevent dangerous fume buildups, and carry fire extinguishers. Never go boating while intoxicated or impaired. Use nautical charts and other reference materials to familiarize yourself with the waters in which you plan to go boating, and take charts with you. Do not speed or create a dangerous wake. Maintain a proper lookout for other boats, swimmers, and obstructions at all times. Be careful to avoid running aground in shallow areas, and to avoid hitting an obstruction, whether submerged, floating, or above the water such as a dock or a bridge abutment. Avoid rocks, dams and weirs, and rapids and waterfalls. Carry effective communications gear, radar, and depth-finding equipment whenever appropriate. Check a detailed weather forecast, and only go out when conditions are safe given your experience and your boat's capabilities.

SOURCE USCG, uscgboating.org/statistics/accident_statistics.php

Fatal Accidents by Vessel Operation

ANNUAL US DEATHS   See table below.

Deaths from vessel operation in 2019:

At anchor	17
Being towed	1
Changing direction	41
Changing speed	26
Cruising	180
Docking/undocking	2
Drifting	125
Idling	3
Launching/loading	1
Rowing/paddling	137
Sailing	6
Tied to dock/moored	1
Towing	1
Trolling	5
Other	4
Unknown	63

Racing a Cherub-class boat

STAY SAFE Always comply with state and federal (US Coast Guard) safety regulations and navigation rules. Make sure the boat's hull is sound and watertight, and all machinery and equipment are in good working order. If the vessel is mechanically powered, be very careful of the propeller(s) or water jet(s). Handle all fuel safely, avoid smoking, prevent dangerous fume buildups, and carry fire extinguishers. Never go boating while intoxicated or impaired. Use nautical charts and other reference materials to familiarize yourself with the waters in which you plan to go boating, and take charts with you. Do not speed or create a dangerous wake. Maintain a proper lookout for other boats, swimmers, and obstructions at all times. Be careful to avoid running aground in shallow areas, and to avoid hitting an obstruction, whether submerged, floating, or above the water such as a dock or a bridge abutment. Avoid rocks, dams and weirs, and rapids and waterfalls. Carry effective communications gear, radar, and depth-finding equipment whenever appropriate. Check a detailed weather forecast, and only go out when conditions are safe given your experience and your boat's capabilities.

SOURCE USCG, uscgboating.org/statistics/accident_statistics.php

Fatal Accidents by Body of Water Type

ANNUAL US DEATHS See table in entry.

Deaths by body of water type in 2019:

Lakes, ponds, reservoirs	294
Bays, sounds, marinas, harbors, canals	188
Ocean, gulf	46
Great Lakes	15

STAY SAFE Always comply with state and federal (US Coast Guard) safety regulations and navigation rules. Make sure the boat's hull is sound and watertight, and all machinery and equipment are in good working order. If the vessel is mechanically powered, be very careful of the propeller(s) or water jet(s). Handle all fuel safely, avoid smoking, prevent dangerous fume buildups, and carry fire extinguishers. Never go boating while intoxicated or impaired. Use nautical charts and other reference materials to familiarize yourself with the waters in which you plan to go boating, and take charts with you. Do not speed or create a dangerous wake. Maintain a proper lookout for other boats, swimmers, and obstructions at all times. Be careful to avoid running aground in shallow areas, and to avoid hitting an obstruction, whether submerged, floating, or above the water such as a dock or a bridge abutment. Avoid rocks, dams and weirs, and rapids and waterfalls. Carry effective communications gear, radar, and depth-finding equipment whenever appropriate. Check a detailed weather forecast, and only go out when conditions are safe given your experience and your boat's capabilities.

SOURCE   USCG, uscgboating.org/statistics/accident_statistics.php

Vessel: Fatal Accidents by Engine Horsepower

ANNUAL US DEATHS   See table in entry.

Most fatal boat accidents in 2019 occurred in boats with no engine.

No engine	180
10 hp or less	30
11 to 25	32
26 to 75	56
76 to 150	77
151 to 250	56
Over 250 hp	55
Unknown	127

STAY SAFE   Always comply with state and federal (US Coast Guard) safety regulations and navigation rules. Make sure the boat's hull is sound and watertight, and all machinery and equipment are in good working order. If the vessel is mechanically powered, be very careful of the propeller(s) or water jet(s). Handle all fuel safely, avoid smoking, prevent dangerous fume buildups, and carry fire extinguishers. Never go boating while intoxicated or impaired. Use nautical charts and other reference materials to familiarize yourself with the waters in which you plan to go boating, and take charts with you. Do not speed or create a dangerous wake. Maintain a proper lookout for other boats, swimmers, and obstructions at all times. Be careful to avoid running aground in shallow areas, and to avoid hitting an obstruction, whether submerged, floating, or above the water such as a dock or a bridge abutment. Avoid rocks, dams and weirs, and rapids and waterfalls. Carry effective communications gear, radar, and depth-finding equipment whenever appropriate. Check a detailed weather forecast, and only go out when conditions are safe given your experience and your boat's capabilities.

SOURCE   USCG, uscgboating.org/statistics/accident_statistics.php

Vessel: Fatal Accidents by Hull Material

ANNUAL US DEATHS   See table in entry.

In 2019, fiberglass hulls were the most common material in fatal boating accidents.

Aluminum	167
Fiberglass	291
Plastic	81
Rubber/vinyl/canvas	29
Steel	5
Wood	4
Other	0
Unknown	36

STAY SAFE   Always comply with state and federal (US Coast Guard) safety regulations and navigation rules. Make sure the boat's hull is sound and watertight, and all machinery and equipment are in good working order. If the vessel is mechanically powered, be very careful of the propeller(s) or water jet(s). Handle all fuel safely, avoid smoking, prevent dangerous fume buildups, and carry fire extinguishers. Never go boating while intoxicated or impaired. Use nautical charts and other reference materials to familiarize yourself with the waters in which you plan to go boating, and take charts with you. Do not speed or create a dangerous wake. Maintain a proper lookout for other boats, swimmers, and obstructions at all times. Be careful to avoid running aground in shallow areas, and to avoid hitting an obstruction, whether submerged, floating, or above the water such as a dock or a bridge abutment. Avoid rocks, dams and weirs, and rapids and waterfalls. Carry effective communications gear, radar, and depth-finding equipment whenever appropriate. Check a detailed weather forecast, and only go out when conditions are safe given your experience and your boat's capabilities.

SOURCE   USCG, uscgboating.org/statistics/accident_statistics.php

Vessel: Fatal Accidents by Length

ANNUAL US DEATHS   See table in entry.

Interpret this data from 2019 cautiously, because smaller vessels tend to be more prone to some types of fatal accidents, such as swamping or person overboard. There are also many more smaller vessels than larger ones on the water.

Less than 16 feet	252
16 to less than 26 feet	243
26 to less than 40 feet	46
40 to 65 feet	11
More than 65 feet	0
Unknown	61

STAY SAFE Always comply with state and federal (US Coast Guard) safety regulations and navigation rules. Make sure the boat's hull is sound and watertight, and all machinery and equipment are in good working order. If the vessel is mechanically powered, be very careful of the propeller(s) or water jet(s). Handle all fuel safely, avoid smoking, prevent dangerous fume buildups, and carry fire extinguishers. Never go boating while intoxicated or impaired. Use nautical charts and other reference materials to familiarize yourself with the waters in which you plan to go boating, and take charts with you. Do not speed or create a dangerous wake. Maintain a proper lookout for other boats, swimmers, and obstructions at all times. Be careful to avoid running aground in shallow areas, and to avoid hitting an obstruction, whether submerged, floating, or above the water such as a dock or a bridge abutment. Avoid rocks, dams and weirs, and rapids and waterfalls. Carry effective communications gear, radar, and depth-finding equipment whenever appropriate. Check a detailed weather forecast, and only go out when conditions are safe given your experience and your boat's capabilities.

SOURCE USCG, uscgboating.org/statistics/accident_statistics.php

Vessel: Fatal Accidents by Rental Status

ANNUAL US DEATHS See data in entry.

Of total recreational boating fatalities in the US in 2019, 418 occurred in vessels that were not rented, 41 occurred in vessels that were rented, and 154 had unknown rental status.

STAY SAFE Always comply with state and federal (US Coast Guard) safety regulations and navigation rules. Make sure the boat's hull is sound and watertight, and all machinery and equipment are in good working order. If the vessel is mechanically powered, be very careful of the propeller(s) or water jet(s). Handle all fuel safely, avoid smoking, prevent dangerous fume buildups, and carry fire extinguishers. Never go boating while intoxicated or impaired. Use nautical charts and other reference materials to familiarize yourself with the waters in which you plan to go boating, and take charts with you. Do not speed or create a dangerous wake. Maintain a proper lookout for other boats, swimmers, and obstructions at all times. Be careful to avoid running aground in shallow areas, and to avoid hitting an obstruction, whether submerged, floating, or above the water such as a dock or a bridge abutment. Avoid rocks, dams and weirs, and rapids and waterfalls. Carry effective communications gear, radar, and depth-finding equipment whenever appropriate. Check a detailed weather forecast, and only go out when conditions are safe given your experience and your boat's capabilities.

SOURCE   USCG, uscgboating.org/statistics/accident_statistics.php

Vessel: Fatal Accidents by Type

ANNUAL US DEATHS   See table in entry.

The relatively large number of fatalities involving open motorboats in 2019 may be in part because this type of vessel has an engine that allows high speed, while providing less enclosure to protect occupants.

Airboat	1
Cabin motorboat	34
Canoe	39
Houseboat	3
Inflatable	12
Kayak	86
Open motorboat	288
Personal watercraft	46
Pontoon	40
Rowboat	18
Sailboat (no engine)	4
Sailboat (with auxiliary engine)	14
Standup paddleboard	12
Other	8
Unknown	8

STAY SAFE   Always comply with state and federal (US Coast Guard) safety regulations and navigation rules. Make sure the boat's hull is sound and watertight, and all machinery and equipment are in good working order. If the vessel is mechanically powered, be very careful of the propeller(s) or water jet(s). Handle all fuel safely, avoid smoking, prevent dangerous fume buildups, and carry fire extinguishers. Never go boating while intoxicated or impaired. Use nautical charts and other reference materials to familiarize yourself with the waters in which you plan to go boating, and take charts with you. Do not speed or create a dangerous wake. Maintain a proper lookout for other boats, swimmers, and obstructions at all times. Be careful to avoid running aground in shallow areas, and to avoid hitting an obstruction, whether submerged, floating, or above the water such as a dock or a bridge abutment. Avoid rocks, dams and weirs, and rapids and waterfalls. Carry effective communications gear, radar, and depth-finding equipment whenever appropriate. Check a detailed weather forecast, and only go out when conditions are safe given your experience and your boat's capabilities.

SOURCE   USCG, uscgboating.org/statistics/accident_statistics.php

Vessel: Fatal Accidents by Year Built

ANNUAL US DEATHS   See table in entry.

This data from 2019 suggests that vessels more than 10 years old tend to be involved in more fatal recreational boating accidents. Deteriorating physical condition, and lack of up-to-date safety equipment, may be contributing factors.

2019	34
2018	31
2016 to 2017	34
2014 to 2015	23
2012 to 2013	8
2006 to 2011	59
Before 2006	245
Unknown	179

STAY SAFE   Always comply with state and federal (US Coast Guard) safety regulations and navigation rules. Make sure the boat's hull is sound and watertight, and all machinery and equipment are in good working order. If the vessel is mechanically powered, be very careful of the propeller(s) or water jet(s). Handle all fuel safely, avoid smoking, prevent dangerous fume buildups, and carry fire extinguishers. Never go boating while intoxicated or impaired. Use nautical charts and other reference materials to familiarize yourself with the waters in which you plan to go boating, and take charts with you. Do not speed or create a dangerous wake. Maintain a proper lookout for other boats, swimmers, and obstructions at all times. Be careful to avoid running aground in

shallow areas, and to avoid hitting an obstruction, whether submerged, floating, or above the water such as a dock or a bridge abutment. Avoid rocks, dams and weirs, and rapids and waterfalls. Carry effective communications gear, radar, and depth-finding equipment whenever appropriate. Check a detailed weather forecast, and only go out when conditions are safe given your experience and your boat's capabilities.

SOURCE   USCG, uscgboating.org/statistics/accident_statistics.php

Fatal Accidents by Visibility

ANNUAL US DEATHS   See table in entry.

In 2019, most boating fatalities occurred in good visibility and in daytime.

Poor – day	17
Poor – night	25
Fair – day	31
Fair – night	24
Good – day	359
Good – night	74
Unknown – day	56
Unknown – night	15
Not reported	12

STAY SAFE   Always comply with state and federal (US Coast Guard) safety regulations and navigation rules. Make sure the boat's hull is sound and watertight, and all machinery and equipment are in good working order. If the vessel is mechanically powered, be very careful of the propeller(s) or water jet(s). Handle all fuel safely, avoid smoking, prevent dangerous fume buildups, and carry fire extinguishers. Never go boating while intoxicated or impaired. Use nautical charts and other reference materials to familiarize yourself with the waters in which you plan to go boating, and take charts with you. Do not speed or create a dangerous wake. Maintain a proper lookout for other boats, swimmers, and obstructions at all times. Be careful to avoid running aground in shallow areas, and to avoid hitting an obstruction, whether submerged, floating, or above the water such as a dock or a bridge abutment. Avoid rocks, dams and weirs, and rapids and waterfalls. Carry effective communications gear, radar, and depth-finding equipment whenever appropriate. Check a detailed weather forecast, and only go out when conditions are safe given your experience and your boat's capabilities.

SOURCE   USCG, uscgboating.org/statistics/accident_statistics.php

Fatal Accidents by Water Temperature

ANNUAL US DEATHS   See table in entry.

Note that most boating fatalities in 2019 occurred in temperatures between 70 and 80 degrees F.

39 degrees F and below	15
40 to 49	39
50 to 59	65
60 to 69	87
70 to 79	137
80 to 89	125
90 degrees F and above	3
Unreported	142

STAY SAFE   Always comply with state and federal (US Coast Guard) safety regulations and navigation rules. Make sure the boat's hull is sound and watertight, and all machinery and equipment are in good working order. If the vessel is mechanically powered, be very careful of the propeller(s) or water jet(s). Handle all fuel safely, avoid smoking, prevent dangerous fume buildups, and carry fire extinguishers. Never go boating while intoxicated or impaired. Use nautical charts and other reference materials to familiarize yourself with the waters in which you plan to go boating, and take charts with you. Do not speed or create a dangerous wake. Maintain a proper lookout for other boats, swimmers, and obstructions at all times. Be careful to avoid running aground in shallow areas, and to avoid hitting an obstruction, whether submerged, floating, or above the water such as a dock or a bridge abutment. Avoid rocks, dams and weirs, and rapids and waterfalls. Carry effective communications gear, radar, and depth-finding equipment whenever appropriate. Check a detailed weather forecast, and only go out when conditions are safe given your experience and your boat's capabilities.

SOURCE   USCG, uscgboating.org/statistics/accident_statistics.php

Fatal Accidents by Wave Conditions

ANNUAL US DEATHS   See table in entry.

Most fatal boating accidents in 2019 occurred in calm water:

Calm	310
Choppy	136
Rough	54
Very rough	21
Not reported	92

STAY SAFE   Always comply with state and federal (US Coast Guard) safety regulations and navigation rules. Make sure the boat's hull is sound and watertight, and all machinery and equipment are in good working order. If the vessel is mechanically powered, be very careful of the propeller(s) or water jet(s). Handle all fuel safely, avoid smoking, prevent dangerous fume buildups, and carry fire extinguishers. Never go boating while intoxicated or impaired. Use nautical charts and other reference materials to familiarize yourself with the waters in which you plan to go boating, and take charts with you. Do not speed or create a dangerous wake. Maintain a proper lookout for other boats, swimmers, and obstructions at all times. Be careful to avoid running aground in shallow areas, and to avoid hitting an obstruction, whether submerged, floating, or above the water such as a dock or a bridge abutment. Avoid rocks, dams and weirs, and rapids and waterfalls. Carry effective communications gear, radar, and depth-finding equipment whenever appropriate. Check a detailed weather forecast, and only go out when conditions are safe given your experience and your boat's capabilities.

SOURCE   USCG, uscgboating.org/statistics/accident_statistics.php

Fatal Accidents by Wind Conditions

ANNUAL US DEATHS   See table in entry.

Most boating fatalities in 2019 occurred in light wind.

None	46
Light (0 to 6 mph)	312
Moderate (7 to 14 mph)	114
Strong (15 to 25 mph)	53
Storm (over 25 mph)	14
Not reported	74

STAY SAFE   Always comply with state and federal (US Coast Guard) safety regulations and navigation rules. Make sure the boat's hull is sound and watertight, and all machinery and equipment are in good working order. If the vessel is mechanically powered, be very careful of the propeller(s) or water jet(s). Handle all fuel safely, avoid smoking, prevent dangerous fume buildups, and carry fire extinguishers. Never go boating while intoxicated or impaired. Use nautical charts and other reference materials to familiarize yourself with the waters in which you plan to go boating, and take charts with you. Do not speed or create a dangerous wake. Maintain a proper lookout for other boats, swimmers, and obstructions at all times. Be careful to avoid running aground in shallow areas, and to avoid hitting an obstruction, whether submerged, floating, or above the water such as a dock or a bridge abutment. Avoid rocks, dams and weirs, and rapids and waterfalls. Carry effective communications gear, radar, and depth-finding equipment whenever appropriate. Check a detailed weather forecast, and only go out when conditions are safe given your experience and your boat's capabilities.

SOURCE   USCG, uscgboating.org/statistics/accident_statistics.php

Fatal Accidents by Worst States

ANNUAL US DEATHS   See data in entry.

In 2019 in the US, Florida had the most recreational boating deaths, with 62. Then came Texas, with 43, California with 39, Alabama with 28, and Washington with 27.

SOURCE   USCG, uscgboating.org/statistics/accident_statistics.php

Spectator and Group Events

Statistics for the entries in this category do *not* reflect the infection risks in mass gatherings posed during times of mass public health emergency, such as the COVID-19 pandemic.

Air Shows

> **ANNUAL US DEATHS**   Average of 5, pilots and spectators

A study for the years 1993 through 2013 found that there was an aircraft crash rate of 31 per 1,000 civil air show events. The study period recorded 174 crashes, of which 91 had at least 1 fatality, with an average of 1.1 deaths per fatal crash.

Overall deaths included both pilots and spectators. The types of flying machines involved included propeller airplanes, jets, helicopters, gyroplanes, gliders, home-builts, and hot-air balloons; no 1 type of aircraft stood out. Causes of fatal mishaps included pilot error, engine failure, and improper maintenance. Crashes were more likely to occur during aerobatics.

The worst single US air show or air race disaster of the last 20 years occurred in 2011 at the Reno Air Races, where an aircraft crashed into VIP seating, killing the pilot and 10 on the ground.

STAY SAFE   Spectators should remain in designated safe viewing areas. Pilots should make sure their aircraft are always properly maintained, and should not attempt maneuvers beyond their skill level.

SOURCES   S.B. Ballard and V.B. Osorio. U.S. Civil Air Show Crashes, 1993 to 2013: Burden, Fatal Risk Factors, and Evaluation of a Risk Index for Aviation Crashes. *Transportation Research Record*, October 2015; FAA, faa.gov/data_research/accident_incident; NTSB, ntsb.gov/safety/data/Pages/Data_Stats.aspx

A USAF pilot ejects just before his F-16 crashes at an Idaho airshow. These events pose some danger to both pilots and spectators.

Bars, Pubs, and Beer Gardens

> **ANNUAL US DEATHS** No data available.

US data on deaths that occur at drinking establishments such as bars, pubs, and beer gardens are generally not tabulated separately from deaths at other locations for the same reasons, such as homicides, alcohol abuse and drunk driving, or natural disease. But in general, these causes of death are among the leading drivers of human mortality.

Drinking establishments can sometimes be particularly exposed to the risks of fights and assaults, robberies and shootings, and drug transactions and abuse. Overcrowded bars and night clubs that have poor fire safety have over the years led to mass deaths from fire, smoke inhalation, and stampedes.

STAY SAFE For occupational death data on bartenders, see **Bartenders**. If you or someone you know has a problem with alcohol or drug abuse, call the SAMHSA national helpline at 1-800-662-4357. Help is free and is available 24/7/365. All calls are strictly confidential.

SOURCE None

Birthday Effect

> **ANNUAL US DEATHS** See data in entry.

Statistically credible studies in several countries, including the US, UK, the Ukraine, and Switzerland, have found that people are more likely to die on or close to their birthday.

The data on 2.4 million Swiss people found that those who were aged 60 or older had 18 percent greater odds of dying on their birthday, compared to other days. Among females, the probability of dying from natural causes was 22 percent higher on their birthday. Among males, the probability of accidental death was 29 percent higher on their birthday.

Excess deaths on birthdays are not offset by fewer deaths in a 10-day range. Excess death rates on birthdays are greater for younger people, and on weekends. For the age range 20 to 29 years, the average excess death rate on a birthday is 25.39.

Theories to explain this birthday effect include the increased emotional stresses about having a birthday, such as a reminder of one's own mortality, or social anxiety and family angst at the birthday party (or the lack of one). There could also be a rise in suicides on birthdays for psychological reasons, analogous to the spike on Mondays (called the new beginnings effect). Another factor for younger victims is an increased intake of celebratory alcohol and/or drugs.

William Shakespeare is traditionally said to have died on his birthday, April 23, in 1616. He was 52.

STAY SAFE   Be cautious getting ready for your next birthday, and during it, to avoid excessive intoxication or reckless behavior. If you are feeling depressed or suicidal, or have a friend or loved one who might be, call the National Suicide Prevention Hotline at 1-800-273-8255 (TALK); help is free and is available 24/7/365.

SOURCES   V. Ajdacic-Gross et al. Death has a preference for birthdays—an analysis of death time series. *Annals of Epidemiology*, August 2012; P.A. Peña. A not so happy day after all: excess death rates on birthdays in the U.S. *Social Science and Medicine*, February 2015.

Casinos

ANNUAL US DEATHS   Few statistics, but an average of 10 lethal heart attacks yearly

Comprehensive national data is not readily available for people who die while at casinos in the US. Casino management is understandably reluctant to report deaths among their guests. Cruise ships with casinos that operate in international waters and foreign territorial waters also are reluctant to report deaths.

Casino patrons do sometimes die of heart attacks, perhaps brought on by the stress of losing or the excitement of winning. Over the 15-year period 1993 through 2008, at Foxwoods and Mohegan Sun casinos in Connecticut, almost 150 people (both patrons and employees) have died of a heart attack on the premises, averaging about 10 people a year. In that same time frame, the casinos also reported a total of 10 suicides and 6 drug-related deaths. For perspective, these same casinos between them entertain more than 25 million guests every year and have trained medical staff and equipment such as AEDs on site for emergencies. Based on this information, the odds of dying while at a casino appear to be under 1 in 1 million annually.

Passive smoke is a potentially fatal issue for casino workers and frequent patrons. Although casinos have nonsmoking sections and run extensive air-ventilation systems, studies have shown that blood levels of cigarette toxins among nonsmoking casino employees are elevated by 3 to 6 times above normal. One study estimated that in Pennsylvania, 6 out of every 10,000 nonsmoking casino workers will die each year due to high exposures to secondhand smoke. Casino patrons presumably have shorter, only intermittent exposures compared to full-time casino employees, but gamblers should be aware of the health risks at gaming establishments, some of which are exempt from state no-smoking laws.

STAY SAFE   Don't smoke. For help with any thoughts of self-harm or severe depression, contact the National Suicide Prevention Lifeline at 1-800-273-8255 (TALK) or suicidepreventionlifeline.org; help is free, 24/7/365. For help with a gambling addiction problem, contact the National Council on Problem Gambling at 1-800-522-4700 or ncpgambling.org/chat; help is free, 24/7/365.

SOURCES   J.L. Repace. Secondhand smoke in Pennsylvania casinos: a study of nonsmokers' exposure, dose, and risk. *American Journal of Public Health,* 2009; D. Trout et al. Exposure of casino employees to environmental tobacco smoke. *Journal of Occupational and Environmental Medicine,* March 1998.

Collapses of Grandstands, Stages, and Arena Roofs

ANNUAL US DEATHS   No cumulative data available.

Although there have been massive death tolls in ancient times, and in other countries, from collapses of grandstands, stages, and arena roofs, such incidents in the US in recent years have been very rare.

The most notable fatal incident of this kind in the US in the past 20 years occurred in 2011 at the Indiana State Fair, before an outdoor concert began. A wind gust from a thunderstorm caused a stage's temporary roof to collapse onto spectators, killing 7 and injuring another 58.

STAY SAFE   If severe weather approaches at an outdoor event, seek shelter away from temporary stages and loudspeaker arrays that might fall over.

SOURCE   The *Indianapolis Star,* indystar.com, "Forever and yesterday": Remembering the Indiana State Fair stage collapse 10 years later.

Country Clubs, Health Spas, and Sweat Lodges

> **ANNUAL US DEATHS** No cumulative data available; see details in entry.

Annual US death statistics for fatalities while engaged in activities that are popular at country clubs, such as golf, tennis, swimming, and fitness training, are not readily available broken out from deaths at other locations (such as public parks, beaches, gyms) during such activities.

While their deaths were due to a mass-shooting hate crime, in April 2021, 8 people were fatally shot at 3 separate massage spas in a single incident in the Atlanta area.

In 2009, 3 people died at an extreme sweat lodge in Arizona, and another 18 were hospitalized; the owner was subsequently convicted of negligent homicide.

STAY SAFE Match your level of exertion to your level of fitness and health. If you are feeling overtaxed, stop at once, rehydrate immediately, and rest. Ignore fanatical group leaders or trainers who might try to intimidate or embarrass you into continuing. If you ever become involved in a mass-shooting incident, quickly evaluate your options and choose the safest course between fleeing the scene, hiding for safety, or in the last extreme fighting for your life.

SOURCES The *New York Times*, nytimes.com, 8 Dead in Atlanta Spa Shootings, With Fears of Anti-Asian Bias; Oxygen, oxygen.com/deadly-cults/crime-news/james-arthur-ray-sweat-lodge-deaths

Crowd Stampedes

> **ANNUAL US DEATHS** No cumulative data available.

Although there have been massive death tolls at crowd stampedes in other countries, such disasters in the US have been very rare in recent years.

The largest fatal crowd stampede in the US not part of a larger disaster such as a fire or explosion occurred in 2003 at a nightclub in Chicago. It was apparently precipitated by security staff using pepper spray to try to break up a fight. The death toll was 21, and more than 50 others were injured. Reportedly, patrons who did not know about the fight detected the smell of the pepper spray, and thought there was a fire or even a terrorist poison gas attack, so they all panicked and ran for the exits. On November 6, 2021, a crowd stampede at a concert in Houston's Astroworld stadium by rapper Travis Scott killed 9 and injured hundreds more.

Four US citizens were among the 45 or more people killed in the crush of a crowd in a narrow, tunnel-like passageway during a religious festival in Israel in April 2021.

STAY SAFE At any crowded gathering, it is important to keep your wits about you, maintain situational awareness, and give a priority to personal safety. Avoid getting into situations that involve overcrowding and/or inadequate exit pathways. Facility managers and security staff must make sure that exits are not blocked, local regulations about maximum occupation capacity are not violated, all mandated fire protection equipment is installed and in good working order, and a public address system is available to issue instructions for crowd control.

SOURCE Reuters, reuters.com/world/ some-worlds-worst-stampedes-2021-11-06

Discos, Raves, Dance Clubs, and Nightclubs

ANNUAL US DEATHS No cumulative data available, but see entry details in entry.

Like bars, pubs, and other drinking establishments, discos, raves, and clubs can be great places to have a good time, but at their worst can sometimes involve alcohol poisonings and drug overdoses, interpersonal violence, and unsafe overcrowding. The main distinction is that discos, raves, and clubs also have large numbers of people dancing, sometimes under the influence of alcohol and drugs such as ecstasy. A few of the worst-ever disaster news items can be illustrative.

The Cocoanut Grove fire on November 28, 1942, in Boston was the deadliest nightclub fire in history. The fire at the popular club killed 492 people. The club was crowded to twice its authorized capacity; flammable draperies and décor ignited, and the flames and smoke spread rapidly. The tragedy led to new fire laws for public establishments, including new regulations requiring emergency doors to stay open and unblocked.

One of the deadliest mass shootings in the US occurred at the Pulse nightclub in Orlando in June 2016, when a domestic terrorist killed 49 people and wounded more than 50 others, before being killed in a shoot-out with police.

In 2003 at a nightclub in West Warwick, Rhode Island, a fire set off by pyrotechnics killed 100 and injured about 230 others.

In 2003 in Chicago, a stampede in a nightclub killed 21 and injured another 50 or more. It was started by security guards using pepper spray to break up a fight.

In 2016 in the Los Angeles area, 3 of 9 people hospitalized for drug overdoses at a rave died.

STAY SAFE   If you or someone you know has a problem with alcohol or drug abuse, call the SAMHSA national helpline at 1-800-662-4357. Help is free and is available 24/7/365. All calls are strictly confidential.

SOURCE   NFPA, nfpa.org/Public-Education/Staying-safe/ Safety-in-living-and-entertainment-spaces

Halftime Shows

ANNUAL US DEATHS   Very rare

Performers being killed in accidents related to sporting event halftime shows are extremely rare. In 1997, an aerialist was killed practicing a bungee jumping routine for that year's Super Bowl halftime show in New Orleans.

STAY SAFE   See **Bungee Jumping**.

SOURCE   The *New York Times*, nytimes.com/1997/01/25/sports/ superdome-death-stirs-questions-on-risks-of-shows.html

Marching Bands and Color Guards

ANNUAL US DEATHS   Possibly 4

A study of mortality among marching band directors in the US, for the years 2001 through 2011, found a total of 40 deaths from all causes. It is unclear how many of these were related to the decedent's occupation. The most common cause of death was a heart attack or stroke, and there were 3 suicides. Eight of the deaths occurred while on duty; many of these occurred while traveling with a group. There have been isolated incidents of deaths due to hazing of marching band participants.

Marching band and color guard performances are physically grueling and usually occur outdoors in all weather conditions. They are sometimes part of parades that also involve potential accidents with motor vehicles and towed floats. Participants in marching bands and color guards are frequently injured, but usually not seriously, and hard data on deaths is scarce.

STAY SAFE Exercise care regarding traffic safety whenever performing on the streets and/or in parades with towed floats. Dress properly for the weather and stay hydrated in hot weather.

SOURCE Texas Music Education Research Reports, Health Risks Faced by Public School Band Directors, 2013.

Parade Floats and Hayrides

ANNUAL US DEATHS Estimated 40 (for hayrides; see entry details for floats).

Parade floats and hay wagons, towed by motor vehicles, have the potential for fatal accidents. Annual US death statistics for parade floats and hayrides are not always broken out from other motor vehicle accident deaths. But anecdotal news reports, and volunteer/community safety groups, can provide a perspective.

For instance, at the 2020 Mardi Gras festivities in New Orleans, 2 bystanders were killed when they were struck by towed parade floats. One study found 8 parade-related accidental deaths in the US over a recent 3-year period. Another 3 fatalities related to motor vehicle and float mishaps occurred at Fourth of July parades in 3 different states (Oklahoma, Maryland, and Maine) in 1 year, 2013.

Hayrides, riding atop hay bales on a towed trailer, when combined with towed parade floats, are estimated to kill an average of about 40 people per year in accidents, and seriously injure another 400 annually.

STAY SAFE If you are a spectator at a parade, stay safely on the sidewalk so as not to get hit by a motor vehicle, towed float, or errant large balloon. If you are driving a motor vehicle in the parade, use extreme

Arlington, Texas, Independence Day Parade float

caution at all times to avoid hitting participants, bystanders, or pedestrians. Make sure the trailer is in good repair, that any auxiliary braking system is properly connected and working, and that any parade structures or displays carried on the float are firmly anchored. If you are riding on a towed float, hayride, or other trailer, sit securely and hold on tight at all times. Get on or off carefully, and only while the trailer is not moving. Similar safety precautions should be applied by all involved in hayrides.

SOURCE ParadeSafety.org, paradesafety.org/Other_Hayride_Accidents.html

Parties

> **ANNUAL US DEATHS** No cumulative data available

Block parties, house parties, and birthday parties can sometimes lead to deaths due to accidents, homicides and suicides, or natural causes. Statistics on deaths specifically at such parties are generally not compiled as a separate category, but anecdotal news items can be illustrative.

In May 2021, in Fairfield Township, New Jersey, at a large house party, 2 people were shot to death and at least 12 others wounded. In May 2021, in Colorado Springs, a gunman killed 6 people, including his girlfriend, and then himself at a birthday party. In June 2020, in Charlotte, North Carolina, 4 were killed and 10 others injured when several perpetrators ran people over on the street and then fired dozens of shots into the crowd at a block party.

Other deaths have at times occurred due to motor vehicle accidents on roads or driveways, dangerous stunts and reckless hijinks during celebrations, heart attacks and strokes, lightning strikes, and drownings in pools and ponds. Sometimes people have been struck and killed by bullets fired by celebrants discharging firearms into the air, or have been fatally injured by party-related explosives, such as fireworks or gender reveal bombs.

STAY SAFE If you or someone you know has a problem with alcohol or drug abuse, call the SAMHSA helpline at 1-800-662-4357. Help is free and is available 24/7/365. All calls are strictly confidential.

SOURCES CNN, cnn.com/2021/11/01/us/mass-shootings-halloween-weekend/index.html; *USA Today*, usatoday.com/story/news/nation/2021/05/11/colorado-springs-shooting-police-release-updates-mass-killing; WBTV, wbtv.com/2020/05/24/officials-year-old-year-old-killed-others-injured-after-gunfire-erupts-party

FLYING CHAMPAGNE CORK

ANNUAL US DEATHS 0

There are no documented cases of death from being hit by a flying champagne cork. Eye injuries do occur, as when reality star Theo Campbell (*Love Island*) was blinded in his right eye in 2019 by a flying champagne cork that split his eyeball in half.

STAY SAFE The pressure within a champagne bottle can be as high as 90 pounds per square inch—the same pressure as a fire hose at full blast. Champagne corks can fly out of the bottle at 50 mph. To avoid injury, chill the champagne to avoid pressure buildup. Don't shake the bottle; point it away from people when opening. Firmly press and twist the cork to open. You should hear a gentle puff, not a loud pop.

SOURCE American Academy of Ophthalmology, aao.org/eye-health/tips-prevention/injuries-champagne

Pool Halls

ANNUAL US DEATHS No cumulative data available

While data on people dying while playing billiards or otherwise spending time in pool halls is not readily available, deaths are not entirely unknown. In 2003 in Texas, famous billiards professional "Fast Eddie" Parker died of a heart attack during a pool tournament. In 1999 in West Los Angeles, a horror film director who suffered from syncope (a blackout disorder) and who had a restricted driver's license blacked out at the wheel and crashed his vehicle through a bus stop and the bar of a pool hall, killing 1 person waiting for a bus and 1 person at the bar. In 1991 in Troy, New York, a man died when he was hit in the head with a billiard ball during a drunken dispute. In 1990, in Maryland, 2 bystanders in a pool hall were killed when 2 intruders sprayed gunfire at a fifth man (who escaped unharmed), in what police described as a drug dispute.

STAY SAFE Use common sense about personal safety in pool halls.

SOURCES The *Baltimore Sun*, baltimoresun.com/news/bs-xpm-1992-11-06-1992311040-story.html; *Los Angeles Times*, latimes.com/archives/la-xpm-2000-jun-02-me-36648-story.html; news10.com , news10.com/news/local-news/45-year-old-man-falls-off-roof-dies-in-construction-accident; News and Record, greensboro.com/man-hit-by-billiard-

Religious Gatherings: Worship, Weddings, Births, and Funerals

ANNUAL US DEATHS 46

Houses of worship in the US are sadly not immune to intentional violence, terrorism, and hate crimes. A study of FBI data for the years 2000 through 2017, scaled up with estimates for jurisdictions not reporting their data to the FBI's database, estimated that there was about 480 incidents of serious violence at churches, synagogues, mosques, and temples in the US per year. The result of this interpersonal violence averaged out at about 46 people killed and 218 seriously injured annually.

Virtually every religion, denomination, and sect were affected. The majority of the incidents during this 17-year period were motivated by domestic conflicts, personal disputes, mental derangement, or robbery, rather than hate or terrorism. But some of the biggest single-incident death tolls were intentional acts of domestic or international terrorism/hate.

On October 27, 2018, an anti-Semitic gunman entered the Tree of Life Congregation synagogue in Pittsburgh and opened fire, killing 11 and wounding 6, including 4 police officers. On November 5, 2017, a gunman killed 26 people and wounded at least 20 others at the First Baptist Church of Sutherland Springs, Texas. On June 17, 2015, Dylann Roof, a 21-year-old White supremacist, killed 9 Black worshippers at historic Emanuel African Methodist Episcopal Church in Charleston, South Carolina.

Interpersonal violence is not the only possible cause of deaths at houses of worship. Accidents such as falls, fires, or structure collapse, and natural causes such as heart attacks, also occur. Detailed, comprehensive national statistics about such fatalities are generally not broken out from broader death data. (See, for instance, **Slips, Trips, and Falls** and **Collapses**.

STAY SAFE Every religious congregation should develop a safety plan well in advance of any problems. Ushers and other staff should be trained in safety measures, including fire safety, first aid, and violent incidents.

SOURCE FBI, fbi.gov/services/cjis/ucr/hate-crime

Rodeos

In the US, an average of 1 or 2 participants a year are killed performing in professional, amateur, and youth rodeos. Annually, several more suffer catastrophic spinal or brain injuries. Nearly 20 of every 100,000 rodeo contestants can expect to suffer a catastrophic injury, meaning the performer will either die or have their life altered in a significant way. Not surprisingly, bull riding is the most dangerous rodeo event, in which the rider can be thrown, gored, and/or stepped on by the massive, angry bull. The incidence of injury is reported at 32.2 per 1,000 competitor-exposures. In 2019, professional bull rider Mason Lowe died after being bucked off his bull at the National Western Stock Show in Denver. The bull, weighing between 1,500 and 2,000 pounds, landed on his chest with its hind legs. Despite wearing a flak jacket, Lowe died of massive injuries to his chest and heart.

STAY SAFE   Rodeo sports are inherently dangerous to both the human participants and the animals. Proceed at your own risk.

SOURCES   D.J. Downey. Rodeo injuries and prevention. *Current Sports Medicine Reports*, October 2007; Professional Rodeo Cowboys Association, prorodeo.com; A.J. Sinclair Elder et al. Analysis of 4 years of injury in professional rodeo. *Clinical Journal of Sport Medicine*, 2020.

Spectator Accidents at Sports Stadiums and Arenas

Accidents in the stands at sporting events in the US that lead to spectator deaths have been very rare in the past 20 years, but they do sometimes occur. Most result from an event attendee leaning too far over a railing, such as to try to catch a hit baseball. Other fatalities occur when an attendee attempts a dangerous stunt, such as sliding down a staircase railing, and slips off and falls to their death. There have also been isolated instances of people falling to their deaths after tripping on steps or escalators. Impairment by alcohol or drugs were found in some of the autopsies. There have also been very rare cases of people committing suicide by jumping from high up at a sports arena, stadium, or concert hall.

STAY SAFE   Do not attempt a stunt, or do anything else reckless, while in the stands at a sporting event, concert, or other public gathering, whether indoors or outdoors. Exercise caution on stairs and escalators.

SOURCE   Ranker, ranker.com/list/people-passed-sporting-events/mark

Water Sports

Jet Skiing

> **ANNUAL US DEATHS**   40 to 50

Every year in the US, about 40 to 50 people die in jet ski accidents. Another 600 to 700 are seriously injured. Causes of death include drowning after falling off and lethal trauma from collisions.

STAY SAFE   Know the limitations of your skill level, be careful not to lose control of the jet ski, and never be reckless. Always wear a personal flotation device. Be especially careful while operating a jet ski near any swimmer, boat, bridge piling, pier, buoy, or around other jet skiers. Never use a jet ski while under the influence of alcohol or drugs. Be wary of bad weather, and do not venture too far from shore. The high-pressure water jet coming out of the back is potentially very dangerous.

SOURCE   USCG, uscgboating.org/statistics/accident_statistics.php

Scuba Diving

> **ANNUAL US DEATHS**   Average of 56.3

From 2006 through 2015 in the US, 563 people were killed in recreational scuba diving accidents. In that same 10-year period, an estimated total of 306 million recreational scuba dives were made. This gives an average annual death toll of 56.3, and a fatality rate of 1.8 per million recreational dives.

STAY SAFE   See **Recreational Boating, Drowning Outdoors**, and **Commercial Divers and Underwater Welders**

SOURCE   Divers Alert Network *2019 Annual Diving Report*, dansa.org/annual-diving-report

Snorkeling and Free Diving

ANNUAL US DEATHS No cumulative national data available;
average of 21 deaths in Hawaii

In competitive free diving around the world, there has only ever been 1
fatality, in an estimated 50,000 such organized diving events.

Recreational snorkeling kills a few dozen people per year in the US
in a variety of ways, some of which are not broken out in water-sports
death statistics as specifically due to snorkeling: heart attacks, drownings, trauma from boating mishaps, and other accidents. One study for
the state of Hawaii for the years 2014 through 2018 found an average
of 21 snorkeling deaths per year in that state's waters; the majority of
these victims were visiting tourists. The study identified full-face snorkel masks as carrying higher risk.

STAY SAFE Snorkelers should comply with recommended best practices: Never snorkel in areas of heavy vessel traffic or with underwater
hazards, avoid snorkeling during lightning storms or heavy seas, always have a lookout on the scene, never free dive beyond 30 feet, and
do not snorkel in water with less than 3 feet of visibility. Do not wear a
full face mask while snorkeling. Beginners should be very careful until
they become familiar with their equipment: Flippers can be dangerously awkward against a shallow bottom, and many snorkeling deaths
occur in water no more than 3 feet deep.

SOURCES Divers Alert Network *2019 Annual Diving Report*,
dansa.org/annual-diving-report; HawaiiNewsNow.com,
hawaiinewsnow.com/2019/09/18/spike-snorkel-related-deaths-
again-highlights-potential-danger-full-face-masks; OpenWaterHQ,
openwaterhq.com/snorkeling

Spearfishing

ANNUAL US DEATHS No data available

Spearfishing combines the risks of scuba diving, free diving, snorkeling, and/or open-water swimming, with added dangers analogous to
those of hunting and archery. Statistics on US deaths in spearfishing
mishaps are not tabulated separately by any government agency.

STAY SAFE Follow safety practices for snorkeling. Never aim a spear-
gun at another person, even as a joke.

SOURCE   SpearfishingWorld.com, spearfishingworld.com/diving-safety.html

Surfing and Paddleboarding

ANNUAL US DEATHS   101 in 2019

In the US in 2019, 101 people were killed while surfing and paddle-boarding. These deaths occurred at beaches in the Atlantic and Pacific Oceans, the Gulf of Mexico, and the Great Lakes. Drownings in rip currents were the leading cause of death, with 59 fatalities. In 2021,130 people were killed in surf zone accidents, mostly by drowning in rip currents.

STAY SAFE   Never go surfing where the surf is larger than your ability.

SOURCES   D. Morgan et al. Descriptive epidemiology of drowning deaths in a surf beach swimmer and surfer population. *Injury Prevention*, February 2008; NWS, weather.gov/safety/ripcurrent-fatalities; USCG, uscgboating.org/statistics/accident_statistics.php; Watersportgeek, watersportgeek.com/surfers-drown

Water-Skiing, Wakeboarding, and Parasailing/Kitesailing

In 2019 in the US, 15 people were killed while engaged in water-skiing or other sports involving being towed behind a motorized water vessel. In 2021, a kitesailing man in Fort Lauderdale, Florida, was killed when a sudden strong wind gust blew him into the side of a house.

The Windsor sisters water-skiing

STAY SAFE   See **Recreational Boating.**

SOURCES   Parasail Safety Council, parasail.org/consumer-alerts.
html; USCG, uscgboating.org/statistics/accident_statistics.php

Winter Sports and Activities

Horse-Drawn Sleighs

ANNUAL US DEATHS   Scant data; 2.3 in Pennsylvania

Data on US deaths from horse-drawn sleigh accidents are not broken
out from other statistics on human deaths due to equestrian (horse-
related) fatalities, such as riding, racing, ranching, and using horse-
drawn work, passenger, or tourist wagons and carriages. By analogy,
a serious enough horse-drawn sleigh accident has the potential to kill
the driver, passengers, and/or bystanders/pedestrians, plus the horse
or horses. In a study of horse-drawn buggy accidents on public roads in
Pennsylvania, 23 people were killed in a recent 10-year period, for an
average of 2.3 deaths per year in 1 state.

STAY SAFE   Make sure the sleigh horse is controlled and driven by a
professional, that the sleigh is in good repair, and the horse is in good
health. Do not drive or ride a horse-drawn sleigh in traffic, on rough terrain, over frozen lakes and rivers, or near cliff and embankment edges.
Be careful getting on and off the sleigh, stay securely seated while in
motion, and be careful to not be kicked by a horse.

SOURCE   Pennsylvania Department of Transportation,
penndot.gov, 2018 Pennsylvania Crash Facts and Statistics

Iceboating

ANNUAL US DEATHS   2 in the winter of 2018 to 2019

In the winter of 2018 to 2019 in the US, 2 people were killed while ice-
boating (ice sailing) on frozen lakes and rivers.

STAY SAFE   Be very wary at all times of falling through the ice. Watch
out for open or flowing water or rotten ice, which are warnings that
nearby ice is thin. If the ice is not so thick that it appears black even
in broad daylight, stay away! Beware of unseasonably warm weather,
which can lead to unexpected thin ice. Obey all posted signs and buoys
warning of dangerous ice. Always wear a helmet and personal flota-

tion device, prioritize personal safety, and stay within the limits of your skills, experience, and equipment.

SOURCES   iceboat.org, iceboat.org/iceboatsafety; lakeice.squarespace.com, lakeice.squarespace.com/ice-safety

Ice Fishing

> **ANNUAL US DEATHS**   6 in the winter of 2018 to 2019

During the winter of 2018 to 2019 in the US, 6 people died from accidents while ice fishing on frozen lakes or rivers.

STAY SAFE   Be very wary at all times of falling through the ice. Watch out for open or flowing water or rotten ice, which are warnings that nearby ice is thin. If the ice is not so thick that it appears black even in broad daylight, stay away! Beware of unseasonably warm weather, which can lead to unexpected thin ice. Obey all posted signs and buoys warning of dangerous ice. Dress for the weather. Don't consume too much alcohol, which can contribute to hypothermia; ice fishing while impaired can lead to taking careless risks that end in tragedy.

SOURCES   lakeice.squarespace.com, lakeice.squarespace.com/ice-safety; Takemefishing.org. takemefishing.org/ice-fishing

Luge, Skeleton, and Bobsled

> **ANNUAL US DEATHS**   Scant data; 2 competitors killed in Olympic luging over a 60-year period

Luge is a very dangerous sport that should only be attempted by trained, adult athletes. It involves lying on your back on a sled with no brakes and sliding feet-first down an icy downhill racing track at speeds in excess of 90 miles per hour. Skeleton is the same idea, but you lie on the sled on your stomach. In Olympic luging in the past 60 years, 2 competitors have been killed in crashes while practicing or competing.

Bobsledding is also a very dangerous sport. Each sled weighs hundreds of pounds and can go up to 125 miles per hour, with riders experience g-forces up to 4 times normal gravity on sharp curves. Although no one was killed bobsledding at the Vancouver 2010 or Sochi 2014 Winter Olympics, almost 20 percent of the bobsled athletes reported being injured, sometimes seriously.

STAY SAFE   Leave luging and bobsledding to Olympic athletes.

SOURCES   Bobskeleton.org.uk, Fatal incidents; the *New York Times*, nytimes.com/2020/07/26/sports/olympics/olympics-bobsled-suicide-brain-injuries.html

Skiing and Snowboarding/Snowtubing

ANNUAL US DEATHS   48 in the winter of 2020 to 2021

In the US during the 2020 to 2021 season, 48 people were killed while skiing, snowboarding, or snowtubing. Most deaths were caused by collisions with objects. The majority occurred on terrain designated more difficult or intermediate (blue square). On average in a typical year, about two-thirds of the deaths are skiers and one-third are snowboarders/snowtubers. By gender, 85 percent were male. By age, 70 percent were between their late teens and their late 30s. Many of those killed were experienced skiers going at high speed on the margins of intermediate trails. About half of them were wearing ski helmets. Another 33 suffered life-changing injuries such as brain damage or paralysis.

Injurious collisions by skiers with other people on the slopes are rare, accounting for only 6 percent of all accidents needing medical aid. Collisions between snowboarders and others are very rare.

STAY SAFE   Be careful while riding ski lifts. Be wary of avalanche risks. Obey all warning signs on ski trails and slopes, and obey all instructions from ski patrols. Be confident in your skills relative to the challenges of the trail or slope you attempt. Avoid collisions with trees and other skiers. Wear a protective helmet. Be courteous of others, look around and uphill, stay in control, stop only in a safe area, and yield the right of way as appropriate.

SOURCE   National Ski Areas Association (NSAA), nsaa.org, Ski Area 2020–21 Season Fatality Fact Sheet

Ski Lifts (Chairlifts)

ANNUAL US DEATHS   0.18 per 100 million skier-miles

The death rate on US ski lifts (chairlifts) is about 0.18 per 100 million skier-miles. In a recent 38-year period, there were about 13 deaths involving roughly 3,500 ski lifts, mostly due to mechanical malfunctions or falls. In the same period, there were more than 14.7 billion ski lift rides, over a cumulative distance of 7.3 billion miles.

STAY SAFE   Pay attention to all safety instructions and signage, sit back and upright and hold on firmly, don't adjust skiing equipment while on the lift, be careful getting on and off, and don't ride while intoxicated, impaired, or if you have a medical problem. If the weather looks threatening, such as high winds, consider not using the lift.

SOURCES   National Ski Area Association (NSAA), nsaa.org, NSAA Aerial Ropeway and Surface Lift Fact Sheet; Ski California, skicalifornia.org/ski-safety/lift-safety

Sleds and Toboggans

> **ANNUAL US DEATHS**   2 in toboggan accidents

A study of US data for 2008 through 2017 found that an average of more than 22,000 people per year needed emergency room treatment due to sled injuries. The most common accident was a collision with a fixed obstacle (e.g., tree or boulder), another sled, or a person. Car accidents involving sledders can produce serious, even fatal injuries, especially if the sled is being towed behind a motor vehicle. Deaths are rare but not unknown: In early 2021, a 7-year-old California boy died when his sled went under a moving car.

On average, 2 people per year are killed in toboggan accidents. Toboggans, which lack steering mechanisms, are more dangerous than steerable snow sleds.

STAY SAFE   Do not sled on public streets, or in or near traffic, or anywhere near trees, rocks, or other obstacles you or your sled might hit at high speed. Wear a helmet. Ride only while sitting up and facing forward. Never face backward and never ride headfirst on your belly. Make sure your sled is in good condition. Never use a sheet of plastic as an improvised sled, as it is impossible to steer or control and can be easily pierced by sharp objects hidden in the snow. Dress for the weather. Adult supervision is advisable.

SOURCES   K.H.C. Evans et al. Sledding-related injuries among children and adults treated in US emergency departments from 2008 to 2017. *Clinical Journal of Sports Medicine*, 2021.

In the Great Outdoors

In this section, the great outdoors means recreational activities that are usually, though not always, enjoyed in the open air.

Archery and Bowhunting

ANNUAL US DEATHS   No cumulative data available

Although hard data on US deaths from recreational archery and bowhunting is scarce, a study of hunting-related deaths from 2002 through 2007 indicated that fewer than 100 people per year were killed in hunting-related accidents of all kinds—primarily due to firearm mishaps or falls from tree stands, not arrow accidents. Injuries from carelessly handled hunting arrows are not uncommon, due to their razor-sharp points with large edges designed to cause large animals to quickly bleed to death.

The sport of archery is reportedly very safe. Many urban and sub-urban jurisdictions have laws against archery except at designated ranges, where very sharp arrow points, such as for hunting, are strictly forbidden.

STAY SAFE   Always be extremely careful while bowhunting and handling, loading, or shooting arrows. Archery should be done only at a designated range if local laws so require and should never be done with arrows having sharp points. Beginning archers of all ages should get professional training on safety and accuracy. Never aim a loaded bow at another person or a pet, even as a joke.

SOURCES   CDC, cdc.gov/injury/wisqars/fatal.html; International Hunter Education Association, ihea-usa.org/hunter-incident-database; Target Tamers, targettamers.com/guides/hunting-accident-statistics

BB Guns, Pellet Guns, and Air Rifles

ANNUAL US DEATHS   About 4

About 4 people per year in the US are killed by BB guns, pellet guns, and air rifles, which use compressed air (by pumping a handle) or compressed carbon dioxide (from a canister) as propellant. The small metal projectiles, whether round (BBs) or bullet shaped (pellets), often have a muzzle velocity of well over 350 feet per second (fps), which can kill a person. In addition, about 30,000 Americans each year need emergency medical care due to nonlethal injuries from these weapons.

STAY SAFE   Obey any local laws banning guns powered by compressed gas. Treat all these weapons as loaded firearms that are potentially lethal due to accidents or unsafe use. They are not toys. Never aim them at a person or a pet, even as a joke.

SOURCES   CDC, cdc.gov/injury/wisqars/fatal.html; D. Laraque, Injury risk of nonpowder guns. *Pediatrics*, November 2004

Beaches

ANNUAL US DEATHS   No cumulative data available

Many millions of Americans enjoy beaches every year without any problems, but deaths from beach-related activities do occur. Aside from natural causes such as heart attacks, perhaps brought on by overexertion, most of the hazards to life and limb come from going into the water (e.g., drowning) or being out in the open air (e.g., getting struck by lightning).

Many beaches ban private motor vehicles. On beaches that permit operation of recreational ATVs and other motor vehicles, always drive safely and never while impaired or distracted. Never overload the vehicle or exceed safe speeds. Beware of extreme maneuvers and unstable terrain that can cause rollovers.

Swimming at the beach can be dangerous when surf conditions are rough or when rip currents (also called, incorrectly, rip tides) are running. In 2021, 130 people were killed in surf zone accidents, most by drowning in rip currents.

STAY SAFE   Keep hydrated in hot weather and avoid overexertion that can cause heatstroke or other medical problems. Swim only in safe areas that have lifeguards. Be aware of surf conditions and rip currents. Stay out of restricted areas on the beach and in the water.

SOURCES   CDC, cdc.gov/injury/wisqars/fatal.html; NWS, weather.gov/safety/ripcurrent-fatalities

Bungee Jumping

ANNUAL US DEATHS 1.2 (worldwide)

A study of data over a recent 23-year period found an average of 1.2 people per year were killed in bungee jumping accidents around the world. Another study reported a fatality rate of 1 in every 500,000 bungee jumps, comparable to the fatality rate of skydiving jumps.

STAY SAFE Staff should ensure that all equipment is in good condition, and cables are of proper length relative to the height of the takeoff point above the ground. Be sure that jumping harnesses are properly fitted and securely fastened to the body.

SOURCES K. Søreide. The epidemiology of injury in bungee jumping, BASE jumping, and skydiving. *Medicine and Sport Science*, 2012; K. Soreide et al. How dangerous is BASE jumping? An analysis of adverse events in 20,850 jumps from the Kjerag Massif, Norway. *Journal of Trauma*, May 2007.

Bungee jumping

Camping

ANNUAL US DEATHS No cumulative data available

Like hiking, camping is enjoyed by huge numbers of Americans but can on rare occasions be fatal. The risks of camping are in many ways similar to those of hiking, as are the difficulties in identifying accurately the number of deaths each year in the US due specifically to camping. (See **Hiking**.)

Camping is analogous to hiking but with increased risk of death (or serious injury), from sharp instruments; open flames, accidental fire, smoke inhalation, and burns; carbon monoxide poisoning; food poisoning and bad water; bad weather; medical crises from overexertion; being hit by a falling tree, a falling rock, or a snow avalanche; or being attacked by bees, snakes, or bears. Those who hunt while camping, or who camp in order to hunt, face the added risks of hunting as well.

STAY SAFE Be very careful with knives and axes, and always practice fire safety. Purify water with tablets, a portable filter, or by boiling. Do not tax yourself beyond the limits of your physical abilities and fitness conditioning. Bring both warm *and* cool clothing, since multiday weather forecasts can be wrong. Be careful to not pitch camp near beehives or hornet nests, animal dens, precipice edges, unstable/crumbling overhanging rocks, or trees that might fall on you.

SOURCE   NPS, nps.gov/orgs/1336/data.htm

Dirt Bikes and Mountain Bikes

ANNUAL US DEATHS   245 dirt bike deaths in 2013

Data on US deaths in mountain bike accidents are not broken out from all bicycling deaths. (See **Bicycles**.) Studies from Canada and Europe indicate that the annual US deaths from bicycling in mountainous terrain is between about 0.2 and 0.35 per 100,000 mountain biking outings. Leading causes of death are brain and spine injuries due to collisions, or accidentally vaulting over the handlebars due to sudden stops while going downhill at high speed. At least 1 person was killed by a bear.

Data on US dirt bike accidents indicate that in 2003, 245 Americans were killed on dirt bikes. During the period 2001 through 2004, about 100,000 people per year were injured badly enough to need hospitalization. There is probably some overlap in the data between mountain bike accidents and dirt bike accidents, and some of the fatalities occurred while riding on roads in traffic, rather than on trails, racetracks, or in the mountains.

STAY SAFE   Stick to riding on trails; riding on racetracks is more dangerous due to competitive pressures, crowding by multiple contestants, and prevalence of daredevil stunts. Always wear a helmet, and make sure your bike's brakes, tires, and other parts are in good condition. Confine your speed and any tricks you perform to your level of skill and experience. In mountainous terrain, be prepared for cold/wet weather. Don't speed downhill or stop too suddenly. Always beware of the dangers of unstable or precipitate terrain and of potential attack by wild animals.

SOURCE   H. Gatterer et al. Mortality in Different Mountain Sports Activities Primarily Practiced in the Summer Season—A Narrative Review. *International Journal of Environmental Research and Public Health*, October 2019.

Drowning Outdoors

ANNUAL US DEATHS 3,868

From 2005 through 2014 in the US, an average of 3,536 people of all ages drowned in nonboating-related accidents, and another 332 drowned in boating incidents. Drowning, especially for young children, can happen anywhere there is water: natural bodies such as the ocean or a lake, or swimming pools (public or at home), or a bathtub, or even a toilet, bucket, or sink. About 20 percent of all drowning deaths are children 14 or younger. Boys are twice as likely to drown as girls. Most drownings occur in late spring or summer.

Nonfatal drownings serious enough to require medical attention injure about 4,000 additional American children in an average year. Some of these involve permanent brain damage. Drowning is one of the leading causes of death for children under age 4; many of these drownings occur at home.

Swimming only at beaches with certified lifeguards on duty greatly improves swimmer safety. One study found the odds of fatal drowning at such a beach were 1 per 18 million beachgoer-days. There were almost 5 times as many fatal drownings at unguarded beaches.

STAY SAFE   Learn to swim. Never leave young children alone in bathtubs or pools. In any natural body of water, such as a lake, river, or the ocean, be very wary of underwater obstructions, unexpected deeps and sudden bottom drop-offs, and treacherous currents and tides. Beware of water that is much colder than current air temperature, especially in early spring; sudden immersion in freezing water contributes to drowning risk. Swim only at beaches that have certified lifeguards on duty. Be especially wary swimming in bodies of water where other people are boating, as collisions or violent motorboat and jet ski wakes can heighten the drowning risk. Swimming while intoxicated also significantly increases the risk of drowning. Always wear a personal flotation device while boating. When fishing from shore or in waders, be careful to not slip and fall in and drown.

SOURCES   CDC, cdc.gov/drowning/facts/index.html; US Lifesaving Association, usla.org/page/STATISTICS

Fireworks

ANNUAL US DEATHS   Average of 7

In 2017 in the US, 8 people were reported killed by amateur fireworks. This compares to 4 in 2016, and an average of 7 per year since 2002. However, the number of fireworks deaths has been rising. Twelve deaths were reported in 2019 and at least 18 in 2020.

In July 2021, Matiss Kivlenieks, an ice hockey goalie for the Columbus Blue Jackets, was killed when he was struck in the chest by a mortar-type firework while in a hot tub nearby. He threw his body in front of his pregnant wife, making the ultimate save.

About 11,000 to 13,000 people per year are injured nonfatally in fireworks accidents; about two-thirds of them happen during the 4 weeks surrounding the Fourth of July.

STAY SAFE   Children should not be allowed to play with fireworks. Exploding fireworks bodies should be made only of paper, to avoid metal fragments forming potentially lethal shrapnel. Skyrockets should be aimed well away from all spectators and carefully supported before lighting the fuse, so they can only go straight into the sky. Never alter or misuse a firework or intentionally aim it at another person. Sparklers and other fireworks are serious burn and fire hazards. Use them only outdoors, well away from structures and property that could get ignited, and have a fire extinguisher nearby.

SOURCES   CPSC, cpsc.gov, 2020 Fireworks Annual Report; *USA Today*, usatoday.com/story/sports/nhl/2021/12/02/ matiss-kivlenieks-death-fireworks-police-report

Fishing

ANNUAL US DEATHS   198 in 2019

In 2019 in the US, at least 198 people were killed while engaged in recreational fishing. This data includes those fatalities reported to US Coast Guard statisticians by way of state regulatory authorities. This total may be incomplete, in that some deaths (from drowning, vehicle mishap, lightning, etc.) while fishing from shore, or from a dock or bridge, might not be included.

SOURCE   USCG, uscgboating.org/statistics/accident_statistics.php

Gender Reveals

ANNUAL US DEATHS 7 from July 2017 to March 2021

In the US, between July 2017 and March 2021, 7 people were reported killed at gender reveal parties. Causes of death included shrapnel from improvised explosive devices, a plane crash, and a shooting.

In September 2020, a couple set off a smoke-generating device in Yucaipa, California. The device started the El Dorado wildfire, which burned 22,744 acres and killed a firefighter.

STAY SAFE Gender reveal methods should never use explosive or incendiary materials or devices.

SOURCE Daily Dot, dailydot.com/irl/gender-reveal-deaths/

Go-Karts

ANNUAL US DEATHS 19.3

A study for the period 1985 through 1996 found that an average of 19.3 people per year were killed in go-kart accidents; on average 11.3 per year of those fatalities were children.

STAY SAFE Always wear a properly fitting crash helmet. Only use go-karts that are in good repair on closed-circuit tracks maintained in good condition and equipped with proper crash barriers and other safety provisions. Do not exceed a go-kart's weight limit or drive faster than your skill level. Children should be closely supervised at all times and should only be permitted to drive after training in safe practices.

SOURCE CPSC report, cpsc.gov, Go-Kart Related Injuries and Deaths to Children

Hiking

ANNUAL US DEATHS 35 to 40 (National Parks only)

Hiking and other nature/wilderness/scenic activities such as camping, cross-country skiing, hunting, rock climbing, backpacking, and mountaineering, are enjoyed every year by immense numbers of Americans. Over 280 million visitors a year enjoy outdoor recreation at US national park sites. Hard data on the number of people who die each year specifically while hiking is not readily available separate from other national park or wilderness fatality totals. Deaths from motor vehicle accidents within park grounds, drownings, and other accidental deaths can't be broken out as part of a hike. A single park visit might combine different activities, further clouding the death classification issue.

An estimate for hiking deaths comes from annual National Park Service (NPS) search-and-rescue incident counts in which a fatality related to hiking was reported. This gives 35 to 40 deaths per year nationally specifically due to hiking. This is certainly an undercount, as non-NPS hiking deaths aren't included.

The primary contributing factors in these deaths were lack of knowledge, lack of experience, and poor judgment. The means by which they occurred were primarily falls from cliffs, rocks, or waterfalls, drownings, dehydration, health issues such as heart attacks, lightning strikes, hyperthermia or hypothermia, and animal attacks.

STAY SAFE   Wear appropriate footgear and clothing, bring plenty of water and a map of the hiking area, be aware of the weather, and be careful on the trail. Inform someone reliable of your route and anticipated return time.

SOURCES   NPS, nps.gov/orgs/1336/data.htm and nps.gov/articles/ hiking-safety.htm; SkyAboveUs, skyaboveus.com/climbing-hiking/ Whats-Killing-Americas-Hikers

Hunting

ANNUAL US DEATHS   Less than 100

A study for 2001 through 2007 found that fewer than 100 people in the US were killed each year in hunting mishaps of all kinds. The exact proportion of deaths due to firearms accidents alone is not readily available, but many human fatalities while hunting are not related to gunshot wounds. Some 15 million or more Americans hunt every year, for enjoyment, for food, for livestock protection, and for income. This implies an annual US deaths of well under 1 per 100,000 hunter-years.

In 2005, 5,686 serious injuries resulted from tree stand use, including some permanent paralysis or even death. Some hunters died of heart attacks, heatstroke and dehydration, or hypothermia. Another frequent cause of serious injury is self-laceration with the knife used to field dress game. There are also many trip-and-fall injuries, even deaths, while hunting, due to the rugged, wooded, slippery terrain often traversed during a hunt. Drowning is also a risk, especially when hunting waterbirds or hunting from a boat.

Ernest Hemingway, novelist and big-game hunter

STAY SAFE Take a hunter safety course even if your state doesn't require one to obtain a hunting license. Whether you hunt with a firearm or a bow, get professional training and practice at a range to learn safe handling of your weapon and responsible, accurate marksmanship. If you hunt from a tree stand, make sure it is in good repair and has a rail or other fall barrier on the platform. Use a full body harness or other fall-restraining device. Be very careful how you handle your weapon when climbing up or down, or whenever going over fences, fallen trees, and other obstacles, or while getting into and out of a vehicle; put the safety on and then unload the gun. Dress for the weather. Tell someone responsible where you are going and when to expect your return. Bring along food and water and basic survival supplies. Stay hydrated in very hot weather. Do not handle weapons if you are intoxicated or impaired. Obey all local laws about allowable weapon types in different hunting seasons for different species, and observe any restrictions about hunting near houses, hunting from motor vehicles, hunting at night, or hunting with dogs. *Always* handle a firearm as if it is loaded, even if you "know" it isn't. Always know your backstop—always be sure the bullet will hit somewhere safe, in case you miss your target or the projectile overpenetrates the game.

SOURCE Target Tamers, targettamers.com/guides/ hunting-accident-statistics

Killfies

ANNUAL US DEATHS   Average of 37 worldwide

From 2011 through 2017, 259 people worldwide died in accidents while taking "killfie" or "selficide" selfie photos, for an average of 37 selfie-linked deaths per year. There has been a significant upward trend, from only 3 in 2011 to 93 in 2017.

At least 36 more people died globally from killfies in 2018, 2019, and 2020. Although this more recent death toll may be incomplete, the decline in the yearly average to 12 implies that public information warnings and news reports about selfie-linked deaths are having some beneficial effect.

Most of the deaths were caused by falls from the edges of cliffs, precipices, dams, or waterfalls, when the victim was either posing for a photo or looking through a camera viewfinder. In other fatal cases, people have been crushed by logs, electrocuted, hit by oncoming trains, drowned in tidal surf, rivers, or capsized boats, dissolved in an acidic hot spring, and mauled by wild animals.

In 2018 in the US, 5 passengers in a sightseeing helicopter died when it crashed into water after 1 of them, while attempting a selfie, snagged the aircraft's fuel shutoff lever.

STAY SAFE   Use basic common sense and avoid attempting to take any selfies in potentially life-threatening situations. Be very cautious near cliffs, waterfalls, and other precipice edges, as well as around moving water, vehicles of all kinds, wild animals, and electric train line trolley wires and third rails.

SOURCES   A. Bansal et al. Selfies: A boon or bane? *India Journal of Family Medicine and Primary Care*, July 2018; Wikipedia, List of selfie-related injuries and deaths

Lightning

ANNUAL US DEATHS   17 in 2020

In 2021, 11 people were killed by lightning, the lowest number ever. In 2020, 17 people were struck and killed by lightning in the US. Seven were under a tree. Between 2006 and 2019, 418 people in the US were struck and killed by lightning, for an average death toll of 30 people per year. By sex, 79 percent were male and 21 percent were female. Deaths occurred in all age groups from young children to octogenarians but were concentrated between ages 20 and 59.

Over 70 percent of all lightning-strike deaths occurred in June, July, and August, although at least a few happened in every calendar month. They occurred throughout the week, with the most on Saturdays. Alabama, Colorado, Florida, Georgia, Missouri, New Jersey, North Carolina, Ohio, Pennsylvania, and Texas have the most lightning deaths and injuries. Florida is considered the lightning capital of the country, with more than 2,000 lightning injuries since 1963.

About 32 percent of lightning injuries take place indoors. A lightning strike can flow through conductive wires, pipes, and framing inside a building, dangerously energizing pipes, faucets, appliances, doorknobs, and the like and electrocuting someone who contacts such metal components at the wrong time. In addition, lightning strikes can start fires, killing people who are unable to escape a burning structure or vehicle.

The odds of being struck by lightning in a given year are around 1 in 500,000. However, some activities put you at greater risk for being struck. Most lightning deaths occurred while the victim was outside. Of the 418 total deaths during the 14 years ending in 2019, 40 victims were fishing from shore or a boat, 25 were at a beach, 20 were camping, and 18 were boating (but not for fishing). In sports, soccer had 12 deaths and golf 10. There were 18 fatalities among people doing yard work. Another 19 were killed while ranching or farming. In 2010, a woman and a man were struck by lightning at the top of Max Patch Bald, a mountain in North Carolina. The woman was killed just as the man was proposing to her.

In 1943, 432 people in the US were killed by lightning. Deaths have dropped considerably since 102 in 1974, the last time more than 100 were killed.

STAY SAFE Always check a local forecast before spending time outdoors, and maintain situational awareness for threatening weather that could include lightning. Get to shelter early and stay there until the lightning is well past your location: Many lightning deaths occur while victims are only steps away from safety, or when victims leave safe areas prematurely. Do not shelter under large trees, as they can attract lightning strikes. Seek low ground, stay low to the ground, discard any umbrella with metal parts, and accept getting wet. When indoors during a lightning storm, do not touch any parts of the plumbing and electrics, and avoid contact with metal doorframes or metal doors. If you are driving, pull over and stay in the car. The outer metal shell of your vehicle provides protection if the windows are closed.

SOURCE National Lightning Safety Council, lightningsafetycouncil. org/LSC-LightningFatalities.html

GOLF

ANNUAL US DEATHS   Average of 1.4

The odds of being struck by lightning on a golf course are 1 in 3,000. About 5 percent of all lightning deaths and injuries in the US happen on golf courses. Between 2005 and 2019, 10 people were killed by lightning while playing golf. In June 2021, a golfer on a course in New Jersey was killed by lightning.

The golfer Lee Trevino was struck by lightning at the 1975 Western Open in Illinois. He suffered burns on his back that needed surgery. He said afterward, "If you are caught on a golf course during a storm and are afraid of lightning, hold up a 1-iron. Not even God can hit a 1-iron."

In 1991, 1 spectator was killed by lightning at the PGA Championship at Crooked Stick Golf Club in Indiana. At the US Open at Hazeltine National Golf Course in Minnesota the same year, 1 spectator was killed and 5 were injured by lightning as they sheltered under a tree.

Most golf courses sound a warning siren if lightning is near. Seek shelter in a safe structure. Golf carts provide no protection and may even increase the danger.

SOURCES   National Lightning Safety Council, lightningsafetycouncil.org/LSC-LightningFatalities.html; NOAA, nssl.noaa.gov/education/svrwx101/lightning; PGA, pga.com/archive/news/golf-buzz/how-stay-safe-in-lightning-golf-course

National Parks

ANNUAL US DEATHS   330

Based on data for 2014 through 2016, the annual US death rate for visitors at all sites of the US National Parks Service (NPS) combined is about 0.1 per 100,000 recreational visits. Males were 79 percent of all deaths. People aged 45 or older were 53 percent of all deaths. There were on average 330 total reported deaths from all causes per year, compared to about 280 million recreational visits per year.

About 55 people per year commit suicide in a national park. About 3 people per year are murdered. The leading cause of accidental death is drowning, with an average of 58 fatalities per year. Then comes motor vehicle accidents, with an average of 54 people killed in crashes per year. Next are falls, with an average of 29 people killed by falling per year. An average of 74 deaths per year are due to natural causes, such as heart attacks; half of these occur while the victim is engaged in strenuous physical activity.

STAY SAFE   Be careful while driving in national parks, as roads are often narrow and winding, with poor visibility around tight curves, no shoulders along steep drops, and fast-moving vehicles from the other direction can appear unexpectedly in seemingly quiet wilderness. Do not overexert yourself if you might not be in the best health. Observe all wildlife respectfully, from a safe distance. Although no one in the US has been reported killed by a bison (buffalo) attack in recent years, some people who harassed them or threatened their calves have been injured. Always stay inside the designated safe viewing areas at geysers and hot springs, and watch young children.

SOURCE   NPS, nps.gov/orgs/1336/data.htm

MOST AND LEAST DANGEROUS NATIONAL PARKS

ANNUAL US DEATHS   See table in entry.

This data should be reviewed cautiously, especially because some national parks have both low numbers of visitors and low numbers of deaths, causing the ratio—their reported visitor death rate—to be mathematically unstable, lacking much statistical credibility.

Here are the 5 most and 5 least deadly US National Park sites from 2007 through 2018, based on cumulative data for visits and visitor deaths.

PARK	DEATHS	VISITS	DEATHS PER 10 MILLION VISITS
Deadliest			
North Cascades	19	291,255	652.35
Denali Preserve	59	5,870,403	100.50
Upper Delaware River	21	3,064,806	68.52
Big Thicket Preserve	11	1,643,769	66.92
Little River Canyon	17	3,199,845	53.13
Least Deadly			
George Washington	16	84,846,911	1.78
Cape Cod	17	53,034,104	3.21
Colonial National	14	39,977,457	3.50
C & O Canal	18	51,351,603	3.51
Golden Gate	86	177,394,212	4.85

STAY SAFE Be careful while driving in national parks, as roads are often narrow and winding, with poor visibility around tight curves, no shoulders along steep drops, and fast-moving vehicles from the other direction can appear unexpectedly in seemingly quiet wilderness. Do not overexert yourself if you might not be in the best health. Observe all wildlife respectfully, from a safe distance. Although no one in the US has been reported killed by a bison (buffalo) attack in recent years, some people who harassed them or threatened their calves have been injured. Always stay inside the designated safe viewing areas at geysers and hot springs, and watch young children.

SOURCE NPS, nps.gov/orgs/1336/data.htm

GEYSERS AND HOT SPRINGS

ANNUAL US DEATHS At least 20 since the late 1800s

Over the decades since the late 1800s, at least 20 people of all ages have been boiled to death or severely burned in geysers, hot springs, mud pots, and steam vents at Yellowstone National Park. The park has more than 10,000 of these geological hot spots. Most recently, in 2016, a man slipped and fell into a thermal pool and was boiled to death. His body was literally dissolved before rescue crews could arrive. All that was left behind were his wallet and flip-flops.

The water temperature in these geological features, fed by underly-

Castle Geyser, Yellowstone National Park

ing volcanic lava streams, can reach 250 degrees F or higher. The most frequently visited sites at Yellowstone are protected by boardwalks, railings, and warning signs to allow sightseeing from a safe distance. But tragically, children and even adults occasionally get too close and fall or even jump in, in at least 1 case when hiking after dark without flashlights.

These often beautiful, always fascinating natural attractions can seem deceptively docile until too late. Scalding geysers can erupt unexpectedly. Standing water can be much, much hotter than it seems from any distance. Walkable terrain can hide areas where only a treacherous, thin crust of earth covers a deep body of boiling water—anyone stepping there will fall in, with horrific and tragic consequences.

STAY SAFE   Stay on designated walkways and behind railings, and obey all safety signage when viewing geysers and hot springs. When in doubt, remain well clear of the area of a suspected geothermal feature, whether at Yellowstone or anywhere else.

SOURCES   NPS, nps.gov/orgs/1336/data.htm; *Outside*, outsideonline.com/outdoor-adventure/exploration-survival/ brief-history-deaths-yellowstones-hot-springs

Picnicking

ANNUAL US DEATHS   No data available

Picnicking carries with it various risks of death, depending on where the picnic takes place and what else the picnickers do on their outing. Basic risk variables, as with any outdoor activity, include the altitude, weather, terrain conditions, local flora and fauna, proximity to water, proximity to moving motor vehicles, proximity to overhanging trees and/or rock outcroppings, accessibility to shelter, sustenance, prompt medical care if needed, and degree of reckless behavior due to impairment from drugs and/or alcohol. Foodborne illness may be an issue.

STAY SAFE   Because picnicking is usually done on warm-to-hot, sunny days, and revolves around having a meal outside away from home, cleanliness and food preservation are important. Keep food and drinks cold and fresh. Bring plenty of drinking water. Check the weather forecast and dress for the weather. Bring sunscreen, first aid kit, insect repellent, trash bags, umbrella, jacket, and some dry things to change into, just in case—especially if you might go for a swim.

SOURCE   Mental Floss, mentalfloss.com/article/550335/ bacteria-could-spoil-your-picnic-foods-less-two-hours

Rock Climbing and Mountain Climbing

ANNUAL US DEATHS   Varies depending on location; see data in entry.

Rock climbing and mountain climbing are related and intertwined sports, with similar risks of injury or death. Overall mortality rates can be significant, and they vary considerably depending on the altitude and the technical difficulty of the climb, as well as the relative isolation of the part of the world where the climbing takes place.

One US study in the late 1990s put the death rate at 3.2 per 100,000 outings. Another, using data from the 1980s and 1990s, put the death rate at 30 per 100,000 outings for climbing Mount Rainier and 60 per 100,000 for climbing in the Grand Tetons.

For comparison, this second study, using data for 1970 through 2010 for climbing in the Himalayas, found death rates per 100,000 outings to be 640 for Cho You, 1,560 for Mount Everest, and 4,500 for Annapurna.

STAY SAFE   Get professional training before attempting any rock climbing or mountain climbing activities. Do not climb beyond your abilities. Always use the full gamut of safety equipment, bring all needed extreme weather clothing and shelter plus adequate survival supplies, and have effective communications gear and avalanche finder beacons in case of any emergencies. Check the weather forecast first, remember it can be much more extreme at higher altitudes, and use common sense to decide whether to start, or delay, your next outing.

SOURCES   S. Rauch et al. Climbing Accidents—Prospective Data Analysis from the International Alpine Trauma Registry and Systematic Review of the Literature. *International Journal of Environmental Research and Public Health*, December 2019; V. Schöffl et al. Evaluation of injury and fatality risk in rock and ice climbing. *Sports Medicine*, August 2010; Tripsavvy, tripsavvy.com/ ways-to-die-climbing-756069

Trapping

ANNUAL US DEATHS   No cumulative data available

Trapping is a specialized type of hunting, in which an animal is captured by a mechanical device so that its fur can be harvested intact. Trap checking intervals can vary between once a day and biweekly. The animal might be killed by the trap, or else be killed after capture by the hunter. The total number of trappers who are active in the US is not readily available, but based on some individual state licensing figures, the number is probably well over 10,000.

Hard data on human fatalities due to trapping activities is scarce, and some trapper deaths due to accidents are probably included in the broader category of hunting. Trappers are exposed to risks of accidental death similar to those that other hunters face.

STAY SAFE Handle traps and equipment carefully. Dress for the weather. Tell someone responsible where you are going and when to expect your return. Bring along food and water and basic survival supplies. Stay hydrated in very hot weather. Obey all state and local laws about allowable trapping methods, species, and seasons.

SOURCE Bass Pro Shops, 1source.basspro.com/news-tips/ habitats-food-plots/4797/animal-trapping-safety

Waterfalls

ANNUAL US DEATHS No cumulative data available, but see details in entry.

Hard data is scarce on overall US deaths from people unintentionally going off the edge of a waterfall. Several deaths have been reported in North America as a result of people falling down a waterfall after losing their footing while taking a selfie. At least 9 have died at Kaaterskill Falls in New York State since 1992; in 2018, 3 vloggers died at Shannon Falls, and 1 freelance photographer at Albion Falls, in British Columbia, Canada.

Since 1990, at least 4 people have been killed while swimming or boating and falling over the edge of Niagara Falls, either by accident, by suicide, or on purpose as a publicity stunt gone wrong. A study of waterfall deaths at popular waterfalls in western North Carolina from 2001 through 2013 found 28 deaths. A study of waterfall deaths in New England for the past several decades found at least 65 deaths at that region's 5 most deadly waterfalls, some while swimming and some after slipping from shore.

STAY SAFE Be wary of vertigo, unsafe footing, or slipping while taking a photo near any dangerous, precipitate edge, including waterfalls. Never let that extraspecial selfie turn into a killfie. When swimming or boating, be aware of the location of any nearby waterfalls and stay away from the edge. Do not inadvertently let water currents carry you too close and then down into mortal danger.

The Yellowstone Falls of the Yellowstone River

SOURCES *Citizen Times*, citizen-times.com/story/news/local/2018/08/01/waterfall-safety-features-barriers-and-signs-widely-vary-across-wnc/848518002; NewEnglandWaterfalls.com. newenglandwaterfalls.com/listofwaterfalldeaths.php; NWS, weather.com/safety/news/2020-06-29-lessons-one-deadliest-waterfalls-northeast

Whitewater Rafting

ANNUAL US DEATHS 53

In the US from 2007 through 2016, there were 530 deaths from whitewater paddling sports such as rafting, kayaking, and boating, which implies an average annual death toll of 53. A 1998 study found an annual US death rate for whitewater sports of 3.8 per 100,000 rafter-days. (There were an estimated 700,000 whitewater paddling enthusiasts in the US in 2019.)

For 2019, US Coast Guard data indicates that 19 people were reported killed during whitewater activities involving water vessels that were required to be registered with states. Note that the USCG data excludes deaths from natural causes, such as heart attacks from overexertion, which are included in some of the other data.

STAY SAFE Consider rafting with professional guides, and always wear a life jacket.

SOURCES American Whitewater, americanwhitewater.org/
content/Accident/view; Tripsavvy, tripsavvy.com/
whitewater-rafting-death-statistics-3969676

Wilderness Areas

ANNUAL US DEATHS No cumulative data available, but see details
in entry.

The following 9 popular wilderness attractions within the US offer
unique, but sometimes deadly, extreme environmental experiences:
intense beauty, extreme isolation, and keen challenges to even the
most experienced adventurers. Many of these sites demand advanced
technical skills in rock climbing, orienteering, or scuba diving. They are
listed here in no particular order.

Grand Canyon National Park, Arizona: About 12 people per year die
at the Grand Canyon.

Hawksbill Crag, Ozark National Forest, Arkansas: Someone fell to
their death in 2021 while taking a selfie. Authorities need to rescue 5
to 10 hikers a year.

Precipice Trail, Acadia National Park, Maine: Climbing the sheer cliff
face here has seen at least 1 death, in 2012.

Whitewater rafting, Colorado River: About 12 people a year are killed.

Eagle's Nest Cave, Florida: Cave diving to depths of 300 feet. About
12 people have died in the past 30 years.

The Maze, Canyonlands National Park, Utah: No reported deaths,
but hikers should beware that temperatures in this remote area can
reach 110 degrees F.

Volusia County, Florida: Includes immensely popular Daytona
Beach, but in both 2017 and 2018 Volusia had the most shark attacks
of any 1 place in the world. None of these, an average of 6 per year,
were fatal.

Angel's Landing, Zion National Park, Utah: Since 2004, at least 6
people have died at the 1,000-foot drop on this extremely narrow hiking trail.

Half Dome, Yosemite National Park, California: Up to 20 people have
died in the past 100 years while scaling this steep cliff.

STAY SAFE Be careful while driving in national parks, as roads are
often narrow and winding, with poor visibility around tight curves, no
shoulders along steep drops, and fast-moving vehicles from the other

direction can appear unexpectedly in seemingly quiet wilderness. Do not overexert yourself if you might not be in the best health. Observe all wildlife respectfully, from a safe distance. Although no one in the US has been reported killed by a bison (buffalo) attack in recent years, some people who harassed them or threatened their calves have been injured. Always stay inside the designated safe viewing areas at geysers and hot springs, and watch young children.

SOURCE   NPS, nps.gov/orgs/1336/data.htm

JUDICIAL AND STATE ACTIONS

Law and Disorder

The statistics for law enforcement and national, border, and homeland security given here are for US civilians (including those incarcerated in the US) who encounter police, corrections, security, and military professionals. The fatalities result from interactions with police, sheriffs, FBI, TSA, corrections officers, prison guards, Coast Guard personnel, ICE, Border Patrol, and military personnel or with various adversaries (criminals, domestic and international terrorists, and enemy combatants).

Military contractors, employees, and consultants who die in conflict zones are also included below as civilians, along with other US non-military personnel who happen to die in foreign conflict zones.

Occupational fatality statistics for US law enforcement and corrections officers, for people serving in the US military, and for people serving in national, border, and homeland security are given in the **On the Job** section of this book.

Police Use of Force

ANNUAL US DEATHS 1,000+ civilians

Every year in the US, police use of force contributes to the death of just over 1,000 civilians. From 2015 through early June 2020, officers of the law shot and killed 5,400 people, for an average of about 995 such deaths per year. In addition, some people have been killed by police via chokeholds, kneeling on necks, tasing, being run over by a police vehicle, and/or other uses of force. These death statistics are approximate, because some deaths in police custody are almost certainly concealed or mischaracterized (for example, as justified shootings, arrestee suicides, resisting arrest, or natural causes). Some local police departments refuse to release their data. A recent study suggests that police killings have been significantly undercounted, perhaps by more than half, between 1980 and 2018. The study showed that about 55 percent of fatal encounters with police in this time, as found through news reports and public records requests, were listed as another cause of death on death certificates. In this time, nearly 31,000 Americans were killed by police, with more than 17,000 of them not accounted for in official police figures.

There can be no doubt that many deaths due to police gunfire and other uses of force are entirely justified. Guns abound in the hands of criminals, and some perpetrators of serious crimes seek to evade and avoid apprehension via deadly resistance. Sometimes, unfortunately for everyone involved, in the tense and confusing environment of a violent encounter, police officers need to make split-second decisions that can lead to mistakes. Most often, the officer mistakes an object in the hands of a civilian for a gun and fires on the person. These incidents can sometimes be seen as tragic accidents. Multiple police officers may feel lives are threatened and open fire in the same critical instant.

Whether the use of lethal force was justified or not can be very controversial and may come down to a matter of opinion. The appearance of police bias in data on the ethnicity or race of killed civilians, significant variations in such death rates by jurisdiction, and significant variation in the treatment of involved officers imply that some departments have a more negative culture than others. More accountability for violent police/civilian interactions is needed. (See the **Police Use of Force** entries below for data by age, sex, and race/ethnicity, and by city and state.)

STAY SAFE Members of the public should avoid getting caught up in active police investigations, pursuits, and apprehension efforts. Cooperate with police officers; don't resist if at all possible. Participate in community efforts to improve local policing.

SOURCES mappingpoliceviolence.org; GBD 2019 Police Violence US Subnational Collaborators. Fatal police violence by race and state in the USA, 1980–2019: a network meta-regression. *Lancet*, 2021; the *Washington Post* Policing in America Project, washingtonpost.com/police-america

By Age, Sex, and Race/Ethnicity

> **ANNUAL US DEATHS**   A complex issue; see table in entry.

A commonly used, revealing measure to quantify and compare exposures to police use of deadly force between different ethnic subgroups—within the same jurisdiction or in separate jurisdictions—is to project the theoretical/implied number of people in each subgroup, per 100,000 members of that subgroup alive today, whose eventual cause of death (however long their lifetimes might be predicted to extend from their current attained ages) will be getting killed by the police. This is a complex actuarial calculation that uses a range of demographic data for the different ethnic groups and for deaths from police use of force.

This rate per 100,000 can then also be turned into an "odds of death-by-cop" by dividing the predicted death toll per 100,000 by 100,000. (Note that every 1 of the 100,000 notional humans, within each ethnic group being compared, will eventually die of something.)

The average lifetime odds of being killed by police in the US, based on police-related deaths from 2013 through 2018, and population data for 2017 (assumed to remain stable into the foreseeable future), was about 1 in 2,000 for all males and about 1 in 33,000 for all females.

The rates per 100,000 males belonging to each ethnic group, of eventually dying from police use of force were:

Black	96 per 100,000 residents
Native American	60
Latino	53
White	39
Asian/Pacific Islander	16

For all ethnic groups, the lifetime odds of dying by police use of force—as opposed to by another cause of (eventually inevitable) death—is at its highest for those currently ages 20 through 30.

STAY SAFE   Police training on proper use of firearms and other means of force, and on de-escalation tactics, is important. Members of the public should avoid getting caught up in active police investigations, pursuits, and apprehension efforts. Cooperate with police officers; don't resist if at all possible. Participate in community efforts to improve local policing.

SOURCES   F. Edwards et al. Risk of being killed by police use of force in the United States by age, race-ethnicity, and sex. *Proceedings of the*

National Academy of Sciences, August 2019; FBI, fbi.gov/services/cjis/
cjis-link/the-national-use-of-force-data-collection; the *Washington Post*
Policing in America Project, washingtonpost.com/police-america

By City and State

ANNUAL US DEATHS   1,021 in 2020

Among the largest US cities, rates of civilian deaths due to police use of
force vary considerably. The city with the highest rate per capita, averaging about 18 people killed per million residents per year from 2013
through 2019, was St. Louis, Missouri. Other cities with comparatively
high rates were Oklahoma City, Phoenix, Tulsa, and Kansas City,
Missouri. The big city with the lowest rate of people killed by police was
New York, with an average of 1.3 per million residents per year. Other
cities with low rates were Virginia Beach, Raleigh, Boston, and Detroit.

Note that the rate for St. Louis is *14 times* the rate for New York
City.

By state, over the period 2013 through 2020, the states with the
highest rates of death per million residents due to police use of force
were New Mexico at 9.8 per million, Alaska at 8.3, Oklahoma at 7.8,
Arizona at 7.1, and Nevada at 6.3. The states with the lowest death
rates per million residents were Rhode Island at 0.7 per million, New
York at 1.1, Massachusetts also at 1.1, Connecticut at 1.4, and New
Jersey at 1.6.

Note that the rate for New Mexico is *14 times* the rate for Rhode
Island.

Between 2015 and 2020, police officers in rural areas shot and killed
about 1,200 people. In the same time frame, police in urban areas shot
and killed at least 2,100 people. The rate of rural police shootings was
about 30 percent lower than the urban rate.

STAY SAFE   Police training on proper use of firearms and other means
of force, and on de-escalation tactics, is important. Members of the public should avoid getting caught up in active police investigations, pursuits, and apprehension efforts. Cooperate with police officers; don't
resist if at all possible. Participate in community efforts to improve local policing.

SOURCES   mappingpoliceviolence.org; GBD 2019 Police Violence
US Subnational Collaborators. Fatal police violence by race and state
in the USA, 1980–2019: a network meta-regression. *Lancet*, 2021; the
Washington Post Policing in America Project, washingtonpost.com/
police-america

Of the 1,126 people killed by police in 2020, the vast majority—96 percent—were killed by shootings. Almost all the other deaths were by Tasers, physical force, and police vehicles.

Tasers, or electroshock weapons, are less lethal than guns, but at least 1,200 deaths have occurred since these weapons came into law enforcement use in the early 2000s. While Tasers are usually safe when used properly, many police officers are not well trained and Tasers are often misused. At least 6 cases of fatal burns caused by electroshock weapons have been documented, most involving a cigarette lighter in the suspect's pocket or being close to a flammable liquid. In 2021, an apparently intoxicated man walked into a police station in Catskill, New York, and doused himself in hand sanitizer. When an officer tried to subdue him with a Taser, the man burst into flames and was critically burned; he later died.

Deaths from being restrained in a choke hold still occur, even though many police departments ban them. Despite the extensive publicity deaths from police choke holds sometimes receive, reliable statistics aren't available. Deaths from crashes during police pursuits killed more than 5,000 bystanders and passengers between 1979 and 2013.

SOURCES M.W. Kroll et al. Fatal and non-fatal burn injuries with electrical weapons and explosive fumes. *Journal of Forensic and Legal Medicine*, 2017; FBI, fbi.gov/services/cjis/cjis-link/the-national-use-of-force-data-collection; the *Washington Post* Policing in America Project, washingtonpost.com/police-america

Political Protests

> **ANNUAL US DEATHS** 25 in 2020

In 2020, at least 25 Americans died violently at political protests. Nine of them were demonstrators taking part in Black Lives Matter protests. In addition to the people killed while demonstrating, at least 14 others were killed in incidents linked to political unrest. Who was responsible and what crimes might have been committed is under investigation.

STAY SAFE Police, marchers, and communities at large should share the job of seeing that demonstrations and protests remain valid expressions of Constitutional free speech, without violence by any group of citizens or by their countermarching or infiltrating opponents, and without excessive use of force by the police. The safest course when things get out of hand is to disperse and reconvene at a more peaceful time and place for further expression of free speech rights. If arrested, comply peacefully, and then arrange good legal counsel.

SOURCE   Armed Conflict Location & Event Data Project (ACLED), acleddata.com/data-export-tool

Prison Populations

ANNUAL US DEATHS   1,088 in 2019

In 2016, about 2.3 million people in the US were incarcerated out of a population of 324.2 million, for an incarceration rate of 0.7 percent.

A breakdown of the prison population by type/jurisdiction is:

State prisons	1,316,000
Local jails	615,000
Federal prisons	225,000
Juvenile facilities	48,000
ICE detention camps	34,000
US territories	11,000
Native American jails	25,000
US military prisons	13,0000
Involuntarily committed	22,000

In 2019, the total incarcerated population was 2,155,000, with 1,435,500 in prisons and 758,419 in jails. Triggered by the COVID-19 pandemic, the number of people incarcerated in state and federal prisons and local jails dropped from around 2.1 million in 2019 to 1.8 million by mid-2020—a 14 percent decrease.

STAY SAFE   If you can't do the time, don't do the crime. Work in your community for a more effective and equitable justice system.

SOURCES   Prison Policy Initiative, prisonpolicy.org/data; Vera Institute of Justice, vera.org/ending-mass-incarceration/ scoping-the-problem

Federal Prisoners: Arrest-Related and In-Custody Deaths

> **ANNUAL US DEATHS**   Average of 495 in 2016 and 2017

Federal law enforcement and detention agencies (various police and criminal investigation forces, courts/jails, and prisons) reported an average of 46 arrest-related deaths and 449 deaths of in-custody/incarcerated prisoners in 2016 and 2017. This was an overall average death rate of about 220 per 100,000 federal prisoners.

Among the arrest-related deaths, 47 percent were homicides (mostly via police justified use of force); 42 percent were suicides. In 78 percent of all arrest-related deaths, the accused had or appeared to have a weapon (almost always a firearm); 55 percent of all arrest-related deaths occurred while a warrant was being served. In about 70 percent of the arrests, the decedent's most serious alleged offense was a violent crime, a weapons charge, or a drug charge.

Almost all the deaths were of males. Of these, 66 percent were White, and 26 percent were Black.

Among the in-custody/incarcerated deaths, almost 90 percent were due to illnesses; about 7 percent were suicides. Almost all the deaths were of males, of which 61 percent were White and 31 percent were Black.

STAY SAFE   If you can't do the time, don't do the crime. Work in your community for a more effective and equitable justice system.

SOURCE   Department of Justice, Bureau of Justice Statistics, bjs.ojp. gov/data/topic

Juveniles Incarcerated or Confined

> **ANNUAL US DEATHS**   8 in 2018

In 2018, 37,500 juvenile offenders (under age 21) were held in 1,510 facilities in the US. By type of facility, they included: 625 detention centers (usually local), 116 shelters, 37 reception and diagnostic centers, 240 group homes (mostly private), 27 ranch or wilderness camps, 164 training schools (mostly state), and 553 residential addiction treatment centers (mostly private).

In 2018, 8 youths died in detention; 6 of these were suicides, 1 was an accident, and 1 was natural causes (illness). This is an overall annual US deaths per 100,000 incarcerated youths of about 21; the death rate from suicide alone was 16 per 100,000.

Philadelphia
County
Prison,
old and
forbidding.

Among these 8 deaths, 5 were male and 3 were female. By race/ethnicity, 4 were White, 1 was Black, and 3 were Latino. These numbers were too small for death rates per 100,000, or by type of facility, to be statistically meaningful.

STAY SAFE   If you can't do the time, don't do the crime. Work in your community for a more effective and equitable juvenile justice system.

SOURCE   US Department of Justice, Office of Juvenile Justice and Delinquency Prevention, ojjdp.ojp.gov/statistics

Local Prisoners

ANNUAL US DEATHS   1,071 in 2016

In 2016, 1,071 inmates died in local jails in the US. This was an annual US deaths of 174 per 100,000 local prisoners. While 2,788 different local jurisdictions contributed to the data, only 572 reported at least 1 prisoner death in 2016.

About half of all prisoner deaths in local jails were from illnesses, mostly heart disease, liver disease, and cancer. Suicides accounted for 31 percent of all deaths; homicides for 3 percent. Deaths from drug or alcohol overdoses were 10 percent of the total local prisoner death count. About 40 percent of all inmate deaths occurred within the first 7 days of admission to the jail.

By race/ethnicity, the death rates per 100,000 prisoners were:

White	240 per 100,000
Black	118
Latino	87

These death rates are not adjusted for differences in age, sex, state of health, reason for incarceration, and so on, which might vary substantially between the different ethnic groups, and which might explain the significant differences in their overall (unadjusted) death rates.

STAY SAFE If you can't do the time, don't do the crime. Work in your community for a more effective and equitable justice system.

SOURCE US Department of Justice, Bureau of Justice Statistics, bjs.ojp.gov/data/topic

Military Prisoners

ANNUAL US DEATHS No data published; assuming deaths are similar to US federal prisoners

In 2019, US military prisons held a total of 1,214 prisoners, convicted by court martial of a serious violation of the Uniform Code of Military Justice (UCMJ). A breakdown of the branches in which they served is as follows:

Army	577
Air Force	228
Marine Corps	261
Navy	140
Coast Guard	8

The only maximum security facility, run by the US Army but holding inmates from all branches, is the United States Disciplinary Barracks (USDB) at Fort Leavenworth, Kansas, which houses males only. All female inmates are incarcerated at the Naval Consolidated Brig, Miramar, California. There are several other, smaller correctional facilities at different locations as well.

As of mid-2016, there were 4 prisoners on death row at the USDB, but the last time the US military executed a service member for a crime (rape and attempted murder of a child) was in 1961. Data is not readily available on deaths of inmates in US military prisons. This entry estimates the death risk as comparable to that for civilian prisons and jails in the US: high.

STAY SAFE   If you can't do the time, don't do the crime.

SOURCE   Department of Justice, Bureau of Justice Statistics, bjs.ojp.gov/library/publications/prisoners-2020-statistical-tables

State Prisoners

ANNUAL US DEATHS   4,135 in 2018

In 2018, 4,135 incarcerated persons died in state prisons in the US. This was a death rate of 344 deaths per 100,000 state prisoners.

Illness-related, or "natural," deaths were 77 percent of state prisoner deaths in 2018, of which almost one-third each were due to cancer and heart disease. Suicides were 7 percent of all deaths. The 2018 suicide rate in state prisons was 26 per 100,000 population, compared to the overall US suicide rate of 14.2 per 100,000 population. Homicides were 2.5 percent of all state prisoner deaths. Males were 96 percent of deaths. White prisoners were 31 percent of the total state prison population but made up 55 percent of all state prisoner deaths.

STAY SAFE   If you can't do the time, don't do the crime. Work in your community for better prison conditions and a more effective and equitable justice system.

SOURCE   Prison Policy Initiative, prisonpolicy.org/blog/2021/06/08/prison_mortality

Probation and Parole

ANNUAL US DEATHS   13,978 adults on probation and 6,712 adults on parole died in 2018.

Probation means a period of community supervision in lieu of being sent to prison, after conviction for a crime. Parole means a conditional release from prison for good behavior, with community supervision, after serving only part of a sentence for a crime.

In the US in 2018, about 3.54 million adults were on probation and about 880,000 were on parole. The number of probationers declined by 17 percent from 2008 to 2018, while parolees increased by 6 percent in that period. These totals are each the net of new individuals going on probation or parole, and old individuals leaving those statuses for reasons such as completion of sentence, return to incarceration (for violation of conditions or for conviction of a new crime), or death.

In 2018, 13,978 adults on probation died, and 6,712 adults on parole died. This gives annual US deaths of 395 per 100,000 probationers and 763 per 100,000 parolees. These death rates are significantly higher than those for federal, state, and local adult incarcerated prisoners and include deaths due to dangerous living conditions such as homelessness, dangerous lifestyles such as drug and alcohol abuse, lack of access to healthcare, succumbing to long-term diseases acquired in prison (such as AIDS), and being victims or perpetrators of violent crimes while released into the community. Persons on probation died at a rate 3.42 times higher than persons in jail, 2.81 times higher than persons in state prison, and 2.10 times higher than the general US population.

STAY SAFE   If you can't do the time, don't do the crime. Work in your community to help individuals on parole and probation have more equitable treatment and access to services.

SOURCES   US Department of Justice, Bureau of Justice Statistics, bjs.ojp.gov/library/publications/probation-and-parole-united-states-2019; Prison Policy Initiative, prisonpolicy.org/data; C. Wildeman et al. Age-Standardized Mortality of Persons on Probation, in Jail, or in State Prison and the General Population, 2001-2012. *Public Health Reports*, November 2019.

Formerly Incarcerated

ANNUAL US DEATHS   Numbers are unavailable, but the death rate of previously incarcerated persons is 2.1 to 2.6 times that of the general population.

An analysis for the period from 1980 to 2020 comparing the annual age-specific mortality rates of Americans who had been incarcerated to the same data for the entire US population found that having been incarcerated increased the annual US deaths by a factor of between 2.1 and 2.6 times. This increase in the odds of dying at every age subsequent to incarceration had the effect of reducing the life expectancy of a formerly incarcerated person at age 45 by 4 or 5 years.

Reasons for this lasting bad effect on mortality rates of having ever been in prison include: selection into the postprison population of individuals who had been exposed, even before prison, to worse lifestyle-related hazards than the overall population (e.g., poverty); persisting socioeconomic and familial disadvantages and stigmas after release from prison; exposure while in prison to severe physical and mental stresses that permanently impaired subsequent life expectancy; acquiring additional life-shortening diseases (including AIDS, hepatitis C, or substance addictions) while in prison, due to boredom, deprivation, depression, overcrowding, unsanitary conditions, and inadequate physical and mental healthcare; causal links between behavioral traits (such as violence) that led to imprisonment, and further life-shortening behaviors postimprisonment; and "criminogenic contagion," or increased criminality postrelease due to having been held in close confines with other criminals.

STAY SAFE   If you can't do the time, don't do the crime. Incarcerated persons should have access to high-quality mental health and medical treatment in prison and after release. Work in your community to improve access to services and opportunities for formerly incarcerated persons.

SOURCE   Prison Policy Initiative, prisonpolicy.org/data

National Security and Insecurity

Customs and Border Protection (CBP) Encounters

ANNUAL US DEATHS Approximately 24.5

The US Customs and Border Protection (CBP) service, which includes the Border Patrol, is in charge of border management and control. The service handles customs, immigration, border security, and agricultural protection. With more than 60,000 employees, it is the largest federal law enforcement agency. Of the nearly 20,000 Border Patrol agents, approximately 17,000 are concentrated along the Southwest border, in California, Arizona, New Mexico, and Texas. Reliable information on deaths as a result of an encounter with a CBP agent are not maintained by the service. However, nonprofit organizations that track border issues, primarily along the US–Mexico border, report that from 2010 to June 2021, at least 130 people died in this manner. Of the deaths, 55 were caused by an on-duty CBP agent's use of force. Another 38 deaths were the result of vehicle collisions involving the Border Patrol. An additional 22 deaths were related to the failure to provide timely medical care. Off-duty officers were responsible for 11 homicides; most of them were later sentenced to prison time. The ACLU of Texas tallies the numbers in a different way that includes deaths from high-speed vehicle pursuits. As of December 13, 2021, the ACLU of Texas tracker had a total death count of 209 since January 2010.

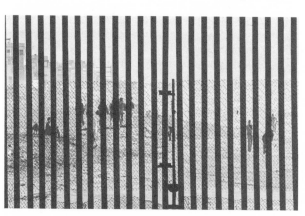

A stretch of fence along the US–Mexico border

At least 5 children died in custody or after being detained by CBP agents at the border during the migrant surge in 2018 and 2019, when as many as 2,600 children were held in border facilities.

STAY SAFE   Cross the border legally. Avoid encounters with CPB. Work in your community for a more equitable immigration system.

SOURCES   ACLU of Texas, aclutx.org/en/cbp-fatal-encounters-tracker; Southern Border Communities Coalition, southernborder.org/deaths_by_border_patrol; US Customs and Border Protection, cbp.gov/newsroom/publications/performance-accountability-financial

Border Patrol Agent Deaths

ANNUAL US DEATHS   2+

Between 2003 and 2019, 35 Border Patrol agents died in the line of duty. The deadliest year was 2004, when 3 agents died out of 10,819, for a mortality rate of 1 in 3,606. From 2003 through 2019, the annual chance of a Border Patrol agent dying in the line of duty was about 1 in 8,628 per year. About half of all agent deaths were from vehicle accidents. Another 34 percent died in other types of on-the-job accidents, mostly drowning, and 17 percent died as a result of assault or murder.

STAY SAFE   Avoid high-speed vehicle chases. Be very cautious and always wear a personal flotation device during water operations.

SOURCES   Cato Institute, cato.org/blog/border-patrol-agent-deaths-line-duty-2003-2019; US Customs and Border Protection, cbp.gov/about/in-memoriam/memoriam-those-who-died-line-duty

Migrant Deaths

ANNUAL US DEATHS   About 372 (probably an undercount)

Between 1998 and 2017, the US Border Patrol recorded over 7,216 migrant deaths along the US–Mexico border. In 2020, 227 deaths were recorded among migrants attempting to cross the border through the desert areas of the Southwest. Humanitarian groups estimate the true number of migrant deaths is higher, because only bodies that have been found are included, and deaths on the Mexican side are excluded. The humanitarian group Border Angels estimates about 10,000 deaths since 1994. Between 1990 and 2020, the remains of at least 3,356 migrants were recovered in southern Arizona. Between 2000 and 2019, the number of recovered undocumented border crosser remains examined by the Pima County Office of the Medical Examiner in Arizona averaged around 162 each year, but in 2020, 220 remains were recovered. In the first half of 2021, 127 remains were recovered. In south Texas, the remains of at least 1,519 undocumented border crossers were recovered between 2012 and 2019. In the first 5 months of 2021, 36 border deaths were reported in this region.

STAY SAFE   Find a reliable guide for illegal border crossings. Work in your community for a more equitable immigration system and better treatment of undocumented immigrants.

SOURCES   Border Angels, borderangels.org; University of Arizona Binational Migration Institute, *Migrant Deaths in Southern Arizona: Recovered Undocumented Border Crosser Remains Investigated by the Pima County Office of the Medical Examiner, 1990–2020*; Southern Border Communities Coalition, southernborder.org/deaths_by_border_patrol

DEATH IN ICE CUSTODY

ANNUAL US DEATHS   9

The US Immigration and Customs Enforcement (ICE) is part of the Department of Homeland Security. ICE is in charge of enforcing immigration law, including apprehending, detaining, and deporting illegal immigrants. ICE holds immigrants at over 200 detention centers across the country. In 2019, the average daily population in ICE custody was 50,165 people. In 2020, nearly 170,000 immigrants were detained.

From 2011 through 2018, 71 people who were deemed to be illegal immigrants died while being held in ICE detention centers, for an average death toll of 9 per year. Based on the 34,000 such detainees in 2016, this was an annual US deaths of about 26 per 100,000 ICE detainees.

Between April 2018 and April 2020, the number of detainees rose substantially to 170,000, due to stepped-up apprehension measures. In this time, 35 individuals died in ICE detention. The death rate in 2018 was 2.3 per 100,000 admissions; in 2019 the death rate was 1.5, and in 2020 it was 10.8. Suicide by hanging was identified as the cause of death in 9 (25.7 percent) cases. Medical causes accounted for the remaining 26 (74.3 percent). Among 26 deaths attributable to medical causes, 8 (30.8 percent) were attributed to COVID-19, representing 72.7 percent of 11 deaths occurring since April 2020.

Among these deaths, 86 percent were male and their mean age at death was 43. The deceased had lived (illegally) in the US for an average of 16 years before entering ICE detention, and they died after an average of 39 days in detention. Among them were citizens of at least 24 foreign countries.

STAY SAFE   Detainees should receive appropriate mental health and medical care. Work in your community and tell your elected representatives to work toward a more equitable immigration system.

SOURCES   M. Grassini et al. Characteristics of Deaths Among Individuals in US Immigration and Customs Enforcement Detention Facilities, 2011–2018. *JAMA Network Open*, July 2021; ICE, ice.gov; S. Terp et al. Deaths in Immigration and Customs Enforcement (ICE) detention: FY2018–2020. *AIMS Public Health*, January 2021.

Terrorism on US Soil

ANNUAL US DEATHS   In 2020, domestic terrorists killed 110 people, international terrorists killed 0. Excluding 9/11, the annual US death rate from domestic and international terrorists through mid-2021 was 25.

Terrorism on US soil can best be understood, and thus more effectively fought, by separating it into domestic versus international. This distinction is based on the locus of origin of the terrorists' ideological motivation, not their place of birth or citizenship.

In the US, international terrorism usually (but not always) means Islamist extremists, such as al Qaeda or ISIS, whose fighters have infiltrated the US from abroad (as was the case on September 11, 2001) or who might be US-born but have become radicalized through in-person and internet-based propaganda and recruitment efforts.

Domestic terrorism in the US consists of homegrown extremists whose ideologies, whether left wing or right wing, focus on resisting America's elected government (e.g., the Oklahoma City bombing in 1995, White supremacists, or various attacks on police) or attacking members of "enemies" within American society (e.g., LGBTQ+ people; supposed Socialists; neo-Nazis; members of racial, ethnic, or religious minorities).

In 2020 in the US, domestic terrorist attacks from far-right groups killed 91 people, while those from far-left groups killed 19 people. Attacks were made with firearms, knives, or motor vehicles. In 2020, there were no deaths in the US from attacks by international terrorists.

Cumulatively from September 12, 2001, through mid-2021 in the US, 250 people were killed by terrorist attacks. There were 107 interna-

The Pentagon under repair after the terrorist attacks of September 11, 2001

tional terrorist attacks—all jihadist. There were 143 domestic terrorist attacks: 114 far-right, 12 far-left, and 17 misogynist/incel.

STAY SAFE If you see something, say something. Potential recruits vulnerable to radicalization must be made to understand that *all* terrorist groups are run by manipulative sociopathic predators, whose true agenda is indiscriminate death and destruction. Radicalized persons require an urgent intervention by a mental health professional or law enforcement.

SOURCES Center for Strategic and International Studies (CSIS), csis.org/analysis/war-comes-home-evolution-domestic-terrorism-united-states; National Consortium for the Study of Terrorism and Responses to Terrorism (START), start.umd.edu/tevus-portal

War

Civilian Deaths in Foreign Wars

ANNUAL US DEATHS 18,055 based on deaths from 2001 to 2019

Cumulatively from late 2001 through mid-2021, at least 22 US Department of Defense (DoD) civilian employees died in overseas military operations in major conflict zones. Of these, 14 were killed by hostile action and the other 8 died from accidents, disease, or suicide. In addition, 7,950 DoD contractors were killed, many of whom performed warfighting duties comparable to those of special ops troops or military police.

From late 2001 through late 2019, about 325,000 non-US noncombatants (civilians) were killed, many of them residents of the conflict zones in Afghanistan, Pakistan, Iraq, Syria, Yemen, and elsewhere. Enemy fighter deaths numbered between 255,000 and 260,000, while US service members killed numbered just over 7,000. Allied troop deaths numbered 12,500, and other military and police deaths in the various conflicts totaled about 175,000.

Although a breakdown by nationality is not readily available, about 536 journalists and media workers were killed, as were about 807 employees of humanitarian and nongovernment organizations (NGOs) such as the Red Cross and Red Crescent and Doctors Without Borders.

A breakdown of total deaths by conflict from 2001 to mid-2021, combining all civilians and all combatants, is as follows:

Afghanistan	157,000
Iraq	308,000
Pakistan	66,000
Syria	179,000
Yemen	90,000

STAY SAFE War is hell.

SOURCE US Department of Defense, defense.gov, Annual Civilian Casualty Report

Nuclear War

ANNUAL US DEATHS The only available statistic is from the 1945 atomic bombings of Hiroshima and Nagasaki, Japan: 110,000–250,000.

Nuclear and thermonuclear weapons are the deadliest and most destructive devices ever created. The effects of a nuclear detonation are many, widespread, and long-lasting: blinding light, intense heat and terrible radiation, and immensely powerful blast waves, plus deadly radioactive fallout that can last for years and circle the globe. A large-enough nuclear attack, or all-out nuclear war, would create so many raging conflagrations, and hurl so much dirt, dust, aerosols, and smoke into the upper atmosphere, that a nuclear winter would occur, plunging the entire planet into a years-long period of killing total darkness and deep freeze. This would be followed by a nuclear summer of deadly over-exposures to hard ultraviolet rays from the sun because the planet's protective ozone layer would have been destroyed by the radiation. The combination of all these catastrophes would certainly destroy human civilization and maybe extinct all life on Earth.

Historically, uses of nuclear weapons in wartime have only occurred in 1945, during World War II, when the US deployed atomic ordnance against the Japanese cities of Hiroshima and Nagasaki to force the imperial Japanese regime to surrender. Estimates of the death toll from these 2 atomic bombs vary from 110,000 to 250,000 or more Japanese. The difference in these estimates depends on how many long-term excess deaths from cancers due to radiation exposure are added to those people killed by the immediate blast effects and the raging conflagrations they caused.

In addition, many people around the world, called downwinders, have been killed by cancer-causing radioactive fallout effects from tests of both atomic bombs and the vastly more powerful and deadly hydrogen bombs during the Cold War. One estimate of this death toll, over the decades since 1945, is about 400,000 excess deaths throughout the US population alone.

STAY SAFE Safety from nuclear weapons is twofold: preventing nuclear war and trying to survive actual nuclear attack. Prevention comes from a combination of effective mutual nuclear deterrence ("mutual assured destruction") and counterterrorism, and international nuclear arms control and arms reduction. Survival comes from preparedness for a nuclear blast, such as stockpiling survival and first aid supplies, owning Geiger counters and gas masks, and building a backyard or basement bomb shelter. See "Nuclear Explosion" at ready.gov for more information.

SOURCES Bulletin of the Atomic Scientists, thebulletin.org/ 2020/08/counting-the-dead-at-hiroshima-and-nagasaki; CDC, cdc.gov/nceh/radiation/emergencies/nuclearfaq.htm; ICAN, icanw.org/new_study_on_us_russia_nuclear_war

Strategic Infrastructure Cyberattacks

ANNUAL US DEATHS No statistics exist, but the *actual* death count is low.

In the context of cybersecurity, a strategic infrastructure cyberattack threatens to cripple the innumerable computer systems (including the Internet), critical industrial process control software, and myriad other applications needed to operate all the vital aspects of a nation's infrastructure. A strategic level of attack by an adversary means one that impairs a country's ability to defend itself, or that threatens its way of life or even its existence.

Up to now, cyberattacks have caused very few deaths, although recent data intrusion and ransomware attacks on hospital systems are thought to have contributed to some heart attack deaths, due to delays in providing needed emergency care.

Cybersecurity experts are concerned that in the future, an adversary might be able to attack the US with an immensely destructive and deadly strategic attack delivered entirely through the cyber realm. In the worst imaginable case, such an attack could have physically destructive effects almost comparable to those of a nuclear war: It could wipe out the national electrical grid, inflict widespread spoilage and pollution of food and drinking water, derange all sorts of highly toxic industrial processes, cause numerous fires, explosions, floods, and train and plane wrecks, and cripple all of the communications, transportation, healthcare, first responder, and other broad systems that are indispensable to sustaining daily life in the US, and that are even essential to our basic human survival.

STAY SAFE   Citizens, corporations and nonprofits, and governments at all levels need to mount and sustain a concerted, coordinated campaign to enhance *defensive* cybersecurity, and also to implement legal and ethical measures to effectively *deter* cyberattacks by adversaries against our strategic infrastructures.

SOURCE   Parachute Technology, parachutetechs.com/
2022-cyber-attack-statistics-data-and-trends

Intentional Harm and Injury

Homicide

H omicide is a human killing of another person by a volitional act. While a murder is a homicide, the terms are not synonymous. Murder is a volitional act committed with intent to cause harm. In contrast, a homicide results from accidental, reckless, or negligent volitional acts in the absence of intent to harm. These distinctions are of legal rather than medical importance.

> **ANNUAL US DEATHS**   19,141 deaths from homicide in 2019

Murder

> **ANNUAL US DEATHS**   16,214 in 2018; 16,425 in 2019; estimated 21,000+ in 2020

Murder is the unlawful and intentional killing of another person. Courts make a distinction as to whether the act of killing was willful and planned in advance, from prior malice, or was spur-of-the-moment, from passion, but this is irrelevant to the victim.

The total number of US murders is on the rise: 16,214 in 2018 and 16,425 in 2019. In 2020, murder and manslaughter jumped 29.4 percent, bringing the estimated total to 21,350, the largest increase on record. Despite the increase, however, the homicide rate in 2020 was

LAPD detectives investigate a murder scene, 1969

7.8 per 100,000 population, lower than the peak of 10 per 100,000 in the early 1980s. Interestingly, the spike in homicides in 2020 and continuing into 2021 wasn't accompanied by a comparable spike in other crimes, such as robbery. In fact, while violent crime rose 5.6 percent in 2020, property crimes such as burglary fell 7.8 percent. The homicide spike is almost certainly related to the COIVD-19 pandemic, although exactly how is unclear.

The FBI's Uniform Crime Reporting (UCR) Program is the primary source for national crime statistics. The program has been providing crime statistics since 1930. The UCR Program includes data from more than 18,000 city, university and college, county, state, tribal, and federal law enforcement agencies. Agencies participate voluntarily and submit their crime data either through a state UCR program or directly to the FBI's UCR Program. While the numbers are good indicators of overall crime and trends, because participation is voluntary, some information never gets submitted. In addition, the crimes are reported on the basis of the charge (first-degree murder, for example) and not the basis of the final disposition of the case (charge reduced to manslaughter, for instance), so the numbers can mislead in the direction of more serious crime than actually occurs.

Accurate demographic details take years to be tallied: In 2010, nationally, 77.4 percent of murder victims were male; 50.4 percent were Black, 47.0 percent were White, and 2.6 percent were other races. Fifty-three percent were killed by someone they knew. In almost half those cases, the murderer was a family member.

The latest year for which breakdowns by method and motive are available (compiled by the FBI) is 2017. In that year, out of 15,129 total murders, 2,236 of them occurred while the murderer was committing another felony.

STAY SAFE Try to avoid getting into serious arguments that escalate to violence (especially while intoxicated), and don't go into known high-crime areas if you can avoid them. Be careful in your interpersonal relationships with family, friends, lovers, co-workers, and rivals to avoid conflicts and enmities that might give someone a reason to want to kill you.

SOURCES CDC, cdc.gov/violenceprevention/datasources/nvdrs/index. html; FBI, crime-data-explorer.app.cloud.gov/pages/explorer/crime/shr

Active Shooter/Mass Killings

ANNUAL US DEATHS 2,128 mass shootings from 2012 to 2017: roughly 1 per day, some resulting in multiple deaths, others in none

Although there's no official definition, most authorities say a mass killing is one that causes the death of 4 or more victims by the same perpetrator(s) at the same or nearby locations and within the same day. In the US, due to the wide availability of firearms, almost all mass killings involve guns, and these are generally recorded by law enforcement as active shooter incidents. Note that not all such incidents involve fatalities; sometimes the shooter is apprehended or otherwise neutralized before anyone gets killed.

Although active shooter/mass killings incidents get extensive media coverage when they occur, the number of active shooter/mass killing events in the US each year is small. In each of 2014, 2015, and 2016, there were 20 such reported incidents. For 2017, there were 30, for 2018 there were 27, and for 2019 there were 28. The number of people killed each year in these shooting rampages peaked at 138 in 2017, compared to 83 in 2016 and 85 in 2018. Between 2009 and mid-2021, there were 264 mass shootings in the US, resulting in 1,485 people shot and killed and 968 people shot and wounded.

Mass shooting events, where at least 4 people (not including the shooter) are shot but not necessarily killed, are dismayingly common in the US. On average since 2014, at least 1 mass shooting occurs in the US every day. In 2019, mass shootings, including active shooter incidents, led to 417 deaths. In 2020, mass shootings led to 611 deaths.

The worst mass shooting in US history was on October 1, 2017, in Las Vegas when Stephen Paddock opened fire from the 32nd floor of the Mandalay Bay Resort on a crowd at a country music festival, killing 58 people and injuring 413. The next worst death toll was on June 12, 2016, at the Pulse nightclub in Orlando, Florida, where the gunman killed 49 victims and then himself.

In part due to societal stresses from COVID-19, FBI data for 2020 shows that overall active shooter incidents were up almost 40 percent over 2019, the previous annual high.

Active shooter incidents can occur almost anywhere. In recent years, they have taken place at houses of worship, in schools, in workplaces, at outdoor music performances, and in retail stores and shopping malls. The single most common mass killing is a family annihilation, usually occurring in the home. Sometimes mass killings are due to gang wars or result from commission of another felony, such as a robbery. There is no distinctive pattern to the demographics of the perpetrators and their victims. Almost all mass killers are male, family annihilators are usually middle-aged White men, and felony-related mass killings are usually done by Black or Hispanic males with extensive prior criminal records.

STAY SAFE   If caught in an active shooter situation, don't panic. Act quickly to protect yourself and others. Use common sense to choose between fleeing, hiding, or (in the last extreme) fighting back for your life. Listen carefully for instructions from police or other first responders. Comply with all safety lockdowns and wait cautiously for the all-clear.

SOURCES   Everytown Research & Policy, everytownresearch.org/ report/gun-violence-in-america; FBI, fbi.gov/services/cjis/ucr; Gun Violence Archive, gunviolencearchive.org

Felony Murder

ANNUAL US DEATHS   15,129 in 2017

Sometimes a victim is murdered by a perpetrator who is in the process of committing some other serious crime. This is felony murder. US FBI data for 2017 showed that out of 15,129 murders, 2,236 (14.8 percent) involved another felony. Record-keeping and reporting by local jurisdictions are imperfect, so 953 of these felonies were not specified by type. The single biggest known felony murder that year was robbery, with 680 people killed in 2017. Second were drug-related felonies, with 533 murders. Burglary had 90, larceny-theft 20, motor vehicle theft 31, prostitution and commercialized vice 15, arson 31, gambling 10, and rape and other sex offenses led to 26 murders in 2017.

STAY SAFE   Try to avoid getting into serious arguments that escalate to violence (especially while intoxicated), and don't go into known high-crime areas if you can avoid them. Be careful in your interpersonal relationships with family, friends, lovers, co-workers, and rivals to avoid conflicts and enmities that might give someone a reason to want to kill you.

SOURCE   FBI, crime-data-explorer.app.cloud.gov/pages/explorer/ crime/shr

SIG Sauer handguns are popular in the US. This one is the company's P239.

Firearms

ANNUAL US DEATHS 14,861 (firearm homicide, including murder) in 2019

Firearms are by far the most frequent weapon used for murder in the US. In 2016, out of 15,070 total murders, 11,004 (73.0 percent) were committed with a gun. Of these, 7,105 (64.6 percent of all firearm murders) were reported to be committed with a handgun. The totals that year for other types of firearms were 374 for rifles and 262 for shotguns; the remainder were tallied only as "other guns" or "firearm type not stated."

STAY SAFE Try to avoid getting into serious arguments that escalate to violence (especially while intoxicated), and don't go into known high-crime areas if you can avoid them. Be careful in your interpersonal relationships with family, friends, lovers, co-workers, and rivals to avoid conflicts and enmities that might give someone a reason to want to kill you.

SOURCE FBI, ucr.fbi.gov/crime-in-the-u.s

Gang-Related

ANNUAL US DEATHS   Average of 2,015 each year from 2007 to 2011

Over 30,000 criminal gangs are known in the US, totaling about 1 million members. Gang-related homicides are estimated at about 13 percent of all homicides nationally. About 67 percent of the roughly 2,000 annual gang killings occurred in big cities.

STAY SAFE   Don't join or associate with criminal gangs or use them to source drugs or other contraband.

SOURCES   FBI, ucr.fbi.gov/crime-in-the-u.s; National Gang Center, nationalgangcenter.ojp.gov/about/surveys-and-analyses

Hate Crimes

ANNUAL US DEATHS   51 in 2019

The criminal justice system defines a hate crime murder as a killing that is motivated by prejudice against the victim(s) based on their race and ethnicity, place/country of origin, faith/religion, disability, sex, sexual orientation/preference, and/or sexual identity/transsexuality. For 2019, 51 murders were reported as hate crimes, but because not all jurisdictions reported data this way, the national total is incomplete.

Demographic breakdowns are only available (and statistically significant) across the much broader category of all hate crimes against persons. Of the 8,552 victims of hate-based intimidation, assault, and homicide in 2019, 57.6 percent were targeted for their race/ethnicity/ancestry, 20.1 percent for their religion, 20.3 percent for sex/gender bigotry, and 2.0 percent because of a disability.

Among the perpetrators of hate/bias offenses, 52.5 percent were White and 23.9 percent were Black; 8.9 percent were of other races/ethnicities (Asian, Native American, and others). In 14.6 percent of cases, the offender's race/ethnicity was unknown. When the offender's age was reported, 84.6 percent were over 18.

STAY SAFE   Work within your community to increase awareness of hate crimes and take preventive measures.

SOURCE   FBI, fbi.gov/hate-crime/2019 and crime-data-explorer.app.cloud.gov/pages/explorer/crime/hate-crime

Hazing

ANNUAL US DEATHS Average of 4.9

Many groups that consider themselves special or elite have rigorous initiation processes for new entrants. Hazing occurs when these rituals become abusive, reckless, and/or violent, often fueled by extreme alcohol consumption. Hazing tends to be focused among males in their late teens and early 20s—high school and college age. Fraternities and marching bands are particularly notorious for dangerous, sometimes lethal, hazing. Between 2008 and 2018, an average of 4.9 deaths per year were related to fraternity hazing, including alcohol poisoning from extreme drinking.

In 2011, a new member of a Florida college's marching band died after a physically violent hazing gauntlet on a band bus used for football away games; the ringleader was subsequently convicted of manslaughter. Early in the 2019–2020 school year, 5 young men in the US died in college fraternity hazing incidents.

Law enforcement and institution crackdowns on the perpetrators of hazing deaths are helping to convey that such life-threatening practices must be discontinued. In Ohio in 2021, 8 young men were indicted on charges including involuntary manslaughter and providing alcohol to minors, after an underage college student died in a fraternity's off-campus hazing incident.

STAY SAFE If you are a student, parent, faculty member or coach, or alumnus/alumna, help to discourage and prevent reckless or toxic behavior in fraternities and sororities, marching bands, and other groups at your school. See "If You Want to Help Stop Hazing" at iup.edu. If you are being hazed or are being pressured to participate in hazing, call the HazingPrevention.Org's antihazing hotline at 1-888-668-4293.

SOURCES HazingPrevention.org, hazingprevention.org/about/hazing-facts; Inside Hazing, insidehazing.com/statistics

Illegal Drugs

ANNUAL US DEATHS 560 in 2018

Illegal drugs are a significant cause of crime, including murder, whether committed to get money to buy drugs, or in the process of trafficking and selling them. About 4 percent of all US homicides are drug related. In 2018, 560 of 14,123 murders involved illegal drugs.

Much greater risk comes from being killed by the drugs themselves.

In 2020, there were more than 100,000 fatal overdoses from abusing drugs.

STAY SAFE   Avoid abusing drugs. Call the Substance Abuse and Mental Health Services Administration's 24/7/365 national helpline at 1-800-662-4357.

SOURCES   CDC NCHS, cdc.gov/nchs/ndi/index.htm; FBI, ucr.fbi.gov/crime-in-the-u.s

Infanticide

ANNUAL US DEATHS   About 500

Infanticide means homicide of a child under 1 year of age, though the term is often used for any murder of a child. The US has the highest rate of child murder of all developed countries.

Hard data is scarce. In some cases, the victim's body disappears; sometimes the cause of death cannot be distinguished from natural causes. When the killing is intentional, either 1 or both parents, or a new lover of one of the parents, is usually the perpetrator, although on rare occasions a babysitter, nanny, or other caregiver is responsible. One notorious case of parental infanticide occurred in 2001, when a Texas mother drowned her 5 young children in a bathtub.

About 2.5 percent of all homicide arrests in the US are for parents killing their children—about 500 cases per year. Motives/reasons include acute psychosis of the parent (including postpartum depression), murder/suicide or family annihilation, indirect revenge after rejection by the other parent, or because the child is seen as an inconvenience or impediment, or to relieve the child's real or imagined physical or spiritual suffering, or from chronic domestic violence that goes too far (battered child syndrome).

STAY SAFE   If you become aware of a new parent struggling to cope, or if you are one yourself, help is available. Call the National Domestic Violence Hotline at 1-800-799-7233. Most local jurisdictions have a department of child protective services. In an emergency, call 911.

SOURCES   CDC, cdc.gov/nchs/nvss/deaths.htm; Child Welfare Information Gateway, childwelfare.gov/topics/can/fatalities

Justifiable Homicide

ANNUAL US DEATHS 665 in 2010 (most reliable statistic available)

"Homicide" is an all-inclusive term for being killed by the direct act of another person. Not all such deaths count as murder or manslaughter, which lead to legal penalties for taking a life. Some killings are deemed by the courts to be justified, necessary, and appropriate to protect oneself or other people from a bad actor who winds up dead. In 2010, there were 665 justifiable homicides in the US. Law enforcement officers killed 387 felons who were committing a crime. Private citizens justifiably killed 278 people who were committing a crime. However, self-defense based on stand-your-ground laws is a controversial claim; it is sometimes used to justify a hate crime.

STAY SAFE Do not commit a crime, resist arrest, or threaten other people.

SOURCE FBI, ucr.fbi.gov/crime-in-the-u.s/2019

Juvenile Killers

ANNUAL US DEATHS About 920 in 2018

Legally, juveniles are youths under age 18. During 2018, an estimated 920 juveniles were arrested for murder and nonnegligent manslaughter. Of these, 79 percent were male, and 21 percent were female. Ten percent were younger than 15; 90 percent were ages 15 to 17.

Data on the race/ethnicity of juvenile killers is not available in much detail, but 40 percent were reported as White. Arrests of juveniles for murder showed an upward trend of 21 percent between 2014 and 2018. While data on the number of people (of all ages) killed by juvenile offenders is not available, it seems reasonable to assume that on average 1 youth murderer accounts for 1 dead victim. The exception would be juvenile school shooters (see **School Shootings** for details).

STAY SAFE Whatever your age, try to minimize contact with potentially violent juveniles.

SOURCE US Department of Justice, Office of Juvenile Justice and Delinquency Prevention, ojjdp.ojp.gov/statistics

Motive or Circumstances

> **ANNUAL US DEATHS**   See explanation in entry.

Data on the motives and circumstances behind total murders each calendar year is incomplete. Available information gives insight into situations to avoid so as not to become the victim of a homicide. In 2017, arguments that escalated into deadly violence accounted for 3,423 of the total 15,129 murders that year (22.6 percent). Gang violence caused 750 deaths. Brawls due to the influence of alcohol and/or drugs killed 247 people in 2017. Romantic triangles triggered 107 killings, whereas 32 children were murdered by their babysitter.

STAY SAFE   Avoid getting into arguments that could escalate into serious violence. Be careful with your own and other people's behavior while under the influence of drugs and alcohol. Be aware of the dangers if you get involved in a love triangle. Be cautious whom you hire as a babysitter.

SOURCE   FBI, ucr.fbi.gov/crime-in-the-u.s/2019

Murder-Suicide

> **ANNUAL US DEATHS**   About 1,300

A murder-suicide is defined statistically as a murder of 1 or more people, followed within 1 week or less by the murderer committing suicide. Murder-suicide incidents are seen by experts as distinct from murders and from suicides; in some cases, the murders precipitate the suicide, while in others the perpetrator's suicide decision triggers a need to kill others, too. The incidence of murder-suicide in the US is about 0.31 per 100,000 of population, or about 1,300 incidents across the US in any 1 year. Of these, 65 percent involve an intimate partner.

A murder-suicide might involve a family annihilation, a failed romantic relationship, or domestic or international terrorism. The event might or might not kill people other than the intended victims; this is especially true when a transportation crash is the method/weapon, such as a motor vehicle or an aircraft. The deadliest ever such incident in the US was the September 11, 2001, terrorist attacks, in which a total of 2,977 died on the 4 planes or on the ground, excluding the 19 terrorist hijackers.

STAY SAFE   Known risk factors for murder-suicide include substance abuse, depression, and the failing health of someone involved. Domestic violence, mental illness, and/or a failing/failed intimate relationship are also risk factors. Seek help at the first sign of domestic abuse, intimate partner violence, excessive jealousy, paranoia, or self-harming thoughts or tendencies by family member or intimate partner. In an emergency, call 911.

SOURCES FBI, ucr.fbi.gov/crime-in-the-u.s/2019; P. Roma et al. Mental illness in homicide-suicide: a review. *Journal of the American Academy of Psychiatry and the Law*, 2012.

Nonfirearm Weapon or Method

ANNUAL US DEATHS 4,727 in 2018

Homicide detectives often say that killing someone is easy; the hard part is disposing of the body. While firearm murders could be called easiest because a mere squeeze of the trigger is all it takes, there are a number of other ways to kill someone. Chief among these nongun homicides in the US is to stab or slash the victim with a knife or other sharp-edged cutting instrument: 1,604 of the total 15,070 murders in 2016 were committed this way. Blunt objects such as clubs or hammers amounted to 472 killings, whereas 656 murders were caused by attacks with parts of the murderer's own body (hands, fists, feet, etc.) Poison was used 11 times; intentional narcotics overdoses were used to kill 114 times. A bomb was only used once in 2016, whereas fire via arson and incendiaries was used 107 times. And cutting off someone's air is a good way to make them stop breathing permanently: Strangulation murders happened 98 times, other asphyxiation (e.g., smothering) murders 91 times, and 9 people were murdered by the killer intentionally drowning them.

Although conspicuous, even spectacular, when it does occur, homicide by bombing is very rare in the US. The Bureau of Alcohol Tobacco Firearms and Explosives (BATFE) indicates that for 2017 and 2018, fewer than 20 people per year in the US were killed by intentional criminal bombings, whether for murder, extortion, revenge, or terrorism.

STAY SAFE Try to avoid getting into serious arguments that escalate to violence (especially while intoxicated), and don't go into known high-crime areas if you can avoid them. Be careful in your interpersonal relationships with family, friends, lovers, co-workers, and rivals to avoid conflicts and enmities that might give someone a reason to want to kill you. In an emergency, call 911.

SOURCES Bureau of Alcohol, Tobacco, Firearms, and Explosives, United States Bomb Data Center (USBDC), atf.gov/resource-center/

united-states-bomb-data-center-usbdc; FBI, ucr.fbi.gov/
crime-in-the-u.s/2019

Rape and Sexual Assault

> **ANNUAL US DEATHS** About 1,000

Hard data on rape and other sexual assault victims who are murdered
is scarce and not up to date. US cases studied in the early 1990s in
which a sex offense was the primary reason for the killing amounted
to a fraction of 1 percent of all murders with known circumstances. If
this rate were to apply more recently, roughly 1,000 people per year are
murdered as the aftermath of a sexual attack.

Sexual assault is a very serious problem in the US. Reliable data is
difficult to come by because most such assaults are never reported to
the police. In 2018, there were an estimated 750,000 attempted or completed rapes. Using these figures, the probability of being murdered if
you are the victim of a sexual attack is about 0.1 percent.

STAY SAFE See the CDC's resource, "Stop SV: A Technical Pack-
age to Prevent Sexual Violence" at cdc.gov/violenceprevention/sexual
violence/prevention.html.

SOURCES CDC, cdc.gov/violenceprevention/sexualviolence; FBI,
ucr.fbi.gov/crime-in-the-u.s/2019; National Sexual Violence Resource
Center, nsvrc.org/statistics; RAINN (Rape, Abuse & Incest National
Network), rainn.org/about-sexual-assault

School Shootings

> **ANNUAL US DEATHS** Average of 11 events annually between 2010
> and 2018

The incidence of US school shootings goes back to the nineteenth century, but until the Columbine High School shooting in April 1999 (15
killed, including the perpetrators), they were rare events. Between
2011 and 2018, an average of 11 school shootings occurred. They have
become more frequent since then. In 2018, there were 30 shootings on
K–12 campuses (33 dead, 55 injured); in 2019, there were 27 (5 killed,
17 injured); in 2020, there were only 9 because schools were mostly
closed due to the COVID-19 pandemic. In 2021, however, the number
jumped to 42 (9 dead, 36 injured).

Looked at more broadly to include all incidents of gunfire on a school

Adam Lanza used this Bushmaster XM 15-E2S assault rifle
(along with other weapons) to kill 28 people, mostly young
children, at Sandy Hook Elementary School, December 14, 2012.

campus, the numbers are higher. In 2021, there were at least 149 incidents of gunfire on school grounds, resulting in 32 deaths and 94 injuries nationally. About 34,000 students were exposed to gun violence in 2021.

The worst high school mass killing was in Parkland, Florida, in 2018, where 17 students and teachers were murdered at Marjory Stoneman Douglas High School; the gunman survived and was arrested. He was later sentenced to death. The worst elementary school shooting was in Newtown, Connecticut, in 2012, where the gunman first killed his mother at home, then 26 children and staff at Sandy Hook Elementary School, and then himself as police arrived. Shootings on college campuses are rarer. The worst ever was at Virginia Tech (in Blacksburg, Virginia), in 2007, where the gunman killed 32 people and then himself.

STAY SAFE Pay attention during active shooter drills. If you hear gunfire on a school campus, seek cover and concealment. Listen for and obey instructions from teachers, school security officials, and police.

SOURCES Everytown for Gun Safety, everytownresearch.org/maps/gunfire-on-school-grounds; Naval Postgraduate School Center for Homeland Defense and Security (CHDS) K-12 School Shooting Database, chds.us/ssdb; NBC News School Shooting Tracker, nbcnews.com/news/us-news/school-shooting-tracker

Serial Killers

ANNUAL US DEATHS   63 from 2015 to 2019

A serial killer is someone who commits 3 or more murders. Motives can be as diverse as repeat "black widows" killing husbands for money, to "angels of death," medical workers who kill patients, to homicidal sexual sadists and various other psychopathic "pattern killers."

Serial killers are a favorite topic of law enforcement dramas, but the odds of being murdered by one are exceedingly low. In fact, the era with the most known serial killers active in the US was the 1980s, with an average of 115 such perps on the prowl per year. (Sometimes 1 individual serial killer was active across more than 1 year.) The number of serial killers and victims has declined steadily since then. From 2015 through 2019, the active serial killer count fell below 30, and the number of victims killed averaged 63 per year. Caution is needed with this data, as there can be a significant time lag before a murder is attributed correctly as a serial killing.

Reasons for this long-term decline include: better technologies to deter or detect and interdict/apprehend serial killers more effectively; stricter parole policies and longer prison sentences, to keep serial killers off the streets once first apprehended; and better public awareness and safety precautions by potential victims. These beneficial trends include: the ever-increasing ubiquity of cellphones to call for help, and of video cameras enabling security surveillance of public areas; fewer people hitchhiking; and fewer children walking to/from school alone or accepting rides from strangers.

In the US in 2015, 80 people were known to be murdered by serial killers, as were 80 people in 2016. In 2017, 77 died this way, followed by 44 in 2018 and 36 in 2019. This is consistent with both the overall multidecade downward trend in serial murders and the multiyear time lag in accurate reporting.

STAY SAFE   Don't expose yourself unduly to being abducted by a serial killer. Don't hitchhike. Make sure your kids don't ever accept rides from strangers or walk to and from school alone. Be careful after dark and along isolated or deserted thoroughfares. Avoid suspicious-looking and out-of-place people. Seventy-nine percent of US serial killers spent time in prison for some other offense before their first murder.

SOURCES   FBI, fbi.gov/stats-services/publications/serial-murder; Radford University Serial Killer Information Center, Serial Killer Statistics, maamodt.asp.radford.edu/Serial%20Killer%20 Information%20Center/Project%20Description.htm

By Sex/Gender

> **ANNUAL US DEATHS** 15,064 male, 1,900 female (based on incomplete FBI 2016 data)

Of the 16,964 homicides committed in the US in 2016 (murders, manslaughters, plus justified killings), the age, sex/gender, and race/ethnicity of the victims was reported to the FBI for only about 66 percent. For sex/gender, 88.8 percent of homicide victims were male. Only 11.2 percent were female.

STAY SAFE Men especially should avoid getting into violence-prone conflicts, especially ones fueled by drugs and/or alcohol, and should try to minimize their presence in high-crime areas.

SOURCE FBI, crime-data-explorer.app.cloud.gov/pages/explorer/crime/shr

By Time of Day and Month

> **ANNUAL US DEATHS** Not applicable

US murders show noticeable patterns of higher or lower probabilities of occurrence across diurnal (time of day or night) and seasonal (month of year) cycles.

Only a minority of all murder statistics reported to the FBI provide time-of-day information. Among the 5,263 homicides of 2018 that did, the deadliest hour was between midnight and 1 a.m., with 415. The wee hours between 1 a.m. and 4 a.m. were noticeably less dangerous, averaging 233 homicides per hour. The period from 4 a.m. to 8 a.m. was the safest, with about 110 killed hourly. Then the average hourly rate climbed gradually, more than doubling to 259 an hour between 4 p.m. and 8 p.m. and climbing higher still to 325 killed per hour between 8 p.m. and midnight.

National data about murder by month of occurrence are more complete. The 15,343 murders in 2015 that had this information were distributed by calendar quarter: From January through March, 3,052 people were murdered; from April through June, 3,724; from July through September, 3,973; from October through December, 4,594. This pattern shows a distinct rise from quarter to quarter, with a peak in the fourth quarter. This peak is due mostly to murders in the month of December, which at 2,161 murders ran 80 percent higher than the average of the other 11 months of 2015. Anecdotally, the year-end holidays are a time of heavy drinking and socializing, leading to a spike in people killing each other. The single month with the fewest murders was February, with 862, partly because this month has the fewest days and partly because it comes in the dead of winter at more northern locations in the US.

STAY SAFE   Be cautious when going out and about, or staying in and socializing, from 9 p.m. to 1 a.m., when more people are intoxicated, and more violent criminals are on the prowl. Also be cautious around the year-end holidays, when intoxication, heavy socializing, and generally raucous behavior cause an annual peak in murders.

SOURCE   FBI, ucr.fbi.gov/crime-in-the-u.s/2019

By Victim Age

ANNUAL US DEATHS   Consistently lowest for those 70 and over; highest for those in their 20s

Of the 16,964 homicides committed in the US in 2016 (murders, manslaughters, plus justified killings), the age, sex/gender, and race/ethnicity of the victims was reported to the FBI for only about 66 percent. For age, 3.5 percent of homicide victims were age 16 or younger, and 12.9 percent were in the age range of 17 through 19. Fully 41.5 percent were in their 20s, 20.7 percent were in their 30s, 10.6 percent were in their 40s, 6.7 percent were in their 50s, 2.8 percent were in their 60s, and 1.3 percent were age 70 and over.

STAY SAFE   By individual age, the early 20s are the most dangerous years, slightly ahead of the late teens, especially for males. If you are a parent, guardian, or counselor to youth, try to impart your greater maturity about caution, risk mitigation, and conflict resolution.

SOURCE   FBI, crime-data-explorer.app.cloud.gov/pages/explorer/crime/shr

By Victim Race or Ethnicity

ANNUAL US DEATHS   Consistently highest among Black people, lower among Whites, and lowest among Native Americans, Asian Americans, and others

Of the 16,964 homicides committed in the US in 2016 (murders, manslaughters, plus justified killings), the age, sex/gender, and race/ethnicity of the victims was reported to the FBI for only about 66 percent.

For race/ethnicity, 43.9 percent of 2016 homicide victims were

White, 53.5 percent were Black, and 2.6 percent were Native American and others. Among White homicide victims, 21.6 percent were Latino.

STAY SAFE Murder is an equal-opportunity destroyer. No race or ethnicity is immune.

SOURCE FBI, crime-data-explorer.app.cloud.gov/pages/explorer/crime/shr and ucr.fbi.gov/crime-in-the-u.s/2019

Worst and Best Cities

ANNUAL US DEATHS See data in entry

To meaningfully compare the level of murder rates in different jurisdictions with significantly different overall populations, actuaries use a statistic that "normalizes" (standardizes) the data. This means figures compared are not distorted by differences in the total number of people living in each place. Murder statistics are normalized to the annual number of deaths per 100,000 of population. (This is calculated as 100,000 × Total Homicides / Total Population). Note that this city-by-city discussion is not 100 percent complete and lags a year or more behind current numbers. Not every city reports their homicide counts for every calendar year, and the statistics take time to compile. In addition, the FBI cautions against using crime statistics to compile rankings, pointing out that the numbers provide no insight into the numerous variables that mold crime in a particular town, city, county, state, tribal area, or region.

The COVID-19 pandemic has caused a sharp uptick in murders in both 2020 and 2021. FBI data shows that homicides rose nationally in 2020 by about 30 percent over 2019, the largest 1-year increase since the FBI began keeping records. Guns were used in more than 75 percent of all murders in 2020.

By normalized actuarial standards, in 2019 the US city with the highest homicide rate of all was Baltimore, Maryland, with 58.27 homicides per 100,000 of population. Next worst was Detroit, Michigan, with 41.45 homicides per 100,000 of population. Third was New Orleans, Louisiana, with 30.67 homicides per 100,000 of population. Fourth was Memphis, Tennessee, with 29.21 homicides per 100,000 of population. Fifth in the top 5 cities for homicide in 2019 was Cleveland, Ohio, with 24.09 homicides per 100,000 of population.

For cities with the best (lowest) murder rates, a criterion for materiality of the sample size is needed, since very small or sparsely populated places will show a lot of meaningless statistical noise (variations) from 1 year to the next. Among cities with a population above 250,000 in

2019, the city with the lowest homicide rate was Irvine, California, with a rate of 0.34 homicides per 100,000 of population. Next was Raleigh, North Carolina, at 1.05 homicides per 100,000 of population, followed by Chula Vista, California, at 1.09 homicides per 100,000 of population, Scottsdale, Arizona, at 1.15 homicides per 100,000 of population, and Laredo, Texas, at 1.51 homicides per 100,000 of population.

STAY SAFE   Try to live in a city with a relatively low homicide rate.

SOURCE   FBI, crime-data-explorer.app.cloud.gov/pages/explorer/crime/shr

Worst and Best States

ANNUAL US DEATHS   See data in entry

To meaningfully compare the level of murder and non-negligent manslaughter rates, in different jurisdictions with significantly different overall populations, actuaries use a statistic that "normalizes" (standardizes) the data, so that the figures compared aren't distorted by differences in the total number of people living in each place. Instead, they rely on the annual number of deaths per 100,000 of population. (This is calculated as 100,000 × Total Deaths / Total Population). The FBI cautions against using crime statistics to compile rankings, pointing out that the numbers provide no insight into the numerous variables that mold crime in a particular town, city, county, state, tribal area, or region.

By the normalized actuarial measure, the US state with the highest homicide rate of all in 2019 was also ranked 6th worst among all US cities: Washington, DC, with a rate per 100,000 of 23.45. Among the actual 50 states, the worst was Mississippi at 15.4 homicides per 100,00 population. Then came Louisiana at 14.7 homicides per 100,000 of population, Alaska at 12.8 homicides per 100,000 of population, New Mexico at 11.8 homicides per 100,000 of population, and South Carolina at 11.0 homicides per 100,000 of population.

The state with the lowest homicide rate in 2019 was Vermont at 0 homicides per 100,000 of population (only 11 people were homicides), followed by Idaho at 1.7 homicides per 100,000 of population, Maine at 1.8 homicides per 100,000 of population, Massachusetts at 2.3 homicides per 100,000 of population, and Hawaii at 2.5 homicides per 100,000 of population.

STAY SAFE   Try to live in a state with a relatively low homicide rate. Whatever state you do live in, try to live outside of cities with high homicide rates.

SOURCES   CDC, cdc.gov/nchs/pressroom/sosmap/homicide_mortality/homicide.htm; FBI, crime-data-explorer.app.cloud.gov/pages/explorer/crime/shr

Suicide

ANNUAL US DEATHS   44,834 in 2020

Suicide means dying as a result of self-harm done with the specific intent to cause one's own death. Suicide is a serious worldwide public health problem. For the US population overall, it is the 10th leading cause of death, taking a toll of 47,511 victims in 2019. The age-adjusted rate has been rising since 2000 and reached 14 per 100,000 in 2019. Just under 70 percent of suicides were White males; 50 percent of all suicides used firearms.

It is estimated that over 5 million Americans have lost a relative or close friend to suicide. Suicide is often an impulsive act, one that can be permanently avoided in many cases if only an in-the-moment intervention lifeline is offered in time. In 2019 an estimated 1.38 million people attempted suicide; 3.4 percent of them actually died.

A *suicide attempt* is an act of self-injury committed with the purported intent to die, but which is nonfatal. *Suicide (or suicidal) ideation* means a person thinks about, talks about, plans, or reads obsessively on the topic of suicide, and/or the ideas of death (afterlife, corpses, funerals, cemeteries, undertakers, etc.) and of killing themselves. Any suicide attempt or suicide ideation should be taken very seriously. Suicide attempts and/or ideation can be precursors to a later self-injurious act that does result in death.

STAY SAFE   If you are aware of someone who is suicidal or might be, tell anyone who might help: If you see something, say/do something. Call the National Suicide Prevention Lifeline at 1-800-273-8255 (TALK). Help is available 24/7/365.

Installing suicide barriers on San Francisco's Golden Gate Bridge

SOURCES   American Foundation for Suicide Prevention (AFSP), afsp. org/suicide-statistics; Behind the Badge, behindthebadge.com/more-police-officers-committed-suicide-last-year-than-were-killed-in-line-of-duty; CDC, cdc.gov/nchs/fastats/suicide.htm; Mental Health America, mhanational.org/issues/2022/mental-health-america-prevalence-data#-four; NIH NIMH, nimh.nih.gov/health/statistics/suicide; SAVE.org, save.org/about-suicide/suicide-facts; TheTrevorProject.org, thetrevor-project.org/resources/category/talking-about-suicide; US Department of Veterans Affairs, mentalhealth.va.gov/suicide_prevention/data.asp; WHO, who.int/news/item/17-06-2021-one-in-100-deaths-is-by-suicide

Bullying and Cyberbullying

ANNUAL US DEATHS   No specifically relevant statistics, but suicides average 15,000 annually for those ages 10 to 34.

The brick-and-mortar and internet/social media worlds are intertwined, especially for young people who participate in bullying and cyberbullying as perpetrators or victims. In the worst cases, enraged and socially isolated victims of bullying and cyberbullying have retaliated with escalating physical violence, even mass shootings in schools.

Bullying and cyberbullying can also drive the victim to commit suicide: Studies of annual youth suicide rates in the US find a drastic rise between 2007 and 2017, consistent with the increase in use of social media such as Instagram; victims of repeated cyberbullying were twice as likely as nonvictims to harm or kill themselves. In 2020, suicide was the 2nd leading cause of death among people ages 10 through 34 in the US, with over 15,000 deaths.

Bullying and cyberbullying, which affect about 1 in every 5 teenagers, increase the risks of developing anxiety and depression, poor grades, delinquency, and dropping out of school, and of having adjustment and relationship problems entering adulthood. Victims of cyberbullying are often also the victims of online identity theft. More than 8 in 10 students who identify as LGBTQ+ experienced harassment or assault at school in 2019.

STAY SAFE   For information about how to discourage, prevent, and remediate bullying and cyberbullying, see "Get Help Now" at stopbullying.gov. Parents of school-age kids who are cyberbullied should not hesitate to inform school authorities and the police; more and more jurisdictions have laws against cyberbullying. For help with any thoughts of self-harm/suicide or severe depression, call the National Suicide Prevention Lifeline at 1-800-273-8255 (TALK); help is free, 24/7/365.

SOURCES CDC, cdc.gov/violenceprevention/youthviolence/ bullyingresearch/fastfact.html; National Center for Education Statistics, nces.ed.gov/fastfacts/display.asp?id=719; stopbullying. gov, stopbullying.gov/bullying/effects; TheTrevorProject.org, thetrevorproject.org/resources/category/talking-about-suicide

Alcohol or Substance Abuse

> **ANNUAL US DEATHS** 33 percent of all suicides involve alcohol

People with alcohol or drug abuse problems are much more likely to commit suicide. The rate of suicide is 2 to 3 times higher for men with a substance abuse disorder, compared to those without. For women, having such an addiction raises suicide rates by a factor of 6 to 9 times.

Mental health disorders, crippling disabilities, and/or serious chronic pain due to an injury or disease are dangerous contributing factors for becoming addicted to alcohol and/or drugs. These precursors of substance abuse also significantly elevate the likelihood of a suicide.

Hard data is scarce, but alcohol and drug addiction are known to significantly raise the risk of suicide. When the substance abuse was brought on by a persisting mental illness, serious disability, or chronic physical pain, the risk of suicide was even higher.

STAY SAFE The Substance Abuse and Mental Health Services Administration's National Helpline is 1-800-662-4357. Help is available 24/7/365. All calls are strictly confidential. The Alcoholics Anonymous Hotline is 1-800-839-1686. If you or someone you know is in a crisis where they might harm themselves, call the toll-free National Suicide Prevention Lifeline at 1-800-273-8255 (TALK).

SOURCE CDC, cdc.gov/nchs/fastats/alcohol.htm and cdc.gov/nchs/fastats/drug-use-illicit.htm

Mass Suicide

> **ANNUAL US DEATHS** Near 0

Mass suicides in the US are exceedingly rare. The worst case ever involving Americans was in Jonestown, Guyana, in 1978, when 917 followers of the Peoples Temple cult, including 276 children, died by drinking Flavor Aid mixed with cyanide; their leader, Jim Jones, shot himself.

In 1997, in Rancho Santa Fe, California, 39 members of the Heaven's

Gate cult killed themselves, believing that their souls would "exit their human vessels" to travel aboard an alien spacecraft supposedly following Comet Hale-Bopp. Two more members later killed themselves, 1 in 1997 and 1 in 1998.

STAY SAFE Try to avoid joining cults with death wishes. If you or someone you know is in a crisis where they might harm themselves, call the toll-free National Suicide Prevention Lifeline at 1-800-273-8255 (TALK). Help is available 24/7/365. All calls are strictly confidential.

SOURCES CNN, cnn.com/US/9703/26/mass.suicide. too/index.html; *Rolling Stone*, rollingstone.com/feature/ jonestown-13-things-you-should-know-about-cult-massacre-121974

Mental Illness

(**ANNUAL US DEATHS** See data in entry.)

An estimated 1 in 5 American adults (20 percent) has a diagnosable mental illness in any given year, and the problem is also serious for children and teenagers. Many such sufferers go unrecognized or untreated.

Depression is a serious, persistent emotional disorder and is known to significantly elevate the risk of suicide. Other mental conditions can cause depression as a side effect, including bipolar disorder, post-traumatic stress disorder (PTSD), and schizophrenia (hallucinations and delusions). While most people with depression never try to kill themselves, an estimated 30 percent to 70 percent of all suicides do suffer from depression or bipolar disorder. Hard data is scarce, but mental health experts state that an emotional or brain chemistry disorder can significantly raise the risk of suicide.

Mental health and physical health are linked, and mental illness can shorten life spans in ways other than suicide. People with schizophrenia, for example, have a death rate that is 3.5 times higher than the general population.

STAY SAFE If you or someone you know is in a crisis where they might harm themselves, call the toll-free National Suicide Prevention Lifeline at 1-800-273-8255 (TALK). Help is available 24/7/365. All calls are strictly confidential.

SOURCES CDC, cdc.gov/nchs/fastats/mental-health.htm; M. Olfson et al. Premature Mortality Among Adults With Schizophrenia in the United States. *JAMA Psychiatry*, October 2015.

Methods

ANNUAL US DEATHS 44,834 in 2020

As with murder, someone who commits suicide needs to pick a method: use a lethal weapon (e.g., gun, noose), consume a lethal substance (e.g., drug overdose, poison), or subject themselves to a lethal environment or experience (e.g., fall from a height, jump under a train). Statistical data on suicide methods are valuable because they point to public policies that can limit availability. Since suicide is often impulsive, denying the means also denies the opportunity. Studies show that restricting access to, say, pesticides, firearms, drugs, train tracks, and dangerous precipices all lead to fewer suicides.

In the US in 2018, firearms were used in 50.5 percent of all suicides. Suffocation, including hanging, accounted for 28.6 percent. Poisoning, including intentional drug overdoses, accounted for 12.9 percent of all suicides. About 8 percent used other methods: Jumping from a high place was under 2 percent, as was drowning. Under 1 percent jumped in front of a train. Suicide carbon monoxide asphyxiation resulting from automobile tailpipe exhaust is not possible for any vehicle new enough to have a catalytic converter. This device removes more than 99 percent of the carbon monoxide in the tailpipe fumes and became mandatory in the US in 1975. However, it is still possible for diverted exhaust emissions to exhaust the available oxygen in a closed space.

STAY SAFE Keep potential suicide methods, such as guns, knives, drugs, and rope, away from potentially suicidal people. Provide effective fencing and signage for suicide help lines on bridges and other high points.

SOURCES CDC, cdc.gov/injury/wisqars/fatal.html; S.R. Kegler et al. Firearm Homicides and Suicides in Major Metropolitan Areas —United States, 2015–2016 and 2018–2019. *MMWR Morbidity and Mortality Weekly Report*, January 2022; Suicide Prevention Resource Center, sprc.org/scope/means-suicide

Suicide by Cop

ANNUAL US DEATHS Estimated 90 to 300

Suicide by cop means someone intentionally acting in a threatening manner designed to provoke police into killing them. There are 2 patterns: a perpetrator-victim who, fleeing after committing a crime, chooses death over incarceration when about to be apprehended; or someone who first decides to kill themselves, then chooses as their method to be shot by a cop. Determining afterward whether an officer-involved shooting was a suicide by cop is difficult, so available data is imprecise.

For each year from 2015 through 2018 in the US, there were about 900 to 1,000 fatal shootings of suspects by police, counting all different circumstances. Experts estimate that between 10 and 30 percent of these were suicides by cop, for an annual total between 90 and 300.

STAY SAFE If you or someone you know is in a crisis where they might harm themselves, call the toll-free National Suicide Prevention Lifeline at 1-800-273-8255 (TALK). Help is available 24/7/365. All calls are strictly confidential.

SOURCES K. Mohandie et al. Suicide by cop among officer-involved shooting cases. *Journal of Forensic Sciences*, 2009; C.L. Patton and W. J. Fremouw. Examining 'suicide by cop': A critical review of the literature. *Aggression and Violent Behavior*, 2016; Police Executive Research Forum, policeforum.org/suicidebycop

Police and Law Enforcement

ANNUAL US DEATHS See data in entry.

Police officers and other law enforcement personnel face unique job and personal pressures that can lead to depression, suicidal thoughts, and suicide. Statistics are scarce, but Blue H.E.L.P., a nonprofit organization that has been collecting law enforcement suicide information since 2016, reported 239 officer suicides in 2019. By contrast, in 2019 only 89 officers died in the line of duty from felonious assault or accidents. The FBI has begun suicide data collection at the federal, state, tribal, and local levels for law enforcement officers, corrections employees, 911 operators, judges, and prosecutors. The information will appear, starting in 2023, in the Law Enforcement Suicide Data Collection (LESDC) database.

STAY SAFE If you are a police officer or know one who might harm themselves, call the toll-free National Suicide Prevention Lifeline at 1-800-273-8255 (TALK); or call Copline, a crisis intervention hotline staffed by retired police officers, at 1-800-267-5463.

SOURCES Blue H.E.L.P, bluehelp.org/resources/statistics; FBI, fbi.gov/services/cjis/ucr/law-enforcement-suicide-data-collection

Veterans and Active Duty Troops

ANNUAL US DEATHS   See data in entry.

Suicide by active duty troops and military veterans is a very serious problem in the US. In 2018 alone, 6,435 veterans killed themselves, compared to a total of 7,032 US troop deaths in world conflict zones over the *entire* period from September 11, 2001, through late 2020. Reasons include postcombat PTSD-related severe depression, persisting postservice unemployment, homelessness, and social isolation, and other difficulties readapting to civilian life.

The age-and-sex-adjusted suicide rate for veterans was 27.5 per 100,000 in 2018, up from 25.8 in 2016, despite increased national attention and funding devoted to mitigating the problem. This rate was fully 50 percent greater than for US adults overall. Suicides by active duty service members add about another 9 percent to the veterans, death toll, for an average of 20 suicides *per day* by current and former US military men and women.

Past or current military service increases the risk of suicide by about 55 percent compared to the overall US population.

STAY SAFE   If you or someone you know is a veteran or active duty service member in a life crisis, seriously depressed, or at risk of suicide, call the Department of Veterans Affairs' Veteran Crisis Line at 1-800-273-8255, or text them at 838255. Help is available 24/7/365 and is completely confidential.

SOURCE   US Department of Veterans Affairs, Office of Mental Health and Suicide Prevention, mentalhealth.va.gov/suicide_prevention

By Demographic Status

By Age

ANNUAL US DEATHS   See data in entry.

The rate of US suicides by age shows different patterns between males and females.

For males, the rate per 100,000 in 2018 was 3.7 for ages 10 to 14, 22.7 for ages 15 to 24, 27.7 for ages 25 to 44, 31.0 for ages 45 to 64, 27.8 for ages 65 to 74, and 39.9 for ages 75 and over.

For females, the rates were 2.0 for ages 10 to 14, 5.8 for ages 15 to 24, 7.9 for ages 25 to 44, 9.8 for ages 45 to 64, 6.2 for ages 65 to 74, and 4.0 for ages 75 and over.

The suicide risk by age is highest for women between 45 and 64; for men it is highest for ages 75 and up.

STAY SAFE   If you or someone you know is in a crisis where they might harm themselves, call the toll-free National Suicide Prevention Lifeline at 1-800-273-8255 (TALK). Help is available 24/7/365. All calls are strictly confidential.

SOURCE   CDC, cdc.gov/suicide/facts/disparities-in-suicide.html

By Race/Ethnicity

ANNUAL US DEATHS   See data in entry.

The rate of suicide per 100,000 people per year during 2018, separated out by race/ethnicity, shows different patterns for males and for females.

Among males, the suicide rate was highest for Whites at 30.4 per 100,000. It was 12.1 for Hispanics, 12.0 for Blacks, and 10.8 for Asians.

Among females, it was 8.3 for Whites, 2.9 for both Hispanics and for Blacks, and 4.1 for Asians.

Among American Indian/Alaska Native young people aged 8 to 24, suicide was the 2nd leading cause of death in 2019.

Your risk remains moderate overall but about 2.5 times as high for non–Hispanic Whites as for others.

STAY SAFE   If you or someone you know is in a crisis where they might harm themselves, call the toll-free National Suicide Prevention Lifeline at 1-800-273-8255 (TALK). Help is available 24/7/365. All calls are strictly confidential.

SOURCE   CDC, cdc.gov/suicide/facts/disparities-in-suicide.html

By Sex

ANNUAL US DEATHS   See data in entry.

In the US, men commit suicide 3.5 times as often as women. For 2016, the total number of males who killed themselves was 34,727; for females it was 10,238. The age-adjusted rate for men was 21.5 suicides per 100,000 of male population; the rate for women was 6.3 per 100,000 of female population. For men, the most common method was a firearm, at almost 70 percent of all male deaths. For women, the method used was divided roughly evenly among firearm, drug overdose, or hanging.

STAY SAFE If you or someone you know is in a crisis where they might harm themselves call the toll-free National Suicide Prevention Lifeline at 1-800-273-8255 (TALK). Help is available 24/7/365/ All calls are strictly confidential.

SOURCES CDC, cdc.gov/nchs/products/databriefs/db398.htm, cdc. gov/nchs/data/hestat/suicide/rates_1999_2017.htm, and cdc.gov/injury/wisqars/fatal.html

By Sexual Orientation or LGBTQ+ Identity

ANNUAL US DEATHS See data in entry.

People who belong to the lesbian, gay, bisexual, transgender, and queer community, or who are confused or ambivalent about their sexual/gender orientation/identity/preference, are especially prone to attempting and committing suicide. Statistics show this risk is significantly increased when the person has been subject to verbal abuse and/or physical assault, and/or to being rejected by their families.

The psychosocial challenges that LGBTQ+ young people face make them especially vulnerable to suicide. LGBTQ+ youths are 3 to 4 times as likely as straight kids to try to kill themselves at some time in their life. They are over 8 times more likely to attempt suicide if their family rejects them than if their family accepts them. LGBTQ+ community members who have been subjected to insults or assaults are 2.5 times as likely to commit self-harm, including self-killing. For transgender people, 41 percent surveyed reported that they attempted suicide at least once; if they were ever the victim of a physical assault because of their identity, this figure rose to 61 percent.

Statistics about suicide ideation and attempted suicide in this population are readily available, but statistics about actual suicides are harder to find. This is probably because for some suicides, sexual orientation isn't known or considered relevant, and for others, family and societal pressures mean sexual orientation is omitted or falsified on the death certificate. In a study of 123,289 suicide decedents from 18 states participating in the CDC's National Violent Death Reporting System, 621 (0.5 percent) were identified as lesbian, gay, bisexual, or transgender. The majority of lesbian, gay, bisexual, or transgender decedents were identified as gay male (53.9 percent), followed by lesbian (28.0 percent), transgender (10.4 percent), and bisexual (7.5 percent).

STAY SAFE   The Trevor Project Lifeline, at 1-866-488-7386, specializes in helping LGBTQ+ young people in crisis. If you or someone you know is in a crisis where they might harm themselves, call the toll-free National Suicide Prevention Lifeline at 1-800-273-8255 (TALK). Help is available 24/7/365. All calls are strictly confidential.

SOURCES   CDC, cdc.gov/suicide/facts/disparities-in-suicide.html and cdc.gov/violenceprevention/datasources/nvdrs; B.H. Lyons et al. Suicides Among Lesbian and Gay Male Individuals: Findings From the National Violent Death Reporting System. *American Journal of Preventive Medicine*, February 2019.

By Time

By Day of the Week

ANNUAL US DEATHS   The new beginnings effect may be responsible for a 12 percent spike in suicides on Sundays or Mondays.

Overall, suicides are most likely to happen on a Sunday or Monday. They are least likely on Fridays and Saturdays. The difference is not great: Based on data on suicide attempts as a proxy for fatalities, about 12 percent more occur at the start of a week than at the end. Experts call this the new beginnings effect: Someone with depression reaches a new beginning, sees that little in their prospects has changed, and lacks the wherewithal to endure what will follow. However, for school-age kids, the suicide peak comes on Monday and Tuesday instead of Sunday and Monday.

STAY SAFE   If you or someone you know is in a crisis where they might harm themselves, call the toll-free National Suicide Prevention Lifeline at 1-800-273-8255 (TALK). Help is available 24/7/365. All calls are strictly confidential.

SOURCES   G.A. Beauchamp et al. Variation in suicide occurrence by day and during major American holidays. *Journal of Emergency Medicine*, June 2014; QuickStats: Average Number of Deaths from Motor Vehicle Injuries, Suicide, and Homicide, by Day of the Week — National Vital Statistics System, United States, 2015. *MMWR, Morbidity and Mortality Weekly Report*, June 2017.

During Holidays

ANNUAL US DEATHS   See data in entry.

The US holiday with the most suicides is New Year's Day, with a peak 14.5 percent above the average day of a calendar year. This is followed by Labor Day and then Memorial Day, at 10.6 percent and 7.3 percent higher than an average day, respectively. Then come Mother's Day at 7.0 percent and Father's Day at 4.2 percent above an average day.

Christmas and Thanksgiving show significant lows in suicide rates, perhaps because people tend to be with their loved ones most on those holidays and so are less likely to kill themselves. The rate is 15.8 percent below average on Thanksgiving Day and 21.7 percent below average on Christmas Day.

Your risk is moderate overall but about one-seventh higher on New Year's Day, and one-fifth lower on Christmas.

STAY SAFE   If you or someone you know is in a crisis where they might harm themselves, call the toll-free National Suicide Prevention Lifeline at 1-800-273-8255 (TALK). Help is available 24/7/365. All calls are strictly confidential.

SOURCES   G.A. Beauchamp et al. Variation in suicide occurrence by day and during major American holidays. *Journal of Emergency Medicine*, June 2014; QuickStats: Average Number of Deaths from Motor Vehicle Injuries, Suicide, and Homicide, by Day of the Week — National Vital Statistics System, United States, 2015. *MMWR, Morbidity and Mortality Weekly Report*, June 2017.

By Lunar Phase

ANNUAL US DEATHS   Lunar phase has no effect on suicide rates.

Despite the persistent belief that suicide attempts peak on nights of the full moon, reinforced by anecdotes from police and paramedics to the same effect, data does not support any pattern to the incidence of suicide by lunar phase. No lunar phase is particularly dangerous.

STAY SAFE   If you or someone you know is in a crisis where they might harm themselves, call the toll-free National Suicide Prevention Lifeline at 1-800-273-8255 (TALK). Help is available 24/7/365. All calls are strictly confidential.

Some say the presence of a full moon provokes suicides. The data does not support this.

SOURCE   T. Biermann et al. Influence of lunar phases on suicide: the end of a myth? A population-based study. *Chronobiology International*, 2005.

By Season

ANNUAL US DEATHS   See data in entry.

Seasonal variations in suicide are small. The rate in spring is highest, being 5 percent higher than the rate in winter (lowest rate). Summer and autumn are less than 2 percent below spring. Experts think the spring peak resembles the weekly Monday peak, due to a new beginnings effect: Someone with depression reaches a new beginning, sees that little in their prospects has changed, and lacks the wherewithal to endure what will follow.

Your risk remains moderate but is slightly higher in springtime.

STAY SAFE   If you or someone you know is in a crisis where they might harm themselves, call the toll-free National Suicide Prevention Lifeline at 1-800-273-8255 (TALK). Help is available 24/7/365. All calls are strictly confidential.

SOURCE   J. Yu et al. Seasonality of suicide: a multi-country multi-community observational study. *Epidemiology and Psychiatric Sciences*, August 2020.

By Location

Rural versus Urban Areas

ANNUAL US DEATHS   See data in entry.

US suicide rates are higher for people living in rural areas than for those living in urban areas, and this disparity has been rising. Reasons include increased isolation, boredom leading to substance abuse, more access to firearms, and a rise in mental illness among rural youths.

Between 2000 and 2018, rural suicide rates increased 48 percent, while rates in cities grew by only 34 percent. By sex, rural suicide rates exceeded city ones by 40 percent for men and by 25 percent for women.

In 2018, the suicide rate for men living in rural areas was 31 per 100,000, compared to 22 per 100,000 in urban areas. For women, the figure was 8 per 100,000 for women living in rural areas compared to 6 per 100,000 in urban areas.

STAY SAFE   If you or someone you know is in a crisis where they might harm themselves, call the toll-free National Suicide Prevention Lifeline at 1-800-273-8255 (TALK). Help is available 24/7/365. All calls are strictly confidential.

SOURCES   CDC, cdc.gov/ruralhealth/Suicide.html; P.S. Nestadt et al. Urban-Rural Differences in Suicide in the State of Maryland: The Role of Firearms. *American Journal of Public Health*, October 2017; D.L. Steelesmith et al. Contextual Factors Associated With County-Level Suicide Rates in the United States, 1999 to 2016. *JAMA Network Open*, September 2019.

Worst and Best Countries

In 2019, the age-adjusted annual suicide rate per 100,000 of population varied considerably by country. The highest rate was Lesotho, at 72.4 per 100,000, followed by Guyana (40.3), Eswatini (29.4), South Korea (28.6), Kiribati (28.3), Federated States of Micronesia (28.2), Lithuania (26.1), Suriname (25.4), Russia (25.1), and South Africa (23.5).

The lowest suicide rates per 100,000 of population in 2019 were Antigua and Barbuda (0.4), Barbados (0.6), Grenada (0.7), Saint Vincent and the Grenadines (1.0), Sao Tome and Principe (1.5), Jordan (1.6), Syria (2.0), Venezuela (2.1), Honduras (2.1), and the Philippines (2.2).

STAY SAFE   If you or someone you know is in a crisis where they might harm themselves, call the toll-free National Suicide Prevention

"Welcome to Fabulous Las Vegas," America's most prolific suicide city.

Lifeline at 1-800-273-8255 (TALK). Help is available 24/7/365. All calls are strictly confidential.

SOURCE World Population Review, worldpopulationreview.com/country-rankings/suicide-rate-by-country

Worst and Best States

Data on US suicides by state for 2019, based on the number of suicides per 100,000 of population, showed that Wyoming was worst, with 29.3 people killing themselves per 100,000 residents. Then came Alaska at 28.5, Montana at 26.2, New Mexico at 24.0, and Colorado at 22.1.

The lowest suicide rate of all was in New Jersey at 8.0 per 100,000 residents, followed by New York at 8.3, Massachusetts at 8.7, Maryland at 10.3, and California at 10.7.

Your risk is moderate, though about 3.5 times as high in the worst states as in the best states.

STAY SAFE If you or someone you know is in a crisis where they might harm themselves, call the toll-free National Suicide Prevention Lifeline at 1-800-273-8255 (TALK). Help is available 24/7/365. All calls are strictly confidential.

SOURCE CDC, cdc.gov/nchs/pressroom/sosmap/suicide-mortality/suicide.htm

Prevention

If you or someone you know is in a crisis where they might harm themselves, call the National Suicide Prevention Lifeline at 1-800-273-8255 (TALK). Help is available 24/7/365. All calls are strictly confidential.

Suicide is a serious public health problem everywhere, and suicide prevention is an important responsibility for every community. Suicide is the 10th leading cause of death overall in the US. For people between age 10 and age 34, it is the 2nd leading cause of death.

Steps that help reduce the death rate for suicide include: training in suicide awareness, depression and wellness screening, and coping/resilience skill training. Training is recommended for healthcare providers, school administrators and teachers, law enforcement and court officers, clergy, counselors and coaches, social workers, tribal elders and other ethnic advocates, military and Veterans Administration officials, and even bartenders and firearms dealers and gun safety instructors.

Limiting the means to kill oneself has also been shown to reduce the suicide rate. This includes use of home gun safes, secure firearm storage outside the home through gun dealers and shooting ranges, and safe storage of prescription medications in the home.

STAY SAFE   If you or someone you know is seriously depressed, feels despair, or is thinking and/or talking about harming themselves or taking their own life, get professional help for them right away.

SOURCES   American Foundation for Suicide Prevention (AFSP); Behind the Badge; CDC; Mental Health America; NIH NIMH; The Samaritans; SAVE.org; TheTrevorProject.org; US Veterans Administration; WHO

Suggested Reading

These books are a good starting point for learning more about statistics and actuarial science:

Joel Best. *Damned Lies and Statistics: Untangling Numbers from the Media, Politicians, and Activists*. University of California Press, 2012.

Émile Borel. *Probabilities and Life*. Dover Publications, 1962.

Robert Matthews. *Chancing It: The Laws of Chance and How They Can Work for You*. Skyhorse, 2017.

John D. McGervey. *Probabilities in Everyday Life*. Ballantine Books, 1986.

Jack Mingo et al. *Cause of Death: A Perfect Little Guide to What Kills Us*. Pocket Books, 2008.

Joseph Newmark. *Statistics and Probability in Modern Life*. Brooks Cole; 6th edition, 1996.

Peter Olofsson. *Probabilities: The Little Numbers that Rule Our Lives*. Wiley, 2006.

Index of Causes of Death

A

Picture Credits

Captions to full-page images
p. 2: Operation Buster-Jangle Dog nuclear weapon test, Nevada, 1951
pp. 4–5: Cloud-to-ground lightning in Poland
pp. 6–7: A Jewish cemetery in Dörzbach, Germany
p. 16: The aftermath of the 1994 Northridge earthquake in Los Angeles
p. 36: A scanning electron microscope image of a cell infected with SARS-CoV-2 (the small spheres)
p. 56: Illustrations of a diseased heart from an 1834 book
p. 241: An alligator in Loxahtchee National Wildlife Refuge, Florida
pp. 242–43: British Airways Flight 38 after crash-landing at London Heathrow Airport on January 17, 2008
p. 244: The eruption of Mount St. Helens on May 18, 1980
pp. 594–95: Minnesota State Patrol troopers in Minneapolis on May 29, 2020, during the protests following the death of George Floyd
p. 596: A cell at Alcatraz

Editor: Alan Axelrod
Proofreader and indexer: Jennifer Dixon
Designers: Misha Beletsky and David Fabricant
Cover designer: David Fabricant
Production manager: Louise Kurtz

First edition
10 9 8 7 6 5 4 3 2 1

Library of Congress Cataloging-in-Publication Data
Names: Buff, Sheila, author. | Buff, Joe, author.
Title: You bet your life : your guide to deadly risk / Sheila Buff & Joe Buff, MS, FSA.
Description: New York : Half Full, [2022] | Includes bibliographical references. | Summary: "A book on the wagers we make with our lives every day"—Provided by publisher.
Identifiers: LCCN 2022010644 (print) | LCCN 2022010645 (ebook) | ISBN 9780789270177 (hardback) | ISBN 9780789260963 (ebook)
Subjects: LCSH: Mortality. | Risk. | Premature death.
Classification: LCC HB1321 .B84 2022 (print) | LCC HB1321 (ebook) | DDC 304.6/4--dc23/eng/20220302
LC record available at https://lccn.loc.gov/2022010644
LC ebook record available at https://lccn.loc.gov/2022010645

For bulk and premium sales and for text adoption procedures, write to Customer Service Manager, Half Full, An Imprint of Abbeville Press, 655 Third Avenue, New York, NY 10017, or call 1-800-ARTBOOK.

Visit Half Full online at www.halffullbooks.com.

CAUSE OF DEATH NAVIGATOR

ACUTE CONDITIONS AND DISEASES

CHRONIC DISEASES

INFECTIOUS DISEASES

MEDICAL MISHAPS, MALPRACTICE, AND MURDER

NOT DEAD YET

OTHER (MORE OR LESS) NATURAL CAUSES

SUBSTANCE ABUSE

CONTRIBUTORY FACTORS

ACTS OF GOD AND NATURE

ANIMAL ATTACKS

AT HOME

ON THE GO

ON THE JOB

AT PLAY

IN THE GREAT OUTDOORS

LAW AND DISORDER

NATIONAL SECURITY AND INSECURITY

HOMICIDE

SUICIDE